POLYCHLOROAROMATIC COMPOUNDS

POLYCHLOROAROMATIC COMPOUNDS

Edited by

H. Suschitzky

Department of Chemistry and Applied Chemistry
University of Salford
Salford, England

Springer Science+Business Media, LLC

Library of Congress Catalog Card Number: 74-1617
ISBN 978-1-4684-2099-9   ISBN 978-1-4684-2097-5 (eBook)
DOI 10.1007/978-1-4684-2097-5
© Springer Science+Business Media New York 1974
Originally published by Plenum Publishing Company Ltd 1974
Softcover reprint of the hardcover 1st edition 1974

U.S. Edition published by
Plenum Publishing Corporation
227 West 17th Street
New York, New York 10011

Contributors

Manuel Ballester   Instituto de Química Orgánica de Barcelona, Patronato Juan de la Cierva, C.S.I.C., Barcelona, Spain.

T. Chivers   Department of Chemistry, University of Calgary, Calgary T2N 1N4, Alberta, Canada.

M. B. Green   Imperial Chemical Industries Ltd, Mond Division, Research and Development Department, Runcorn Heath Laboratory, Runcorn, Cheshire, England.

B. Iddon   The Ramage Laboratories, Department of Chemistry and Applied Chemistry, University of Salford, Salford M5 4WT, Lancashire, England.

Santiago Olivella   Instituto de Química Orgánica de Barcelona, Patronato Juan de la Cierva, C.S.I.C., Barcelona, Spain.

H. Suschitzky   The Ramage Laboratories, Department of Chemistry and Applied Chemistry, University of Salford, Salford M5 4WT, Lancashire, England.

B. Wakefield   The Ramage Laboratories, Department of Chemistry and Applied Chemistry, University of Salford, Salford M5 4WT, Lancashire, England.

Preface

The chlorination of aromatic and heteroaromatic compounds is one of the oldest and also one of the most important reactions since it heralded the rise of the heavy organic chemicals industry. Reports of polychlorobenzenes and polychloropyridines date back well into the last century and it is of historical interest that Kekulé was probably the first to have prepared pentachloropyridine.

Yet, in spite of the ancient lineage and the practical as well as theoretical significance of polychloroaromatics, there has existed so far no review exclusively devoted to their chemistry. Three international symposia on polychloro-chemistry since 1968, in which polychloroaromatics figured prominently in discussions amongst chemists and biologists, are evidence for the growing importance of this subject. Information on the chemistry and uses of the numerous polychloroaromatic compounds has remained buried largely in the journal and patent literature and only occasionally receives sparse and incidental treatment in an article primarily dealing with a specific ring system.

Our objective which seems more than timely under the circumstances is to provide a manageable and simply written reference book on the present state of knowledge in the field of polychloro-aromatic and -heteroaromatic compounds. A very liberal view of the prefix 'poly' was taken by including compounds with as few as two chlorine atoms on the same ring whenever it was felt that the continuity of the discussion would suffer unless this was done. Linguistically, our interpretation is, of course, open to criticism. A truly comprehensive treatment of the subject would have produced an unwieldy monograph and so we have tried and hopefully succeeded in steering our discussion between over-detailed exposition and superficial treatment. The specific problem of the specialist may be followed up by recourse to the judicially selected list of references. The field has been surveyed in four chapters by different authors with the advantage of an authoritative treatment of each section which far outweighs the possible disadvantage of less uniformity. The chapters vary by necessity in style and scope since I endeavoured to preserve the author's view and personal choice of material. I tolerated a certain amount of overlap between chapters so that readers interested only in a certain section would be made aware of related concepts. Incidental mention of known biological activity and applications of the polychloro compounds as well as comparisons with their more glamorous relatives, namely the analogous fluorine compounds, are made. A full chapter on established industrial applications will draw attention to patent literature which is often overlooked by chemists who rely primarily on periodical journal literature.

As editor whose fortune it has been to enlist an international team of authors who are leading in their field, I wish to thank each of my contributors for giving so unstintingly of their time and talent. My thanks also go to numerous colleagues for their critical advice and co-operation in making unpublished material available.

It is hoped that the selective presentation of the material will provide all readers concerned with the chemistry, application, or environmental aspects of polychloro compounds with a ready reference work and with new ideas.

H. Suschitzky,
Salford, Lancs., England.
April 1974.

ERRATA

Page 183, line 8: for '*para* and *ortho*'
read '*bridgehead* and *ortho*'

Page 183, line 12: for '*para* carbon'
read '*bridgehead* carbon'

POLYCHLOROAROMATIC COMPOUNDS

Contents

CHAPTER 1

Aromatic and Alkaromatic Chlorocarbons

Manuel Ballester and Santiago Olivella

Instituto de Química Orgánica de Barcelona, Patronato
Juan de la Cierva, C.S.I.C., Barcelona, Spain

1

1.1 INTRODUCTION

Chemical exploration of the chlorocarbons was initiated by the almost simultaneous discovery of three chlorocompounds. Perchlorobenzene,† the simplest aromatic chlorocarbon, was reported by Julin in 1821 [2], following the announcement of the synthesis of perchloroethane and perchloroethylene by Faraday [3, 4-6].

Müller, about forty years later [7], while studying the chlorination of benzene in the presence of iodine, observed the formation of 'a beautiful crystalline compound' identical to Julin's chlorocarbon, and concluded that it was hexachlorobenzene [8].

In spite of its early start, the chemistry of the aromatic chlorocarbons remained dormant until the nineteen-fifties, when it received a considerable impetus with the discovery of a powerful perchlorination agent: reagent BMC [9, 10]. It is prepared as a complex solution from sulphuryl chloride, aluminium chloride and sulphur monochloride [9]. It is, therefore, similar to Silberrad's chlorinating agent [11] which is much less effective.

Doorenbos has recently suggested that the active species of reagent BMC is the trichlorosulphonium tetrachloroaluminate [12, 13]. This is, however, doubtful since it is only slightly soluble in sulphuryl chloride. Nevertheless, Doorenbos has described a new chlorinating reagent consisting of a mixture of that salt, chlorine and thionyl chloride which in some cases performs like reagent BMC [12]. This reagent may advantageously be used in large scale preparations because of its much more limited solvent requirements. Ballester and co-workers have recently found two new reagents (BMC-S and BMC-P) which, in some cases, need much less sulphuryl chloride [14].

Reagent BMC has allowed the preparation of sterically strained chlorocarbons such as perchlorotoluene [15], which had already been attempted unsuccessfully by Beilstein [16, 17] in 1869. Its efficiency is due to the high concentration of the active species present and the relatively low temperature (usually below 100°) which minimizes chlorinolysis, i.e., fragmentation of the carbon framework.

Other aromatic chlorinating methods, such as the use of chlorine in the presence of iodine, [7, 8, 18, 19], antimony pentachloride [7, 8, 16, 18, 19], ferric chloride [20, 21], or aluminium chloride [22], are either not powerful enough to chlorinate exhaustively or lead to products of chlorinolysis. In some cases, the 'chlorine carriers' (iodine trichloride [23, 24], antimony pentachloride [8, 17] and ferric chloride [21]) have been used without chlorine, but, except in the case of iodine chloride, significantly higher temperatures are required for comparable results.

† 'Per' denotes substitution of all hydrogen atoms attached to carbon [1].

West and co-workers have recently found new routes to some aromatic chlorocarbons based on the rearrangement of certain aliphatic chlorocarbons.

The chemistry of the aromatic and alkaromatic chlorocarbons is remarkable. Steric effects arising from the space-filling capacity of the chlorine atom and the aromatic character are responsible for physical and chemical phenomena of considerable interest, some of which are unexpected, such as the existence of inert chlorocarbon free radicals.

The strength of the carbon(sp^2)-chlorine bond, the inertness of the non-bonding p-electrons of the chlorine, combined with steric effects, confer a considerable thermal stability and chemical inertness on some chlorocarbons. By contrast, chlorocarbons with carbon-(sp^3)-chlorine bonds and possessing steric strain are reactive and thermally unstable, even at moderate temperatures.

Since in principle every hydrocarbon has a corresponding chlorocarbon, the aromatic chlorocarbons can be grouped under arenes, arylalkanes, arylethylenes, arylacetylenes, macromolecules, free radicals, carbanions, carbonium ions, etc. However, as Wheland [25], Tatlow [26], and Ballester [27] have pointed out, geometrical limitations due to the unusual space requirements of the chlorine atoms could be expected, which are seldom found in the hydrocarbon chemistry. Molecules of relative simplicity such as perchlorotoluene or perchlorophenanthrene cannot be assembled with the Stuart-Briegleb space-filling atomic models. A chlorocarbon might be 'geometrically allowed' and yet the transition state of the reaction that would normally lead to it might be prohibitively overcrowded (steric hindrance). Electronic deactivation due to accumulation of chlorines might also play a decisive role.

The present review describes many unpublished results observed by its authors and their co-workers. The authors wish to acknowledge their indebtedness to their numerous past and present co-workers at the Instituto de Química Orgánica de Barcelona, especially Professors Juan Castañer and Juan Riera, for their outstanding contribution to the development of the aromatic and alkaromatic chlorocarbons. They also wish to express their sincere gratitude to the Aerospace Research Laboratories (Wright-Patterson AFB, Ohio) for continued generous moral and financial support, and to their scientist, Dr. Leonard Spialter, for his enthusiastic advice and encouragement.

1.2   PERCHLOROARENES

In this section perchlorobenzene and non-fused and fused polycyclic aromatic chlorocarbons are described.

1.2.1   Perchlorobenzene: Syntheses

Perchlorobenzene (I) has been obtained under drastic reaction conditions (red heat, electric discharge, etc.) starting from the elements, chloroform, carbon tetrachloride, chlorinated ethanes, tetrachloroethylene, etc. However, the syntheses here described start from organic compounds possessing five or more carbon atoms per molecule.

The numerous syntheses of perchlorobenzene are here classified into different categories, according to the nature of the starting material and the type of reaction.

(a)   From Benzene and Chlorobenzenes
The reported syntheses from benzene are carried out by chlorine and iodine [7, 8, 19]; chlorine and ferric chloride [20]; chlorine and antimony pentachloride [8]; chlorine and stannic chloride (or tin) [28]; chlorine and aluminium chloride [22]; antimony pentachloride [8, 18, 29]; ferric chloride [21]; iodine trichloride [23, 24]; Silberrad's reagent [11]; electrochemical chlorination [30]; and vapour phase chlorination on alumina with iodine trichloride (180-300°) or rhodium trichloride (300°) [31]. The by-products are usually polychlorobenzenes.

$$C_6H_6 \rightarrow C_6Cl_6 + HCl$$
$$(I)$$

When iodine is used as a carrier, it becomes strongly bonded to the benzene ring and its elimination is extremely difficult [22, 28].

Other similar syntheses of perchlorobenzene start from chlorobenzenes [11, 18, 32]. The chlorination of 1,2,3,4-; 1,2,3,5- and 1,2,4,5-tetrachlorobenzenes to give a mixture of penta- and perchlorobenzene has been effected with chlorine and a Al-Hg couple [33].

Chlorosulphonic acid can act at temperatures 140-220° as chlorinating agent of polychlorobenzenes [34-36]. If temperature or degree of chlorination of the starting polychlorobenzene are not high enough, sulphochlorides become the predominant products. Chlorination of polychlorobenzenes with chlorine without catalyst does not usually take place. However, with sunlight, at room temperature, they are converted into addition products which on further treatment yield perchlorobenzene [19, 37-40]. In the case of the three trichlorobenzenes, such treatment is accompanied by elimination of hydrogen chloride [37].

$$C_6H_3Cl_3 \xrightarrow[h\nu]{Cl_2} C_6H_3Cl_9 \xrightarrow{HO^-} C_6Cl_6$$

In the three isomeric tetrachlorobenzenes [38] and in pentachlorobenzene [39, 40] the addition products are frequently cyclohexenes and eventually some perchlorobenzene. This is possibly due to repulsions hindering either the formation of cyclohexanes or assisting elimination. Perchlorobenzene has been obtained from these adducts by further eliminations either with ethanolic hydroxide or zinc in ethanol.

Ladenburg reported that some perchlorobenzene is formed by chlorination of chlorobenzene with wet chlorine [41].

(b) *From Other Substituted Benzenes*
Riemschneider reported chlorinations of fluorobenzene [42]. With chlorine and iron at 180-200° a mixture of fluoropentachloro- and perchlorobenzene is obtained. The chlorination could be performed at 100° with ultraviolet light. The iron playes a decisive and

complex role in preventing chlorine addition and helping the replacement of the hydrogens and the fluorine.

$$C_6H_5F \rightarrow C_6Cl_6$$

Perchlorobenzene is also formed by treating phenol, resorcinol, anisol, and anethol [43] at 350° with chlorine and iodine.

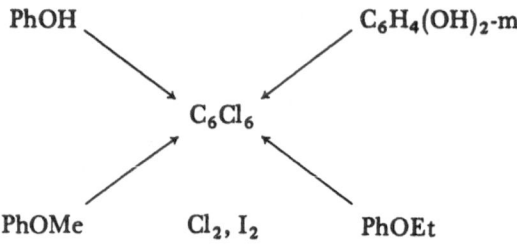

Under milder conditions, pentachlorophenol is obtained from phenol, probably by way of chloroiodophenols [44]. Diphenyl ether with chlorine, in the presence of both iron and iodine, gives perchlorobenzene [45].

Pentachlorophenol [46], trichlorohydroquinone [47], and tetrachlorohydroquinone [47] are converted into perchlorobenzene by heating with phosphorus pentachloride. Hydroquinone-2,6-disulphonyl dichloride behaves similarly and both the hydroxyl and the chlorosulphonyl groups are replaced.

The conversion of phenols such as resorcinol [48], phloroglucinol [49], and pyrocatechol [49] into perchlorobenzene can also be performed by heating them in chlorosulphonic acid (160-250°). In the case of phloroglucinol, if the temperature is low or the reaction time insufficient, pentachlorophenol or its mixtures with perchlorobenzene are obtained. At still lower temperatures, sulphonation occurs.

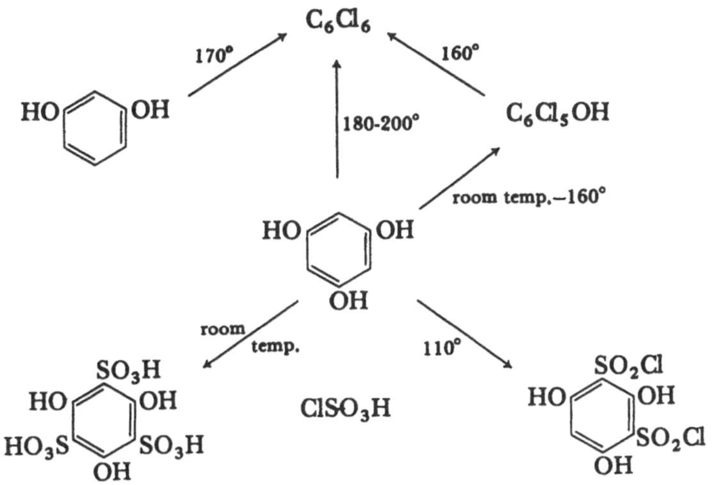

$$C_6Cl_6$$

170° ↗　　160° ↖

HO–⬡–OH　　180-200° ↑　　C_6Cl_5OH

room temp.–160°　↗

HO–⬡–OH

OH

room temp. ↗　　110° ↘

SO$_3$H
HO–⬡–OH
HO$_3$S–⬡–SO$_3$H
OH

ClSO$_3$H

SO$_2$Cl
HO–⬡–OH
⬡–SO$_2$Cl
OH

Pyrocatechol behaves similarly, sulphonation occurring at 100° and formation of pentachlorophenol at 150°. However, in this case the reaction at higher temperatures giving perchlorobenzene has not yet been described, but it is obtained by treatment of pyrocatechol-disulphonic acid or its dichloride with thionyl chloride at 250°.

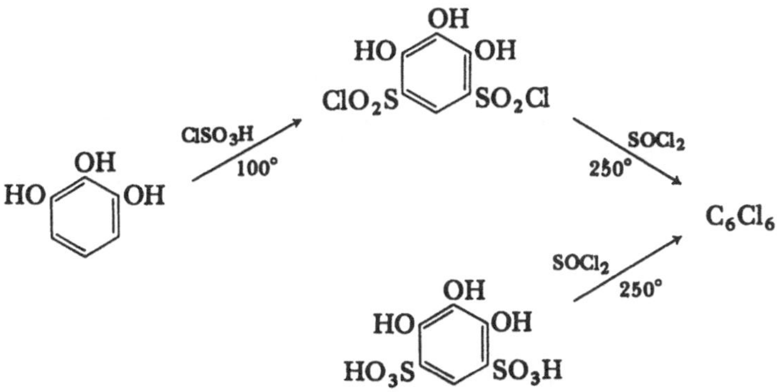

OH
HO–⬡–OH
ClO$_2$S–⬡–SO$_2$Cl

ClSO$_3$H
100° ↗

HO–⬡–OH (with OH)
OH

SOCl$_2$
250° ↘

C_6Cl_6

SOCl$_2$
250° ↗

OH
HO–⬡–OH
HO$_3$S–⬡–SO$_3$H

Chlorination of nitrobenzene with chlorine in the presence of iron, at 100-125°, gives perchlorobenzene [20, 51-53]. At lower temperature (40°) a mixture of mono- and dichloronitrobenzenes is obtained [52], indicating that the nitro group is probably replaced at a later stage of chlorination. When nitrobenzene is submitted to photochlorination with chlorine and sunlight and the resulting complex mixture is then treated with methanolic sodium hydroxide, penta- and perchlorobenzene are obtained [54], presumably through the intermediacy of polychlorocyclohexanes. While chlorination of nitrobenzene with sulphuryl chloride at 200° is reported to give a mixture of 1,2,4,5-tetrachloro- and pentachlorobenzene, that of m-dinitro- and 1,3,5-trinitrobenzene gives perchlorobenzene [55].

Aniline, diphenylamine and triphenylamine when treated with chlorine in the presence of iodine, at high temperature (350°) give perchlorobenzene [43]. Tribenzylamine and azobenzene behave similarly [43].

A remarkable formation of perchlorobenzene occurs in the chlorination of N-nitroaminobenzenes with hydrogen chloride in refluxing acetic acid or acetyl chloride [56].

Perchlorobenzene is formed from pentachlorophenylmetallic compounds such as pentachlorophenylmagnesium chloride [57] or pentachlorophenyllithium [58] by chlorine–metal exchange with chlorosilanes, such as silicon tetrachloride [57], phenyltrichlorosilane [57, 58], and triphenylchlorosilane [57]. There seems to be considerable steric hindrance to the desired formation of polypentachlorophenylsilanes and, consequently, the exchange takes place preferentially.

$$C_6Cl_5MgCl \xrightarrow[\text{or SiCl}_4]{\text{PhSiCl}_3} C_6Cl_6 \xleftarrow{\text{SiCl}_4} C_6Cl_5Li$$

The reaction of phenylmagnesium chloride with stannic chloride gives analogous results.

(c)   From Non-aromatic, Six-carbon Ring Compounds

It has already been mentioned (section 1.2.1a) that some poly-chlorinated cyclohexanes and cyclohexenes give perchlorobenzene with base or with zinc [37-40, 59]. It must be added that perchloro-cyclohexane (II) and perchlorocyclohexene (III), which are obtained by extended photochlorination of perchlorobenzene (or pentachloro-benzene) with chlorine and sunlight, by treatment with zinc give back perchlorobenzene [39, 40].

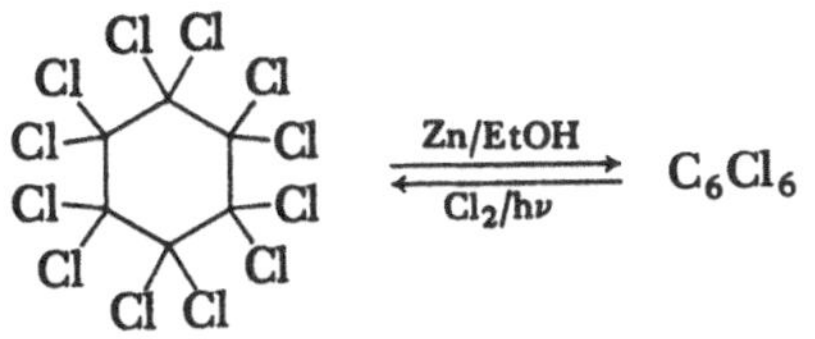

$$\text{(II)} \xrightleftharpoons[\text{Cl}_2/\text{h}\nu]{\text{Zn/EtOH}} \text{C}_6\text{Cl}_6$$

(II)

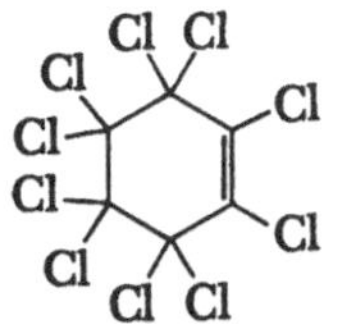

(III)

Benzene hexachloride (γ-isomer) gives perchlorobenzene with chlorine and sunlight [59]. Oxidation of a benzene hexachloride to perchlorobenzene with chlorosulphonic acid or sulphur trioxide has been reported [60]. Other polychlorocyclohexanes react with chlor-ine at 820°, using charcoal as the catalyst, giving excellent yields of perchlorobenzene [61].

Perchlorocyclohexa-1,4-diene (IV; Barral's chlorocarbon) can be dechlorinated to perchlorobenzene by sodium amalgam [62], ethanolic sodium or potassium hydroxide [38, 62], oleum [62, 63], thermolysis above 200° or below with phosphorus pentachloride and chlorine [62].

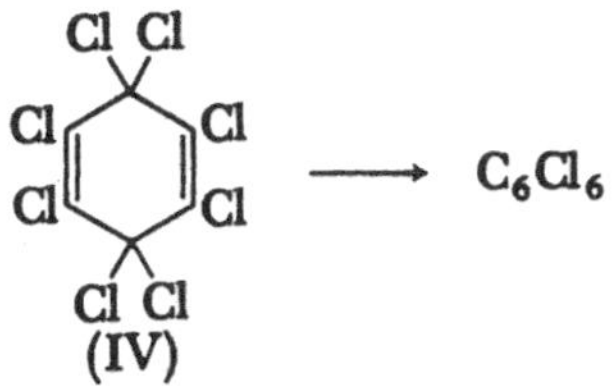

$$\text{(IV)} \longrightarrow \text{C}_6\text{Cl}_6$$

(IV)

The dechlorination with oleum is particularly interesting since it takes place stepwise. The following mechanism is consistent with the facts.

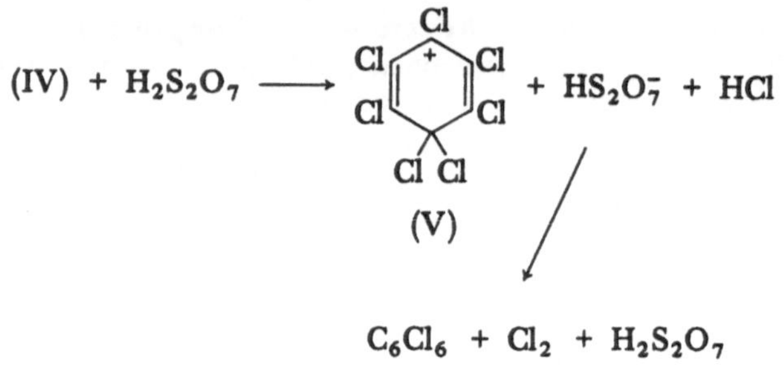

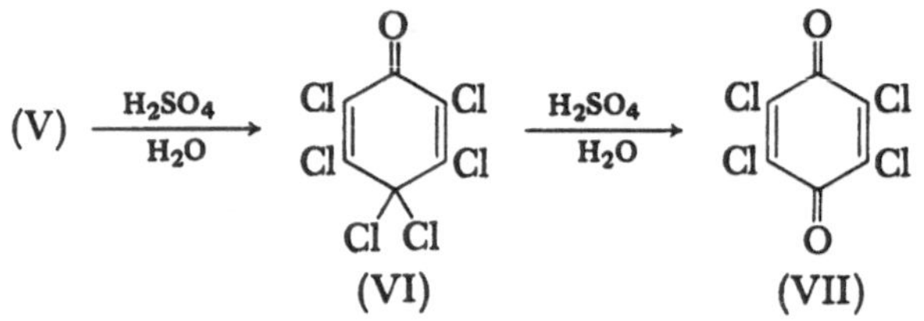

$$(IV) + H_2S_2O_7 \longrightarrow \text{[structure (V)]} + HS_2O_7^- + HCl$$

(V)

$$C_6Cl_6 + Cl_2 + H_2S_2O_7$$

$$(V) \xrightarrow[H_2O]{H_2SO_4} \text{[structure (VI)]} \xrightarrow[H_2O]{H_2SO_4} \text{[structure (VII)]}$$

(VI)   (VII)

Accordingly, chloranil (VII) and hydrogen chloride are also formed [62, 63]. The deep red-violet colour appearing immediately after the reaction components are mixed is an indication of carbonium ion formation. Since neither perchlorocyclohexa-2,5-dienone (VI; hexachlorophenol) nor (VII) give deep colours with oleum it is concluded that carbonium ion (V) is an intermediate [64]. (V) becomes stable in the absence of hydrogen chloride. In this connection it is mentioned that hydrogen chloride is evolved first and chlorine later [62, 63].

Perfluorocyclohexene reacts with aluminium chloride at mild temperatures yielding perchlorobenzene [65].

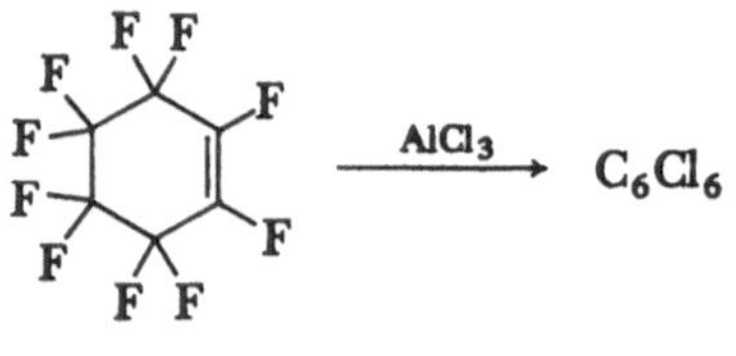

$$\text{[perfluorocyclohexene structure]} \xrightarrow{AlCl_3} C_6Cl_6$$

Heating (VI) at 110-160° with phosphorus pentachloride yields a mixture of perchlorobenzene, a pentachlorophenyl phosphate, and Barral's chlorocarbon (IV) [62]. If the temperature is too high (> 170°) (IV) is not obtained [62, 66]. A reasonable mechanism for this process involves the formation of phosphonium salts (VIII) and (IX). Accordingly, it has been reported that (IV) decomposes above 200° giving perchlorobenzene and chlorine [62]. It has recently been

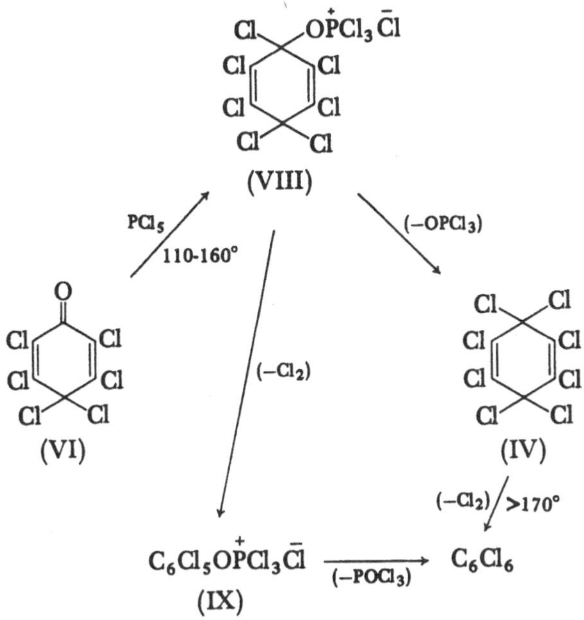

$$\overset{+}{\text{OPCl}_3}\,\bar{\text{Cl}} \qquad \text{(VIII)}$$

$$\xrightarrow[\text{(IX)}]{(-\text{POCl}_3)} \quad C_6Cl_6$$

$C_6Cl_5\overset{+}{\text{OPCl}}_3\bar{\text{Cl}}$

(VI)   (IV)

PCl_5, 110-160°

(-OPCl$_3$)

(-Cl$_2$)

(-Cl$_2$), >170°

found that phosphonium salts of type (X) decompose thermally as follows [67].

$$(C_6Cl_5\overset{+}{\text{OP}}Ph_3)X^- \xrightarrow{\Delta} C_6Cl_5X + Ph_3PO$$

(X)

(X = Cl, Br, I)

Heating at 170-175° an equimolecular mixture of hexachlorophenol (VI) and aluminium chloride results in the formation of perchlorobenzene and phosgene [68].

Trichloro-p-quinone [47], chloranil [47, 62, 66] and chloranilic acid (dichloro-2,6-dihydroxy-p-quinone) [47] give perchlorobenzene when heated (180-200°) with phosphorus pentachloride. At lower temperatures (135-140°) chloranil gives Barral's chlorocarbon (IV) instead, along with a pentachlorophenyl phosphate [62]. Presumably, (VI) is an intermediate. Conversion of chloranil can also be effected with chlorine in the presence of iodine, at 350°.

(*d*)   *By Chlorinolysis of Alkylbenzenes*

The alkylbenzenes are particularly sensitive to chlorinolysis when submitted to exhaustive chlorination.

With chlorine and ultraviolet light αH-heptachlorotoluene [69], $\alpha H \alpha' H$-octachloro-o-xylene [69], $\alpha H, \alpha' H, \alpha'' H$-nonachloromesitylene [69], and a mixture of bisdichloromethylbenzenes [70] are converted into perchlorobenzene [69-72].

With chlorine and white incandescent light the dichloromethyl polychlorobenzenes remain unaltered [69-72]. However, ethylpentachlorobenzene gives some perchlorobenzene, although $\alpha H, \alpha H, \beta H, \beta H$-hexachloroethylbenzene is the major product [70a] indicating a greater resistance of the methyl group to chlorinolysis.

It has been found that chlorine and white light converts readily and quantitatively perchlorotoluene and perchloro-p-xylene into perchlorobenzene [9], the process being assisted by steric strain. Extended chlorination of 2,4,6-trichloromesitylene under the same conditions gives $\alpha H, \alpha' H, \alpha'' H$-nonachloromesitylene, $\alpha H, \alpha' H$-octachlor-m-xylene, αH-heptachlorotoluene and perchlorobenzene; i.e., all products of gradual chlorinolysis [9].

Because of chlorinolysis, Beilstein and Kuhlberg were unable to obtain perchlorotoluene from $\alpha H, \alpha H$-hexachlorotoluene, αH-heptachlorotoluene, and $o H$-heptachlorotoluene by heating with antimony pentachloride, perchlorobenzene being the sole product [16]. The chlorination of bibenzyl and triphenylmethane under similar conditions (350°) yields also perchlorobenzene [73].

Chlorination of isopropylbenzene [29], p-cymene [29], diphenylmethane [46], and α-trichloromethyldiphenylmethane [43] with chlorine and iodine, at 200-350°, converts them into perchlorobenzene.

Chlorination with chlorine, ferric chloride and/or iron of ethylbenzene [74, 75], isopropylbenzene [74], and 1,4-dibutylbenzene [74] gives perchlorobenzene, the conversion being favoured by longer reaction times and/or higher temperatures [74]. With ethylbenzene, ethyl chloride is formed, indicating that chlorinolysis occurs prior to side-chain chlorination. p-Cymene yields, however, $\alpha H, \alpha H, \alpha H$-pentachlorotoluene, no perchlorobenzene being detected. This again indicates that methyl is the alkyl group most difficult to replace by chlorine.

Ethylpentachlorobenzene, 1,4-diethyltetrachlorobenzene and isopropylpentachlorobenzene react with sulphuryl chloride, sulphur monochloride and iron to give 82, 57, and 45% of perchlorobenzene [76] respectively. From 1,4-diethyltetrachlorobenzene some ethylpentachlorobenzene is also isolated. It is pointed out, however, that under chlorination conditions $\alpha H, \alpha H, \alpha H$-pentachlorotoluene is stable, so confirming once more the remarkable resistance of the methyl group to chlorinolysis.

The conversion into perchlorobenzene of benzotrichloride [15, 69, 77], oH-heptachlorotoluene [9], and perchlorotoluene [9] with reagent BMC under forcing conditions has also been reported.

The remarkable chlorinolyses of alkylbenzenes mentioned above are set out:

$$C_6H_5R \xrightarrow{\text{(1) (4)}} C_6Cl_6 \xleftarrow{\text{(2) (3) (4) (5) (6)}} C_6Cl_5R'$$

(1) Cl_2, $FeCl_3/Fe$; R = Et, i–Pr

(2) Cl_2 hν; R' = $CHCl_2$

(3) Cl_2 hν (white); R' = Et, CCl_3

(4) $SbCl_5$, Δ; R = CH_2CH_2Ph, $CHPh_2$; R' = CH_2Cl, $CHCl_2$

(5) SO_2Cl_2, S_2Cl_2, Fe; R' = Et, i–Pr

(6) BMC, Δ; R = CCl_3, R' = CCl_3

$$C_6Cl_5CHCl_2$$

$$Cl_2 \Big| h\nu$$

$$C_6Cl_6$$

$$Me_2CH \langle \bigcirc \rangle Me \xrightarrow{Cl_2, \ FeCl_3/Fe} C_6Cl_5Me$$

Perhalo compounds of type $C_6Cl_5C_2F_xCl_{5-x}$ when chlorinated with chlorine and ferric chloride or treated with antimony pentafluoride give perchlorobenzene [75]. In the latter case, it is assumed that antimony chlorofluorides are formed by chlorine-fluorine exchange, causing subsequent chlorinolysis.

Perchlorobenzene is also formed from ethylpentachlorobenzene with Kharash's chlorinating agent [70], by electro-chemical chlorination of toluene in hydrochloric acid–acetic acid [30], and with chlorine and iodine on tribenzylamine [43].

(e) By Chlorinolysis of Fused Polycyclic Aromatic Hydrocarbons
High temperature chlorination of these hydrocarbons with antimony pentachloride or with chlorine in the presence of iodine yields

usually perchlorobenzene. Naphthalene gives a mixture of perchloro-benzene, perchloroethane and carbon tetrachloride [43, 78, 79]. Anthracene behaves similarly, carbon tetrachloride being the only by-product [43, 80].

These reactions proceed most probably through perchlorodihydro-arenes, such as perchloro-1,4-dihydronaphthalene (XI) and per-chloro-9,10-dihydroanthracene (XII), which undergo subsequent chlorolytic fragmentation. Milder chlorination conditions allow, in some cases, the isolation of such intermediates.

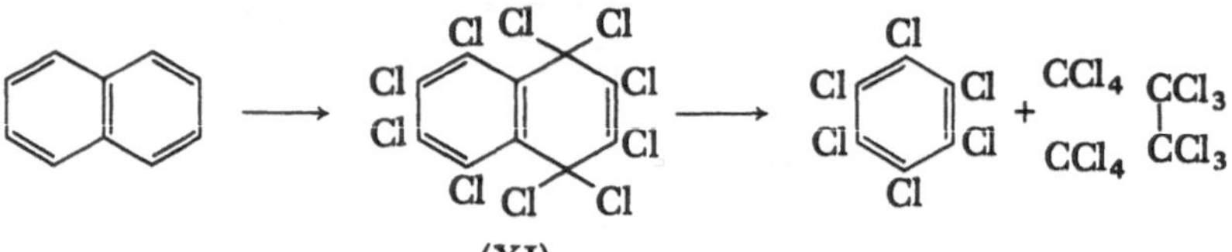

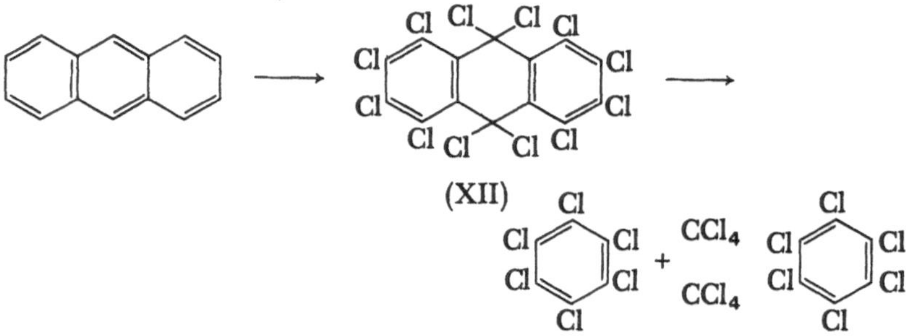

(XI)

(XII)

Compounds possessing a biphenyl system give perchlorobiphenyl as the end-product rather than perchlorobenzene (see 1.2.1f). For example, biphenyl itself gives perchlorobiphenyl exclusively when submitted to exhaustive chlorination [43]. However, phenanthrene, with a built-in biphenyl system, gives perchlorobenzene and carbon tetrachloride [43, 81]. Similarly, chrysene gives a mixture of perchlorobenzene, perchloroethane, and carbon tetrachloride [73]. This remarkable behaviour is partly due to steric strain as chlorine substitution increases which is relieved by chlorine addition. Chlorine addition to the central carbon–carbon bond of the 'built-in' biphenyl is favoured on account of its high double bond character and followed by chlorinolysis to perchlorobenzene. Pyrene, by contrast, does not give perchlorobenzene [73]. Presumably, bond constraints play a decisive role here. Chlorinolysis in naphthalene, chrysene and pyrene also proceed, presumably, via chlorine addition to the naphthalene system.

(f) By Chlorinolysis of Benzenoid Carbonyl and Carboxyl
Derivatives

Perchlorobenzene among other products is found in the following
chlorinations: Antimony pentachloride with benzaldehyde in the
presence of iodine [82]; chlorine and aluminium trichloride with
4-methylacetophenone [73]; antimony pentachloride with benzophenone [84], benzil [84], benzoic acid [84], benzoyl chloride
[84], phthalic acid [84], and o-benzoylbenzoic acid [84-86]; chlorine in chlorosulphonic acid with benzoic acid in the presence of
iodine [72-87]; antimony pentachloride with o-naphthoquinone
[76], anthraquinone [84-86, 88], xanthone [85, 86], and acridone
[85, 86]; chlorine and ferric chloride with phthalic anhydride or
perchlorophthalide [89]; chlorine with phthalic anhydride at
360-400°, using metal chlorides and alumina as catalysts [31]; and
phosphorus pentachloride with perchlorophthalide [90].

Some of the preceding chlorinolyses deserve comment.

The chlorination of benzoic acid in chlorosulphonic acid takes
place through sulphonic acids and perchlorobenzoic acid, the latter
being actually the major reaction product (78% yield).

$$PhCO_2H \xrightarrow{ClSO_3H} C_6H_4(SO_3H)CO_2H \xrightarrow[(ClSO_3H)]{Cl_2,I_2} C_6Cl_5CO_2H$$

$$\downarrow$$

$$C_6Cl_6$$

In the exhaustive chlorination of anthraquinone with antimony
pentachloride some intermediates of considerable mechanistic interest have been isolated. Their formation is interpreted as follows:

The exhaustive chlorinations of xanthone and acridone occur also, presumably, through the corresponding perchloro-aromatic compounds which have been isolated under the actual chlorination conditions.

$$C_6Cl_5CO_2H + C_6Cl_6$$

$$C_6Cl_6$$

It is also remarkable that the chlorinolysis of phthalic anhydride with chlorine in the presence of ferric chloride gives carbon dioxide and phosgene. A plausible reaction mechanism is set out:

$$C_6Cl_6 + CO_2 + COCl_2$$

Biphenyl systems with carbonyl groups give little or no perchlorobenzene when submitted to chlorolytic conditions, the major product being perchlorobiphenyl. Some perchlorobenzene has been reported in the exhaustive chlorination with antimony chloride of biphenyleneglycollic acid [73], phenanthraquinone [73], 2,2'-biphenylenedicarboxylic acid [73], and fluorenone [85].

(g) *By Chlorinolysis of Other Compounds*

Exhaustive chlorination of benzidine with antimony pentachloride gives much perchlorobenzene and little perchlorobiphenyl [73]. This result is related to the high activation of the *para* position of anilines towards electrophilic attack, particularly chlorination, giving quinonoid structures [91] which then chlorinolyse.

$$C_6Cl_6$$

Carbazole, another aromatic amine, yields a mixture of perchloro-benzene and perchlorobiphenyl [73].

$$C_6Cl_6 + C_6Cl_5C_6Cl_5$$

It has been found that while 2-cyanonaphthalene with antimony pentachloride gives perchlorobenzene, its 1-cyano isomer does not [73]. Rosaniline yields a mixture of perchlorobenzene and carbon tetrachloride [73].

Some aliphatic compounds give perchlorobenzene when chlorin-ated exhaustively: hexyl iodide [93], cetyl iodide [94], di-idobutyl [94], di-isoamyl [94], palmitic acid [94], camphor [43], and cycloheptatriene hexachloride [95].

(h)   By Thermolysis of Chlorinated Aliphatic Compounds
Disregarding those compounds containing less than five carbon atoms the following behave similarly: perchloropenta-1,3-diene [96], dodecachlorohexane [97], isomeric perchlorohexatrienes [98], per-chlorohexa-1,5-diene [98], a decachlorohexene [99], *cis,cis*-1-carboxyheptachloro-1,3,5-triene [100], perchlorocyclopentene [101], perchlorocyclopentadiene [102], perchlorofulvene [98, 103], isomers of perchloromethylcyclopentadiene [96], perchloro-tropone [104], and perchlorodicyclopentenyl [105].

(i)   By Thermolysis and Photolysis of Benzenoid Perchloro-Compounds
The thermolysis of pentachlorophenol at 300° gives perchloro-benzene (54.5%) along with perchlorodibenzdioxane (XIII) [45].

(XIII)

The following mechanism has been suggested [50]:

$$C_6Cl_5OH \xrightarrow[(-H_2O)]{\Delta} (C_6Cl_5)_2O \xrightarrow{HCl} C_6Cl_6 + C_6Cl_5OH$$

The required hydrogen chloride would result from the formation of (XIII) and other by-products. Accordingly, when calcium carbon-ate is added the yield of perchlorobenzene decreases sharply.

It has already been mentioned (see section 1.2.1c) that perchlorobenzene is obtained by thermolysis of triphenylpentachlorophenoxyphosphonium chloride (XIV). This salt is formed in the reaction of pentachlorophenol with triphenylphosphine dichloride and it is decomposed in boiling benzene in an overall yield of 79% [67].

$$C_6Cl_5OH + Ph_3PCl_2 \xrightarrow{\text{(–HCl)}} \{Ph_3C_6Cl_5O\overset{+}{P}\}\ \overset{-}{Cl}$$

$$(XIV)$$

$$C_6Cl_6 + Ph_3PO$$

It has recently been reported [106] that perchloro-*N*-methyleneaniline (XV) obtained by chlorination from various anilines, at 400-700° is converted partly into perchlorobenzene.

$$\text{(XV)} \qquad \xrightarrow{\Delta} \quad C_6Cl_6 + C_6Cl_5CN + CNCl + Cl$$

The thermolysis of perchlorotoluene at 310-320° gives, among other chlorocarbons, a 66% yield of perchlorobenzene [107]. This product is also obtained (9.7%) by irradiating a solution of the toluene in carbon tetrachloride [108] with u.v. light. These two processes will be considered later in some detail.

It has been found that the pyrolysis of perchlorophthalic anhydride at 800° gives 34% yield of perchlorobenzene [109, 110]. The following mechanism involving the intermediacy of perchlorobenzyne (XVI) has been suggested [109].

1.2.2   Perchlorobenzene: Properties

Perchlorobenzene is a sublimable, colourless chlorocarbon melting at 227°. The distance between two vicinal chlorine nuclei (3.12 Å) is shorter than twice the value accepted for the van der Waals radius (3.60 Å) and it is reasonable, therefore, to assume that the ring of perchlorobenzene is puckered, with the C–Cl bonds being bent alternatively upwards and downwards from its mean plane.

Controversy has been raging during the past twenty-five years as to whether perchlorobenzene was a distorted molecule. Coulson and Stocker [111, 112] have pointed out that significant electrostatic forces oppose the repulsions between vicinal chlorines. Although their own quantum calculations favour the corrugated conformation, nevertheless they regard them as inconclusive. Scherer, on the basis of Urey-Bradley and Valence Force Fields, has calculated the vibrational frequencies for both the planar (D_{6h}) and the distorted (D_{3d}; θ, $10°$) forms [113]. Comparison with the observed (infrared and Raman) data does not allow any definite conclusion either. Delorme and co-workers have recently studied the vibrational spectrum of perchlorobenzene but are unable to find any evidence for out-of-plane bendings [114].

Other conflicting or inconclusive physical measurements [115] include X-ray [116-122], electron diffraction [123-126], infrared [127, 128], quadrupole resonance [129], and ultraviolet [130] spectra. However, the most recent electron diffraction [126] and X-ray [122] data indicate that perchlorobenzene is planar.

The ultraviolet spectrum of perchlorobenzene has been reported by Conrad-Billroth [131], Kopelman [130], Ballester et al. [132, 133] and Hammond [134]. Its infrared spectrum has been recorded by Parodi [135], Young et al. [136], Plyler et al. [137], Kopelman and Schnepp [127], Saeki [138], Scherer and Evans [139], Chambers et al. [140], Delorme et al. [114], and Darmon et al. [141]. It has been studied by Murray et al. [142], Lecompte [143], Nonnenmacher and Mecke [144], Saeki [138], Scherer et al. [113, 139, 145-149], and Delorme et al. [114, 128]. The Raman spectrum has been studied by Dadieu et al. [150], Truchet and Chapron [151], Murray and Andrews [152], Saeki [138], Scherer [139], Ziegler [153], Delorme et al. [114], and Sirkar and Bishui [154].

Chlorine nuclear quadrupole resonance studies on perchlorobenzene have been carried out by Weatherley et al. [155], Duchesne and Monfils [129], Bray et al. [156, 157], Richardson [158], and Dixon and Bloembergen [159]. Krishnan and Banerjee [160], and Lasheen [161] have studied its diamagnetism.

The vapour density of perchlorobenzene was first measured by Basset in 1867 [162], and its value confirmed it as a C_6Cl_6 chlorocarbon.

1.2.3   Perchlorobenzene: Reactions

(a) *Stability*

Julin [2], and Phillips and Faraday [6] reported in 1821 the exceptionally high thermal stability and chemical inertness of perchlorobenzene. It can be submitted to temperatures around $500°$ [163, 164] and its vapours withstand red-heat temperatures with little decomposition [6]. It is significant, for example, that perchlorobenzene is formed by heating chloroform vapour at bright-red heat [162]. Perchlorobenzene has been reported to be inert towards concentrated hydrochloric acid [2], boiling concentrated nitric acid [2], concentrated sulphuric acid [2], boiling potassium permanganate solutions [162], chlorine at red-heat [6] or in sunlight [19], bromine and sunlight [19], and hot alkalies, either as solid [32] or in solution [19, 162]. However, boiling, fuming nitric acid and, under certain circumstances, bases attack it (cf. below).

(b) Fragmentation

The fragmentation of perchlorobenzene upon pyrolysis and electron impact in the mass spectrometer has been studied by Mayerson and Fields [166], and Schäfer [167].

(c) Fluorination

Whearty and Bancroft [168, 169] found that fluorine reacts explosively with perchlorobenzene giving a liquid mixture of fluorochlorinated compounds with liberation of chlorine and heat. Bigelow and Pearson showed that this reaction, when carried out in carbon tetrachloride at $0°$-to-room temperature, consists of an addition of fluorine to the benzene ring [170]. This result was confirmed later by Musgrave and co-workers using 1,2,2-trifluorotrichloroethane as a solvent [171]. They obtained in an excellent yield perhalocyclohexanes $C_6Cl_xF_{12-x}$, where x depended on the temperature at which the reaction was carried out. Subsequent treatment with iron at $300°$ gave perfluorobenzene and various perfluorochlorobenzenes.

$$C_6Cl_6 \xrightarrow{F_2} C_6Cl_xF_{12-x} \xrightarrow{Fe, \Delta} C_6F_6 + C_6Cl_yF_{6-y}$$
(mixture)

Although other reactions with fluorine gas have been reported [172], most fluorinations have been performed employing milder agents, for instance with antimony pentafluoride [173-176]. The products are, predominantly, mixtures of perfluorochlorocyclohexenes:

There exists a remarkable inertness of the dichlorovinylene system.

The use of antimony pentachloride as a solvent has been recommended since it does not react with antimony pentafluoride [176]. In this way, it has been possible to isolate 3,3,6,6-tetrafluorotetrachlorocyclohexa-1,4-diene.

Fluorinations with halogen fluorides have also been reported [177-179]. McBee *et al.* employed bromine trifluoride so obtaining a mixture of perfluorochlorobromocyclohexanes [177]. Treatment of this mixture with antimony pentafluoride cannot effect replacement of the bromine completely. The product resulting from the last treatment when submitted to dehalogenation with zinc in ethanol gives a mixture containing perfluorobenzene, perfluorocyclohexa-1,3-diene and perfluorocyclohexa-1,3-dienes and -cyclohexenes [177].

Chambers and co-workers used chlorine trifluoride instead, operating at rather high temperature (240°) [140]. They obtained mixtures of perfluorochlorocyclohexanes and -cyclohexenes, which could be aromatized by heating with iron, in excellent conversion yields. These authors confirmed the reluctance of the dichlorovinylene group to undergo fluorination. They proposed two simultaneously operating mechanisms involving 1,2- and 1,4-addition of either chlorine monofluoride or fluorine, and allylic substitutions. The initial 1,4-addition had been proposed in the fluorination with antimony pentafluoride [174, 175].

Musgrave *et al.* have found that fluorination of perchlorobenzene with cobaltic fluoride gives a mixture of perfluorocyclohexanes [179].

It has been shown so far that substitution of chlorine by fluorine in perchlorobenzene occurs in two steps: (1) fluorine addition, and (2) halogen elimination. Fluorine substitution can, however, be performed directly by means of alkali fluorides in dipolar aprotic solvents, as reported by Finger *et al.* [180]. Thus, with potassium fluoride in *N*-methyl-2-pyrrolidone (NMP), at 200°, perfluorochlorobenzenes are obtained [181, 182]. While rubidium and caesium fluorides can replace, even advantageously, potassium fluoride, sodium fluoride is ineffective [181]. Other solvents such as dimethylformamide [180, 183], dimethylsulphoxide [180, 181, 183], and tetramethylenesulphone (sulpholane) [183], have also been successfully employed. With the latter solvent, and potassium fluoride, at 230-240°, a mixture containing a very small proportion of perfluorobenzene is obtained while, with caesium fluoride, even at lower temperatures, the latter turned out to be the major product (42% yield) [183].

$$C_6Cl_6 \xrightarrow[\text{NMP}]{KF/200°} \begin{array}{c} Cl \\ F \diagup\!\!\!\diagdown F \\ Cl \diagdown\!\!\!\diagup Cl \\ F \end{array} + C_6F_4Cl_2 + C_6F_5Cl$$

$$C_6Cl_6 \xrightarrow[\text{sulpholane}]{CsF/160\text{-}190°} C_6F_6 + C_6F_5Cl$$

Vorozhtsov *et al.* have reported that the reaction of perchloro-benzene with potassium fluoride at 450-500°, in the absence of solvent, results in a mixture of perhalobenzenes containing a substantial proportion of perfluorobenzene (20%) [184].

It is known that accumulation of chlorines in perchlorobenzene activates the aromatic ring towards nucleophilic displacement reactions [185-188]. Consequently, in media enhancing the nucleophilicity of the attacking species, fluoride ion attack on perchloro-benzene becomes possible. The strong (−I) effect of the first introduced fluorine and its (+T) effect direct the second and third into the *meta* positions [182] yielding the *sym*-trifluorotrichloro-benzene as the predominant reaction product [181].

$$C_6Cl_6 \xrightarrow[\text{F}]{\text{F}^-} \quad \underset{Cl}{\overset{F}{\underset{Cl}{\bigcirc}}} \quad + \quad \underset{Cl}{\overset{F}{\underset{Cl}{\bigcirc}}} \quad + \quad \underset{Cl}{\overset{Cl}{\underset{F}{\bigcirc}}}$$

(d)   *Chlorination*

Riemschneider and Oswald chlorinated perchlorobenzene with liquid chlorine and sunlight [40]. The reaction is so slow that it takes about half a year to go to completion. The products, as it has already been advanced (Section 1.2.1c), are perchlorocyclohexane (II) and perchlorocyclohexene (III).

(e)   *Oxidations*

Perchlorobenzene reacts sluggishly with cupric oxide at red-heat temperatures [6]. In general, the oxidation of aromatic chloro-carbons is an arduous process, a fact that accounts for the difficulties usually encountered in their elemental quantitative analysis when based on oxidative methods. However, perchlorobenzene is attacked readily with boiling red fuming nitric acid or its mixtures with concentrated sulphuric acid giving chloranil, nitrogen oxides and hydrogen chloride [165].

(f)   *Reductions*

The reduction of perchlorobenzene with hydrogen at red heat and with hydrogen iodide at high temperatures had been carried out by Berthelot and Jungfleisch in connection with their early structural studies [32, 189, 191]. In some cases they obtained benzene and other—presumably aliphatic—hydrocarbons. Dechlorinations with alkali metals, such as lithium [192], sodium [192, 193], sodium amalgam [193], and potassium [19], yield carbonaceous materials. It has been reported that with sodium and sodium chloride at 300°, a highly impure, insoluble material is obtained to which the structure of a poly-*p*-phenylene has tentatively been assigned [194]. Perchlorobenzene reacts with magnesium at 600° giving magnesium chloride [195]. A method of analysis of chlorine in highly chlorinated compounds based on this reaction has been proposed [195]. Although it gives reproducible data its procedure is cumbersome. As far as the authors know, this method has been abandoned for routine analyses. Perchlorobenzene is dechlorinated by red phosphorus. The nature of the resulting product remains unknown [196].

(g) *Metallations†*
    The use of tetrahydrofuran (THF) as a solvent in the preparation of arylmagnesium chlorides from aryl chlorides and magnesium [197] has been applied successfully to perchlorobenzene [198-204, 57] using halides, such as ethyl bromide and others. The product is pentachlorophenylmagnesium chloride, although in one case some (~5%) 1,4-dimetallation was reported [57]. According to Pearson and co-workers, a good yield of pentachlorophenylmagnesium chloride (XVII) can be obtained in refluxing ethyl ether–benzene instead of THF when ethylene bromide is used as an entrainer [205, 206].

$$C_6Cl_6 + Mg \xrightarrow[\text{(THF)}]{\text{CH}_2\text{BrCH}_2\text{Br}} C_6Cl_5MgCl$$
$$(XVII)$$

Perchlorobenzene was reported to be inert towards alkylmagnesium halides in boiling ethyl ether [207]. However, in ether–THF a metal-halogen exchange reaction with benzylmagnesium chloride giving pentachlorophenylmagnesium chloride [57] takes place.
    Pentachlorophenyllithium (XVIII) has been prepared by Rausch and co-workers [208], and other authors [58, 203, 209-216], also by metal-halogen exchange with n-butyllithium at low temperatures (~−70°C), generally in THF:

$$C_6Cl_6 + n\text{-BuLi} \xrightarrow[\text{(THF)}]{-78°} C_6Cl_5Li \text{ or } C_6Cl_4Li_2\text{-}p$$
$$(XVIII) \qquad (XIX)$$

An excess of alkyllithium leads to tetrachloro-p-phenylenedilithium (XIX) [216]. These metallations are of synthetic value, particularly in the formation of perchlorobenzyne.
    Perchlorobenzene reacts at −70°, in THF, with triphenylsilyllithium giving, presumably, pentachlorophenyllithium which reacts subsequently with an active chlorosilane yielding the corresponding pentachlorophenylsilane [217].

$$C_6Cl_6 + Ph_3SiLi \rightarrow C_6Cl_5Li + Ph_3SiCl$$
$$C_6Cl_5Li + Me_3SiCl \rightarrow C_6Cl_5Me_3Si$$

† For detailed discussion see Chapter 3.

Similarly, the reaction of perchlorobenzene with other silyllithium and -sodium derivatives yields pentachlorosilanes and, depending on the relative amount of the components, even 1,4-bis-silyltetrachlorobenzenes [218].

The reaction between perchlorobenzene, lithium dimethylcopper and benzoyl chloride in ethyl ether–THF gives $\alpha H,\alpha H,\alpha H$-pentachloroacetophenone, presumably via pentachlorophenylcopper compounds resulting from metal-halogen exchanges [219]. A remarkable reaction takes place when perchlorobenzene, trimethylsilyl chloride and lithium metal are mixed in THF or 1,2-dimethoxyethane [220, 221]. The product is tetrakistrimethylsilylallene (XX). Presumably, the benzene ring is split into two persilylated moieties [221] by an unknown mechanism.

$$C_6Cl_6 \longrightarrow$$

(XX)

(R = Me$_3$Si)

It is significant that if 1,4-bistrimethylsilyltetrachlorobenzene is substituted for perchlorobenzene the yield of that product nearly doubles (52%) [221], which suggests that this silane might be an intermediate.

(h) *Other Nucleophilic Substitutions*

It has already been mentioned (see section 1.2.3a) that perchlorobenzene is stable towards hot alkalies. However, Weber and Söllscher reported as early as 1883 that ethanolic sodium hydroxide at 180-200° displaces almost all chlorine atoms [222]. Similar results were reported for sodium and ethanol [223] and sodium methoxide in methanol [224]. The reactions with ethanolic sodium ethoxide [225] and methanolic sodium methoxide [226] at high temperatures give mixtures of pentachlorophenol and their corresponding ethers.

A number of such nucleophilic displacements giving pentachlorophenol have been performed with alkaline hydroxides in alcoholic media: sodium hydroxide in glycerol at 250° [227], sodium hydroxide in methanol at 200° [228] and various alkaline solutions at 150-190° [229]. In other cases, pentachlorophenol ethers are obtained; for example, boiling potassium hydroxide in ethanol gives a 95% yield of the ethyl ether [74]. Similar results are obtained with sodium hydroxide in triethyleneglycol at 200° [230] or in methanol at 120° [186] giving, respectively, the hydroxyphenol ether and pentachloroanisole. It has recently been found that when the reaction with sodium hydroxide is carried out in dimethylsulphoxide (DMSO) at 110-120° a 97% of pentachlorophenol is obtained [231]. It has already been mentioned (see section 1.2.3c) that persubstitution by chlorine enhances the susceptibility of the aromatic systems towards nucleophilic substitution reactions owing to the negative inductive effect of the chlorines [186]. Accordingly, it has been found that the rate of basic hydrolysis is about 5×10^4 faster than that of chlorobenzene [232]. The results just mentioned indicate, however, that perchlorobenzene is not so reactive as expected. Part of this abnormal behaviour is probably due to its low solubility in polar solvents [186]. Rocklin claimed that nucleophilic substitution is greatly accelerated when pyridine is used as a solvent [186], producing pentachloroanisole with methanolic sodium hydroxide, and pentachlorophenylisopropyl ether with base in isopropanol in yields above 90% [186]. That pyridine plays an additional role besides being the solvent is shown by the fact that while picolines and lutidines are also effective the 2,6-lutidine [186] is not. Steric shielding preventing the approach of the reagent to a transition complex may be responsible.

Pentachlorothiophenol and its thioethers can be prepared in excellent yields by the pyridine procedure using sodium sulphide or a mercaptide, respectively, as the nucleophile [186]. The latter represents a practical alternative to the Williamson synthesis of thioethers from pentachlorothiophenoxide and alkyl halides.

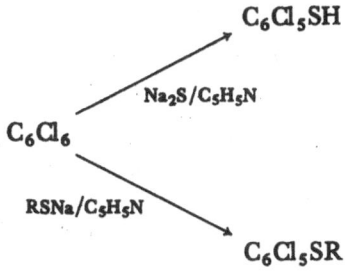

$$C_6Cl_5SH$$

$$Na_2S/C_5H_5N$$

$$C_6Cl_6$$

$$RSNa/C_5H_5N$$

$$C_6Cl_5SR$$

Mercaptides are very strong nucleophiles and react readily with perchlorobenzene even in the absence of pyridine giving frequently p-disubstitution [36] and even 1,2,4,5-tetrasubstitution.

Perchlorobenzene reacts with aqueous ammonia at 250°, methylamine at 150°, boiling ethylenediamine, and dimethylamine at 160° yielding pentachloroaniline (95%), N-methylpentachloroaniline (74%), N-pentachlorophenylethylenediamine (95%), and N,N-dimethylpentachloroaniline (69%), respectively [186]. At lower temperatures the yields decrease sharply [233].

28   MANUEL BALLESTER AND SANTIAGO OLIVELLA

1.2.4 Perchlorobiphenyl: Syntheses

In 1876, Ruoff reported the first synthesis of perchlorobiphenyl (XXI) by chlorination of biphenyl with chlorine, using iodine as a carrier [43]. Other syntheses starting also from biphenyl have been carried out with antimony pentachloride at high (up to 300°) temperature [73, 222] and with reagent BMC (99.6% yield) [234, 234a]. Adrianov, using chlorine and iron at 100-200°, claims to have obtained a mixture of perchlorobiphenyl, 4H-nonachlorobiphenyl and 4H,4'H-octachlorobiphenyl [235]. The work, however, needs verification in view of the reported melting points. Perchlorobiphenyl has been prepared from commercial pentachlorobiphenyl by chlorination with chlorine at high temperature (up to 300°), using a mixture of ferric chloride and iodine as a catalyst [236].

$$\text{Ph--Ph} \rightarrow C_6Cl_5\text{--}C_6Cl_5$$

(XXI)

Perchlorobiphenyl is also formed in the chlorination of some substituted biphenyls: by exhaustive chlorination of benzidine and carbazole with antimony pentachloride [73] and from 4,4'-diamino-octachlorobiphenyl (octachlorobenzidine) by a Sandmeyer reaction

$$p\text{-}H_2NC_6Cl_4C_6Cl_4NH_2\text{-}p \xrightarrow[\text{(b) } Cu_2Cl_2, \text{ HCl}]{\text{(a) } H_2SO_4, NaNO_2} C_6Cl_5C_6Cl_5$$

[237]. Chlorination of 4,4'-dihydroxyoctachlorobiphenyl to perchlorobiphenyl with phosphorus pentachloride at 230° was unsuccessful [222].

Exhaustive chlorination of bi-p-tolyl with antimony pentachloride, at 360° [73], 4,4'-dichloro-3,3'-bistrichloromethylbi-p-tolyl with reagent BMC [234], and photochlorination of perchlorobi-p-tolyl with chlorine and white light [234] give perchlorobiphenyl. Other compounds possessing a built-in biphenyl system, such as phenanthraquinone [73, 85, 86], fluorenone [85, 86], 2,2'-dicarboxybiphenyl [73], and 2,2'-biphenyleneglycollic acid [73], when heated with antimony pentachloride, give mixtures of perchlorobiphenyl and perchlorobenzene (Section 1,2,1f).

Evidence concerning the mechanism of the chlorinolysis of phenanthraquinone and fluorenone includes the isolation of the intermediates shown below when the reaction temperature is lowered [85, 86].

Exhaustive chlorination of cetyl iodide with antimony pentachloride in the presence of iodine, at high temperature, seems to give some perchlorobiphenyl among other chlorocarbons (cf. Section 1.2.1g) [94]. Oxidations of pentachlorophenyllithium involving metal compounds such as titanium tetrachloride give some perchlorobiphenyl [204, 209]. It is unlikely that this reaction proceeds via

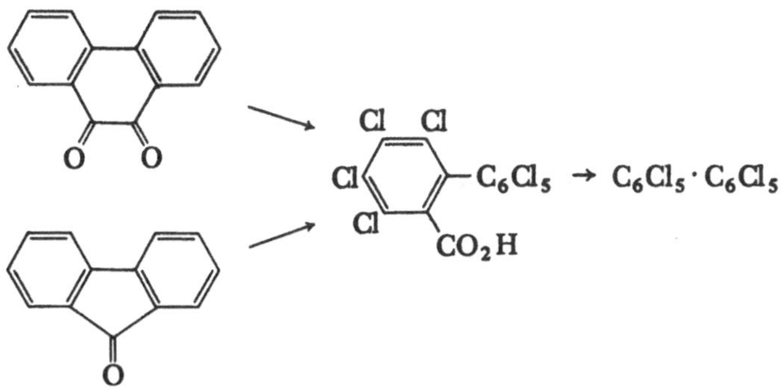

$$Cl_2 \quad Cl$$
$$Cl \diagdown \diagup - C_6Cl_5 \rightarrow C_6Cl_5 \cdot C_6Cl_5$$
$$Cl \diagup \diagdown CO_2H$$

perchlorophenyl radicals since no pentachlorobenzene results which would be expected from radical attack on the solvent [209].

Perchlorobiphenyl is formed in the thermolyses of perchlorobenzylidenecyclohexa-1,4-diene and perchlorodiphenylmethyl radical, to be considered later [238].

1.2.5   Perchlorobiphenyl: Properties and Reactions

(a) *Properties*

Perchlorobiphenyl is a colourless, sublimable solid melting at 309°. Its infrared [234, 239] and ultraviolet [234, 239, 240] spectra have been recorded by Ballester and co-workers. Its mass spectrum has recently been published by Safe and Hutzinger [234a, 241].

(b) *Stability*

Ruoff, in 1876, had already emphasized the high stability of perchlorobiphenyl. Its vapours withstand red heat temperatures remarkably well [43]. Its thermal stability has recently been studied by Johns et al. [163, 164]. They found that it starts to decompose at 510-520°. It does not undergo significant change when treated by antimony pentachloride at 350° [73]. Thus its stability is, in a general sense, comparable to that of perchlorobenzene (see section 1.2.3a).

(c) *Reductions*

Perchlorobiphenyl is stable towards sodium, magnesium and aluminium at temperatures up to 200° [73]. However, sodium in benzene at 250-300° converts it into a carbonaceous material and sodium chloride [73]. The reduction of perchlorobiphenyl with lithium aluminium hydride in THF yields a mixture of 4H-nonachlorobiphenyl and 4H,4'H-octachlorobiphenyl, their relative amounts depending upon the proportion of hydride used [242].

(d) Metallations†

According to Binns and Suschitzky, perchlorobiphenyl reacts with magnesium in THF with the assistance of iodine and ethylene bromide to give a mixture of nonachloro-4-biphenylylmagnesium chloride (XXII) and octachloro-4,4'-biphenylenebismagnesium dichloride (XXIII) [242].

$$C_6Cl_5C_6Cl_5 \xrightarrow[C_2H_4Br_2(THF)]{Mg, I_2} C_6Cl_5C_6Cl_4MgCl\text{-}p + p\text{-}ClMgC_6Cl_4C_6Cl_4MgCl\text{-}p$$
$$\text{(XXII)} \qquad\qquad\qquad \text{(XXIII)}$$

n-Butyllithium exchanges with perchlorobiphenyl in ethyl ether, at low temperatures, giving, analogously, a mixture of 4-mono- (XXIV) and 4,4'-dilithium (XXV) derivatives, depending upon the relative amounts of the reaction components [242].

$$C_6Cl_5C_6Cl_5 \xrightarrow[(ether)]{n\text{-}Bu} C_6Cl_5C_6Cl_4Li\text{-}p + p\text{Li}C_6Cl_5C_6Cl_5Li\text{-}p$$
$$\text{(XXIV)} \qquad\qquad \text{(XXV)}$$

As it occurs in the case of perchlorobenzene, perchlorobiphenyl can be converted into the corresponding mono- and di-arynes by elimination of lithium chloride.

(e) Nucleophilic Substitutions

Like perchlorobenzene and for the same reasons (see sections 1.2.3c and 1.2.3h), perchlorobiphenyl undergoes a variety of nucleophilic reactions. Ethanolic sodium hydroxide at 140-160° yields, presumably, 4,4'-dihydroxyoctachlorobiphenyl ('perchlorodiphenol') [222]. The first systematic study of these processes has recently been published by Binns and Suschitzky [242]. Reaction with sodium methoxide according to Rocklin's pyridine procedure (see section 1.2.3h) gave 4-methoxynonachlorobiphenyl or 4,4'-dimethoxyoctachlorobiphenyl in 90% yields. This reaction has been studied independently by Ballester and co-workers, in methanolic dioxane, establishing unambiguously by an independent synthesis the structure of the dimethoxy derivative [231, 243].

Perchlorobiphenyl reacts with piperidine in dimethylformamide to give mixtures of 4N-piperidylnonachlorobiphenyl and 4,4'-bis-N-

† cf. Chapter 3.

piperidyloctachlorobiphenyl [242]. In the absence of solvent, at 175°, the product is the bispiperidyl derivative [242].

1.2.6 Perchloropolyphenyls

Merz and Weith reported the preparation of the first perchloropolyphenyls [73]. Perchloro-p-terphenyl (XXVI) is obtained from the corresponding hydrocarbon by means of antimony pentachloride at 360° [73, 234a]. Its mass spectrum has been reported [234a].

$$p\text{-}C_6H_5\text{–}C_6H_4\text{–}C_6H_5 \;\rightarrow\; p\text{-}C_6Cl_5\text{–}C_6Cl_4\text{–}C_6Cl_5$$
$$(XXVI)$$

1,3,5-Triphenylbenzene is similarly converted into perchloro-1,3,5-triphenylbenzene (XXVII) [73]. Ballester and co-workers have prepared the three isomeric perchlorotriphenylbenzenes by chlorination of the hydrocarbons with reagent BMC followed by antimony pentachloride at high temperature [234] in yields of 55% (XXVII), 36% (XXVIII), and 60% (XXIX). Their melting points are: 255-9° (XXVII), 302-4° (XXVIII), and 306-8° (XXIX).

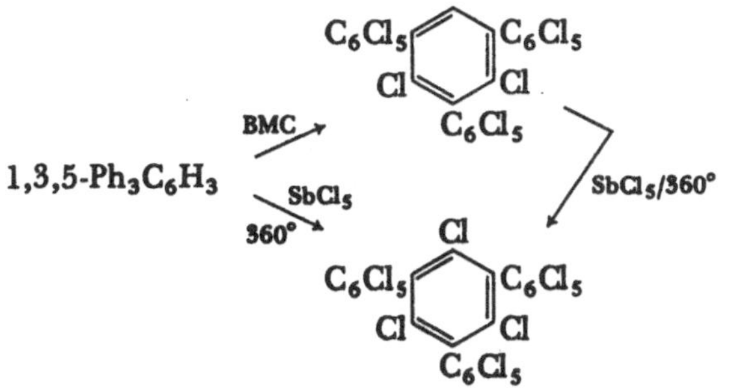

(XXVII)

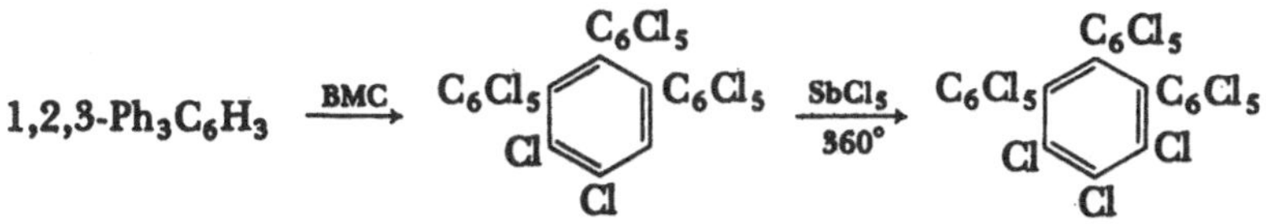

(XXVIII)

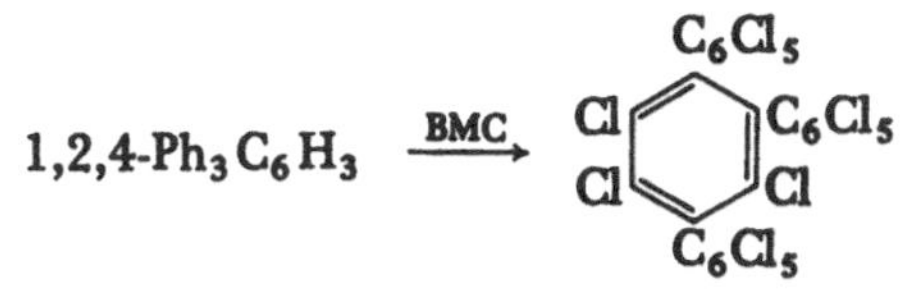

(XXIX)

The 1,2,3- and 1,2,4-isomers are formed, among other chlorocarbons, by thermal oligomerization of perchlorophenylacetylene (XXX) [244] (cf. later).

$$C_6Cl_5C{\equiv}CCl \xrightarrow{\Delta} (XXVII) + (XXIX)$$
(XXX)

Wibaut *et al.* [236] performed the perchlorination of *p*-quaterphenyl (4,4′-diphenylbiphenyl) with antimony pentachloride at 220-270° obtaining perchloro-*p*-quaterphenyl (XXXI), m.p. 364-65°.

$$p\text{-}Ph{-}C_6H_4C_6H_4{-}Ph\text{-}p \rightarrow p\text{-}C_6Cl_5{-}C_6Cl_4C_6Cl_4{-}C_6Cl_5\text{-}p$$
(XXXI)

These perchloropolyphenyls possess the high thermal stability and chemical inertness of perchlorobenzene and perchlorobiphenyl. Concentrated nitric acid at 300-350° attacks them only with great difficulty [73] and sodium in benzene at these temperatures leaves a carbonaceous material from perchloro-1,3,5-triphenylbenzene [73].

1.2.7 Perchlorobiphenylene

Perchlorobiphenylene (XXII) has been reported simultaneously by Brown [110, 245] and Cava [109] and co-workers, to be formed in ca. 30% by pyrolysis of tetrachlorophthalic anhydride, under vacuum at 450-1100°. According to Cava *et al.*, perchlorobenzene is also obtained.

(XVI)   (XXXII)

The proposed mechanism of this reaction involves the intermediate formation of perchlorobenzyne (XVI) (see section 1.2.1j). Perchlorobiphenylene is a yellowish-green solid, m.p. 292-4° [109] (or 296-8° [110]). The fact that this chlorocarbon is obtained at very high temperatures speaks for its excellent thermal stability. Its ultra-violet spectrum has been reported [109, 110].

1.2.8 Perchloronaphthalene: Syntheses

Perchloronaphthalene (XXXIV) was first prepared by Berthelot and Jungfleisch in 1868 [32, 190] by the exhaustive chlorination of naphthalene with chlorine and antimony pentachloride to give presumably a mixture of perchloronaphthalene and perchloro-1,4-dihydronaphthalene (XXXIII). The latter could not be fully converted into perchloronaphthalene by distillation under atmospheric pressure as they claimed.† Ruoff succeeded in preparing it pure using chlorine and antimony pentachloride [43]. Schwemberger and Gordon performed the chlorination with chlorine, iron and iodine at 100-150° in about 90% yields [78, 246]. Care had to be taken to avoid chlorinolysis to perchloroindan which occurs easily at higher temperatures [246]. Goubeau used Schwemberger's method, completing the chlorination by treatment with antimony pentachloride [247]. Brintzinger and Orth realized the perchlorination in the vapour phase using charcoal (400°), rhodium trichloride-silica (400-480°), or rhodium trichloride-alumina (400°) [31] in nearly quantitative yields. Ballester *et al.* chlorinated naphthalene using reagent BMC [248-250] and obtained perchloro-1,4-dihydronaphthalene which is easily converted into perchloronaphthalene in boiling isopropyl ether (89% yield) [248-250].

Perchloro-1,4-dihydronaphthalene seems to be, in general, the precursor of perchloronaphthalene, which decomposes when the temperature is high enough.

(XXXIII)   (XXXIV)

Claus and co-workers found that $2H$-pentachloro-1,4-naphthoquinone [251], $2H$,3-hydroxytetrachloro-1,4-naphthoquinone [251], and 2,3,4-trichlorosulphonyl-α-naphthol [252] react with phosphorus pentachloride at 250° giving perchloronaphthalene. Similarly, perchloronaphthalene results from the reaction of bis-chlorosulphonyl-β-naphthols [253] or 1-naphthylamine-3,6,8-tri-sulphonic acid [78] with phosphorus pentachloride.

Perchloro-1,4-naphthoquinone (XXXV) reacts with phosphorus pentachloride to give perchloronaphthalene [254].

An example of chlorinolysis leading to perchloronaphthalene is the chlorination of methyl α-naphthoate with chlorine at 110° in the presence of iron and ferric chloride, and using perchloroethylene as a solvent [255]. Its isomer, the methyl β-naphthoate, under similar conditions, seems to give perchloro-1,4-dihydronaphthalene

† The melting point reported by Berthelot was about 70° lower than its actual value. Laurent had claimed much earlier the isolation of perchloronaphthalene, although his melting point was 30° too low [43].

(XXXIII) instead, which it was claimed is hydrolyzed to perchloro-1,4-naphthoquinone during the workup of the product [255]. At higher temperatures the product resulting from both esters is perchloroindane [255].

A most unusual, high-yield synthesis of perchloronaphthalene consists of the vapour-phase pyrolysis of perchlorocyclopentadiene [256]. Some (~5%) perchlorobiscyclopentadienyl is also formed which does not seem to be an intermediate, since under the reaction conditions it carbonizes without forming any perchloronaphthalene [256].

1.2.9 Perchloronaphthalene: Properties and Reactions

(a) *Properties*

Perchloronaphthalene is a yellowish solid, m.p. 198°. Its ultraviolet [250, 257], infrared [250, 255, 258], and Raman [247] and mass [234a] spectra have been studied.

X-ray data on perchloronaphthalene have been taken by Gafner and Herbstein [259, 260]. The molecule is warped because repulsions between the perichlorines cause out-of-plane bending of the C–Cl bonds in opposite senses, the deviations from the planar arrangement being 0.54 to 0.79 Å. The α-chlorines cause their vicinal β-chlorines to deviate in the same sense, although to a smaller extent (0.37 to 0.47 Å). Also, the carbon atoms suffer displacements from their mean plane which is about one third that of the chlorine atom to which they are attached. In addition, there is an in-plane bending of the α-chlorine-carbon bonds. This type of distortion is believed to be more favourable than alternating out-of-plane displacements.

(b) *Stability*

The thermal stability of perchloronaphthalene is high, although, presumably, less than that of perchlorobenzene and perchlorobiphenyl. According to Berthelot and Jungfleisch, perchloronaphthalene loses chlorine when distilled under atmospheric pressure [32] which is at least partly [78] due to the presence of perchloro-1,4-dihydronaphthalene (see section 1.2.7). Stability towards aqueous sodium or potassium hydroxide and cold ethanolic potassium hydroxide [78] has been reported.

(c) *Fluorination*

Fuller has found that perchloronaphthalene reacts with potassium fluoride in sulpholane at 230-240°, giving a 50-60% yield of perfluoronaphthalene [183].

$$(XXXIV) \quad \xrightarrow[\text{sulpholane}]{KF\,(230\text{-}40°)}$$

(d) *Chlorination*

Schwemberger *et al.* have found that when perchloronaphthalene is submitted to exhaustive chlorination it fragments to perchloroindane and carbon tetrachloride [246], or even to perchlorobenzene, perchloroethane and carbon tetrachloride (1.2.1e) [78]. With antimony pentachloride in carbon tetrachloride a red colour, due possibly to formation of a carbonium ion (XXVI) [78], is observed (cf Section 1.2.1c).

(XXXV)

(e) Oxidation

Perchloronaphthalene reacts with hot concentrated nitric acid giving perchloronaphthoquinone and tetrachlorophthalic acid [78, 254].

(f) Reductions

Berthelot performed the reduction of perchloronaphthalene to naphthalene at high temperature with hydrogen over pumice [32, 261]. With hydrogen iodide at 275° the reduction proceeded further to give saturated hydrocarbons [190].

(g) Other Nucleophilic Substitutions

Berthelot observed the formation of violet vapours and an acidic substance, possibly naphthols, when perchloronaphthalene was heated with solid sodium or potassium hydroxide [32, 190]. Treatment with ethanolic potassium hydroxide yields a resinous material [78]. Ballester and co-workers have recently found that under similar conditions perchlorinated naphthols [262] are formed.

1.2.10   Perchlorophenylnaphthalenes and Perchlorofluoranthene

Ballester and co-workers have performed the perchlorination of 1- and 2-phenylnaphthalene with reagent BMC [250]. As in other fused polycyclic aromatic hydrocarbons, this moderate temperature reagent gives perchlorodihydrophenylnaphthalenes (see sections 1.4.23 and 1.4.24) which can subsequently be dechlorinated by either heat or stannous chloride in dioxane to the corresponding perchloro-phenylnaphthalenes. The overall yield in the dechlorination of (XXVI) is 72.2% and that of (XXXVIII) 87 or 93.6%, depending on

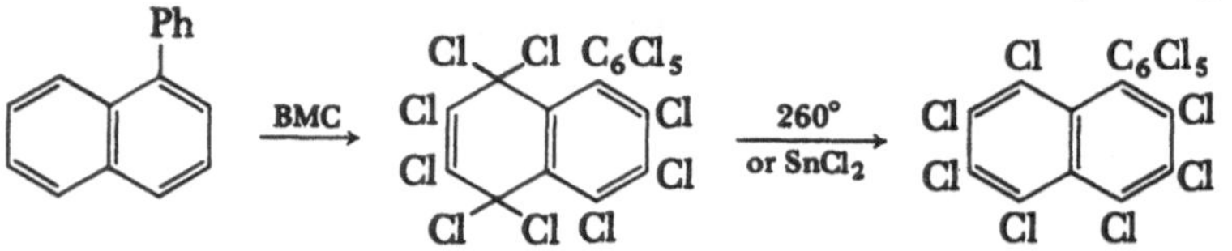

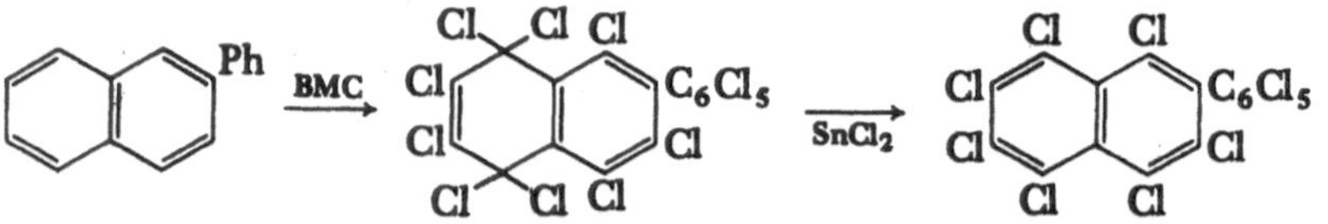

whether it is performed by stannous chloride or thermally. Perchloro-1-phenylnaphthalene is also formed among other chloro-carbons by a thermal dimerization of perchlorophenylacetylene (XXX; Section 1.5.1).

The compounds (XXXVII) and (XXXIX) are yellow solids melting at 276-8° and 238-41° respectively.

Perchlorofluoranthene (XL) is obtained among other chlorocarbons (XXXVI and XXXVII) in a 15% yield in the chlorination of 1-phenyl-naphthalene with reagent BMC (Section 1.4.23) [250]. It is an intense-yellow solid, melting at 263-5° with known infrared and electronic spectra [250]. The formation of (XL) is accounted for in the following manner (cf. Section 1.2.1c and 1.2.9d):

(XL)

The infrared and electronic spectra of these compounds are known [250].

1.2.11 Perchloroanthracene

Ruoff attempted unsuccessfully the synthesis of perchloroanthracene (XLI) from anthracene by high temperature chlorination, obtaining perchlorobenzene and carbon tetrachloride [43], presumably because of chlorinolysis of the intermediate perchloro-9,10-dihydroanthracene (XII) (Section 1.2.1e). Antimony pentachloride at 280-290° gave the same result [80] and at lower temperatures an unidentified octachloroanthracene was isolated. Perchloroanthracene has been prepared by Ballester and co-workers from perchloro-9,10-dihydroanthracene by dechlorination with stannous chloride in chloroform [248-250] in 45% yield.

(XII)   (XLI)

The anthracene (XLI) is a yellow solid m.p. 238-43°. Its molecular shape has not yet been studied but there is no doubt that it is probably more distorted than perchloronaphthalene. Its infrared and electronic absorption spectra have been taken [250]. In the solid state it presents thermochromy being orange-yellow at room temperature and changing to orange-red when heated to 100° [250].

1.2.12   Perchlorophenanthrene

The synthesis of perchlorophenanthrene (XLII) from phenanthrene has unsuccessfully been attempted by Ruoff with chlorine and iodine at 350° [81], and Grinbaum and Marchlewski with chlorine and sunlight, in chloroform and the presence of red phosphorus [263]. Ruoff and Zetter obtained perchlorobenzene and carbon tetrachloride, although the latter author was able to isolate an intermediate octachlorophenanthrene at lower temperature (180-200°). Grinbaum and Marchlewski obtained, presumably, the same octachlorophenanthrene (similar melting points) as a final product, namely the 4H,5H-octachloroisomer.

Perchlorophenanthrene has been reported by Ballester [248] and Brooks et al. [264]. The latter authors obtained it from phenanthrene in low yield through a cumbersome sequence with chlorine, iodine and chlorosulphonic acid, followed by thermolysis at 330°. The former workers prepared it in two steps, namely, chlorination with reagent BMC and dechlorination with isopropyl ether in 44% yield. The first step gives perchloro-9,10-dihydrophenanthrene (XLIII) which undergoes conversion in a 78.4% yield [250]. Recently, Ishimori, West et al. have described an elegant high-yield synthesis of perchlorophenanthrene by thermolysis of perchloroheptafulvalene (XLIV) at 340° [265].

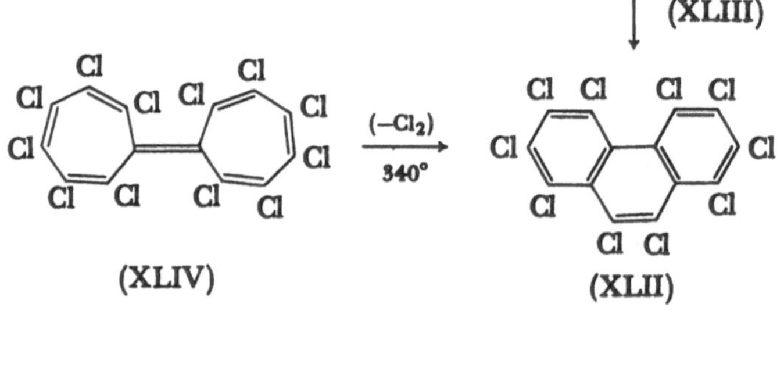

(XLIII)

(XLIV)   (XLII)

Perchlorophenanthrene is a yellow solid m.p. 252.0–3.5° of great thermal stability. The steric interaction between the chlorines of positions 4 and 5 is necessarily very high, causing the molecule to be warped. The fact that the chlorination of phenanthrene stops when two hydrogens are still unreplaced indicates the existence of steric shielding. The carbonization of perchlorophenanthrene with alkali metals has been studied by Brooks et al. [192]. Its photo-chlorination yields perchloro-9,10-dihydrophenanthrene [250].

The infrared and ultraviolet-visible spectra of this chlorocarbon have been recorded [250].

1.2.13   Perchloropyrene

By chlorination of pyrene with antimony pentachloride at 360° Merz and Weith obtained two unidentified chlorocarbons—$C_{14}Cl_{10}$ and $C_{15}Cl_{10}$—and carbon tetrachloride as the products of chlorinolysis [73]. Vollman et al. obtained perchloropyrene (XLV) from pyrene in a multi-step chlorination [266]. Chlorination with chlorine in hot trichlorobenzene gives $2H,4H,5H,7H,9H,10H$-octachloro-4,5,9,10-

Hexachloropyrenes

(XLVI)   (XLV)

tetrahydropyrene. This compound when dechlorinated by thermolysis at 292° or by ethanolic base at 100° yields an unresolvable mixture of two hexachloro derivatives tentatively regarded as 2H,5H,7H,9H- and 2H,5H,7H,10H-octachloropyrenes, which by further chlorination followed by thermolysis of the product at 260° gives XLV.

Reimlinger and King repeated the chlorination of the mixture of hexachloropyrenes with chlorine, chlorosulphonic acid and iodine at 0°, obtaining perchloro-4,5,9,10-tetrahydropyrene (XLVI) which on heating above its melting point (312°) eliminates chlorine yielding perchloropyrene [267].

The melting point of perchloropyrene has not been reported. Its electronic absorption [257] and infrared [268] spectra have been described and studied. Brooks et al. have studied the carbonization of perchloropyrene with alkali metals [192]. It undergoes oxidation with concentrated nitric acid to give perchloropyrene-3,8-quinone (XLVII) [255]. However when heated with oleum and nitric acid it gives perchloronaphthalene-1,4,5,8-tetracarboxylic dianhydride (XLVIII) instead [266, 267]. Presumably (XLVII) is a precursor for (XLVIII).

(XLVII)   (XLVIII)

1.2.14   Perchlorochrysene

An unsuccessful attempt to synthesize the title compound is recorded (1.2.1e) [73].

1.2.15   Final Remarks Concerning Fused Polycyclic Aromatic Chlorocarbons

In the perchlorination of fused polycyclic aromatic hydrocarbons, alkaromatic chlorocarbons, such as perchloro-1,4-dihydroanthracene or perchloro-9,10-dihydrophenanthrene, are formed. They can be isolated when the reaction temperature is low enough, e.g., with reagent BMC. Their formation accounts for the relatively easy chlorinolysis to perchlorobenzene or perchlorobiphenyl. The tendency of the fused polycyclic aromatic chlorocarbons to add chlorine is due to the well-known reactivity of certain of their bonds and to the fact that the products although more overcrowded with chlorines are less strained.

Polycyclic chlorocarbons show three important types of steric interactions between two vicinal chlorines: (1) 1–2, as in perchlorobenzene; (2) 1–3, as in perchloronaphthalene; and (3) 1–4, as in phenanthrene and the strain increases in this order.

The approximate repulsions between vicinal chlorines in some aromatic chlorocarbons is given by the Δ value in Table 1.1. $\Delta = 2R_{Cl} - d$ is the difference between twice the effective radius of chlorine (R_{Cl}) and the calculated distance (d) between the chlorine nuclei in the all-planar conformation. The actual distances between any pair of neighbouring chlorine nuclei in distorted perchloronaphthalene have been found to be 3.00–3.08 Å (Section 1.2.9a). Consequently, half this value (1.50 Å) is the effective chlorine radius to be used in predicting molecular distortions and the Van der Waals radius adopted in some space-filling models is incorrect. The very small negative Δ value for perchlorobenzene indicates that it is a borderline case.

The strain of a 1-3 steric interaction is released when one of the in-plane (sp^2)-carbon-chlorine bonds is replaced by two out-of-plane (sp^3)-carbon-chlorine bonds, as in the conversion of the aromatic chlorocarbon into the corresponding alkaromatic chlorocarbon (e.g., perchloro-1,4-dihydronaphthalene, -9,10-dihydroanthracene, -9,10-dihydrophenanthrene, -4,5,9,10-tetrahydropyrene). The strain, as measured by Δ, becomes then insignificant (0.07 Å). The considerable 1-4 repulsion in perchlorophenanthrene is lessened in perchloro-9,10-dihydrophenanthrene on account of the rotation about the biphenylic bond of one benzene ring in respect to the other.

TABLE 1.1

Chlorocarbon	Type of interaction	d(Å)	Strain $\Delta = 2R_{Cl} - d$ (Å)
Perchlorobenzene	1–2	3.10	−0.10
Perchloronaphthalene	1–3	2.45	0.55
Perchloroanthracene	1–3	2.45	0.55
Perchlorophenanthrene	1–3	2.45	0.55
Perchlorophenanthrene	1–4	1.12	1.88
Perchloropyrene	1–3	2.45	0.55

1.3 PERCHLOROARYLALKANES

In perchlorotoluene, the simplest perchloroarylalkane, the steric interactions among the *alpha* and the *ortho* chlorine atoms are very high [9]. The least strained arrangement appears to be one in which two *alpha* chlorines are located below while the other is positioned above the plane of the benzene ring.

The alternative conformation where an *alpha* chlorine is lying in the plane of the ring while the other two are placed symmetrically above and below it should be ruled out because of its greater steric strain. In this case, one *ortho* chlorine is locked between the two out-of-plane chlorines and consequently steric repulsions between the in-plane *alpha* and *ortho* chlorines cannot be released by rotation of the trichloromethyl group about its bond with the phenyl. Its strain would therefore be of type 1-3 (Section 1.2.15).

Even in the first, more favourable conformation the repulsions between chlorines must be very high since chlorine is similar in size to the methyl group. Therefore, perchlorotoluene can be regarded as a 'homomorph' of 2,6-dimethyl-*t*-butylbenzene for which Brown and co-workers estimated the strain of about 24 kcal/mole [269, 270], a substantial fraction of bond dissociation energy.

It is not surprising, therefore, that in spite of many previous attempts as far back as 1869 [16] no alkylaromatic chlorocarbon with a side-chain flanked by two *ortho* chlorines was synthesized until 1954 [15]. The steric strain not only hinders the formation of the alkylaromatic chlorocarbons (or perchloroarylalkanes) but also assists their fragmentation (chlorinolysis and thermolysis).

The appearance of chlorinolysis products does not necessarily indicate that the alkylaromatic chlorocarbon is an intermediate since decomposition may occur before perchlorination. For this reason reagent BMC [9], which requires a relatively low reaction temperature, is so successful in the synthesis of such compounds.

1.3.1 Perchlorotoluene: Syntheses

The synthesis of perchlorotoluene (IL) by exhaustive chlorination of toluene or polychlorotoluenes had repeatedly been attempted.

Beilstein and Kuhlberg [16] by chlorination of benzyl chloride with chlorine in the presence of iodine, followed by distillation with antimony pentachloride, obtained a mixture of two heptachlorotoluenes, presumably *oH*- and *αH*-heptachlorotoluene.

oH-Heptachlorotoluene seems to be the usual product when chlorinated toluenes are treated with chlorine at high temperature and/or distilled with antimony pentachloride [16]. Attempts to replace the last hydrogen by further treatment with the latter chlorinating agent gave, if the temperature was high enough (280°) perchlorobenzene.

Consequently, Beilstein and Kuhlberg concluded that perchlorotoluene could not exist [16]. Ballester [72, 77] attempted the synthesis of perchlorotoluene starting from benzotrichloride, using Silberrad's chlorinating agent [11] which was claimed [271] to preserve side-chains from chlorinolysis. The product was oH-heptachlorotoluene (ca. 70%) which was characterized by hydrolysis to 2,3,4,5-tetrachlorobenzoic acid with sulphuric acid followed by thermal decarboxylation to $1H,2H$-tetrachlorobenzene. Under stronger chlorination conditions some perchlorobenzene was obtained [271]. Harvey et al., carrying out the chlorination of benzotrichloride with chlorine in the presence of iron and ferric chloride in carbon tetrachloride, obtained also a 45% yield of oH-heptachlorotoluene [74]. It was concluded, therefore, that steric hindrance [72, 74, 77] and, probably, some electronic deactivation [77] prevented substitution of the last o-hydrogen by chlorine. The most favourable conformation of oH-heptachlorotoluene is predicted [9] to be one in which an α-chlorine is in the plane of the benzene ring, facing the hydrogen. This type of conformation has recently been confirmed by X-ray [271a] and n.m.r. [271b] studies.

Lock performed an exhaustive chlorination of $\alpha H,\alpha H,\alpha H$-pentachlorotoluene with chlorine and white light, at 210-230°, producing αH-heptachlorotoluene in a 70-80% yield [71, 272]. The structure of this compound followed from hydrolysis with sulphuric acid [71, 272], or oleum [70a].

$$C_6Cl_5Me \xrightarrow{Cl_2} C_6Cl_5CHCl_2 \xrightarrow[\Delta]{H_2SO_4} C_6Cl_5CHO$$

This remarkable result confirmed by Ross and co-workers [70a] and by Ballester [72], could be accounted for either in terms of shielding of the α-hydrogen in the heptachlorotoluene by the two ortho chlorines towards the attacking chlorine, or by electronic deactivation (−I effect) due to the chlorine atoms. As shown later, the perchlorobenzyl radical—the pertinent intermediate in the substitution of the last hydrogen—can easily be formed from perchlorotoluene, and it is sufficiently stable to give a strong esr signal [273] and to dimerize in solution, even in the presence of toluene [274]. Therefore, steric inhibition of resonance in the radical does not seem to be important.

In this connection it should be mentioned that Brimelow et al. found that while methyltrichlorobenzenes under exhaustive chlorination with chlorine at 180-220° give trichloromethyltrichlorotoluenes the 2,3,6-trichlorotoluene yields $\alpha,\alpha,2,3,6$-pentachlorotoluene [275]. Photochlorination of $\alpha H,\alpha H,\alpha H$-pentachlorotoluene with white light at lower temperatures has been effected by Harvey and co-workers with the same result (87% yield) [69]. This indicates that steric shielding is the predominant hindering cause. In fact the α-hydrogen is surrounded by three (one *ortho* and two *alpha*) chlorine atoms in these α,α,2,6-tetrachlorotoluenes in their energetically most favourable conformation where that hydrogen lies in the plane of the benzene ring [9]. When the photochlorination is performed with ultraviolet light it gives a mixture of α*H*-heptachlorotoluene and perchlorobenzene [69], probably *via* a photoexcited state.

The synthesis of perchlorotoluene was announced by Ballester and Molinet in 1954 [15] as a result of an unsuccessful experiment [249] and with it reagent BMC was discovered [9, 10] (Section 1.1). The chlorination of benzotrichloride with that reagent gives perchlorobenzene. However, under milder conditions it is possible to isolate perchlorotoluene as an intermediate [15]. When the reaction conditions are too mild *oH*-heptachlorotoluene is the major product. In general, it is not possible to avoid the presence of either the heptachlorotoluene or perchlorobenzene. Nevertheless, since the latter can more easily be separated from perchlorotoluene the reaction is usually conducted beyond the optimal yield. The synthesis of perchlorotoluene in a 53% yield has been reported starting from *oH*-heptachlorotoluene [9], which is prepared from benzotrichloride by means of Silberrad's chlorinating agent [11], i.e., a mixture of sulphur monochloride, aluminium trichloride and sulphuryl chloride [9, 72, 73]. From the resulting mixture *pH*-heptachlorotoluene is also isolated [9] which resembles perchlorotoluene in some of its properties.

The structure of *pH*-heptachlorotoluene was confirmed by hydrolysis to 2,3,5,6-tetrachlorobenzoic acid with sulphuric acid, and decarboxylation of this acid to 1,2,4,5-tetrachlorobenzene. The synthesis of this heptachlorotoluene had previously been reported by

Nicodemus by pyrocondensation of trichloroethylene [276]. However, its melting point was about $40°$ higher and the substance he obtained was probably impure αH-heptachlorotoluene, since heating it with ethanolic potassium hydroxide gave a triethoxytetrachlorotoluene which yielded chloranil by oxidation with chromic acid or nitric acid. In fact, Ross and Markarian, unaware of Nicodemus's work, found that under similar conditions αH-heptachlorotoluene can be converted into chloranil [277].

$$p\text{-MeOC}_6\text{Cl}_4\text{CH(OMe)}_2$$

NaOMe ↙   ↘ H⁺

$$\text{C}_6\text{Cl}_5\text{CHCl}_2 \qquad\qquad p\text{-MeOC}_6\text{Cl}_4\text{CHO}$$

| Fuming HNO₃

CHLORANIL

More recently, Ballester *et al.* have employed a modified version of reagent BMC (BMC-S) (1.1) to perform a high yield synthesis of perchlorotoluene directly from benzotrichloride [14]. They also chlorinated benzotrifluoride with still another version of reagent BMC (BMC-P) obtaining a mixture of perchlorotoluene (57%) and perchlorobenzene (42%) [278]. Under comparable conditions:

$$\text{PhCF}_3 \xrightarrow{\text{BMC-P}} \text{C}_6\text{Cl}_5\text{CCl}_3 + \text{C}_6\text{Cl}_6$$

McBee's group have obtained a mixture of oH-heptachlorotoluene (65%) pH-heptachlorotoluene (3%), the rest being mostly perchlorobenzene [279].

West and co-workers have described an elegant synthesis of perchlorotoluene by thermal isomerization of perchlorocycloheptatriene at $185\text{-}190°$ [104, 280]. This rearrangement is similar to that observed with the corresponding hydrocarbons.

1.3.2   Perchlorotoluene: Properties

Perchlorotoluene is a white solid m.p. $71.5\text{-}72.5°$. No data concerning its precise molecular shape have been reported. It may be assumed that on account of steric strain the molecule has a distorted, chair-like conformation with the substituents located, alternatively, above and below the mean plane [9]. There is compelling spectral (ultraviolet) evidence for the existence of ring distortion [108, 132]. The infrared spectrum has been reported by Ballester and Castañer [108].

1.3.3   Perchlorotoluene: Reactions

(a) *Hydrolyses*

The hydrolysis of perchlorotoluene to pentachlorobenzoic acid occurs easily around 100° in sulphuric [9, 15] and even in acetic acid [274, 281], most likely through a perchlorophenylcarbonium ion.

$$C_6Cl_5CCl_3 \rightarrow C_6Cl_5\overset{+}{C}Cl_2 \rightarrow C_6Cl_5COCl \rightarrow C_6Cl_5CO_2H$$

(L)

While the negative inductive character of the pentachlorophenyl does not favour the formation of a carbonium ion, the positive tautomeric effect of the α-chlorines should assist ionization as borne out by the series benzyl chloride, benzal dichloride and benzotrichloride [282]. Moreover, steric assistance of heterolysis of the carbon-chlorine bond is a major factor. It is well known that acid media favour carbon-chlorine bond heterolysis, probably by proton co-ordination with the leaving chlorine atom [283]. There is evidence that the hydrolysis in acetic acid is assisted by iodine. The major products of the reaction at 100° (two days) are perchlorobenzoyl chloride (65%) and pentachlorobenzoic acid (20.5%) [278] for which the following mechanism is suggested [278]:

$$C_6Cl_5CCl_3 + I_2 \longrightarrow C_6Cl_5\overset{+}{C}Cl_2 + ClI_2^-$$

$$C_6Cl_5\overset{+}{C}Cl_2 + MeCO_2H \xrightarrow{(-H^+)}$$

$$C_6Cl_5CCl_2OCOMe \longrightarrow C_6Cl_5CO_2H + MeCOCl$$

(diacyl ψ-dichloride)

$$C_6Cl_5COCl + MeCO_2H \longrightarrow C_6Cl_5CO_2H + MeCOCl$$

In this connection it is noteworthy that acyl ψ-dichlorides decompose giving acyl n-dichlorides [284].

Perchlorotoluene is hydrolysed with ferric chloride in refluxing methylene chloride under controlled humidity giving a 92% yield of perchlorobenzoyl chloride [283] presumably via a perchlorophenylcarbonium ion.

(b) *Reductions and Reductive Condensations*

Perchlorotoluene reacts apparently as a positive chlorine compound with iodide ion in acetic acid at 100°. The products are

cis-(LI) (47.5%) and $trans$-perchlorostilbene (LII) (36.6%) and αH-heptachlorotoluene [274].

$$C_6Cl_5CCl_3 \xrightarrow[\text{MeCO}_2\text{H}]{\text{I}^-}$$

$$C_6Cl_5CCl=CClC_6H_5 \, (LI=cis + LII=trans) \; + \; C_6Cl_5CHCl_2$$

It has been assumed that this reaction involves the perchlorobenzyl radical (LIII) which dimerizes to give perchlorobibenzyl (LIV) and then, by vicinal dechlorination, a mixture of the two isomeric perchlorostilbenes [274].

$$C_6Cl_5CCl_3 + I^- \longrightarrow C_6Cl_5\overset{\cdot}{C}Cl_2 + Cl^- + I_2$$
$$(LIII)$$

$$2 \times (LIII) \rightleftharpoons C_6Cl_5CCl_2CCl_2C_6Cl_5$$
$$(LIV)$$

$$C_6Cl_5\,CCl=CClCl_6\,Cl_5$$
$$cis \text{ and } trans$$

It will be shown (section 1.7.1) that perchlorobenzyl radicals are present in substantial concentration during that reaction [273], and that they are in equilibrium with the dimer (LIV). This fact accounts for the failure of previous attempts to isolate (LIV) even under mild conditions. It is remarkable that the dimer (LIV) should so easily undergo vicinal dechlorination. Its steric strain, which is probably even greater than that of perchlorotoluene, is relieved by formation of the less strained perchlorostilbene which, in addition, also benefits from some resonance stabilization. Similarly, oH-heptachlorotoluene reacts with iodide ion under the same conditions giving a mixture of $2H,2'H$-decachlorostilbenes [274]. Under these conditions the

cis and $trans$

dodecachlorobibenzyl is gradually converted into a mixture of the decachlorostilbenes.

The conversion of oH-heptachlorotoluene is a much slower process than that of perchlorotoluene which was anticipated since the formation of the corresponding hexachlorobenzyl radical is not assisted sterically. It is reasonable to assume that perchlorophenyl-carbonium ion (L) is an immediate precursor to the perchlorobenzyl radical, followed possibly by a transient, unstable perchlorobenzyl

$$C_6Cl_5CCl_3 \rightarrow C_6Cl_5CCl_2^+ + Cl^-$$

$$C_6Cl_5\overset{+}{C}Cl_2 + I^- \rightarrow C_6Cl_5\overset{\cdot}{C}Cl_2 + I$$

iodide which readily dissociates. The fact that no reaction between perchlorotoluene and alkali iodides takes place in ethanol rules out an alternative, e.g., direct attack of iodide ion on the *alpha* chlorines. In Sections 1.11.3 and 1.11.4 dealing with stable perchloroaryl-carbonium ions their reactions with iodide (bromide) ions is described.

The presence of αH-heptachlorotoluene poses an interesting question concerning its formation: A priori, it could result from the reaction of hydrogen donors either with perchlorobenzyl radical or perchlorophenylcarbonium ion (hydride shift). If the reaction of perchlorotoluene with iodide ion in acetic acid is performed in the presence of toluene—a reasonably good hydrogen donor—the yield of αH-heptachlorotoluene, normally under 10%, increases sharply, becoming the main product (~50%) [279] even with added sodium acetate to neutralize the hydriodic acid present. This indicates that at least in this case the free radical mechanism is operative as set out,

$$C_6Cl_5\overset{\cdot}{C}Cl_2 + PhMe \longrightarrow C_6Cl_5CHCl_2 + Ph\overset{\cdot}{C}H_2 \overset{I_2}{\longrightarrow} PhCH_2I$$

since acetic acid cannot function as a hydride donor. Accordingly, benzyl iodide has been isolated in a 55% yield [274].

A similar increase in the yield (42%) of αH-heptachlorotoluene is achieved by addition of trifluoroacetic acid [279]. The main product is, however, pentachlorobenzoic acid (45%) which points to hydrogen iodide as the donor.

$$C_6Cl_5CCl_3 \longrightarrow C_6Cl_5CHCl_2 + C_6Cl_5CO_2H$$

While in low-acidity media (aqueous acetic acid or acetic acid-sodium acetate) the accompanying formation of the perchloro-stilbenes is favoured, in more acidic conditions (anhydrous acetic acid or acetic acid-trifluoroacetic acid) it is suppressed. In their place substantial (30-45%) yields of pentachlorobenzoic acid are isolated [279] as mentioned above. These results can be rationalized by assuming an equilibrium between two different reactive species resulting from perchlorotoluene such as ion pairs (A) and independent ions (B), the proportion of the latter increasing with the acidity of the medium:

$$C_6Cl_5\overset{+}{C}Cl_2\overset{-}{Cl} \underset{-H^+}{\overset{+H^+}{\rightleftharpoons}} C_6Cl_5\overset{+}{C}Cl_2 + Cl^-$$

$$\text{(A)} \qquad\qquad\qquad \text{(B)}$$

Species B is more readily attacked by the solvent than by iodide ion (and vice versa) and more active in hydride transfer reactions.

Dimethylsulphoxide instead of acetic acid leads to a mixture of cis- (93%) and trans-perchlorostilbene (6.2%) [281]. This steric predominance can be interpreted in terms of the high dielectric constant of DMSO (ϵ, 46.4), favouring the formation of the isomer of higher dipole moment, i.e., the cis (Section 1.4.10).

The condensation can also be effected with ferrous chloride in acetic acid (cis, 32.5; trans, 49.3%) [274] or stannous chloride in diexane (49.3; 49%) [273] or by Karrer's method (copper and pyridine) [274].

The reaction of 4H-heptachlorotoluene with iodide ion in acetic acid at 100° gives a mixture of cis- and trans-4H,4'H-decachlorostilbene, no 4H,4'-H-dodecachlorobibenzyl being detected [287]. This is not unexpected since the latter compound, like perchlorodibenzyl, suffers from steric strain assisting vicinal dechlorination. The yield in perchlorostilbene is, however, moderate, probably due to extensive solvolysis. Direct evidence for the presence of 4H-hexachlorobenzyl radicals has been adduced (Section 1.7.2) [273].

Perchlorotoluene reacts with triphenylphosphine at 80° in benzene giving after treatment with water αH-heptachlorotoluene, cis-perchlorostilbene (Section 1.4.7) and triphenylphosphine oxide [281]. Eventually, some trans-isomer is obtained. The sole formation at lower temperature indicates that an intermediate phosphonium salt, probably (LV) is formed first.

$$C_6Cl_5CCl_3 + Ph_3P \longrightarrow$$

$$C_6Cl_5\overset{+}{C}Cl_2PPh_3Cl^- \overset{H_2O}{\longrightarrow} C_6Cl_5CHCl_2 + Ph_2PO + HCl$$

$$\text{(LV)}$$

Since substitution on the trichloromethyl carbon is subject to steric hindrance, the reaction of perchlorotoluene with triphenylphosphine proceeds by nucleophilic attack on the α-chlorine [281], i.e., a so-called second class phosphonium salt formation [288]. This type of reaction occurs also with carbon tetrachloride [289, 290].

$$C_6Cl_5CCl_2-Cl + \overset{..}{P}Ph_3 \rightarrow C_6Cl_5\overset{-}{C}Cl_2Ph_3\overset{+}{P}Cl \rightleftharpoons C_6Cl_5CCl_2\overset{+}{P}Ph_3Cl^-$$

$$\text{(LVI)} \qquad\qquad\qquad \text{(LV)}$$

According to Ballester and co-workers, the decomposition of the phosphonium salt to give the perchlorostilbenes takes place as follows [281]:

$$(LVI) \rightarrow C_6Cl_5\overset{.}{C}Cl_2 + Ph_3\overset{.}{P}Cl$$

$$C_6Cl_5\overset{.}{C}Cl_2 \rightarrow C_6Cl_5CCl_2CCl_2C_6Cl_5$$

$$\Big| \begin{matrix} Ph_3\overset{.}{P}Cl \\ and/or \\ Ph_3P \end{matrix}$$

$$\downarrow$$

$$C_6Cl_5CCl=CClC_6Cl_5 + Ph_3PCl_2$$

$$Ph_3PCl_2 \xrightarrow{H_2O} Ph_3PO + HCl$$

Epr evidence for perchlorobenzyl radical in this reaction has been obtained. An alternative (carbene) mechanism:

$$(LVI) \rightarrow C_6Cl_5CCl + Ph_3PCl_2$$

$$C_6Cl_5CCl + C_6Cl_5\overset{-}{C}Cl_2Ph_3\overset{+}{P}Cl \rightarrow C_6Cl_5CCl=CClC_6Cl_5 + Ph_3PCl_2$$

Since, in no case, $\alpha H,\alpha H$-hexachlorotoluene has been detected, this is evidence against the formation of the overcrowded phosphorane (LVII) as an intermediate, as it occurs in the reaction with carbon tetrachloride [289, 290].

$$C_6Cl_5CCl=PPh_3 \xrightarrow{H_2O} C_6Cl_5CH_2Cl + Ph_3PO$$
$$(LVII)$$

When the reaction is performed in ethyl ether or tetrahydrofurane αH-heptachlorotoluene and triphenylphosphine oxide are the sole products (71-95 and 88-100% yields respectively). This fact shows that those phosphonium salts are remarkably stable in polar solvents.

Perchlorotoluene reacts with potassium ferrocyanide in boiling dioxane to give a 93% yield of αH-heptachlorotoluene [291], and

$$C_6Cl_5CCl_3 \xrightarrow[\text{(dioxane)}]{K_4Fe(CN)_6} C_6Cl_5CHCl_2$$

with acetic acid containing either cyclohexene (20%) or hydrogen bromide, at 100°, this product is obtained almost quantitatively [248, 292]. The reaction may involve hydride shifts to the perchlorophenylcarbonium ion:

$$C_6Cl_5CCl_3 \longrightarrow C_6Cl_5\overset{+}{C}Cl_2 \xrightarrow{(H^-)} C_6Cl_5CHCl_2$$

Ballester and Molinet found that perchlorotoluene reacts with toluene at 250° giving a complex mixture of αH-heptachlorotoluene, $\alpha H,\alpha H$-hexachlorotoluene, $\alpha H,\alpha H,\alpha H$-pentachlorotoluene, and a condensation product which analyzes for $\alpha H,\alpha(p$-tolyl)hexachlorotoluene [293]. The yields are 21.8, 5.5, 3.1 and 31% respectively and the rest is made up by polychlorinated resins. Two hydrocarbons, namely benzyltoluene and dibenzyl are also isolated. The formation of the former hydrocarbon has been attributed to chlorine-hydrogen exchange with toluene, followed by hydrogen chloride catalyzed Friedel–Crafts condensation of the benzyl chloride with toluene. This was confirmed by a blank experiment. Formation of a small amount of dibenzyl has been observed after merely heating toluene at 250° so that no special significance can be attached to its presence. The $\alpha H,\alpha H$-hexa- and $\alpha H,\alpha H,\alpha H$-pentachlorotoluene are regarded as products of consecutive chlorine–hydrogen exchange with toluene. Although, possibly, perchlorotoluene cannot react with toluene in a Friedel–Crafts condensation because of steric strain of the resulting product, the αH-heptachloro- and $\alpha H,\alpha H$-hexachlorotoluene—mainly the latter—can. This accounts for the formation of that $C_{14}H_4Cl_6$ and other unidentified condensation products. The proportion of hydrogen chloride formed (1.9 moles) is consistent with this assumption.

Although at temperatures above 200° perchlorotoluene gives perchlorobenzyl radicals which might account for the observed reaction products, the authors prefer a polar (hydride shift) rather than a radical mechanism. Accordingly, Ballester and Rosa [292] have found that such an exchange occurs also at low temperature when aluminium chloride instead of hydrogen chloride is used as a catalyst.

$$C_6Cl_5CCl_3 + AlCl_3 \longrightarrow$$

$$C_6Cl_5\overset{+}{C}Cl_2 AlCl_4^- \xrightarrow{\ PhMe\ } C_6Cl_5CHCl_2 + Ph\overset{+}{C}H_2 AlCl_4^-$$

$$C_6H_5\overset{+}{C}H_2 AlCl_4^- + PhMe \longrightarrow PhCH_2C_6H_4Me + AlCl_3 + HCl$$

(c) *Chlorinolysis*

Perchlorotoluene readily undergoes quantitative chlorinolysis to perchlorobenzene with chlorine and white light in carbon tetrachloride solution [9], according to the following mechanism:

Step 1 $Cl_2 \xrightarrow{\ h\nu\ } 2\ Cl$

Step 2 $Cl + C_6Cl_5CCl_3 \longrightarrow C_6Cl_6 + Cl_3\overset{\cdot}{C}$

Step 3 $Cl_3\overset{\cdot}{C} + Cl_2 \longrightarrow CCl_4 + Cl$

Clearly, step 2 is sterically assisted by about 20 kcal/mole. (Perchlorotoluene is stable in white light.)

Perchlorotoluene also chlorolyses quantitatively on prolonged treatment with reagent BMC. This is an electrophilic displacement of the side chain assisted by steric strain.

Step 1 $(BMC)Cl^+ + C_6Cl_5CCl_3 \rightarrow C_6Cl_6 + Cl_3C^+$

Step 2 $Cl_3C^+ + SO_2Cl_2 \rightarrow (BMC)Cl^+ + CCl_4 + SO_2$

(d) Thermolysis

While non-strained polychloroalkylbenzenes are stable at temperatures at 300° and above, perchlorotoluene starts to decompose with gaseous evolution around 200°. This unusually facile thermal decomposition is undoubtedly caused by its high steric strain. The thermolysis has been studied at 310-320° in an inert atmosphere by Ballester and co-workers. The products are almost exclusively perchlorobenzene, perchlorostyrene (Section 1.4.1) and carbon tetrachloride [107] as shown:

$$C_6Cl_5CCl_3 \rightarrow C_6Cl_6 + C_6Cl_5CCl=CCl_2 + CCl_4$$

The first step of the thermolysis is doubtless the homolysis of the trichloromethyl carbon-chlorine bond giving perchlorobenzyl radical as is evidenced by epr in a partly thermolysed sample. The chlorine formed attacks the perchlorotoluene (second step) giving trichloromethyl radical which reacts with the perchlorobenzyl radical with elimination of chlorine (step 3). This step, although sterically unfavourable, is assisted by the formation of molecular chlorine in a concerted manner, and also by the fact that the coupling of radicals occurs as soon as they are formed [294]. The molecular chlorine then chlorolyses the remaining perchlorotoluene.

Step 1 $C_6Cl_5CCl_3 \rightarrow C_6Cl_5\dot{C}Cl_2 + Cl$

Step 2 $Cl + C_6Cl_5CCl_3 \rightarrow C_6Cl_6 + Cl_3\dot{C}$

Step 3 $Cl_3\dot{C} + C_6Cl_5\dot{C}Cl_2 \rightarrow C_6Cl_5CCl=CCl_2 + Cl_2$

Step 4 $Cl_2 + C_6Cl_5CCl_3 \rightarrow C_6Cl_6 + CCl_4$

When the thermolysis is carried out in the presence of oxygen, phosgene is formed which competes with step 3.

The fact that neither perchloroethane nor perchloroethylene were detected is an indication of their easy capture of the trichloromethyl radicals by other species of the medium. Nevertheless, some

perchlorobenzyl radicals escape from the 'reaction cage' giving
trans-(1.4% yield) and *cis*-perchlorostilbene (traces).

$$C_6Cl_5\dot{C}Cl_2 \rightarrow C_6Cl_5CCl=CClC_6Cl_5 + Cl_2$$

Perchlorobibenzyl is a much strained chlorocarbon. Consequently,
as in the formation of perchlorostyrene, the dimerization of
perchlorobenzyl radical above 300° cannot occur unless concerted
molecular chlorine elimination takes place. In this connection, it is
mentioned that $2H,2'H$-dodecachlorobibenzyl thermolyses [107]
also at 300°. The main product (55% yield) is a mixture of *cis*- and
trans-decachlorostilbenes. It is reasonable to assume that the process
is initiated mainly by vicinal dechlorination. Probably, the chlorine

$$C_6HCl_4CCl_2CCl_2C_6HCl_4 \rightarrow C_6HCl_4CCl=CClC_6HCl_4 + Cl_2$$

formed attacks the dodecachlorobibenzyl to give, ultimately, carbon
tetrachloride and pentachlorobenzene in comparable yields (~17%),
and some molecular chlorine.

$$Cl_2 + C_6HCl_4CCl_2CCl_2C_6HCl_4 \rightarrow C_6HCl_4CCl_3$$

$$Cl_2 + C_6HCl_4CCl_3 \rightarrow C_6HCl_5 + CCl_4$$

$$Cl_2 + C_6HCl_4CCl_2CCl_2C_6HCl_4 \rightarrow C_6HCl_5 + Cl_2C=CClC_6HCl_4 + Cl_2$$

(*e*)   *Photolysis*
According to Ballester and Castañer, perchlorotoluene in carbon
tetrachloride undergoes photolysis with ultraviolet light, at room
temperature. The products are perchloroethylbenzene (LVIII)
(46.1% yield) perchlorobenzene (9.7%), *trans*-perchlorostilbene
(5.9%), *cis*-perchlorostilbene (1.1%) and perchlorostyrene (LIX)
(2.1%) [108]. It has been shown by means of epr that perchlorobenzyl radicals are formed initially (Section 1.7.1) as set out:

Step 1 $C_6Cl_5CCl_3 \xrightarrow{h\nu} C_6Cl_5\dot{C}Cl_2 + Cl$

Step 2 $Cl + CCl_4 \rightleftharpoons Cl_2 + Cl_3\dot{C}$

Step 3 $Cl_3\dot{C} + C_6Cl_5CCl_3 \longrightarrow CCl_4 + C_6Cl_5\dot{C}Cl_2$

Step 4 $Cl_3\dot{C} + C_6Cl_5\dot{C}Cl_2 \longrightarrow C_6Cl_5C_2Cl_5$
                                                    (LVIII)

Step 5 $C_6Cl_5\dot{C}Cl_2 \rightleftharpoons C_6Cl_5CCl_2CCl_2C_6Cl_5$

$Step\ 6$ $C_6Cl_5CCl_2CCl_2C_6Cl_5 \xrightarrow[\overset{\cdot}{C}Cl_3]{h\nu} C_6Cl_5\overset{\cdot}{C}Cl-CCl_2C_6Cl_5 + Cl$

$Step\ 6a$ $\longrightarrow C_6Cl_5\overset{\cdot}{C}Cl-CCl_2C_6Cl_5 + Cl_4$

$Step\ 7$ $C_6Cl_5\overset{\cdot}{C}Cl-CCl_2C_6Cl_5 \longrightarrow C_6Cl_5CCl=CClC_6Cl_5 + Cl$

$Step\ 8$ $C_6Cl_5C_2Cl_5 \xrightarrow[\overset{\cdot}{C}Cl_3]{h\nu} C_6Cl_5\overset{\cdot}{C}Cl-CCl_3 + Cl$

$Step\ 8a$ $\longrightarrow C_6Cl_5\overset{\cdot}{C}Cl-CCl_3 + CCl_4$

$Step\ 9$ $C_6Cl_5\overset{\cdot}{C}Cl-CCl_3 \longrightarrow C_6Cl_5CCl=CCl_2 + Cl$
(LIX)

Step 2 is a well-known equilibrium which is responsible for the formation of perchloroethane in the photochlorinations with chlorine in carbon tetrachloride. Steps 3, 6a and 8a must be exothermic on account of the increased electron delocalisation and the release of steric strain. Step 4 is a reasonable one since the two radicals involved are the most stable among those present and evidence for Step 5 comes from epr. Steps 6 and 8 are analogous to step 1 in the sense that they release steric strain and form resonance stabilized benzyl-type radicals which further 'relax' (Steps 7 and 9) by ejection of a chlorine atom and π-bond formation. Perchloroethylbenzene is the major product here, while it is absent in the thermolysis of perchlorotoluene (section 1.3.3d). This result is due to the relatively low temperature under which the photolysis is carried out, avoiding chlorine elimination.

(f) *Condensations*—See sections 1.4.2, 1.4.4 and 1.4.13.

1.3.4   Perchloroethylbenzene

Several attempts to perchlorinate ethylbenzene have been reported. McBee and co-workers obtained, by perchlorination with chlorine and white light, phenylpentachloroethane and, at higher temperatures (130°), a pentachloroethylchlorobenzene [75]. Phenylpentachloroethane when submitted to aromatic perchlorination with chlorine and ferric chloride at 140-160° gives, presumably, *oH*-nonachloroethylbenzene which, at higher temperatures, yields perchlorobenzene.

Harvey *et al.* performed this chlorination in the presence of iron-ferric chloride at 100-110° and, accordingly, obtained a 95.5% yield of *oH*-nonachloroethylbenzene. They suggested that steric hindrance is the reason for the failure to obtain perchloroethylbenzene since an *ortho* hydrogen resists replacement. Hydrolysis of the nonachloroethylbenzene with sulphuric acid at 260-280° followed by oxidation with alkaline aqueous potassium permanganate gives 2,3,4,5-tetrachlorobenzoic acid, presumably according to the following sequence:

$$2,3,4,5\text{-}Cl_4C_6HC_2Cl_5 \xrightarrow[260-280°]{H_2SO_4} 2,3,4,5\text{-}Cl_4C_6HCOCl_3$$

$$\xrightarrow{KMnO_4} 2,3,4,5\text{-}Cl_4C_6HCO_2H$$

The oH-nonachloroethylbenzene has been obtained along with perchlorostyrene by chlorination of α,β,β-trichlorostyrene with reagent BMC [108]. Forcing reaction conditions lead slowly to perchlorostyrene (LIX), indicating that perchloroethylbenzene (LVIII), if formed, dechlorinates readily. oH-Heptachlorostyrene behaves similarly [108].

Ross and co-workers carried out an exhaustive chlorination of ethylpentachlorobenzene with chlorine and white light at 200° [70a]. The only products are perchlorobenzene and 1-pentachlorophenyl-2-chloroethane, even with Kharash's chlorination reagent (sulphuryl chloride and benzoyl peroxide). This result, originally attributed to steric inhibition of resonance in the methylpentachlorophenyl radical [70a], is very likely due to preferential attack on the *beta* carbon (steric shielding of the *alpha* carbon) [295] as well as to a relatively high electron density on the 1-carbon of the ring because of the alkyl substituent. According to Stacey and co-workers, chlorination of ethylpentachlorobenzene in carbon tetrachloride furnishes a mixture of octachloroethylbenzenes [69]. With ultraviolet light an excellent yield of αH-nonachloroethylbenzene is obtained whose structure was inferred from the steric shielding of the *alpha* carbon [69].

$$C_6Cl_5Et \xrightarrow[\text{light}]{Cl_2(CCl_4)} C_6Cl_5C_2H_2Cl_3 \xrightarrow[UV]{Cl_2(CCl_4)} C_6Cl_5CHClCCl_3$$

This result is strikingly different from that obtained with αH-heptachlorotoluene which gives a mixture containing perchlorobenzene [69], indicative of a better shielding or electronic deactivation of the 1-carbon due to the trichloromethyl group.

The synthesis of perchloroethylbenzene is achieved in a 46% yield by photolysis of perchlorotoluene with ultraviolet light in carbon tetrachloride, at room temperature, as mentioned previously (Section 1.3.3e). It results from the coupling of perchlorobenzyl and trichloromethyl radicals.

$$C_6Cl_5CCl_3 \xrightarrow{h\nu} C_6Cl_5\overset{.}{C}Cl_2 \xrightarrow{Cl_3\overset{.}{C}} C_6Cl_5CCl_2CCl_3$$
$$\text{(LVIII)}$$

Perchloroethylbenzene (LVIII) is a colourless solid, melting at 223·4°. Its ultraviolet spectrum [108, 132, 133] indicates that the molecule is *less* distorted than that of perchlorotoluene. This surprising result is due to the fact that in certain molecules strain and distortion are not necessarily combined. The repulsions between the *ortho* chlorines and the two *alpha* chlorines are partly cancelled by interaction from the top by the trichloromethyl group [108]. The infrared spectrum provides, however, evidence that the molecule is strained [108].

Perchloroethylbenzene does not hydrolyze to perchloroacetophenone by treatment with sulphuric acid or oleum, a behaviour contrasting with that of perchlorotoluene (Section 1.3.3a) and *oH*-heptachlorotoluene. Such inertness has been attributed to both the low electron-releasing capacity of the trichloromethyl group as compared to chlorine, and the greater steric inhibition of resonance in the intermediate carbonium ion; because of steric interaction between the α-chlorines and the *ortho* chlorines the plane of the sp^2-hybridized *alpha* carbon should be almost perpendicular to the benzene ring and, consequently, no stabilizing π-electron delocalization can occur. By the same token the stability of perchloroethylbenzene towards ultraviolet light is greater than that of perchlorotoluene; the formation of perchloro-α-methylbenzyl radicals by elimination of an *alpha* chlorine is energetically less favourable on account of the greater steric inhibition of resonance.

Reaction of perchloroethylbenzene with iodide ion takes place readily. While, in the case of perchlorotoluene, this reaction occurs through the relevant carbonium ion, in the present case the elimination of two vicinal chlorines is most probably a concerted process assisted by molecular compression. The product is, therefore, perchlorostyrene (98% yield).

$$C_6Cl_5CCl_2CCl_3 \xrightarrow{\quad I^- \quad} C_6Cl_5CCl=CCl_2$$

Perchloroethylbenzene when treated with reagent BMC gives perchlorobenzene and perchlorostyrene in 95.3 and 2.9% yields respectively. This remarkable result indicates that in the chlorination of *oH*-nonachloroethylbenzene with BMC perchlorobenzene is *not* an intermediate, i.e., vicinal dechlorination takes place *prior* to replacement of the *ortho* hydrogen.

1.3.5   Perchloro-*n*-propylbenzene

Attempts to synthesize perchloro-*n*-propylbenzene by multi-step chlorination of propiophenone were abortive [296] resulting in cyclization to chlorinated indanes (Section 1.4.2).

1.3.6   Perchloro-o-xylene

The synthesis of perchloro-o-xylene has been attempted starting from 3,4,5,6-tetrachloro-o-xylene by chlorination with chlorine and incandescent white light in carbon tetrachloride [69]. The product is $\alpha H, \alpha' H$-octachloro-o-xylene. Chlorination of this compound with ultraviolet light gives an unknown product (probably αH-heptachlorotoluene) and perchlorobenzene [69]. The failure to substitute the last α-hydrogens is due to the causes mentioned in the photochlorination of $\alpha H, \alpha H, \alpha H$-pentachlorotoluene (Section 1.3.1).

Perchlorination of o-xylene with chlorine and incandescent light, in carbon tetrachloride, gives 1-dichloromethyl-2-trichloromethylbenzene [69]. Extreme overcrowding must be the reason for stopping replacement of the last *alpha* hydrogen. However, McBee's group has been able to prepare o-bistrichloromethylbenzene from bistrifluoromethylbenzene with aluminium chloride [297, 298]. The yields are 34 or 51% using refluxing chloroform [297] or methylene chloride at 0° [298] respectively.

The suggested general mechanism for such halogen exchanges is as follows [298]:

$$\text{Ar}-\overset{|}{\underset{|}{\text{C}}}-\text{F} + \text{AlCl}_3 \;\rightarrow\; \text{Ar}-\overset{|}{\underset{|}{\text{C}}}{}^+ \quad \text{AlCl}_3\text{F}^- \;\rightarrow\; \text{Ar}-\overset{|}{\underset{|}{\text{C}}}-\text{Cl} + \text{AlCl}_2\text{F}$$

In this connection, it is noteworthy that 1-dichloromethyl-2-trichloromethylbenzene does not undergo ring chlorination with chlorine, iron, ferric chloride and iodine in carbon tetrachloride [74]. In this case, the deactivation must be electronic rather than steric (section 1.3.7).

1.3.7   Perchloro-m-xylene

Tawildarow [16] performed the first recorded chlorination of xylene as described by Beilstein and Kuhlberg [16], producing perchlorobenzene. McBee and co-workers performed the chlorination with iron as a catalyst and obtained a tetrachloro derivative which, on photochlorination, gave an octachloroxylene. Although no properties were given and the number of chlorines was estimated from the increase in weight, it was probably m-bisdichloromethyltetrachlorobenzene. Further photochlorination causes chlorinolysis to perchlorobenzene.

$$\text{C}_6\text{H}_4\text{Me}_2\text{-}m \;\xrightarrow[\text{Fe}]{\text{Cl}_2}\; \text{C}_6\text{Cl}_4\text{Me}_2\text{-}m \;\xrightarrow[h\nu]{\text{Cl}_2}\;$$

$$\text{C}_6\text{Cl}_4(\text{CHCl}_2)_2\text{-}m \;\xrightarrow[h\nu]{\text{Cl}_2}\; \text{C}_6\text{Cl}_6$$

These results were confirmed by Stacey *et al.* starting from pure tetrachloro-*m*-xylene in carbon tetrachloride [69]. The reasons for the failure to obtain the chlorocarbon are the same as cited for the photochlorination of pentachlorotoluene (Section 1.3.1).

McBee *et al.* obtained from bistrichloromethylbenzene a solid C_8HCl_9 which, because of resistance to chlorinolysis, was possibly 2*H*-nonachloro-*m*-xylene.

$$C_6H_4Me_2 \xrightarrow[h\nu]{Cl_2} C_6H_4(CCl_3)\text{-}m \xrightarrow[FeCl_3]{Cl_2} 4,5,6\text{-}Cl_3\,C_6\,H(CCl_3)_2\text{-}m$$

Chlorination of *m*-bistrichloromethylbenzene with reagent BMC gives 2*H*,4*H*-octachloro-*m*-xylene (42% yield) which is hydrolyzed to 4,5-dichloroisophthalic acid with concentrated sulphuric acid [248].

$$C_6H_4(CCl_3)_2 \xrightarrow{BMC} 4,5\text{-}Cl_2\,C_6\,H_2\,(CCl_3)_2\text{-}m \xrightarrow{H_2SO_4}$$

$$4,5\text{-}Cl_2\,C_6\,H_2\,(CO_2\,H)_2\text{-}m$$

A similar chlorination of 2-chloro-*m*-bistrichloromethylbenzene gave 4*H*,6*H*-octachloro-*m*-xylene, which on treatment with sulphuric acid yielded 2,4-dichloroisophthalic acid [248]. The failure to synthesize perchloro-*m*-xylene is due not so much to steric strain as to the electronic deactivation by trichloromethyl groups towards electrophilic substitution of the last hydrogens.

1.3.8   Perchloro-*p*-xylene: Syntheses

An abortive attempt to synthesize perchloro-*p*-xylene (LX) by chlorination of *p*-bistrichloromethylbenzene with chlorine in carbon tetrachloride at 75-80° in the presence of iron, ferric chloride and iodine [74] was reported. This result clearly shows the high deactivation of the ring on account of the two trichloromethyl groups. Photochlorination of *p*-dimethyltetrachlorobenzene with chlorine and white light in boiling carbon tetrachloride gives a 64% yield of α*H*,α′*H*-octachloro-*p*-xylene [69]. When this product is submitted to the same conditions under ultraviolet light it affords perchlorobenzene [69]. Again, the reasons for the failure to replace the last hydrogens are those already mentioned in Section 1.3.1. Also, syntheses of perchloro-*p*-xylene from tetrachloroterephthaloyl dichloride by reaction with phosphorus pentachloride have been unsuccessful [12]. There is no reaction at 200° and at 300° chlorinolysis occurs. These failures are due to steric hindrance, although the strained perchloro-*p*-xylene would not have survived such drastic chlorination conditions (see section 1.3.9).

Perchloro-*p*-xylene (LX) was synthesized by Ballester *et al.* from

$$C_6H_4(CCl_3)_2\text{-}p \quad \xrightarrow[\text{(CCl}_4)]{Cl_2,\ Fe,\ FeCl_3,\ I_2} \!\!\!\!\!\times\!\!\!\!\!\longrightarrow \quad C_6Cl_4(CCl_3)_2\text{-}p$$
$$\text{(LX)}$$

$$Cl_2(CCl_4) \nearrow\!\!\!\!\!\times$$

$$C_6Cl_4Me_2\text{-}p \quad \xrightarrow[h\nu]{Cl_2(CCl_4)} \quad C_6Cl_4(CHCl_2)_2\text{-}p$$

$$\searrow \underset{h\nu}{Cl_2(CCl_4)}$$

$$C_6Cl_6$$

p-bistrichloromethylbenzene by means of reagent BMC [9]. $2H,5H$-octachloro-p-xylene is obtained as a by-product. The yields are 71 and 12%, respectively, and short reaction times alter their relative amounts. These results have been confirmed by Doorenbos [12].

$$C_6H_4(CCl_3)_2\text{-}p \xrightarrow{\text{BMC}} \text{(LX)} + 2,5\text{-}Cl_2C_6H_2(CCl_3)_2\text{-}p$$

When the reaction is carried out in a pressurized vessel tetrachloro-terephthaloyl chloride is formed quantitatively, presumably through an aluminium chloride catalyzed oxygen–chlorine exchange between perchloro-p-xylene and the sulphuryl chloride. This is substantiated by the fact that perchloro-p-xylene can indeed be transformed under those conditions.

$$C_6H_4(CCl_3)_2\text{-}p \xrightarrow{AlCl_3,\ SO_2Cl_2} C_6Cl_4(COCl)_2\text{-}p$$

Working on the hypothesis that trichlorosulphonium tetrachloro-aluminate $SCl_3^+\ AlCl_4^-$ is the active chlorinating species in reagent BMC, Doorenbos found that this salt, at least in two cases, can effect perchlorination of trichloromethylaromatic compounds [12]. Under pressure 1,4-bistrichloromethylbenzene reacts with chlorine in thionyl chloride containing that sulphonium salt at about 80°, to give 76-82% of perchloro-p-xylene and some octachloro-p-xylene. At atmospheric pressure, the yield is significantly lower [12]. It is remarkable that in these syntheses no $2H,3H$-octachloro-, $2H,6H$-octachloro-, or $2H$-nonachloro-p-xylene are apparently formed. The fact that the $2H,6H$-octachloro-p-xylene gives perchloro-p-xylene rather sluggishly seems to indicate that it is not the main intermediate. It is doubtful, however, that the reaction conditions are strictly comparable.

Perchloro-*p*-xylene is also obtained from perchloro-*p*-xylylene (LXI) in a good (75.5%) yield by chlorination with chlorine and white, incandescent light at room temperature in acetic acid containing hydrogen chloride [299]. This interesting synthesis has, however, no practical value since (LXI) is obtained from (LX).

$$\underset{\text{(LXI)}}{Cl_2C=C_6Cl_4=CCl_2} \xrightarrow[\text{HAc, HCl}]{\text{Cl}_2,\ h\nu} \underset{\text{(LX)}}{C_6Cl_4(CCl_3)_2\text{-}p}$$

1.3.9   Perchloro-*p*-xylene: Properties and Reactions

(*a*)   *Properties*

Perchloro-*p*-xylene is a pale, greenish-yellow solid melting at 153.0-4.5°. Its infrared [12, 281] and electronic [132, 133, 240, 301] spectra† have been recorded and studied. Perchloro-*p*-xylene exhibits both as a solid and in solution thermochromism [249, 281]. At −20° it becomes colourless.

The electronic spectrum affords evidence that the molecule is much distorted, adopting probably a chair-like conformation; i.e., the substituents being alternatively above and below the main plane of the benzene ring [132]. While the distortion due to each trichloromethyl group and its *ortho* chlorines is as important as in perchlorotoluene, the resulting steric strain is less than twice that of the latter compound since the twisting caused by the first trichloromethyl group favours the introduction of the second trichloromethyl group in the *para* position. Consequently, perchloro-*p*-xylene should be more stable than perchlorotoluene both thermally and chemically, particularly in those reactions that are sterically assisted.

Although a comparison of the ultraviolet spectra of perchlorotoluene and perchloro-*p*-xylene indicates that the latter is more than twice as strained as the former [301] this correlation is of limited value since the spectra refer both to the ground and activated state.

It has been mentioned that perchloro-*p*-xylene can be synthesized by photochlorination in acetic acid (section 1.3.8) and also with reagent BMC without a trace of perchlorobenzene. These results are consistent with its expected higher stability. Nevertheless, electronic effects may also play an important role, since chlorinolysis is favoured by electron availability at the carbon bearing the side chain.

The dielectric constant and conductivity of perchloro-*p*-xylene

† Smith and Turton reported the ultraviolet spectrum of perchloro-*p*-xylene in 1955 [300]. There is no doubt whatever that this compound had not been synthesized by the Birmingham group [9]. The confusion was very likely caused by a misprint.

have been measured by Martinez [301a] and Freeburger has studied its mass spectrum [301b].

(b) *Hydrolyses*

Perchloro-p-xylene and 2H,5H-octachloro-p-xylene hydrolyze quantitatively to the corresponding terephthalic acids by treatment with sulphuric acid at 100° [9]. This reaction, followed by decarboxylation, established the structure of the octachloro-p-xylene.

$$C_6Cl_4(CCl_3)_2\text{-}p \xrightarrow[100°]{H_2SO_4} C_6Cl_4(CO_2H)_2$$

$$1,4\text{-}Cl_2C_6H_2(CCl_3)_2\text{-}p \xrightarrow[100°]{H_2SO_4} 1,4\text{-}Cl_2C_6H_2(CO_2H)\text{-}p$$

$$\xrightarrow[HAC]{350°} C_6H_4Cl_2\text{-}p$$

Perchloro-p-xylene does not hydrolyze in acetic acid at 100°, even with added hydrogen chloride (saturation) or trifluoroacetic acid (10%) [281]. This striking difference from perchlorotoluene is due to the strong negative inductive effect of the trichloromethyl groups which hinder the heterolysis of the α-carbon-chlorine bond. Hydrolysis, however, occurs in acetic acid containing iodine, at 100°, giving perchloroterephthaloyl dichloride in an excellent yield [281]. Comparable results are obtained by chlorine-oxygen exchange with sulphuryl chloride (section 1.3.8) [12], and by treatment in boiling methylene chloride under carefully controlled humidity; higher humidity yields mixtures of that dichloride and tetrachloroterephthalic acid [278].

For the mechanism of the hydrolysis in the presence of iodine, see Section 1.3.3a.

(c) *Reductions and Reductive Condensations*

Perchloro-p-xylene reacts with potassium or sodium iodide in acetic acid at 100° to give a nearly quantitative yield of perchloro-p-xylylene [299]. In the presence of hydrogen chloride or trifluoroacetic

$$C_6Cl_4(CCl_3)_2\text{-}p \xrightarrow[AcOH]{I^-,\,100°} Cl_2C{=}C_6Cl_4{=}CCl_2$$

acid the yield diminishes sharply (~40%) and αH,αH'-octachloro-p-xylene is formed (33-42%) [281]. As in the case of the reaction with perchlorotoluene (section 1.3.3b) the hydrogen iodide is most probably the reducing species. The reaction can, therefore, be accounted for by the following sequence:

$$C_6Cl_4(CCl_3)_2\text{-}p \xrightarrow{I_2} Cl_3CC_6Cl_4\overset{+}{C}Cl_2\text{-}p \searrow^{I^-}$$

$$Cl_3C\text{--}C_6Cl_4\text{--}\dot{C}Cl_2\text{-}p$$

$$\downarrow I^-$$

$$Cl_2C=C_6Cl_4=CCl_2\text{-}p$$

$$Cl_3CC_6Cl_4CHCl_2\text{-}p \xrightarrow{I^-} Cl_2\dot{C}C_6Cl_4CHCl_2\text{-}p$$

$$\searrow^{HI}$$

$$Cl_2CHC_6Cl_4CHCl_2\text{-}p$$

A concerted elimination by attack on an *alpha* chlorine cannot account solely for the effect of added acids. Toluene is not as

effective a hydrogen donor as hydriodic acid [281], yet when this hydrocarbon is added the yield of perchloro-*p*-xylylene stays high (72%), while that of *αH,αH*-octachloro-*p*-xylene is only 4%, and compound (LXIII) is also formed (14%). Benzyl acetate is also isolated (0.15 moles per mole of XL). These results are consistent with the following mechanism (on page 63) involving the formation of the perchloro-*p*-xylyl radical (LXIII).

Perchloro-*p*-xylylene is under no circumstances reduced to *αH,αH*-octachloro-*p*-xylene.

Although no reaction between perchloro-*p*-xylene and iodide ion ensues in ethanol, in acetone perchloro-*p*-xylylene is produced almost quantitatively. It is, therefore, likely that in the latter solvent a concerted mechanism operates. When the reaction with iodide ion is performed in dimethylsulphoxide at 100° a 95% yield of *cis*-4H,4H-dodecachlorostilbene (LXVI) is obtained (section 1.3.3b)

$$Cl_3 CC_6 Cl_4 \dot{C}Cl_2\text{-}p + PhMe \longrightarrow Cl_3 CC_6 Cl_4 CHCl_2\text{-}p + Ph\dot{C}H_2$$
(LXII)

$$Cl_3 CC_6 Cl_4 CHCl_2\text{-}p \xrightarrow{\ I^-\ } Cl_2 \dot{C}\text{-}C_6 Cl_4 CHCl_2\text{-}p \searrow PhMe$$

$$Cl_2 CHC_6 Cl_4 CHCl_2\text{-}p$$

$$p\text{-}Cl_2 CHC_6 Cl_4 CCl_2 CCl_2 C_6 Cl_4 CHCl_2\text{-}p \searrow I_2$$

$$p\text{-}Cl_2 CHC_6 Cl_4 CCl{=}ClCC_6 Cl_4 CHCl_2\text{-}p$$
(LXIII)

$$Ph\dot{C}H_2 \xrightarrow{\ I_2\ } PhCH_2 I \xrightarrow{\ HAc\ } PhCH_2 OCOMe$$

[281]. This reaction is probably initiated by a nucleophilic displacement on an *alpha* chlorine atom by iodide ion giving perchloro-carbanion (LXIV) (Step 1), the formation of which is assisted by both the strong electronegative effect of the other trichloromethyl group and the high polarity of the solvent.† (LXIV) is then attacked by the solvent and dichlorocarbene is presumably formed (Step 2). In the authors' opinion, the strong steric inhibition of resonance in carbanion (LXIV) due to the two *ortho* chlorines is essential for Step 2 (carbon-carbon bond fission). These steps are consistent with the degradations referred to in Section 1.3.9e.

Step 1 $Cl_3 CC_6 Cl_4 CCl_3\text{-}p + I^- \rightarrow Cl_3 CC_6 Cl_4 \bar{C}Cl_2\text{-}p + ICl$
(LXIV)

$$Cl_2 C{=}C_6 Cl_4{=}CCl_2$$

Step 2   (LXIV) $\xrightarrow[\ (+H^+)\ ]{(-Cl^-)}$

DMSO $2,3,5,6\text{-}Cl_4 C_6 HCCl_3 + {:}CCl_2$
(LXV)

The reluctance of carbonium ion formation and activation towards nucleophilic displacement on chlorine in perchloro-*p*-xylene are not shown by 4*H*-heptachlorotoluene (LXV). Consequently, the formation of the *cis*-stilbene (LXVI) via 4*H*-hexachlorobenzyl radicals takes place as usual (Section 1.3.3b).

† In a medium of low dielectric constant such as acetone the concerted mechanism yielding perchloro-*p*-xylylene is favoured (see preceding page).

$$2,3,5,6\text{-}Cl_4\,C_6\,CCl=ClCC_6\,Cl_4\text{-}2,3,5,6$$

cis-   (LXVI)

Concerning the stereospecificity of this reaction see Section 1.4.10. The reaction of perchloro-p-xylene with other reducing agents such as stannous chloride or ferrous chloride in dioxane at 100° gives a polymeric chlorocarbon [302] (cf. Section 1.6.1). Other dechlorinations include the reaction with triphenylphosphine, analogous to that of perchlorotoluene (section 1.3.3b), which in benzene at 80° yields perchloro-p-xylylene (74% yield) and triphenylphosphine oxide. A related reaction is that of 1,4-bistrichloro-

$$Cl_3\,CC_6\,Cl_4\,CCl_3\text{-}p \;+\; PPh_3 \;\rightarrow\; Cl_3\,CC_6\,Cl_4\,\bar{C}Cl_2\,Ph_3\,\overset{+}{P}Cl$$

$$Cl_2C=C_6Cl_4=CCl_2 \;+\; Ph_3PCl_2$$

$$Ph_3PCl_2 \;\xrightarrow{\;H_2O\;}\; Ph_3PO$$

methylbenzene with triethylphosphite giving a polymeric material, probably through unstable $\alpha,\alpha,\alpha',\alpha'$-tetrachloro-$p$-xylylene [303].

(d)   Chlorinolysis

The chlorinolysis of perchloro-p-xylene with chlorine and white light in carbon tetrachloride occurs very readily, giving a 72% yield of perchlorobenzene (Section 1.2.1d) [9]. (cf. the chlorination in acetic acid, Section 1.3.8.)

(e)   Dealkylations

In the preceding section dealkylation by iodide ion in dimethylsulphoxide has been considered. The fact that similar dealkylations occur with perchloro-p-xylene by means of non-reducing nucleophiles indicates that the iodide ion acts as both a nucleophile and a reducing agent.

The reaction of perchloro-p-xylene with piperidine under increasingly stronger reaction conditions gives 4H-heptachlorotoluene, αH,4H-hexachlorotoluene, and 1,2,4,5-tetrachlorobenzene [248]. The following mechanism is proposed, involving elimination of dichlorocarbene as with iodide ion:

$$Cl_3C_6Cl_4CCl_3\text{-}p \;\xrightarrow{\;C_5H_{11}N\;}\; Cl_3\,CC_6\,Cl_4\,\bar{C}Cl_2\text{-}p \;\rightarrow\; Cl_3\,CC_6\,HCl_4\text{-}2,3,5,6 \;+$$

$$:CCl_2$$

$$1,2,4,5\text{-}Cl_4C_6H_2 \;+\; :CCl_2 \;\xleftarrow{\;C_5H_{11}N\;}$$

The initiation step of these dealkylations consists of the extraction of a positive chlorine by the nucleophile (or base). Accordingly, the reaction can also be performed starting from αH-compounds such as $\alpha H,\alpha' H$-octachloro-p-xylene ($\sim$80% yield) by removing the proton with a base [248].

$$Cl_2CHC_6Cl_4CHCl_2\text{-}p \xrightarrow[\text{or OH}^-]{C_5H_{11}N} Cl_2CHC_6Cl_4-\bar{C}Cl_2\text{-}p \longrightarrow$$

$$CCl_2 + 2,3,5,6\text{-}C_6HCl_4CHCl_2$$

An interesting reaction was reported in 1933 by Lock [71]. Benzaldehydes, when treated with strong base such as ethanolic potassium hydroxide, undergo the Cannizzaro reaction which involves a hydride shift initiated by addition of the base to the carbonyl group. When this addition is blocked by bulky substituents such as two *ortho* chlorines (2,6-dichloro-, 2,3,6-trichloro-, and pentachlorobenzaldehyde) degradation to the corresponding m-dichlorobenzenes with formation of formate ion takes place. It is reasonable to assume that in this case the base can only abstract a proton from the aldehyde group. The resulting anion is then decomposed by the solvent. These changes are summarized in Sequence 1.

Sequence 1

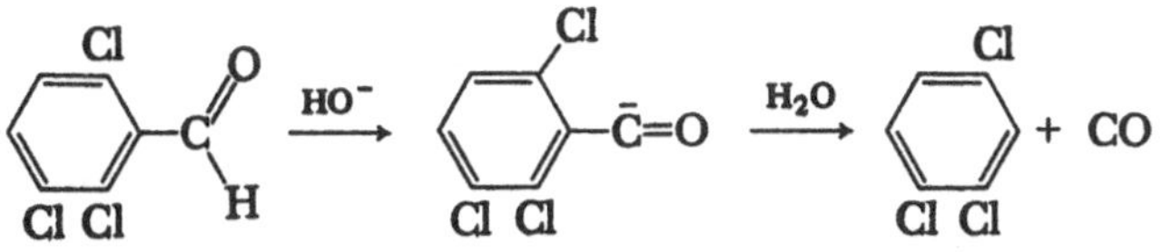

Sequence 2

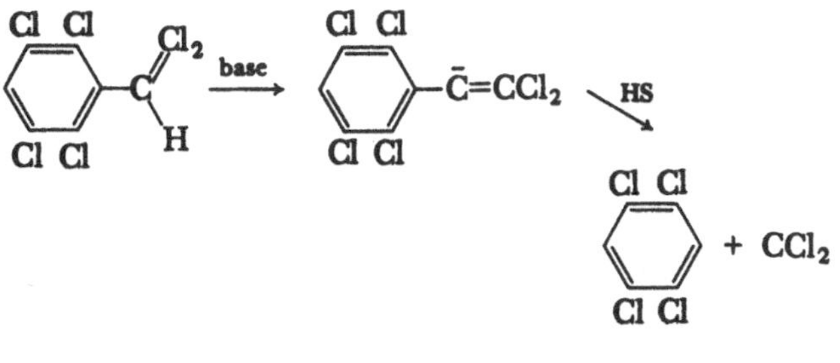

Notice the analogy with Sequence 2 describing the above-mentioned dealkylations.

(f)   Condensations

Perchloro-p-xylene condenses with trichloroethylene. This reaction will be studied in Section 1.4.5.

1.3.10   Perchloromesitylene

According to Harvey *et al.* [74], the chlorination of mesitylene with chlorine, iron and ferric chloride in boiling carbon tetrachloride gives 2,4,6-trichloromesitylene. Photochlorination with chlorine and white light in the same solvent converts this compound into $\alpha H,\alpha'H,\alpha''H$-nonachloromesitylene (91% yield) [69]. With chlorine and ultraviolet light it yields perchlorobenzene (96%) [69].

McBee and Leech had reported that photochlorination of 2,4,6-trichloromesitylene afforded a 37.2% yield of a chlorocarbon melting at 178-179°, assumed to be perchloromesitylene since it gave the correct chlorine analysis [304]. McBee's exhaustive chlorination was repeated independently by Tatlow and by Ballester and co-workers [9] who were unable to repeat it. The latter authors obtained $\alpha H,\alpha'H,\alpha''H$-nonachloromesitylene melting at 178.5-180.5° (68.5% yield), and $\alpha H,\alpha'H$-octachloro-m-xylene (25.2%). Extended reaction times result also in substantial yields of αH-heptachlorotoluene and perchlorobenzene.

$$C_6Cl_6 \xleftarrow{Cl_2}_{h\nu} C_6Cl_3Me_3\text{-}1,3,5 \xrightarrow{Cl_2}_{h\nu} C_6Cl_3(CHCl_2)_3\text{-}1,3,5$$

$$\downarrow Cl_2h\nu$$

$$C_6Cl_4(CHCl_2)_2\text{-}m + C_6Cl_5CHCl_2 + C_6Cl_6$$

It should be emphasized that *photochlorination of a side chain has so far never led directly to alkylaromatic chlorocarbons.*

Ballester *et al.* have found that $2H,4H,6H$-nonachloromesitylene is inert towards reagent BMC [9]. This failure is due not only to steric shielding of the hydrogens by the flanking trichloromethyl groups but also to electronic deactivation of the ring towards electrophilic aromatic substitution.

1.3.11   Other Perchloroalkylbenzenes

Unsuccessful attempts to synthesize other perchloroalkylbenzenes have been recorded in the chemical literature. Among them, the exhaustive chlorination of n-propylbenzene [296], cumene [74], p-menthane, and 1,4-di-n-butylbenzene [74].

1.3.12   Perchlorobenzocyclobutene

Roedig and Kohlhaupt [305] reported that perchlorobutenyne dimerizes at 160-180° giving perchlorobenzocyclobutene (LXVIII).

At 80-100° a stable intermediate (LXVII) is isolated [305], the structure of which was established by spectral evidence (^{13}C-NMR) [306]. The following mechanism has been suggested for this unusual cycloaddition reaction [306, 307].

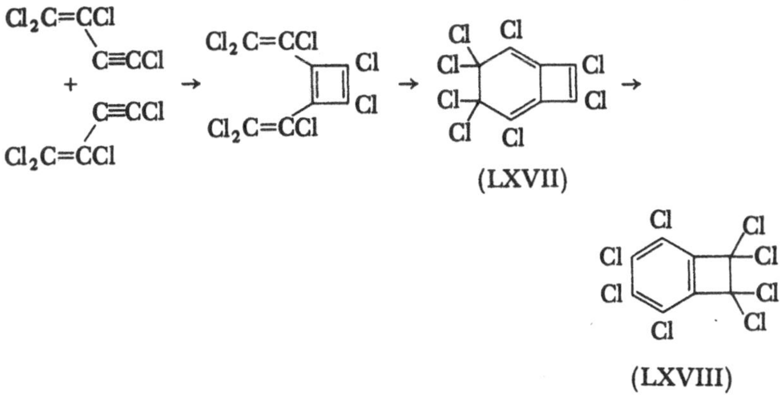

(LXVII)

(LXVIII)

The isomerization of (LXVII) into (LXVIII) also takes place quantitatively at room temperature in polar solvents such as dimethylformamide, dimethylsulphoxide or acetonitrile.

Perchlorobenzocyclobutene is a colourless solid melting at 136-138°. Its infrared [306], ultraviolet [306] and nuclear quadrupole resonance spectra [308] have been reported. Its molecule is bond-strained. However, since its four *alpha* chlorines are located above and below the plane of the benzene ring their steric interactions with the vicinal in-plane aromatic chlorines are minimal, as in *αH*-heptachlorotoluene.

Perchlorobenzocyclobutene is hydrolysed by concentrated sulphuric acid at 160° giving perchlorobenzocyclobutenedione, (LXIX), which is converted into tetrachlorophthalic anhydride with perhydrol-acetic acid [306]. With sodium hydroxide in methanol at

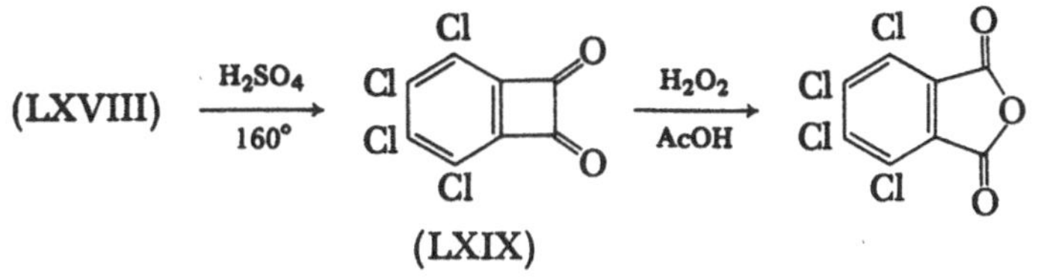

(LXIX)

room temperature a monomethoxy derivative is obtained. At higher temperature the product is a trimethoxy derivative (73% yield). According to Roedig [306] the products are 1-methoxy- (A), and 1,1,2-trimethoxyheptachlorobenzocyclobutene (B).

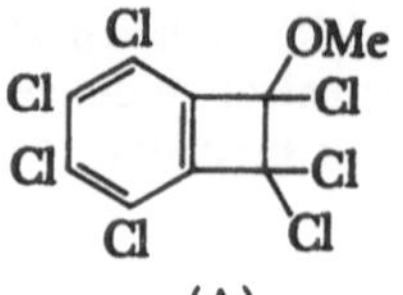

(A) 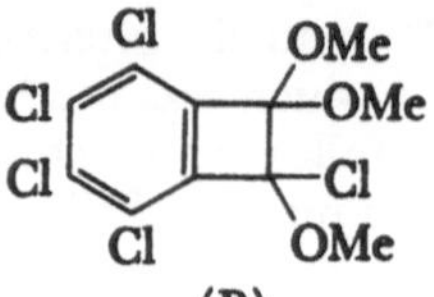   (B)

In our view, these reactions should be interpreted differently.†
It has already been pointed out (Section 1.2.3h) that accumulation of aromatic chlorines facilitates nucleophilic substitution. For instance, Ross and Markarian have shown that αH-heptachlorotoluene reacts with sodium alkoxide to give $\alpha,\alpha,4$-trialkoxy-2,3,5,6-tetrachlorotoluene [277].

$$C_6Cl_5CHCl_2$$

$$\Big\backslash \, RO^-$$

$$p\text{-}RO-C_6Cl_4CH(OR)_2$$

Another example is the basic alcoholysis of oH-heptachlorotoluene reported by Harvey and co-workers [74]. Here, the absence of one o-chlorine atom is compensated by the powerfully electron-attracting trichloromethyl group.

By analogy, it is reasonable to assume that the first step in the reaction of perchlorocyclobutene with methoxide ion leads to (LXX):

$$\text{(LXVIII)} \rightarrow$$

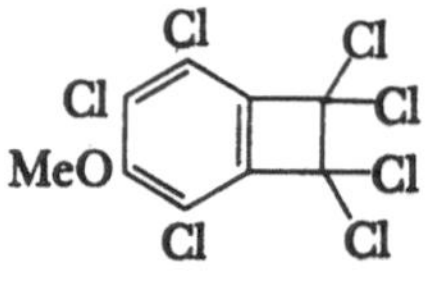

(LXX)

This nucleophilic displacement is favoured by the strong electron-attracting effect of the dichloromethylene groups. Compound A is, therefore, most probably (LXX). The presence of the methoxy group in the benzene ring assists the heterolysis of the two chlorines in

† We have discussed the interpretation of this and other nucleophilic displacements with Professor Roedig and wish to acknowledge his kindness in making available to us some unpublished data at the time of writing this review.

position 1 and consequently, their nucleophilic substitution as shown:

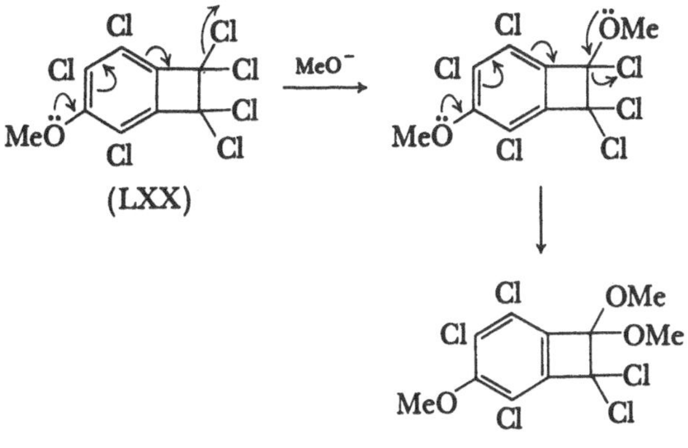

(LXX)

(LXXI)

Compound B should, therefore, be (LXXI). It is emphasized that the introduction of the third methoxy group is the easiest step since it is assisted by the positive tautomeric effect of the two methoxy substituents already present.

The reported stability of the trimethoxy derivative towards boiling methanolic hydroxide [305] is due essentially to two causes: (1) The presence of one methoxy group in the benzene ring deactivates it towards further nucleophilic substitution. (2) Because this methoxy group is *meta* to the dichloromethylene group in (LXXI) its alcoholysis is not assisted.

Roedig and co-workers have also described the reaction of phenylhydrazine with perchlorocyclobutene [305]:

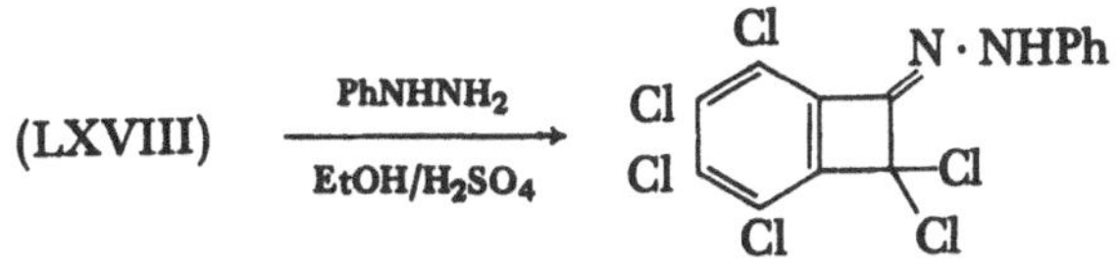

1.3.13   Perchloroindane

Zincke and Meyer synthesized perchloroindane (LXXII) for the first time from the dimer (LXXIII; 57% yield) or from perchloroindanone (LXXIV) with phosphorus pentachloride at 280° [309].

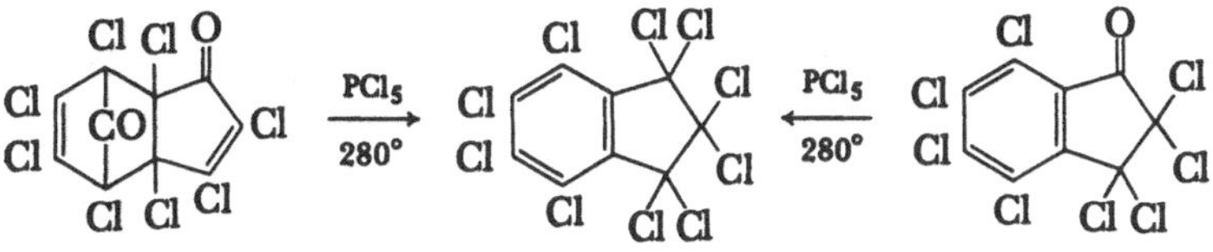

(LXXIII)   (LXXII)   (LXXIV)

Schwemberger obtained it by heating perchloronaphthalene with antimony pentachloride and iodine chloride [78], as well as by destructive chlorination of naphthalene at 200-250° [79].

Roedig produced the compound (LXXII) by destructive chlorination of commercial hexachloronaphthalene with chlorine, iron and iodine, at 160-180° in 70% yield [310]. Bernimolin started from α-chloronaphthalene [311] and also found that chlorination with the Kharash chlorinating agent did not perchlorinate the molecular complex consisting of 4,5,6,7-tetrachloroindane 4,5,6,7-tetrachloro-indene but produced 1H,2H,3H-heptachloroindane [311].

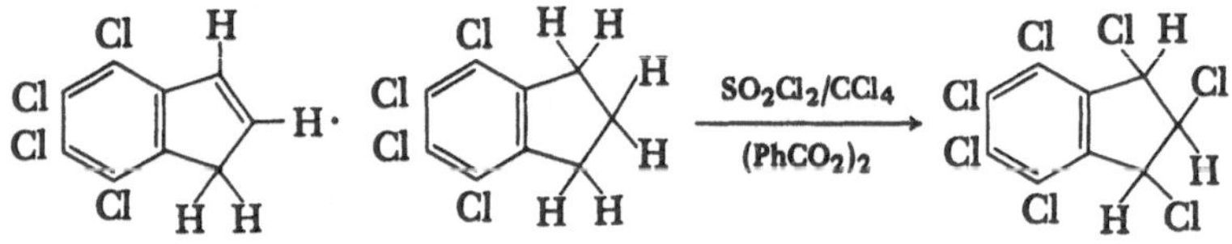

Eaton and co-workers obtained perchloroindane (81%) from perchloroindene by chlorination with chlorine in carbon tetrachloride at 50° and also by destructive chlorination of naphthalene as described by Vollmann [312]. McBee *et al.* found that perchloro-indenone reacts with phosphorus pentachloride giving, among other products, some perchloroindane [102] (see Section 1.4.21).

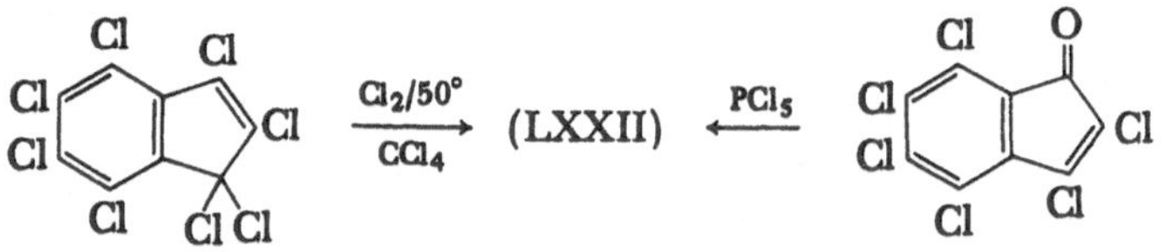

Ballester and co-workers obtained perchlorobenzene in a 96% yield by photochlorination of 1H-nonachloroindane with chlorine and white light in carbon tetrachloride [296]. Photochlorination of γH,γH,2H-nonachloropropylbenzene under the same conditions gives, in addition, perchloroindane [296]:

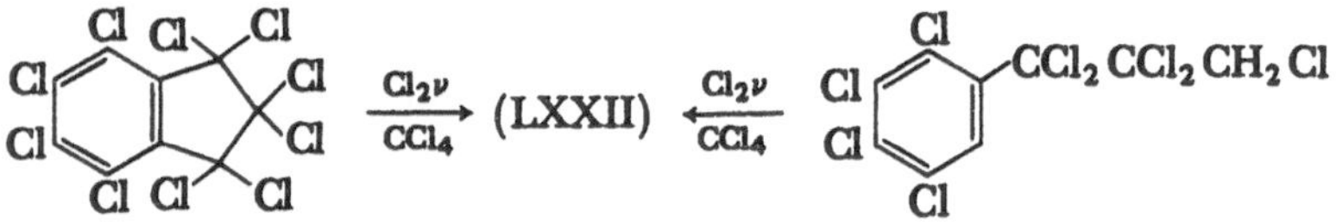

Perchloroindane is also a major product of the aluminium chloride condensation of perchlorotoluene with trichloroethylene (Section 1.4.2) [248, 313].

Perchloroindane is a colourless solid melting at 135.5-136.5°. Its infrared spectrum has been reported [255, 313]. Its ultraviolet spectrum shows the absence of strain as expected [296, 313], since, as in perchlorobenzocyclobutene (Section 1.3.12), the chlorines of positions 1 and 3 are located above and below the plane of the benzene ring and therefore steric interactions with the neighbouring 'in-plane' chlorines are insignificant. The fact that this chlorocarbon can be prepared by destructive chlorination of naphthalene speaks for its high thermal and chemical stability.

Roedig found that warm fuming nitric acid hydrolyzes perchloro-indane to a mixture of perchloroindanone (LXXIII), perchloro-indanedione (LXXIV) and αH-hexachloro-o-acetylbenzoic acid [310]. Presumably, these three products are formed consecutively [310].

This hydrolysis has also been effected in boiling anhydrous nitric acid by Bernimolin, yielding 93% of perchloroindanedione [311] and by Ballester and Riera with sulphuric acid at 100° [296].

According to Roedig, the reaction of perchloroindane with ethan-olic potassium hydroxide yields a monoethoxy derivative (A), along with a methoxyvinylbenzoic (B) acid [310] and with ethanolic sodium ethoxide a triethoxy derivative (C). As in the case of the perchlorobenzocyclobutene (Section 1.3.12) it was assumed that replacement occurred in the aliphatic ring only, which is unlikely.

Ballester *et al.* have recently established that these reactions proceed as follows [231]:

(LXXII) $\xrightarrow{\text{EtO}^-}$ (LXXVI) $\xrightarrow{\text{EtO}^-}$ (LXXVII)

(LXXVI) $\xrightarrow{\bar{\text{O}}\text{H}}$ (LXXIX)

(LXXVII) $\xrightarrow{\text{fum. HNO}_3}$ (LXXVIII)

(LXXIX) $\xrightarrow{\bar{\text{O}}\text{H}}$

$\xrightarrow[\text{(b) } H_3O^+]{\text{(a) } \bar{\text{O}}\text{H}}$ (LXXX)

In order to establish the structures of the ethoxy derivatives Ballester and co-workers performed some additional reactions and studied the spectra involved [231].

Hydrolysis of both ethoxy derivatives with oleum at room temperature give (LXXXI) which rules out structures A and C in favour of (LXXVI) and (LXXVII).

(LXXVI) $\longrightarrow$ (LXXXI) $\longleftarrow$ (LXXVII)

The ketal character of (LXXVII) is established by hydrolysis in hydrochloric acid–dioxane to give the parent ketone (LXXIX). Its hydrolysis with boiling fuming nitric acid gives the diketone (LXXVIII) which on treatment with potassium hydroxide followed by alkaline potassium permanganate yields 4-ethoxytrichlorophthalic acid (LXXXII) as conclusive proof that one methoxy group is attached to the benzene ring.

(LXXVII) $\xrightarrow{H_3O^+}$ (LXXIX)

$\Big\downarrow$ fum. HNO₃

(LXXVIII) $\xrightarrow[\text{(b) kMnO}_4]{\text{(a) } \bar{O}H}$

![structure](Cl substituted benzene with CO₂H groups and EtO group) (LXXXII)

The fact that compound B is obtained from ketone (LXXIX) with sodium hydroxide in aqueous dioxane shows it to be (LXXX) as indicated previously.

1.3.14   Perchlorobenzocyclohexene (Perchlorotetralin)

Berthelot and Jungfleisch found that chlorination of naphthalene with chlorine and antimony pentachloride gives a product containing more chlorine than perchloronaphthalene [190]. (See also Sections 1.2.8 and 1.2.9d.) Other authors reported that exhaustive chlorination of naphthalene yields perchloroindane (Section 1.3.13). The photochlorination of 5,6,7,8-tetrachlorotetralin with chlorine and white light in carbon tetrachloride gives, presumably, a mixture of isomers $C_{10}H_4Cl_8$ [69].

Ballester et al. attempted the chlorination of perchloro-1,4-dihydronaphthalene to perchlorotetralin (LXXXIII) with white light in carbon tetrachloride [250] but obtained only starting material (cf. Section 1.3.13). The synthesis of the title compound appears to be an unusually difficult process (cf. Sections 1.2.1c and 1.2.3d).

1.3.15   Perchlorobi-*m*-Tolyl

All attempts to synthesize this chlorocarbon have been unsuccessful. Chlorination of bi-*m*-tolyl with chlorine and white light in carbon tetrachloride did not give the expected biphenyl but a complex mixture of chlorine addition products. In order to deactivate the rings, the photochlorination was attempted starting from 4,4'-dichloro-3,3'-bistrichloromethylbiphenyl (LXXXIV) and, as expected, 4,4'-dichloro-3,3'-bistrichloromethylbiphenyl (LXXXV) was obtained in a 80.5% yield [234]. The latter with reagent BMC gives a relatively small amount of, presumably, 2*H*,2'*H*-dodecachlorobitolyl (LXXXVI) [234], and under forcing conditions perchlorobiphenyl (LXXXVII) is formed. No perchlorobi-*m*-tolyl has ever been detected [234].

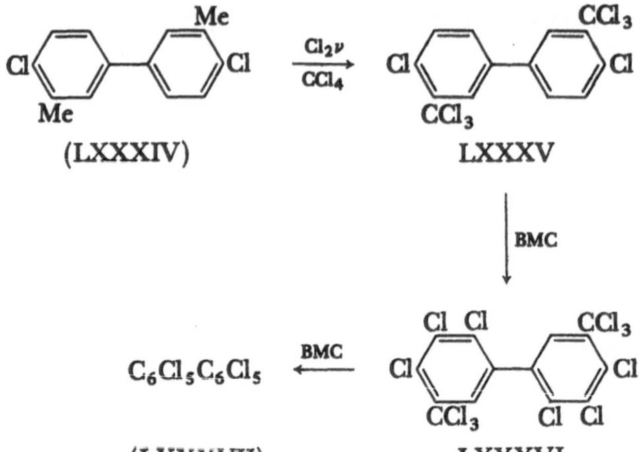

(LXXXIV)   LXXXV

$C_6Cl_5C_6Cl_5$ $\xleftarrow{\text{BMC}}$

(LXXXVII)   LXXXVI

1.3.16 Perchlorobi-p-tolyl

The synthesis of perchlorobi-p-tolyl (LXXXIX) has been effected by chlorination of 4,4'-bistrichloromethylbiphenyl (LXXXVIII) with reagent BMC, the yield being about 60% of theory [234, 248, 249]. The latter is obtained by photochlorination of bi-p-tolyl with chlorine and white light in carbon tetrachloride at 80°.

$$p\text{-}Cl_3CC_6H_4C_6H_4CCl_3\text{-}p \;\rightarrow\; p\text{-}Cl_3CC_6Cl_4C_6Cl_4CCl_3\text{-}p$$

(LXXXVIII)   (LXXXIX)

$$+\, p\text{-}Cl_3CC_6Cl_4C_6Cl_4COCl\text{-}p$$

(XC)

Doorenbos carried out the chlorination with chlorine and trichlorosulphonium tetrachloroaluminate in thionyl chloride, improving the yields (75%) [12].

When reagent BMC is used perchloro-4-p-tolylbenzoyl chloride (XC) is also formed and, under certain reaction conditions, its yield can be as high as 75.5% [234]. Its formation is probably due to chlorine-oxygen exchange of a type already described (Section 1.3.8).

Perchlorobi-p-tolyl is a white solid melting at 233.0-5.5°. Its infrared [12, 234], mass [301b] and ultraviolet [234, 240, 249] spectra have been recorded. The latter shows molecular distortion of each ring as well as inhibition of resonance between them [234, 249]. The steric distortion is similar to that found in perchlorotoluene (Section 1.3.2). Both perchloro-p-tolyl and perchloro-4p-tolylbenzoyl chloride hydrolyse quantitatively with hot sulphuric acid [234, 248, 249] or oleum [278] to octachlorobiphenyl-4,4'-dicarboxylic acid. Perchlorobi-p-tolyl, when treated with anhydrous ferric chloride in refluxing methylene chloride under carefully controlled humidity conditions, gives (XC) (71%). If the treatment is performed in humid air perchlorobiphenyl-4,4'-dicarboxyl dichloride (XCI) is the product (75.5%). In wet methylene chloride octachloro-4,4'-biphenyl-dicarboxylic acid is obtained (67.8%) [278].

$$p\text{-ClOCC}_6\text{Cl}_4\text{C}_6\text{Cl}_4\text{COCl-}p$$

$$\text{(XCI)}$$

Perchlorobi-p-tolyl reacts with cyclohexene in acetic acid in the same manner as perchlorotoluene (Section 1.3.3b) giving $\alpha H,\alpha' H$-octachlorobi-p-tolyl (39% yield) [234].

$$p\text{-Cl}_3\text{CC}_6\text{Cl}_4\text{C}_6\text{Cl}_4\text{CCl}_3\text{-}p \xrightarrow{\text{C}_6\text{H}_{10}} \text{Cl}_2\text{CHC}_6\text{Cl}_4\text{C}_6\text{Cl}_4\text{CHCl}_2$$

Perchlorobi-p-tolyl undergoes reductive polycondensation with stannous chloride or ferric chloride yielding macromolecular chlorocarbons (cf. Sections 1.6.2 and 1.6.5). It condenses with trichloroethylene in the presence of catalytic mixtures containing aluminium trichloride (cf. Section 1.4.6).

1.3.17   Perchlorodiphenylmethane

Ruoff reported that chlorination of diphenylmethane with chlorine and iodine at 150° yields perchlorobenzene, carbon tetrachloride, and hydrogen chloride [43]. Ballester and Riera made perchlorodiphenylmethane (XCII) in low yield (9%) by chlorination of diphenyldichloromethane with reagent BMC, and studied some of its properties [248, 249, 314]. $2H,2'H$-decachlorodiphenylmethane (8.6%) and perchlorofluorene (XCIII) (6.9%) were also isolated [249, 315]. It was found later that under certain conditions $4H$-tridecachloro-3-benzylidenecyclohexene (XCIV) can be formed in 39% yield [238].

Compound (XCIV) reacts with basic alumina to give perchlorobenzylidenecyclohexa-1,4-diene (XCV) (Section 1.4.6). By means of stannous chloride it is converted into $\alpha H,oH$-decachlorodiphenylmethane [238].

Perchlorodiphenylmethane has been obtained in a substantial (48.8%) yield by chlorination of 4,4'-bistrichloromethyldiphenyldichloromethane with reagent BMC [291]. Under milder than usual

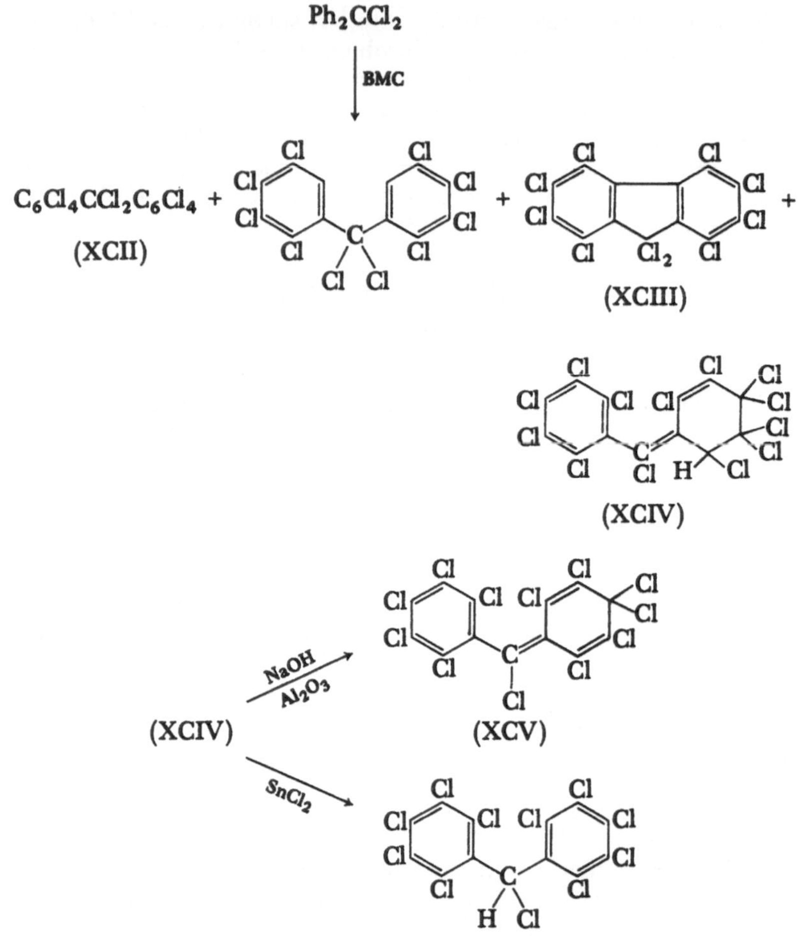

$$Ph_2CCl_2$$

$$C_6Cl_4CCl_2C_6Cl_4 \quad (\text{XCII})$$

(XCIII)

(XCIV)

(XCIV) $\xrightarrow[\text{Al}_2\text{O}_3]{\text{NaOH}}$ (XCV)

(XCIV) $\xrightarrow{SnCl_2}$

conditions an intermediate, presumably 2H,5H,2'H,5'H-dodeca-chlorodi-p-tolylmethane, can be isolated, in which chlorinolysis has affected the trichloromethyl groups.

A third synthesis starts from perchlorodiphenylmethyl (PDM) radical which on chlorination, either under ionic conditions or with bromine or iodine chlorides, affords the product in an 87.1 or a 93.3% yield, respectively (see Section 1.8.1a).

$$(C_6Cl_5)_2\overset{\cdot}{C}Cl \xrightarrow{(+Cl)} (C_6Cl_5)_2CCl_2 \xleftarrow{BMC} (p\text{-}Cl_3CC_6H_4)_2 \quad CCl_2$$
(PDM)

It results also from either thermal (>180°) or catalytic isomerization of perchlorobenzylidenecyclohexa-1,4-diene (see Section

1.4.14) and in a small yield by thermolysis of perchlorodiphenyl-methyl radical (see Section 1.8.1a).

Perchlorodiphenylmethane is a colourless solid melting at $213\text{-}6°$. Its infrared [315] and ultraviolet [315] spectra have been reported showing molecular distortion.†

Perchlorodiphenylmethane   and $2H,2'H$-decachlorodiphenyl-methane are hydrolyzed by oleum to the corresponding benzo-phenones [248, 315], and the former at lower temperature also gives some perchlorobenzylidenecyclohexadienone (see Section 1.11.3). Perchlorodiphenylmethane reacts with stannous chloride in ethyl ether or chloroform [314, 315], with ferrous chloride in ethyl ether [315], or with mercury in ethyl ether dimethylsulphoxide and ultrasonics [315] to give PDM radicals. Sunlight or ultraviolet light can also effect the conversion [238] (Section 1.8.1a).

The reaction with triphenylphosphine gives an acceptable (46.7%) yield of $1H,2H$-icosachlorotetraphenylethane [316] by a mechanism involving the formation of probably perchlorodiphenylcarbene (XCVI), and αH-decachlorodiphenylmethyl radicals [316].

$$(C_6Cl_5)_2CCl_2 + Ph_3P \longrightarrow (C_6Cl_5)_2\bar{C}Cl \ \ Ph_3\overset{+}{P}Cl$$

$$(C_6Cl_5)_2\dot{C}Cl + Ph_3\dot{P}Cl \qquad (C_6Cl_5)_2C: + Ph_3PCl_2$$

$$(XCVI)$$

$$\downarrow \text{solvent}$$

$$(C_6Cl_5)_2CHCH(C_6Cl_5)_2 \longleftarrow (C_6Cl_5)_2\dot{C}H$$

It is noteworthy that the $1H,2H$-icosachlorotetraphenylethane also results from the reaction of αH-undecachlorodiphenylmethane with triphenylphosphine [316] by the following radical mechanism:

$$(C_6Cl_5)_2CHCl + Ph_3P \longrightarrow (C_6Cl_5)_2\bar{C}H \ \ Ph_3\overset{+}{P}Cl$$

$$\downarrow$$

$$(C_6Cl_5)_2CHCH(C_6Cl_5)_2 \longleftarrow (C_6Cl_5)_2\dot{C}H + Ph_3\dot{P}Cl$$

Perchlorodiphenylmethane thermolyses at about $350°$ giving perchlorobenzene, perchlorobiphenyl, perchlorofluorene (XCIII) and

† This chlorocarbon may be assumed to derive from strained perchloro-toluene by substituting an *alpha* chlorine atom, by a bulky pentachlorophenyl group.

chlorine (see also Section 1.4.14). The first step of this process is most probably the formation of PDM radical (see Section 1.8.1a). The chlorine liberated causes the chlorinolysis to perchlorobenzene and perchlorobiphenyl.

It should be pointed out that some thermal processes of perchlorobenzylidenecyclohexa-1,4-diene (XCV) are actually due to perchlorodiphenylmethane (see Section 1.4.14). Perchlorodiphenylmethane reacts with solutions of aluminium chloride or antimony pentachloride to give perchlorodiphenylcarbonium salts (Section 1.11.3).

$$(C_6Cl_5)_2CCl_2 + MCl_x \rightarrow (C_6Cl_5)_2 \overset{+}{C}Cl \overset{-}{M}Cl_{x+1}$$

1.3.18   Perchloro-4-phenyldiphenylmethane

Perchloro-4-phenyldiphenylmethane (XCVII) has been synthesized by interaction of perchloro-4-phenyldiphenylmethyl (PPDM) radicals with chlorine and iodine in carbon tetrachloride (70.7% yield) (see Section 1.8.1b), or with a solution of aluminium chloride in sulphuryl chloride, followed by treatment with water (71.1%) (Section 1.11.3) [317].

$$p\text{-}C_6Cl_5C_6Cl_4\overset{\cdot}{C}ClC_6Cl_5 \xrightarrow{(+Cl)} p\text{-}C_6Cl_5C_6Cl_4CCl_2C_6Cl_5$$

(PPDM)   (XCVII)

Compound (XCVII) is a colourless solid melting at 247-54° (dec.). Its infrared and ultraviolet spectra have been recorded [317] and the latter shows molecular distortion as in perchlorodiphenylmethane (Section 1.3.17).

1.3.19   Perchloro-2-phenyldiphenylmethane

Perchloro-2-phenyldiphenylmethane (XCVIII) is obtained in the same way as perchloro-4-phenyldiphenylmethane (Section 1.3.18) from perchloro-2-phenyldiphenylmethyl (PODM) radical (Section 1.8.1c). The yields of the reactions with chlorine and iodine, and with sulphuryl chloride and aluminium chloride, are 87.8 and 61%, respectively [318].

$$o\text{-}C_6Cl_5C_6Cl_4\overset{\cdot}{C}ClC_6Cl_5 \xrightarrow{(+Cl)} o\text{-}C_6Cl_5C_6Cl_4CCl_2C_6Cl_5$$

(PODM)   (XCVIII)

Perchloro-2-phenyldiphenylmethane is a colourless solid melting at 215° (dec.). Its infrared and ultraviolet spectra have been measured [318]. It is hydrolyzed by oleum followed by addition of water to give perchloro-3-phenyl-4-benzylidenecyclohexa-2,5-dienone (IC) (69.9% yield) [318] (Section 1.11.3). It can be reduced to give the original PODM radical (87.7% yield) (Section 1.8.1c) [318] and at 260-270° it decomposes to perchloro-9-phenylfluorenyl (PPF) radical (73%) (Section 1.8.3a) [318].

1.3.20 Perchloro-4,4'-diphenyldiphenylmethane

There are three ways to synthesize perchloro-4,4'-diphenyldiphenyl-methane (C), all of them starting from perchloro-4,4'-diphenyl-diphenylmethyl (PDDM) radical and operating at room or lower temperatures [317]: (a) Reaction with chlorine and iodine in carbon tetrachloride (70.7% yield); (b) Reaction with a solution of aluminium chloride in sulphuryl chloride, followed by treatment with water (51.7%); (c) Treatment with chlorine and white light in carbon tetrachloride (94%).

$$\{p\text{-}C_6Cl_5C_6Cl_4\}_2 \, \dot{C}Cl \xrightarrow{(+Cl)} \{p\text{-}C_6Cl_5C_6Cl_4\}_2 \, CCl_2$$

(PDDM)   (C)

Perchloro-4,4'-diphenyldiphenylmethane is a colourless solid melting at 304-12° (dec.) with known infrared and ultraviolet spectra [317].

It is remarkable that in spite of steric strain it withstands photochlorination conditions. (Perchlorotoluene (Section 1.3.3e) and perchloro-p-xylene (Section 1.3.9d) undergo facile photo-chlorinolysis). This stability is probably due to the extra shielding afforded by the second tetrachloro-p-phenylene group at positions 1 and 1' of the first one. Perchloro-4,4'-diphenyldiphenylmethane is hydrolyzed to perchloro-4,4'-diphenylbenzophenone (52% yield) as usual (Section 1.11.3) [291, 317] and can easily be converted into the corresponding carbonium ion (Section 1.1.3) [291] or radical (Section 1.8.1d) [291, 317].

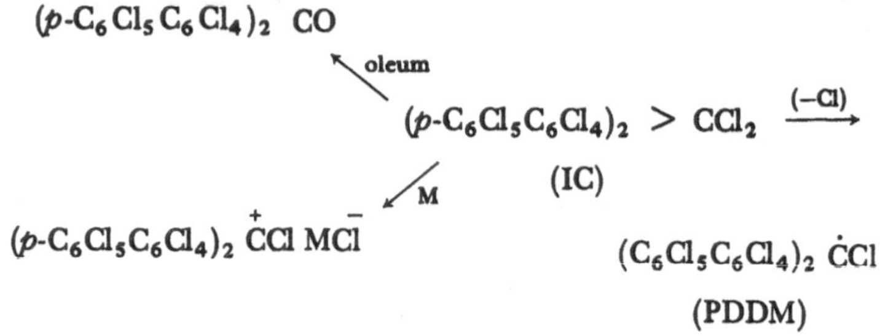

$(p\text{-}C_6Cl_5C_6Cl_4)_2$ CO

$(p\text{-}C_6Cl_5C_6Cl_4)_2 > CCl_2 \xrightarrow{(-Cl)}$

(IC)

$(p\text{-}C_6Cl_5C_6Cl_4)_2 \overset{+}{C}Cl\ M\overset{-}{Cl}$

$(C_6Cl_5C_6Cl_4)_2 \overset{\cdot}{C}Cl$

(PDDM)

1.3.21   Perchlorofluorene

Perchlorofluorene (XCIII) is isolated in a small (6.9%) yield in the chlorination of benzophenone chloride with reagent BMC (Section 1.3.17) [248, 249, 315]. It can be prepared from perchloro-fluorenone with phosphorus pentachloride at 200° (85% yield) [248] or by chlorination of 9,9-dichlorofluorene with reagent BMC-P (71.8%) [64]. It is also found among the thermolysis products of perchlorobenzylidenecyclohexa-1,4-diene (XCV) (350°;

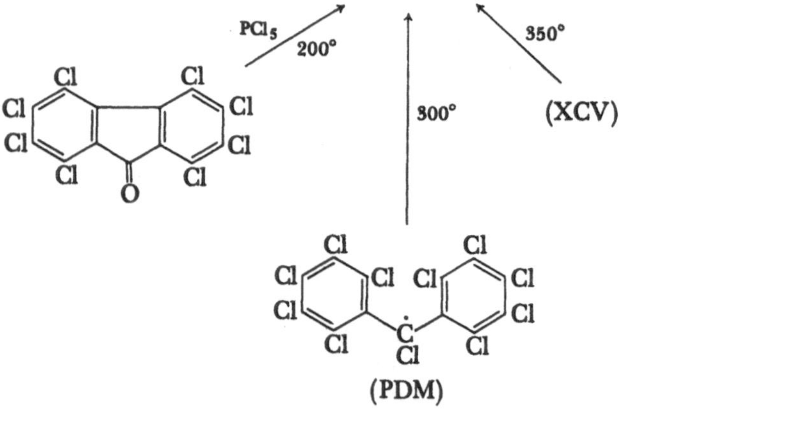

68.6% yield) (Section 1.4.14) [238], or perchlorodiphenylmethyl (PDM) radical (300°; 61.3%) (Section 1.8.1a) [238].

It also results from thermolysis at 350° of αH-undecachlorodiphenylmethane, along with perchlorobenzene and $\alpha H,\alpha H$-decachlorodiphenylmethane [238].

$$(C_6Cl_5)CHCl \xrightarrow{350°} C_6Cl_6 + (C_6Cl_5)_2CH_2 + (XCIII)$$

Perchlorofluorene is a colourless solid melting at 258-9° whose infrared and ultraviolet spectra have been reported [315]. Space filling models indicate that its molecule is not significantly strained since the two 9-chlorines are located above and below the mean plane of the biphenyl system and, therefore, do not interact with the 1- and 8-chlorines. This lack of strain accounts for its easy preparation. The thermal stability of perchlorofluorene is remarkable since it can be heated to 500° in air without significant decomposition [238]. This unexpected behaviour is due to the fact that the usual steric situation is reversed. Normally the alkylaromatic chlorocarbons are strained, while the derived radicals resulting from carbon—chlorine bond homolysis are non-strained. By contrast, perchlorofluorene is a non-strained chlorocarbon but its radical perchlorofluorenyl is strained because of interactions of the chlorines in the three peri-positions.

Perchlorofluorene is hydrolyzed by oleum at room temperature giving perchlorofluorenone (77%), presumably through the carbonium ion [248, 315]. In sunlight, it gives radicals presently under study. Photolysis in the presence of oxygen yields fluorenone (78%), presumably through radical involvement [238]. When perchlorofluorene is heated at 300° with copper dust a 95% yield of blue perchlorobifluorenylene (CI) is obtained (Section 1.4.26) [238].

(CI)

1.3.22   Perchloro-9-phenylfluorene

Perchloro-9-phenylfluorene (CII) has been obtained by Ballester *et al.* from the perchloro-9-phenylfluorenyl (PPF) radical either by chlorination with sulphuryl chloride and aluminium chloride or by reaction with chlorine in carbon tetrachloride in the presence of iodine (Section 1.8.3a) [318]. It is isolated among other chlorocarbons in the photochlorination of the perchlorotriphenylmethyl (PTM) radical in carbon tetrachloride (Section 1.8.2a) [319]. It is also the main product of the reaction of 'black' perchloroketone (CIII) with phosphorus pentachloride [318].

Perchloro-9-phenylfluorene is a white solid, melting with decomposition at 180-210°. Its infrared and ultraviolet spectra are known [318], the latter spectrum clearly showing the existence of molecular distortion due to repulsions of the *ortho* chlorines in the pentachlorophenyl group and those at positions 1, 8 and 9 of the fluorene system (cf. perchlorofluorene; Section 1.3.2).

The thermal decomposition of perchloro-9-phenylfluorene gives PPF radical and chlorine [318]. The elimination of chlorine can be effected at low temperatures with ferrous chloride in ethyl ether but extended treatment with this reducing agent, or with stannous chloride, yields 9*H*-tridecachloro-9-phenylfluorene, presumably via a PPF radical [318] (Section 1.8.3a).

Treatment of perchloro-9-phenylfluorene with oleum and then with water at room temperature gives a mixture of perchloro-9-

phenyl-3-fluorenone (CIV) (22.9% yield) and perchloro-4-(9-fluorenylidene)cyclohexa-2,5-dienone (CIII) (35.1%) [318]. The following mechanism is suggested:

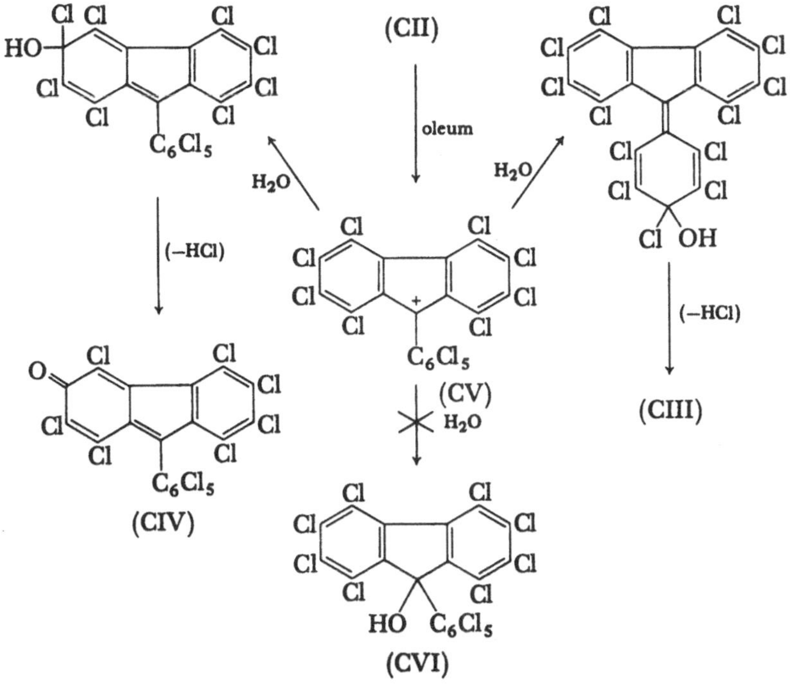

(CII)

oleum

H_2O

(−HCl)

(−HCl)

(CIII)

(CV) H_2O

(CIV)

(CVI)

The absence of alcohol (CVI) is undoubtedly due to steric shielding of the 9-carbon atom of the carbonium ion (CV).

1.3.23 Perchloro-9,9′-spirobifluorene

The title compound (CVII) can be made directly from 9,9′-spirobifluorene with reagent BMC according to Ballester and co-workers [318] in an 84% yield. Another synthesis (82%) is achieved by boiling a solution of perchloro-9-(o-biphenylyl)fluorenyl radical (CVIII) in carbon tetrachloride (Section 1.8.3c) [318]. This chlorocarbon is also obtained (21.6% yield) along with αH-heptadodecachloro-9-(o-biphenylyl)fluorene (CIX) (58.5%) in the perchlorination of 9-o-biphenylylfluorene with the same reagent [318]. Spiran formation occurs at some stage of the perchlorination since (CIX) is quite stable under the reaction conditions [318].

Perchloro-9,9′-spirobifluorene is a colourless solid melting at 388-97° of high thermal stability (up to 400°). Its infrared and ultraviolet spectra have been measured [318].

(CVII)   (CVIII)

+   (CVII)

1.3.24   Perchloro-9,10-dihydroanthracenes

Ballester and co-workers synthesized perchloro-9,10-dihydroanthracene (XII) by chlorination of anthracene with reagent BMC [248-250] (24.2%).

In an attempt to synthesize αH-pentadecachloro-9-benzylanthracene with reagent BMC chlorinolysis occurred giving perchloro-9,10-dihydroanthracene and αH-heptachlorotoluene [250]. This result is related to both the reactivity of the anthracene system at the 9-carbon and the stability of the phenylcarbonium ions.

Perchloro-9,10-dihydroanthracene is a white solid melting at 355-7° with decomposition. Its formation in preference to that of perchloroanthracene has already been discussed in general terms

(Section (1.2.15). Its ultraviolet and infrared spectra have been recorded [250]. The former does not indicate significant molecular distortion.

It has already been mentioned that perchloro-9,10-dihydroanthracene dechlorinates easily to perchloroanthracene (Section 1.2.11). With oleum, at 100°, it gives an 80% yield of perchloroanthraquinone, presumably through carbonium ions (see Sections 1.2.1c and 1.2.9d).

1.3.25   Perchloro-9,10-dihydrophenanthrene

Perchloro-9,10-dihydrophenanthrene (XLIII) was first made by Ballester [248] in a 56.5% yield by chlorination of phenanthrene with reagent BMC [250]. Brooks and co-workers obtained it in a multi-step exhaustive chlorination of phenanthrene (21% yield) [264]. It can also be obtained by chlorination of perchlorophenanthrene with chlorine in white light in carbon tetrachloride (40% yield) [250]. This reaction has, however, no practical value since perchlorophenanthrene is obtained from the title compound.

(XLIII)

Perchloro-9,10-dihydrophenanthrene is a colourless solid melting at 299-302°. Spectral evidence [250] indicates it to be strain-free. Consequently, to avoid the strong 1:4 interaction between chlorines the molecule should be warped (see Section 1.2.15). As mentioned earlier, perchloro-9,10-dihydrophenanthrene yields perchlorophenanthrene [248-250] with isopropyl ether or reducing agents.

An interesting reaction occurs with oleum. At 100°, a mixture of perchlorobiphenyl-2,2'-dicarboxylic acid (CXI) and perchlorofluorenone is obtained, the yields varying from 71% to 10% for the former and 16.4% to 74% for the latter compound. At room temperature, the products are perchlorophenanthrenequinone (CX;

55.5%) and acid (CXI; 5.7%) [248-250]. It is assumed that these products are formed in the following sequence:

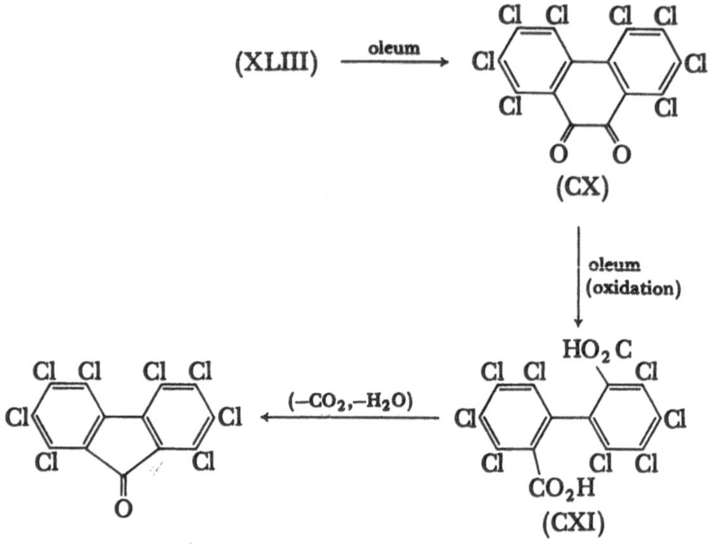

Accordingly, it has been found that acid (CXI) under the reaction conditions gives perchlorofluorenone in a 64% yield [248-250]. It is probable that the acylium ion Ar—Ċ=O plays an essential role in the ring closure, either by displacement of carboxyl group or a hydrogen (after decarboxylation).

The reaction with alkali metals to give a carbonaceous material has been reported [192].

1.3.26   Perchloroacenaphthene

A chlorination leading to a hexachloroacenaphthene has been described by Dashevskii and Petrenko [320].

Perchloroacenaphthene (CXII) was first prepared by Ballester [249] by chlorination of acenaphthylene with reagent BMC in a 71.7% yield [249, 250, 291]. Later, Mack reported a synthesis by a two-step chlorination of the same hydrocarbon with chlorine in boiling perchlorobutadiene and then in the presence of aluminium chloride in 21% yield [321].

Perchloroacenaphthene is a colourless solid melting at 250-5° (dec.) with known infrared and ultraviolet spectra [250, 321]. Its dechlorination to perchloroacenaphthylene has been performed [249, 250, 291, 321]. Conversion into perchloro-1,2-dioxoacenaphene occurs with oleum at 100° in 62% yield.

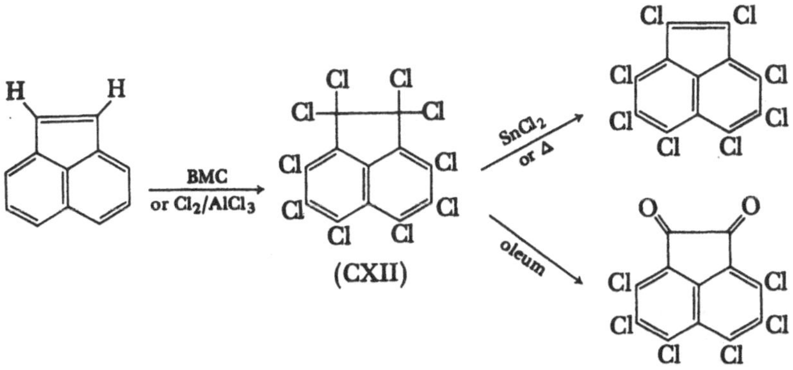

(CXII)

1.3.27   Perchloro-4,5,9,10-tetrahydropyrene

The title compound (XLVI) has been synthesized by Reimlinger and King by a multi-step chlorination of pyrene [267]. No data on this compound were, however, reported.

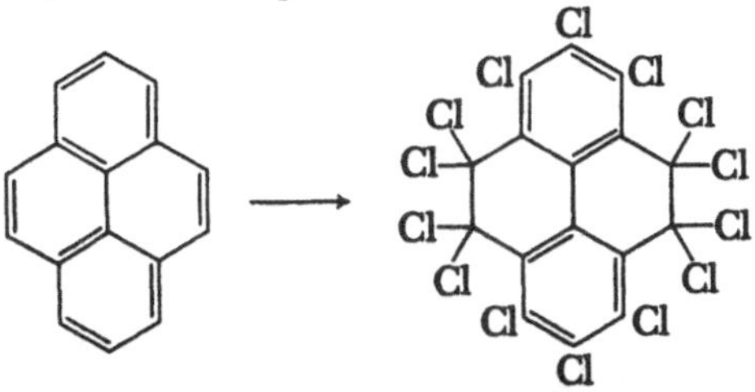

(XLVI)

Merz and Weith performed the chlorination of pyrene with antimony pentachloride at 350° obtaining products of chlorinolysis, namely two yellow solid chlorocarbons infusible up to 300°, and carbon tetrachloride (Section 1.2.13) [46]. It is reasonable to assume that perchloro-4,5,9,10-tetrahydropyrene is an intermediate. Since those solids have formulae $C_{14}Cl_{10}$ and $C_{15}Cl_{10}$, and the stability of perchlorofluorene is known to be extremely high (Section 1.3.21), it is tempting to assume that they are (CXIII) and

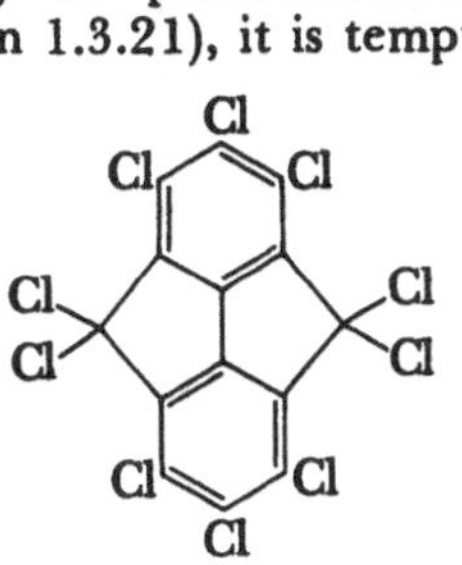

(CXIII)

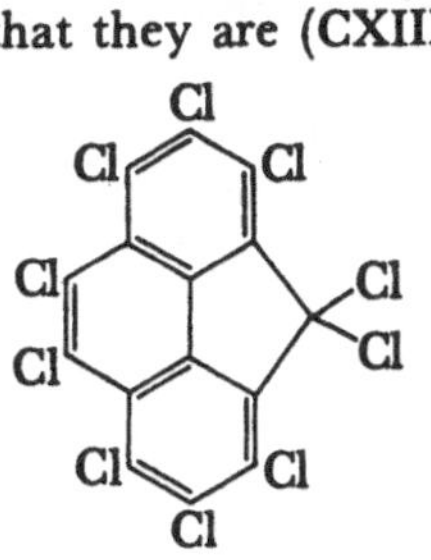

(CXIV)

(CXIV) respectively. However, since such chlorocarbons are rare, no reliable conclusion can be drawn concerning their structures.

1.4   PERCHLOROARYLETHYLENES

Two general characteristics of the perchloroarylethylenes are their extremely high thermal stability and chemical inertness which are due to three factors: (1) Steric effects, such as shielding of the ethylene bond and molecular strain of the would-be adduct; (2) Electronic deactivation caused by the powerful negative inductive effect of the chlorines; (3) The greater strength of the carbon (sp^2)-chlorine bonds, as it occurs in the perchloroarenes (see Introduction).

1.4.1   Perchlorostyrene

Perchlorostyrene (LIX), the simplest aromatic ethylene, was obtained with other products by Ballester and co-workers from thermolysis of perchlorotoluene (XLIX) (Section 1.3.3d) [107]. Although the yield was quantitative only one-third of the starting material was transformed into (LIX).

The most convenient (70% yield) method on a laboratory scale is the chlorination of β,β-dichlorostyrene with reagent BMC or, preferably, BMC-P, some $\alpha H,oH$-octachloroethylbenzene also being obtained (13%) [64].

Perchlorostyrene is also formed by dechlorination of perchloroethylbenzene with potassium iodide in warm acetic acid (98% yield) (Section 1.3.4) [293]. This, however, is not practical since perchloroethylbenzene is rather inaccessible (Section 1.3.4).

It is also formed with reagent BMC from α,β,β-trichlorostyrene (29.2% yield), oH-heptachlorostyrene (42.8%), oH-nonachloroethylbenzene, or perchloroethylbenzene (LVIII) (2.9%) [64]; in the photolysis of perchlorotoluene with ultraviolet light in carbon tetrachloride (2.1%) (Section 1.3.3e) [108] and in excellent yield, from perchlorophenylacetylene (XXX) by addition of chlorine in carbon tetrachloride under illumination [64], or by chlorination with carbon tetrachloride in sunlight (Section 1.5.1) [64]. All these processes are shown in the chart at top of page 89.

The substitution of the hydrogen in oH-nonachloroethylbenzene with reagent BMC is not only sterically hindered but competes with vicinal dechlorination to oH-heptachlorostyrene. Starting from heptachlorostyrene the addition of chlorine to the ethylene bond takes place preferentially and, therefore, the perchlorination is sterically hindered. This is also the case with α,β,β-trichlorostyrene.

At some stage in the reaction of α,α-dichlorostyrene with reagent BMC there is preferential chlorine addition. The resulting precursor,

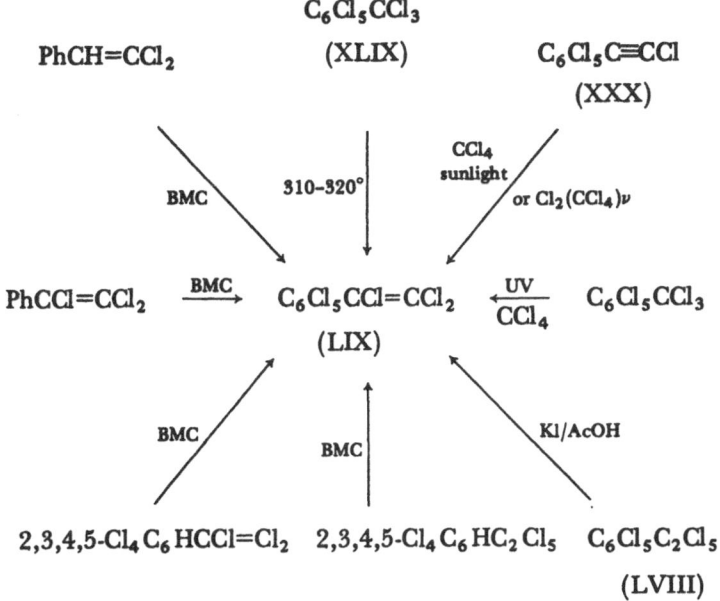

$$C_6Cl_5CCl_3$$

$$PhCH=CCl_2 \qquad (XLIX) \qquad C_6Cl_5C\equiv CCl$$
$$(XXX)$$

$$PhCCl=CCl_2 \xrightarrow{BMC} C_6Cl_5CCl=CCl_2 \xleftarrow[CCl_4]{UV} C_6Cl_5CCl_3$$
$$(LIX)$$

$$2,3,4,5\text{-}Cl_4C_6HCCl=Cl_2 \quad 2,3,4,5\text{-}Cl_4C_6HC_2Cl_5 \quad C_6Cl_5C_2Cl_5$$
$$(LVIII)$$

presumably αH-nonachloroethylbenzene—the formation of which is not significantly hindered like perchlorotoluene (Section 1.3.1)— dechlorinates, however, under the reaction conditions giving the expected perchlorostyrene in good yield.

The dechlorination of carbon tetrachloride by perchlorophenyl-acetylene in sunlight may take place by the following mechanism:

$$C_6Cl_5C\equiv CCl \xrightarrow{\text{sunlight}} C_6Cl_5C\equiv C^{\cdot} + Cl$$

$$Cl + CCl_4 \longrightarrow Cl_2 + Cl_3C^{\cdot}$$

$$Cl_2 + C_6Cl_5C\equiv CCl \xrightarrow{\text{sunlight}} C_6Cl_5CCl=CCl_2$$

Perchlorostyrene is a white solid melting at 99.5-101.0°. Its infrared [108] and ultraviolet [133, 322] spectra have been studied and the latter indicates the absence of steric strain and the presence of inhibition of resonance because the planes of the trichlorovinyl group and the benzene ring are at a significant angle.

Perchlorostyrene has a very high thermal stability and a remarkable inertness towards electrophilic attack. For example, it does not add chlorine either with reagent BMC or with chlorine in light. It is reduced to ωH-pentachlorophenylacetylene with zinc in refluxing dioxane in 62.4% yield [323].

$$C_6Cl_5CCl=CCl_2 \xrightarrow[\text{(dioxane)}]{Zn,\ 100°} C_6Cl_5C\equiv CH$$

Seiber has recently reported its electrolytic reduction in methanol-DME using a mercury cathode, obtaining ωH-pentachlorophenyl-acetylene (21%), αH-heptachlorostyrene (22.8%), and $\omega H,4H$-tetra-chlorophenylacetylene (9.9%) [324]. Using a lead cathode and under varying conditions the yields of ωH-pentachlorophenylacetylene or $\omega H,4H$-tetrachlorophenylacetylene can be greatly increased (up to 80.2 and 40.7%, respectively) [324]. The following mechanism, postulating the intermediacy of perchlorophenylacetylene (XXX), has been proposed [324]:

$$C_6Cl_5CCl=CCl_2 \xrightarrow{(+2e, -Cl^-)} C_6Cl_5\bar{C}=CCl_2 \xrightarrow{(+H^+)} C_6Cl_5CH=CCl_2$$

$$\xrightarrow{(-Cl^-)}$$

$$C_6Cl_5\equiv CCl \xrightarrow[(+2e, +H^+, -Cl^-)]{} C_6Cl_5C\equiv CH \xrightarrow{(+2e, +H^+, -Cl^-)}$$

$$2,3,5,6\text{-}Cl_4C_6HC\equiv CH$$

$$\Big/ (+2e, +H^+)$$

$$2,3,5,6\text{-}Cl_4C_6HCH=CH_2$$

1.4.2 Perchloropropenylbenzenes

An unsuccessful synthesis of perchloropropenylbenzenes has been reported [296]:

$$PhCCl=CClMe \xrightarrow{BMC} C_6Cl_5CCl=CClMe \xrightarrow[h\nu]{Cl_2} C_6Cl_5CCl=CClCCl_3$$

. However, the chlorination of α,β-dichloropropenylbenzene with reagent BMC gives either $1H$-nonachloroindane or, under milder conditions, $\gamma H,\gamma H,2H$-nonachloropropylbenzene. Both compounds perchlorinate to perchloroindane with chlorine in white light (Section 1.3.13) [296].

It appears, therefore, that just as with α,β,β-trichlorostyrene (Section 1.4.1), chlorine addition to the ethylene bond precludes substitution of the last *ortho* hydrogen. The unexpected substitution of the terminal methyl group creates a second reaction centre available for cyclization. The synthesis of the perchloropropenyl-benzenes could possibly be effected starting from the isomeric β-chloropropenylbenzenes (see Section 1.4.1).

Ballester and co-workers succeeded in preparing the cis-(CXV) and trans-perchloropropenylbenzene (CXVI) by condensation of perchlorotoluene with trichloroethylene in methylene chloride in the presence of aluminium chloride-containing catalysts [248, 249, 313]. This remarkable reaction deserves some comment. In principle, this is a Prins-type reaction. It was foreseen, however, that because of the high steric strain of the normal adduct (βH-undecachloropropylbenzene) hydrogen chloride elimination would immediately follow. Nevertheless, the desired condensation took place provided that in addition to the aluminium chloride both calcium chloride and hydrogen chloride were present. The yields were 66% for the cis-isomer and 19% for the trans. A 9.6% yield of 4H-tridecachloro-1-phenylpent-1-ene (CXVII) was also obtained.

$$C_6Cl_5CCl_3 \ + \ CHCl{=}CCl_2$$

$$(CH_2Cl_2) \left| \begin{array}{l} AlCl_3 \\ CaCl_2 \\ HCl \end{array} \right. \downarrow$$

$$\underset{(CXV)}{\overset{C_6Cl_5}{\underset{Cl}{>}}C{=}C\overset{CCl_3}{\underset{Cl}{<}}} \ + \ \underset{(CXVI)}{\overset{C_6Cl_5}{\underset{Cl}{>}}C{=}C\overset{Cl}{\underset{CCl_3}{<}}} \ + \ \underset{(CXVII)}{C_6Cl_5CCl{=}CClCCl_2CHClCCl_3}$$

When calcium chloride and hydrogen chloride are absent a mixture of perchloroindane (Section 1.3.13), perchloro-1-phenylcyclopentene (Section 1.4.13), and perchloroindene (Section 1.4.21) is formed. It has been suggested that those inorganic chlorides form a complex with the aluminium chloride thereby ensuring the right reaction path. As expected, lithium chloride has a similar effect [248, 249, 313]. As far as the mechanism of this condensation is concerned, it is assumed that a catalytic complex W† causes the formation of a perchlorophenylcarbonium ion loosely bonded to its counter-ion through a chlorine bridge (CXVIII), a process which is assisted by the high steric strain of perchlorotoluene (compare with Section 1.3.3b). The cationoid species then attacks the least shielded carbon atom of trichloroethylene to give a strained adduct (CXIX) which is stabilized immediately by loss of hydrogen chloride.

The resulting isomeric perchloropropenylbenzenes associate somehow with the catalytic complex W since they interconvert under the reaction conditions. They are unable to condense with a second molecule of trichloroethylene under those conditions. Nevertheless, as it has already been indicated, a product of condensation (CXVII) with two molecules of trichloroethylene is formed. Therefore, a

† It is known that aluminium chloride forms complexes with chlorides such as lithium chloride ($AlCl_3 \cdot LiCl$) and calcium chloride ($4AlCl_3 \cdot 3CaCl_2$) [313].

$$C_6Cl_5CCl_3 + W \rightarrow C_6Cl_5\overset{+}{C}Cl_2 \; Cl\overline{W}$$
$$(CXVIII)$$

$$C_6Cl_5\overset{+}{C}Cl_2 \; Cl\overline{W} + CHCl{=}CCl_2 \rightarrow C_6Cl_5CCl_2CHCl\overset{+}{C}Cl_2 \; Cl\overline{W}$$
$$(CXIX)$$

$$(-HCl)$$

$$C_6Cl_5\,CCl{=}CCl\overset{+}{C}Cl_2 \; Cl\overline{W} \rightarrow C_6Cl_5CCl{=}CClCCl_3 + W$$
$$(CXX) \qquad\qquad (CXV \; and \; CXVI)$$

small fraction of a cationoid intermediate, possibly (XIX) or (CXX), is captured by the trichloroethylene.

$$\left.\begin{array}{c} C_6Cl_5CCl_2CHCl\overset{+}{C}Cl_2 \; Cl\overline{W} \\ (CXIX) \\ or \\ C_6Cl_5CCl{=}CCl\overset{+}{C}Cl_2 \; Cl\overline{W} \\ (CXX) \end{array}\right\} \; \begin{array}{c} CHCl{=}CCl_2 \\ \longrightarrow \end{array} \; \begin{array}{c} C_6Cl_5\,CCl{=}CClCCl_2\,CHClCCl_3 \\ (CXVII) \end{array}$$

In the absence of both calcium chloride and hydrogen chloride the cationoid species are less associated with their counter-ion (presumably tetrachloroaluminate) and therefore the formation of perchloropropenylbenzenes cannot compete effectively with other reactions, such as polycondensation with trichloroethylene (Section 1.4.13) or cyclization to perchloroindane (Section 1.3.13).

In fact, *trans*-perchloropropenylbenzene cyclizes in the absence of calcium chloride and hydrogen chloride while the *cis*-isomer, although it possesses the favourable configuration, surprisingly does not cyclize. This may be due to a much easier carbonium ion formation from the *trans*-isomer because of steric causes.

The perchloropropenylbenzenes are colourless solids melting at
146-8° (cis) and 163-5° (trans) with known ultraviolet and infrared
spectra [313]. Their configuration has been established on the basis
of melting point and infrared spectrum. The ultraviolet data are
ambiguous, although they afford excellent evidence for steric
inhibition of resonance between the vinyl group and the benzene
ring, as well as the absence of benzene ring distortion.

Steric interconversion occurs in methylene chloride in the presence of aluminium chloride, as indicated before. Regardless of the
starting isomer an equilibrium mixture in the ratio of cis:trans = 7:2
is obtained at room temperature. Ultraviolet light on a carbon
tetrachloride solution has a similar effect. It is noteworthy that the
cis-isomer be favoured in both cases, since its steric inhibition of
resonance should be higher. It has been suggested that its apparently
higher thermodynamic stability might be due to bonding involving
the π-orbitals of the benzene ring and the non-bonding p-orbitals of
the trichloromethyl chlorines [313].

The perchloropropenylbenzenes are hydrolyzed with oleum at
room temperature giving the same—presumably trans-perchloro-
cinnamic acid in yields of 62.5% (cis) and 78.1% (trans) [313] via a
common intermediate perchlorophenylpropenium ion.

The perchloropropenylbenzenes undergo reductive condensation
with ferrous chloride or stannous chloride to give perchloro-1,6-
diphenylhexatrienes (Section 1.4.10).

1.4.3   Perchloro-1-phenylbutadiene

Perchloro-1-phenylbutadiene (CXXI) is obtained in an 88% yield by
thermolysis of perchloro-1-phenylcyclobutene (Section 1.4.12) at
300° [323]. This reaction may proceed by a concerted mechanism
(ring opening and allylic chlorine shift). A two-step process is
unlikely because of the steric strain of the resulting hypothetical
intermediate.

$$C_6Cl_5-\underset{\underset{CCl_2-CCl_2}{|}}{\overset{\overset{}{|}}{C}}=CCl \xrightarrow{\ 300°\ } C_6Cl_5CCl=CClCCl=CCl_2$$

$$(CXXI)$$

Compound (CXXI) is a white solid melting at 137-8° and its
infrared and ultraviolet spectra have been measured [323].

1.4.4   Perchloro-1-phenylpenta-1,3-diene

In the condensation of perchlorotoluene with trichloroethylene a
small proportion of 4H-tridecachloro-1-phenylpent-1-ene (CXVII) is
formed (Section 1.4.2). This compound when submitted to de-

chlorination with basic alumina gives perchloro-1-phenylpenta-1,3-diene (CXXII) in 85% yield [313].

$$C_6Cl_5CCl=CClCCl_2CHClCCl_3 \xrightarrow{\text{NaOH, Al}_2\text{O}_3} C_6Cl_5CCl=CClCCl=CClCCl_3$$

(CXVII)   (CXXII)

It is a colourless chlorocarbon melting at 156-8° which liberates iodine from alkaline iodides in acetic acid giving, presumably, unstable chlorocarbon free radicals (cf. Section 1.3.3b).

1.4.5   Perchloro-1,4-dipropenylbenzenes

Four perchloro-1,4-dipropenylbenzenes (CXXIII) are known [248, 249, 325]. They are obtained in a low (32.5%) yield by condensation of perchloro-p-xylene with trichloroethylene in the manner described for the perchloropropenylbenzenes (Section 1.4.2). A mixture containing polycondensation products is also formed.

$$Cl_3CC_6Cl_4CCl_3\text{-}p + CHCl=CCl_2 \rightarrow Cl_3CCCl=CClC_6Cl_4CCl=CClCCl_3\text{-}p$$

(CXXIII)

The title compounds melt at 281-4° (a; 23%), 184-6° (b; 8%), 208-10° (c; 0.75%), and 218-25° (d; 0.7%) (cf. Table).

The mechanism of this condensation is discussed in Section 1.4.2.

Since cis-$trans$ isomerism allows for three isomers only (cis-cis, cis-$trans$, $trans$-$trans$) the fact that four exist is regarded as evidence for syn-$anti$ isomerism. On the basis of the position of the C=C stretching peaks, the intensity of the peaks around 1525 cm^{-1} and the melting point the following assignments have tentatively been proposed [248] (cf. Table).

Ultraviolet light isomerizes (CXXIIIb) in carbon tetrachloride to give a mixture from which a, c, and d are isolated in conversion yields of 18, 24 and 3% respectively [248, 325]. The stability of the isomers is due to the high steric energy barriers preventing the flipping from syn to $anti$ configurations, and vice versa. They are due to repulsions between the $ortho$ chlorines in the transition states of the syn-$anti$ conversions which require planar conformations.

The hydrolysis with oleum at room temperature of isomer a gives a 90.5% yield of a perchloro-p-phenylenediacrylic acid (CXXIV), and b and c yield CXXIV and an isomer [248, 249, 325]. Consequently, stereomutations occur during the hydrolysis, as in the case of the perchloropropenylbenzenes (Section 1.4.2). The formation of an isomeric acid casts some doubts upon the correctness of the given configurational assignments [325].

Compounds CXXIII	Melting point	Configuration
a	281-4°	cis, anti, cis
b	184-6	cis, syn, trans
c	208-10	trans, anti, cis
d	218-25	trans, syn, trans

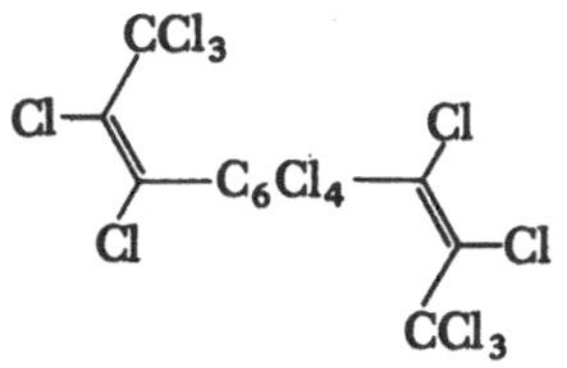

(CXXIIIa)

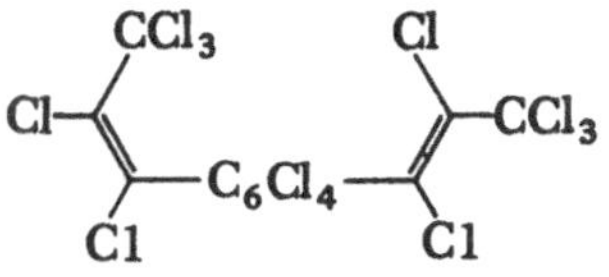

(CXXIIIb)

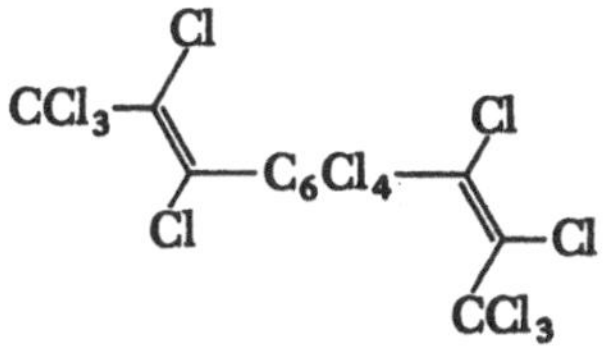

(CXXIIIc)

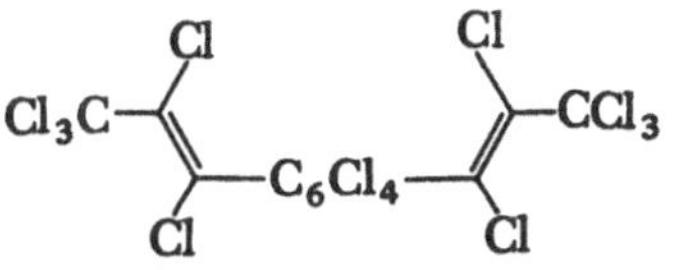

(CXXIIId)

$$(CXXIII) \rightarrow HO_2 CCCl=CClC_6 Cl_4 CCl=CClCO_2 H\text{-}p$$

(CXXIV)

The perchloro-1,4-dipropenylbenzenes undergo polymerization under reductive condensation conditions (Section 1.6.3).

1.4.6   Perchloro-4,4'-dipropenylbiphenyls

Perchlorobi-p-tolyl condenses with trichloroethylene in the presence of aluminium chloride, calcium chloride and hydrogen chloride in methylene chloride, at room temperature, to give a resinous material from which two solid perchloro-4,4'-dipropenylbiphenyls (CXXV), melting at 231-4° and 210-2.5°, are isolated [248, 249, 325] (4.5 and 3.2% respectively). The third possible isomer has not been detected. These results have been confirmed by Doorenbos [12].

$$p\text{-}Cl_3CC_6Cl_4 - C_6Cl_4CCl_3\text{-}p + CHCl=CCl_2 \rightarrow$$

$$p\text{-}Cl_3 CCCl=CClC_6 Cl_4 C_6 Cl_4 CCl=CClCCl_3\text{-}p$$

(CXXV)

The infrared and ultraviolet spectra of these chlorocarbons have been measured [12, 325] to assign their configurations [325]. The higher melting isomer is *trans,trans* (CXXVa), while the other is *cis,cis* (CXXVb) which isomerizes partly to the former in ultraviolet light [325].

Both *a* and *b* give the same perchloro-4,4'-biphenylenediacrylic acid (CXXVI) when treated with oleum at room temperature in 85.4 and 97.4% yields respectively [248, 249, 325].

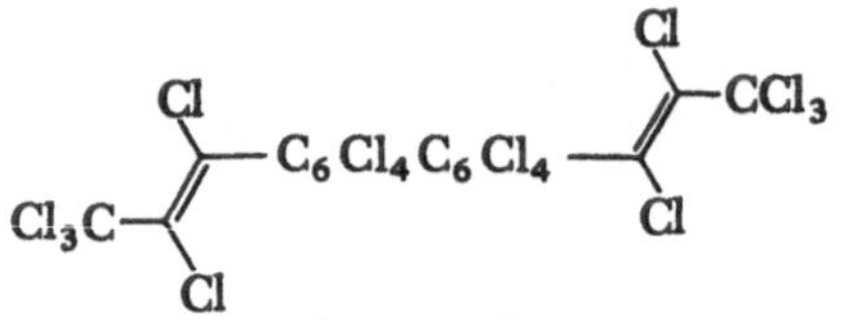

(CXXVa)

p-HO$_2$C–ClCClCC$_6$Cl$_4$C$_6$Cl$_4$CCl=CCl–CO$_2$H-p

(CXXVI)

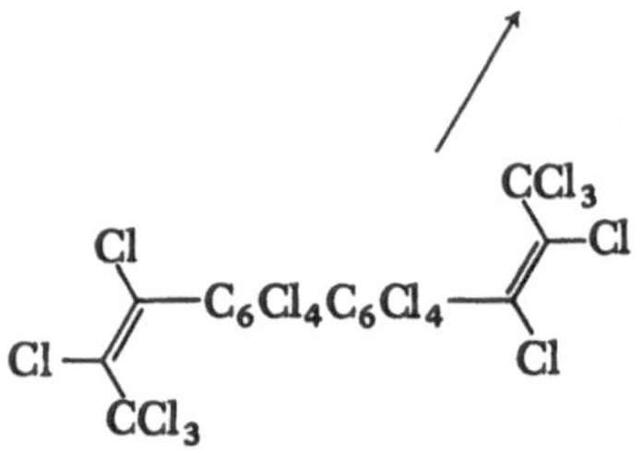

(CXXVb)

The perchloro-4,4'-dipropenylbiphenyls polymerize under reductive condensation conditions (Section 1.6.4).

1.4.7   The Perchlorostilbenes

The reactions of perchlorotoluene with a number of reducing agents such as alkaline iodides—in acetic acid or DMSO, stannous chloride or ferrous chloride, or with triphenylphosphine—have been described in Section 1.3.3b. These reactions give perchlorostilbenes (LI) and (LII). The yields are summarized in Table 1.2.

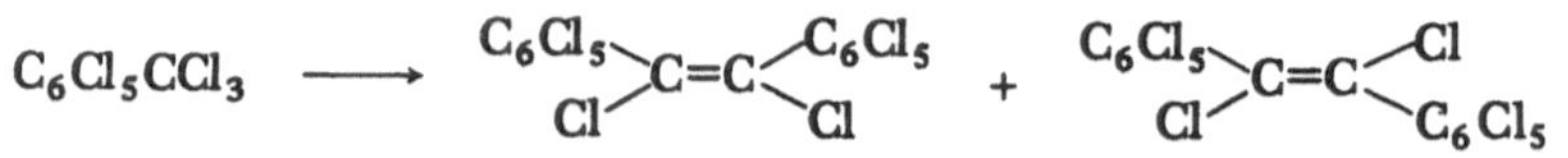

(LI; *cis*)   (LII; *trans*)

TABLE 1.2   Perchlorostilbenes from Perchlorotoluene

Method	Yield %		Reference
	cis (LI)	*trans* (LII)	
Alkaline iodide (acetic acid)	47.5	36.6	274
Alkaline iodide (DMSO)	93	6.2	281
Stannous chloride (dioxane)	40	49	239
Ferrous chloride (dioxane)	32.5	49.3	274
Triphenylphosphine (benzene)	72	8.5	281

Perchlorostilbenes are also formed by chlorination with reagent BMC of *trans*-(*trans*, 45% yield) and *cis*-α,α'-dichlorostilbene (*cis*, 33.6%; *trans*, 41.5%) [274], *trans*-$2H,2'H$-dodecachlorostilbene (*trans*, 99.1%) [326]; or $2H,2'H$-dodecachlorobibenzyl (*cis*, 3.2%; *trans*, 83.8%) [274] (cf. also 1.3.3 and 1.7.1).

Trans-perchlorostilbene (78% yield) along with perchloro-1,1-diphenylethylene (CXXVII; 1.4.8) is also formed by chlorination of 1,1-dichloro-2,2-di-p-chlorophenylethylene with reagent BMC [291] in 78% yield.

$$(p\text{-ClC}_6\text{H}_4)_2 \quad \text{C=CCl}_2 \xrightarrow{\text{BMC}} \underset{\text{C}_6\text{Cl}_5}{\overset{\text{C}_6\text{Cl}_5}{\diagdown}}\text{C=CCl}_2 + \underset{\text{Cl}}{\overset{\text{C}_6\text{Cl}_5}{\diagdown}}\text{C=C}\underset{\text{C}_6\text{Cl}_5}{\overset{\text{Cl}}{\diagup}}$$

$$\text{(DDE)} \qquad\qquad \text{(CXXVII)} \qquad \text{(LII)}$$

The rearrangement, most probably assisted by the repulsion between the phenyl rings, precedes ring-perchlorination since (CXXVII) is inert under the chlorination conditions. The following carbonium ion mechanism is suggested:

$$\text{(LII)}$$

The postulated addition of Cl^+ to the *alpha* ethylene carbon is sterically prevented in perchloro-1,1-diphenylethylene.

The perchlorostilbenes are white solids. The *cis*-isomer melts at 230.5-1.5° while the *trans*-isomer melts and sublimes around 400°. The latter is practically insoluble and of high thermal stability and chemical inertness [274]. Configurations were assigned on the basis of melting point, solubility and infrared and ultraviolet spectra [274] which indicate steric inhibition of resonance.

While thermal isomerization of stilbenes takes place around 200°, [326, 327], perchlorostilbenes behave as if they were non-conjugated ethylenes and isomerize not even at 300°. Steric inhibition of benzylic resonance in the transition state of the stereomutation appears to be responsible for this behaviour. Repulsions between the two *ortho* chlorines of each benzene ring and their *alpha* chlorines prevent the benzylic systems from becoming planar. The activation energy for the perchlorostilbenes is, consequently, high and stereomutation occurs with great difficulty. In *cis*-2H,2'H-decachlorostilbene, where those restrictions are much reduced, isomerization takes place normally.

As expected, the *trans*-isomer is formed from the *cis*-isomer with ultraviolet light in dioxane (62%) while white light has no effect [326]. This result is remarkable since conversion is usually from *trans* to the *cis* form. Although solubilities may play an important role by shifting the photoequilibrium towards the *trans*-isomer, steric repulsions favouring the latter are also important. The 2H,2'H-deca-chlorostilbenes behave similarly, the conversion into the *trans*-isomer being 66.2% [274].

The *cis*-stilbenes isomerize under electrophilic chlorination or photochlorination conditions. In either case, an adduct with positive chlorine or atomic chlorine is probably formed, thereby decreasing the activation energy required for the rotation of one ethylene carbon in respect to the other.

If the formation of such adducts is sterically hindered or the adducts are sterically strained, no stereomutation is expected. Accordingly, the perchlorostilbenes do not isomerize under the

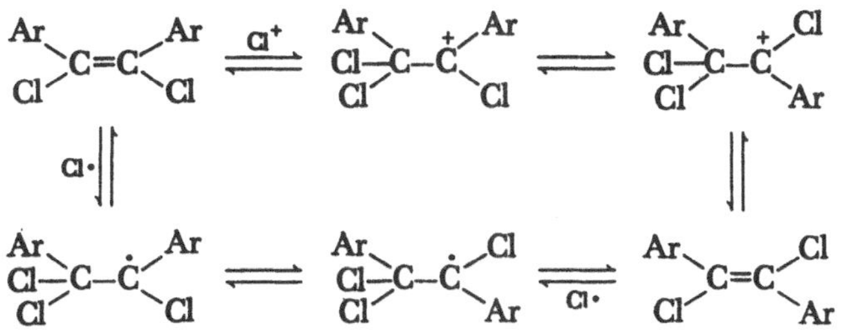

$$\underset{Cl}{\overset{Ar}{>}}C=C\underset{Cl}{\overset{Ar}{<}} \overset{Cl^+}{\rightleftharpoons} \underset{Cl}{\overset{Ar}{>}}Cl-\overset{+}{C}-\overset{+}{C}\underset{Cl}{\overset{Ar}{<}} \rightleftharpoons \underset{Cl}{\overset{Ar}{>}}Cl-\overset{+}{C}-\overset{+}{C}\underset{Ar}{\overset{Cl}{<}}$$

$$Cl\cdot \updownarrow \qquad\qquad\qquad\qquad \updownarrow$$

$$\underset{Cl}{\overset{Ar}{>}}Cl-C-\overset{\cdot}{C}\underset{Cl}{\overset{Ar}{<}} \rightleftharpoons \underset{Cl}{\overset{Ar}{>}}Cl-C-\overset{\cdot}{C}\underset{Ar}{\overset{Cl}{<}} \overset{Cl\cdot}{\rightleftharpoons} \underset{Cl}{\overset{Ar}{>}}C=C\underset{Ar}{\overset{Cl}{<}}$$

above reaction conditions. Since, however, ring chlorination of cis-α,α'-dichlorostilbene gives trans-perchlorostilbene steric effects at the intermediate stages are relatively unimportant.

The perchlorostilbenes are resistant towards oxidation such as boiling fuming nitric acid in the presence of mercury ions. The trans-isomer reacts with zinc in boiling dioxane to give perchlorotolane (Section 1.5.2), while the cis-isomer, under the same conditions, gives a compound melting at 260°, the composition of which is consistent with formula $C_{14}H_4Cl_6$. This different behaviour is possibly connected with solubilities [274].

1.4.8   Perchloro-1,1-diphenylethylene

It has been mentioned in the preceding section that chlorination of 1,1-dichloro-2,2-di-p-chlorophenylethylene with reagent BMC yields perchloro-1,1-diphenylethylene (CXXVII; 21.6% yield), along with trans-perchlorostilbene [234a, 291, 319].

Perchloro-1,1-diphenylethylene is a white solid melting at 203-5° of known infrared and ultraviolet spectra [319] with no evidence for significant ring-distortion. In fact, its ultraviolet spectrum resembles that of cis-perchlorostilbene. Its mass spectrum has been reported [234a]. As indicated previously, this chlorocarbon does not isomerize under the conditions leading to its formation, nor when submitted to chlorination under white light in carbon tetrachloride [274, 319]. It is inert towards concentrated sulphochromic mixture at 100°, and towards hot 60% fuming sulphuric acid.

1.4.9   Perchloro-1,4-diphenylbutadiene

Out of three possible perchloro-1,4-diphenylbutadienes (CXXVIII) only one, probably the trans-trans isomer, is known [323]. It is prepared (79.3%) from perchloro-1,4-diphenylbutadi-ine (CXXIX) with chlorine in carbon tetrachloride under incandescent white light or sunlight:

$$C_6Cl_5C\equiv CC\equiv CC_6Cl_5 \xrightarrow[(CCl_4)]{Cl_2, h\nu} C_6Cl_5CCl=CClCCl=CClC_6Cl_5$$

$$\text{(CXXIX)} \qquad\qquad\qquad\qquad \text{(CXXVIII)}$$

It is a white compound melting at 238-41° and its ultraviolet and infrared spectra have been recorded [323].

1.4.10   The Perchloro-1,6-diphenylhexatrienes

In theory, six isomeric perchloro-1,6-diphenylhexatrienes (CXXX) are possible. The reaction of *cis*- or *trans*-perchloropropenylbenzene (Section 1.4.2) with ferrous chloride in aqueous dioxane at 100° gives the same mixture of two perchloro-1,6-diphenylhexatrienes melting at 141-6° (CXXX*a*) and 304.0-4.5° (CXXX*b*) [248, 249, 325].

The yields, starting from the *cis*-isomer, are 58.5 and 25.2%, respectively, from the *trans* 57.8 and 23.9%. These self-condensations can also be performed with stannous chloride in boiling dioxane [248, 325] giving yields of 59.0 and 27.2% from the *cis*-isomer.

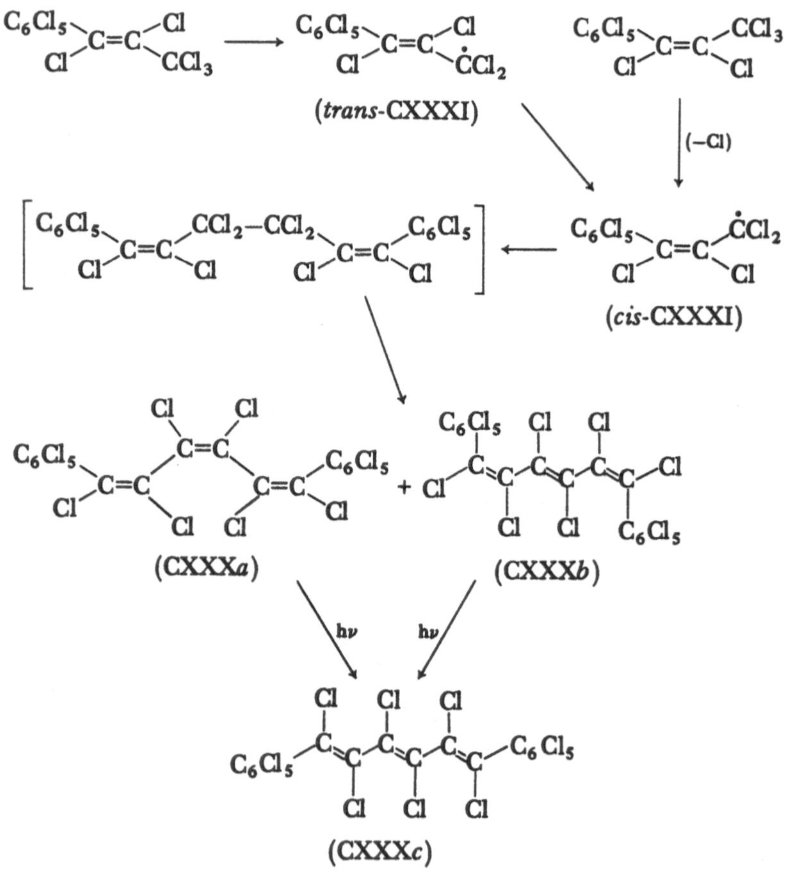

(*trans*-CXXXI)

(*cis*-CXXXI)

(CXXX*a*)   (CXXX*b*)

(CXXX*c*)

Obviously, the same intermediate is involved regardless of the starting material and the dechlorinating agent. It is probably the perchloro-3-phenylallyl radical (CXXXI) and, accordingly, a transient signal is detected by epr spectrometry. The fact that only two perchlorodiphenylhexatrienes are formed indicates reaction of this radical in either its *cis* or *trans* form, exclusively.

Irradiation of a carbon tetrachloride solution of either isomer with ultraviolet light, or heating at 500°, results in about 60% yield of a third isomer (CXXXc) [248, 249, 325]. Its very high melting point (335-6°) indicates that it possesses the all-*trans* configuration. The configurational assignments are therefore: *a, cis,cis,cis; b, cis,trans, cis; c, trans,trans,trans*. Consequently, *cis*-perchloro-3-phenylallyl is the reacting radical. The infrared and ultraviolet spectra of these compounds have been measured [325].

1.4.11   Perchlorotetraphenylethylene

The synthesis of perchlorotetraphenylethylene (CXXXII) was announced erroneously [291, 294]. The compound described was actually $\alpha H,\alpha' H$-icosachlorotetraphenylethane.

Compound (CXXXII) is obtained in a 53% yield by heating a mixture of perchloro-3-benzylidenecyclohexa-1,4-diene (CXX; Section 1.4.14) at 250° with copper [238]. The reaction takes place probably through perchlorodiphenylmethyl (PDM) radical (Section 1.8.1a), and a perchlorodiphenylcarbenoid (possibly XCVI; Section 1.3.17) intermediate.

$$(CXX) \xrightarrow[250-60°]{} (C_6Cl_5)_2CCl_2 \xrightarrow[250-60°]{Cu} (C_6Cl_5)_2\overset{\cdot}{C}Cl$$

$$\Big/ \genfrac{}{}{0pt}{}{Cu}{250-60°}$$

$$(C_6Cl_5)_2C{=}C(C_6Cl_5)_2 \longleftarrow (C_6Cl_5)_2C:$$
$$(CXXXII) \qquad\qquad (XCVI)$$

Perchlorotetraphenylethylene is a white chlorocarbon infusible up to 450°. Infrared and ultraviolet data have been registered [238].

1.4.12   Perchloro-1-phenylcyclobutene

Perchlorophenylacetylene (XXX; Section 1.5.1) reacts with perchloroethylene under the influence of ultraviolet light, at room temperature, giving perchloro-1-phenylcyclobutene (CXXXIII) [323]. The yield is 64.3% of theory. Perchloro-1-phenylcyclobutene

(CXXXIII) [323] is a colourless solid melting at 163-5.5°. Its infrared, ultraviolet and mass spectra have been measured. It is stable towards ultraviolet light in carbon tetrachloride, does not suffer dechlorination with stannous chloride in refluxing dioxane or with alkaline iodide in refluxing acetic acid. However, it reacts with oleum at 100° giving perchlorophenylmaleic acid (CXXXV) (53.4%), probably through the diketone (CXXXIV).

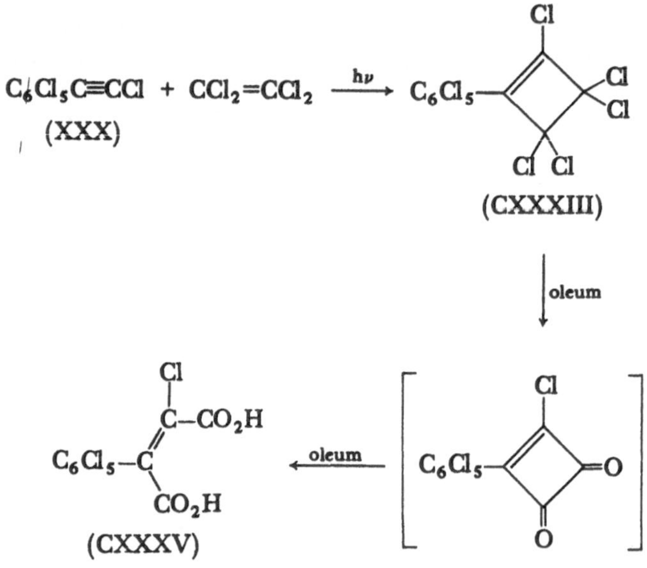

(XXX)

(CXXXIII)

oleum

(CXXXV)

The thermolysis of perchloro-1-phenylcyclobutene yields perchloro-1-phenylbutadiene (Section 1.4.3).

1.4.13   Perchloro-1-phenylcyclopentene

When the condensation of perchlorotoluene with trichloroethylene is carried out in the presence of aluminium chloride, hydrogen chloride being absent (with or without calcium chloride), perchloroindane (>30%) and perchloro-1-phenylcyclopentene (CXXXVI; >10%) are obtained (Section 1.4.2) [248, 313].† The following mechanism (as shown on page 103) is proposed.

Perchloro-1-phenylcyclopentene is a white solid melting at 173-5°. Its infrared and ultraviolet data are published [313]. Its structure is, however, not fully established. It undergoes hydrolysis with oleum at

---

† Minor components are: cis-perchloropropenylbenzene and perchlorobenzene, with calcium chloride; perchloroindane, without calcium chloride.

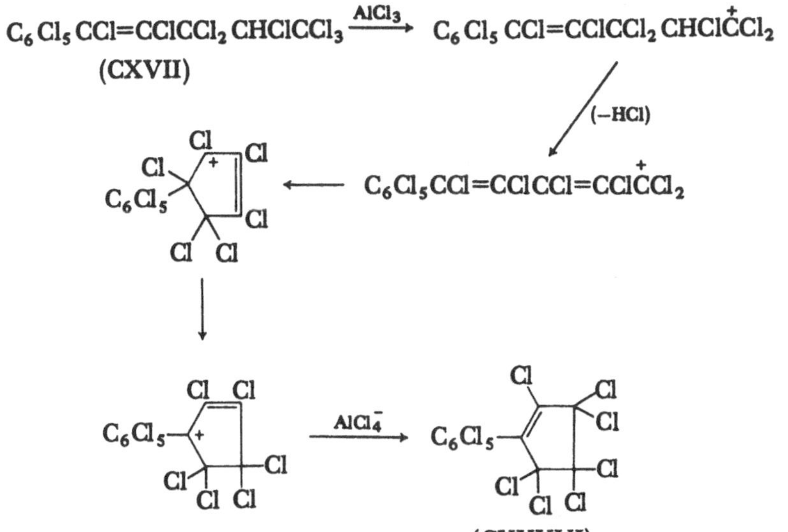

$$C_6 Cl_5 CCl{=}CClCCl_2 CHClCCl_3 \xrightarrow{\text{AlCl}_3} C_6 Cl_5 CCl{=}CClCCl_2 CHCl\overset{+}{C}Cl_2$$

(CXVII)

$$\Big/ \text{(−HCl)}$$

$$\longleftarrow C_6Cl_5CCl{=}CClCCl{=}CCl\overset{+}{C}Cl_2$$

$$\xrightarrow{\text{AlCl}_4^-}$$

(CXXXVI)

room temperature to a perchloroketone melting at 175-8°, which has presumably either structure

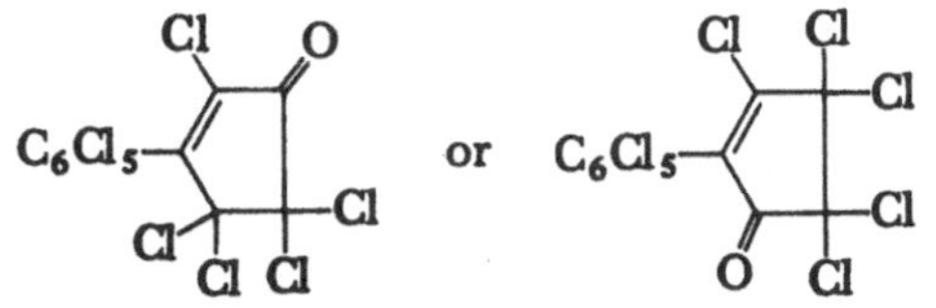

or

1.4.14   Perchlorobenzylidenecyclohexa-1,4-diene

Chlorination of benzophenone chloride by means of reagent BMC yields a mixture containing 4*H*-tridecachloro-3-benzylidenecyclo-hexene (XCIV; Section 1.3.17). This compound gives perchloro-benzylidenecyclohexa-1,4-diene (XCV) when treated with basic alumina in cyclohexane–carbon tetrachloride [238] in 99% yield. This chlorocarbon is also obtained in moderate (~50%) yield from perchlorodiphenylmethane (XCII) or perchlorodiphenylmethyl (PDM) radical when treated below zero with aluminium chloride in sulphuryl chloride [317], or in methylene chloride containing chromic acid [238], and then with water. The dark-green colour appearing before the addition of water indicates the formation of a carbonium ion (Section 1.11.3). At higher temperatures perchloro-diphenylmethane is formed instead (Section 1.3.17).

Perchlorobenzylidenecyclohexa-1,4-diene is a yellowish substance melting at 173.5-6.0°. Its infrared [238, 317], ultraviolet [238,

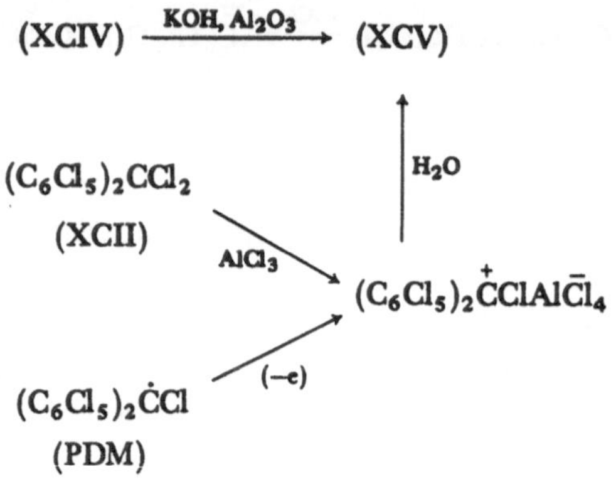
(XCIV)
KOH, Al_2O_3
(XCV)
H_2O
$(C_6Cl_5)_2CCl_2$
(XCII)
$AlCl_3$
$(C_6Cl_5)_2\overset{+}{C}ClAl\overset{-}{Cl}_4$
(−e)
$(C_6Cl_5)_2\dot{C}Cl$
(PDM)

317], and mass [301b] spectra have been recorded. It isomerizes to perchlorodiphenylmethane (70%) above its melting point (Section 1.3.17). The same conversion can be performed by passing its hexane solution through silica gel or by treatment with sulphuryl chloride at room temperature. With oleum (or sulphuric acid) at room temperature, followed by dilution, perchlorobenzophenone is obtained almost quantitatively. However, when the dilution is carried out with ice-dry ice mixture, perchlorobenzophenone is obtained and also

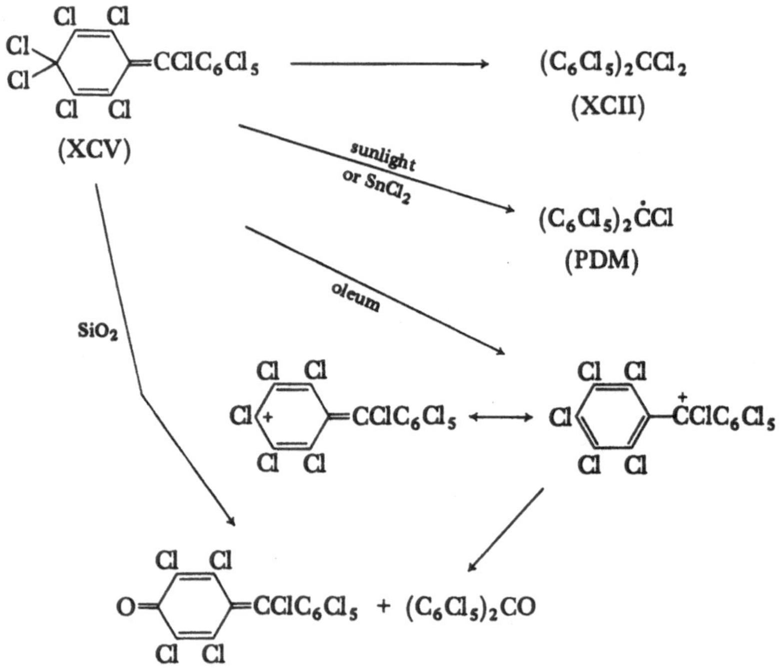
Cl Cl
Cl
Cl
$=CClC_6Cl_5$
Cl Cl
(XCV)
$(C_6Cl_5)_2CCl_2$
(XCII)
sunlight
or $SnCl_2$
$(C_6Cl_5)_2\dot{C}Cl$
(PDM)
oleum
SiO_2
Cl Cl
Cl
$=CClC_6Cl_5$
Cl Cl
Cl Cl
Cl
$\overset{+}{C}ClC_6Cl_5$
Cl Cl
Cl Cl
O
$=CClC_6Cl_5$ + $(C_6Cl_5)_2CO$
Cl Cl

perchlorobenzylidenecyclohexa-2,5-dienone. The green colour of the oleum solution indicates the formation of carbonium salt complexes (Section 1.11.3). Hydrolysis also occurs in carbon tetrachloride solution on silica gel at room temperature producing the two ketones in yields of 47 and 51%, respectively.

On heating the chlorocarbon (XCV) to 250-300° perchloro-diphenylmethyl (PDM) radicals are detected (epr) and at 350° perchlorofluorene (68.6% yield), perchlorobenzene (17.1%), per-chlorobiphenyl (6.6%) and chlorine are formed, undoubtedly via perchlorodiphenylmethane and perchlorodiphenylmethyl radicals [238] (Section 1.3.17). Irradiation with solar or ultraviolet light in carbon tetrachloride or by treatment with stannous chloride in dioxane or ethyl ether generates perchlorodiphenylmethyl (PDM) radicals (Section 1.8.1a).

1.4.15 Perchloro-p-biphenylylmethylenecyclohexa-1,4-diene

Perchloro-p-biphenylylmethylenecyclohexa-1,4-diene (CXXXVII) is obtained in a 34.5% yield from the perchloro-p-phenyldiphenyl-methyl (PPDM) radical by treatment first with a solution of aluminium chloride in sulphuryl chloride and then with water at −15° [317]. The dark-green sulphuryl chloride solution indicates the presence of the corresponding carbonium ion species (Sections 1.8.1b and 1.11.3).

p-$C_6Cl_5C_6Cl_4\dot{C}ClC_6Cl_5$ $\xrightarrow[\text{(b) H}_2\text{O}]{\text{(a) AlCl}_3,\ \text{SO}_2\text{Cl}_2}$

(PPDM)

p-$C_6Cl_5\,C_6\,Cl_4\,CCl\!=\!$

(CXXXVII)

The title compound is a yellowish solid melting at 200-3° with known infrared and electronic spectra [317].

1.4.16 Perchloro-p-Biphenylylmethylene-3-phenylcyclohexa-1,4-diene

The synthesis of perchloro-p-biphenylylmethylene-3-phenylcyclo-hexa-1,4-diene (CXXXVIII) from perchloro-4,4′-diphenyldiphenyl-methyl (PPDM) radicals at low temperatures was unsuccessful [317] yielding perchloro-4,4′-diphenyldiphenylmethane (Sections 1.8.1d and 1.11.3).

1.4.17 Perchlorodiphenylmethylenecyclohexa-1,4-diene

Perchlorodiphenylmethylenecyclohexa-1,4-diene† can be obtained (95%) by interaction of perchlorotriphenylmethyl (PTM) radicals and a mixture of aluminium chloride and sulphuryl chloride followed by addition of water to the resulting green-blue solution at −14° [317].

This reaction takes place probably through triphenylcarbonium tetrachloroaluminate (Section 1.11.4) and unlike in the case of perchlorodiphenylmethyl (PDM) radicals (Sections 1.3.17 and 1.4.14), the isomeric chlorocarbon is not formed. Steric strain and shielding of the central carbon in the intermediate carbonium ion are preventing isomerization.

The chlorocarbon (CXXXIX) is also formed (57.8%) in a time-controlled reaction of PTM with chlorine and white light in carbon tetrachloride [291, 317, 318].

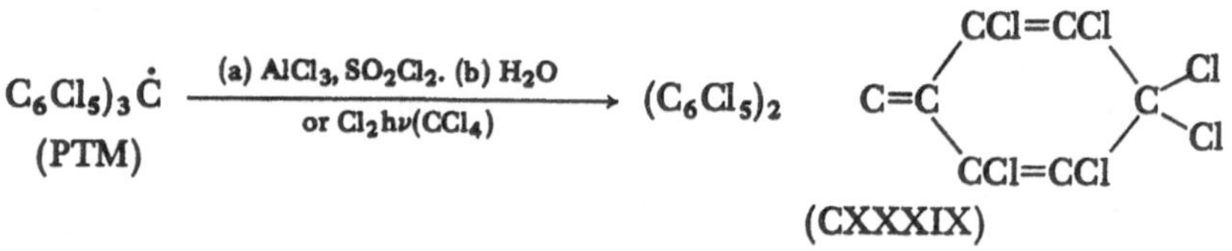

(a) AlCl3, SO2Cl2. (b) H2O
or Cl2hν(CCl4)
(C6Cl5)3Ċ
(PTM)
(C6Cl5)2
C=C
CCl=CCl
C
Cl
(CXXXIX)

Perchlorodiphenylmethylenecyclohexa-1,4-diene is a reddish solid melting at 218-220° (dec.). Its infrared and electronic spectra have been measured [317]. The molecule is extremely twisted at the *exo* ethylene bond and, consequently, unstable. It is converted into PTM in boiling solvents (hexane, carbon tetrachloride, etc.) or by dechlorination with stannous chloride in ethyl ether at room temperature (83.4%) and also by incandescent white light in carbon tetrachloride (95.5%) (Section 1.8.2a) [291, 317, 318].

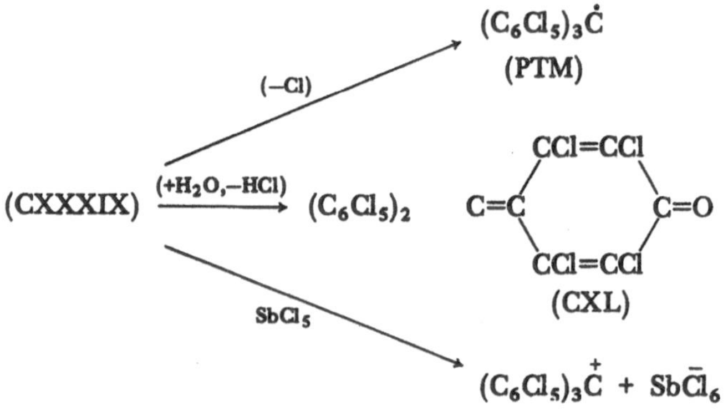

(C6Cl5)3Ċ
(PTM)
(−Cl)
(CXXXIX)
(+H2O,−HCl)
(C6Cl5)2
C=C
CCl=CCl
C=O
(CXL)
SbCl5
(C6Cl5)3C+ + SbCl6−

† Originally assumed to be the perchlorotriphenylmethane [294].

With antimony pentachloride in either sulphuryl chloride or in carbon tetrachloride (CXXXIX) gives perchlorotriphenylcarbonium hexachloroantimonate (Section 1.11.4). In solution, it reacts with silica gel giving perchloro-4-diphenylmethylenecyclohexa-2,5-dienone (CXL) [317]. This hydrolysis can also be accomplished in good yield with oleum [317].

1.4.18 Perchlorodi-p-biphenylylmethylene-3-phenylcyclohexa-1,4-diene

Perchloro-4,4',4''-triphenyltriphenylmethyl (PTTM) radical gives a dark blue-green solution with anhydrous aluminium chloride in sulphuryl chloride, owing to the formation of perchloro-4,4',4''-triphenylcarbonium tetrachloroaluminate. However, by treatment with water at $-15°$ the PTTM radical is recovered [317].

The failure to obtain (CXLI) is due to steric strain and to the shielding by three p-pentachlorophenyl groups in the carbonium ion.

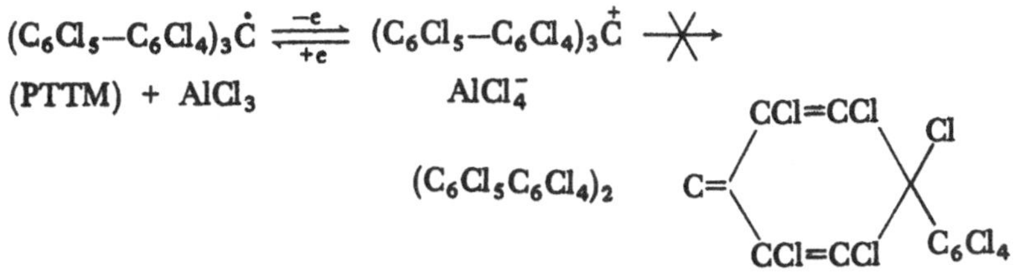

$$(C_6Cl_5\text{-}C_6Cl_4)_3\overset{\bullet}{C} \underset{+e}{\overset{-e}{\rightleftharpoons}} (C_6Cl_5\text{-}C_6Cl_4)_3\overset{+}{C} \longrightarrow\!\!\!\!\times\!\!\!\longrightarrow$$

$$(PTTM) + AlCl_3 \qquad\qquad AlCl_4^-$$

$$(C_6Cl_5C_6Cl_4)_2 \qquad C=\!\!<$$

(CXLI) + AlCl_3

1.4.19 Perchloro-3-diphenylmethylenecyclohexene

Treatment of the perchlorotriphenylmethyl (PTM) radical with chlorine and white light in carbon tetrachloride gives, among other chlorocarbons (Section 1.4.17), some perchloro-3-diphenylmethylenecyclohexene (CXLII) as the major product [291, 319]. It must be formed through perchlorodiphenylmethylenecyclohexa-1,4-diene (CXXXIX).

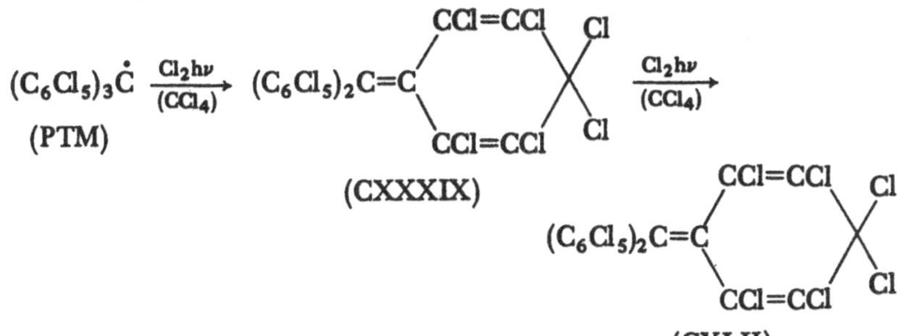

$$(C_6Cl_5)_3\overset{\bullet}{C} \xrightarrow[(CCl_4)]{Cl_2h\nu} (C_6Cl_5)_2C=C\!\!< \xrightarrow[(CCl_4)]{Cl_2h\nu}$$

(PTM)

(CXXXIX)

$$(C_6Cl_5)_2C=C\!\!<$$

(CXLII)

Perchloro-3-diphenylmethylenecyclohexene is a white solid melting at 272-4° (dec.). Its infrared and ultraviolet spectra have been recorded [319]. It reacts with stannous chloride in ethyl ether giving αH-pentadecachlorotriphenylmethane and some PTM. Dechlorination with potassium iodide in acetic acid at 100° gives a good yield of the αH-quasi-perchloro compound only. However, it is remarkable that when the dechlorination is performed with sodium iodide under the same conditions, $9H$-tridecachloro-9-phenylfluorene is obtained in an excellent yield [318, 319].

1.4.20   Perchlorophenylcyclobutadiene

The synthesis of perchlorophenylcyclobutadiene (CXLIV)† has unsuccessfully been attempted by vicinal dechlorination of perchloro-1-phenylcyclobutene (CXXXIII) with stannous chloride in boiling dioxane or by heating above 300° (Section 1.4.12) [323], and by dehydrochlorination of $4H$-nonachloro-1-phenylcyclobutene [323].

The thermal treatment of (CXXXIII) gives an excellent yield of perchloro-1-phenylbutadiene (Section 1.4.3).

† Other electronic structures for this hypothetical chlorocarbon could be given, among them that of perchlorophenyltetrahedrane.

1.4.21 Perchloroindene

Zincke and Günther reported the synthesis of perchloroindene (CXLIII) in 1892 by the reaction of phosphorus pentachloride with perchloroindenone at about 200° [328]. The synthesis was confirmed by Yates and co-workers and the yield reported as 79% [329]. Flash pyrolysis at 475-500° of the cage-ketone perchloropentacyclo-(5.3.0.0[2, 6].0[4, 10].0[5, 9])decan-3-one gave a yield of 90-95% [329]. Initial decarbonylation followed by intra-molecular rearrangement and aromatization by chlorine elimination [102] is the suggested pathway.

A good preparative method is to heat perchloroindane (LXXII; Section 1.3.13) well above 200° [312]. Treatment of naphthalene with chlorine and iron, raising the temperature from 180 to 280°, gives a mixture containing (LXXII) and (CXLIII), which by thermolysis at 330-340° is converted mostly into the latter chlorocarbon [312].

Mcbee and co-workers observed formation of (CXLIII) in a 9% yield among other chlorocarbons in the thermolysis of perchlorocyclopentadiene at 240-250° (Section 1.2.1h) [102].

Perchloroindene is a white solid melting at 138.0-8.5°, of high thermal stability and light-sensitivity [329]. Its spectra (i.r. and u.v.) have been measured [313, 329].

(CXLIII) adds chlorine in warm carbon tetrachloride giving perchloroindane (Section 1.3.13). By treatment with sulphur trioxide and then with water it gives a nearly quantitative yield of perchloroindenone [329].

1.4.22   Perchloro-1,4-dihydronaphthalene

A mixture of perchloro-1,4-dihydronaphthalene (XI; Sections 1.2.1e and 1.2.8) and perchloronaphthalene (XXXIV; Section 1.2.8) was probably obtained by Berthelot and Jungfleisch by exhaustive chlorination of naphthalene at high temperature [190]. Vollman prepared (XI) by chlorination of naphthalene with chlorine in a solvent, in the presence of ferric chloride, antimony pentachloride or iodine [330], or in sulphuryl chloride in the presence of iron at 50-60° [330]. It can also be prepared in a 67.8% yield by treatment of naphthalene with reagent BMC with other minor products, e.g., an octachlorotetralin (4.3%) and a decachlorotetralin (3.5%) [234a, 248-250].

The title compound is a white solid melting at 207-9° with decomposition. Its infrared and ultraviolet spectra [250] confirm the structure and the absence of ring distortion. Its mass spectrum has been registered [234a].

In view of the results obtained by Ruoff [43], it appears to break down with chlorine at 300° in the presence of iodine, giving perchlorobenzene, perchloroethane and carbon tetrachloride (Section 1.2.1e). Illumination does not induce chlorine addition. It reacts with boiling isopropyl ether (88.9% yield), or with stannous chloride in dioxane at 100° (97.0%), giving perchloronaphthalene (XXXIV; Section 1.2.8) [248-250]. When treated with oleum at 100° it gives a 68.5% yield of perchloro-1,4-naphthoquinone [248-250].

1.4.23   Perchloro-1,4-dihydro-5-phenylnaphthalene

The chlorination of 1-phenylnaphthalene with reagent BMC gives a mixture which, according to the spectral data, contains perchloro-1,4-dihydro-5-phenylnaphthalene   (XXXVI),   perchloro-1-phenyl-naphthalene (XXXVII) and perchlorofluoranthene (XL; Section 1.2.10) [250]. The isolation of the title compound has not been effected.

1.4.24   Perchloro-1,4-dihydro-6-phenylnaphthalene

Perchloro-1,4-dihydro-6-phenylnaphthalene   (XXXVIII),   obtained from 2-phenylnaphthalene by means of reagent BMC (76.4%) (see Section 1.2.10) [250], is a pale yellow solid melting at 253-8° (dec.). Its infrared and ultraviolet spectra have been taken [250]. The possibility   that   it   is   actually   perchloro-1,4-dihydro-2-phenyl-naphthalene cannot be ruled out although, on steric grounds, it is less probable. On heating to 260° (XXXVIII) loses chlorine giving 93.6% perchloro-2-phenylnaphthalene (Section 1.2.10). This dechlorination also occurs with stannous chloride in dioxane at 100° (87% yield) (Section 1.2.10).

1.4.25   Perchloroacenaphthylene

The synthesis of perchloroacenaphthylene (CXLV) was reported almost simultaneously by Ballester [249] and Mack [321]. Dechlorination of perchloroacenaphthene (CXLIV; Section 1.3.26) with stannous chloride occurs in a 40% yield [249, 250, 291, 321]. This conversion also takes place by thermolysis above 260° (88%) [321]. Perchloroacenaphthylene is a red solid melting at 380-5° (Ballester) or 375° (Mack). Its infrared, ultraviolet [250, 321], and mass [321] spectra have been recorded.

It reacts with concentrated nitric acid at 180° giving an 82% yield of trichlorohemimellitic acid [321].

(CXLIV)   (CXLV)

1.4.26   Perchlorobi-9-fluorenylene

On heating a mixture of perchlorofluorene with copper dust, a 95% yield of perchlorobi-9-fluorenylene (CI) [238] is obtained (see Section 1.3.21) as a deep-blue solid. Its infrared and electronic spectra have been measured [238]. As a result of steric repulsions between the chlorines the ethylene bond of (CI) is extremely twisted, yet it does not give an epr signal. The study of this chlorocarbon is in progress.

1.4.27   Miscellaneous Perchloroarylethylenes

Perchloro-α,α,α',α'-tetraphenyl-p-xylylene is described in Section 1.8.4a. Related macromolecular chlorocarbons are described in Section 1.6, and a chlorocarbon being both ethylene and free radical is mentioned in Section 1.8.4e.

1.5   PERCHLOROARYLACETYLENES

In Section 1.3 the importance of the perchloroarylalkanes, particularly the trichloromethyl chlorocarbons, in the development of the alkaromatic chlorocarbons was mentioned. It was emphasized in the Introduction and in Sections 1.3.3 that steric compression and sp^3-hybridization are major factors causing their unusual reactivity. By contrast, perchloroarylethylenes which possess sp^2-carbons exclusively, are characterized by high thermal stability and chemical inertness (Section 1.4).

In the perchloroarylacetylenes the reactivity and thermal lability are restored. This is due not only to the inherent reactivity of the triple bond but also to the negligible steric shielding of their sp-carbons, even when they are attached to an aromatic system with blocking *ortho* chlorines. Thus, the perchloroarylacetylenes are a major source for a variety of new aromatic and alkaromatic chlorocarbons. Their chemistry is being developed systematically by the Barcelona research group.

1.5.1   Perchlorophenylacetylene

ωH-pentachlorophenylacetylene (Section 1.4.1) reacts with silver nitrate in aqueous tetrahydrofurane to give a silver salt which, on treatment with chlorine in carbon tetrachloride at room temperature, is rapidly converted to perchlorophenylacetylene (XXX) in a 91.7% yield.

$$C_6Cl_5C{\equiv}CH \xrightarrow[\text{(THF-H}_2\text{O)}]{Ag^+} C_6Cl_5C{\equiv}CAg \xrightarrow[\text{(CCl}_4)]{Cl_2} C_6Cl_5C{\equiv}CCl$$

$$(XXX)$$

Perchlorophenylacetylene is a white solid melting at 132-6°. Its infrared and ultraviolet spectra have been recorded [323]. Reaction with chlorine in carbon tetrachloride in white incandescent light at room temperature produces perchlorostyrene (LIX; Section 1.4.1) quantitatively. The same conversion takes place by exposing its carbon tetrachloride solution to sunlight (see Section 1.4.1). It also reacts with perchloroethylene in ultraviolet light, at room temperature, yielding perchloro-1-phenylcyclobutene (60.3%; Section 1.4.12) and under the same conditions with trichloroethylene to give 4H-nonachloro-1-phenylcyclobutene (56.1%; Section 1.4.20). At 190-195°, perchlorophenylacetylene is converted into a mixture of perchloro-1-phenylnaphthalene, perchloro-1,2,3-triphenylbenzene, and perchloro-1,2,4-triphenylbenzene [244]. No evidence for the formation of perchloro-1,3,5-triphenylbenzene or perchloro-2-phenylnaphthalene, which is the least strained isomer, was found. A similar conversion occurs at lower temperatures (110°) with perchlorostyrene as a solvent [244].

The preceding results are rationalized by two rules concerning the nature of the *reaction intermediates and transitions states*: (1) the steric repulsions should be minimal, and (2) the benzylic resonance should be maximal. Accordingly, the following mechanism involving the formation of perchlorodiphenyltetrahedrane is suggested:

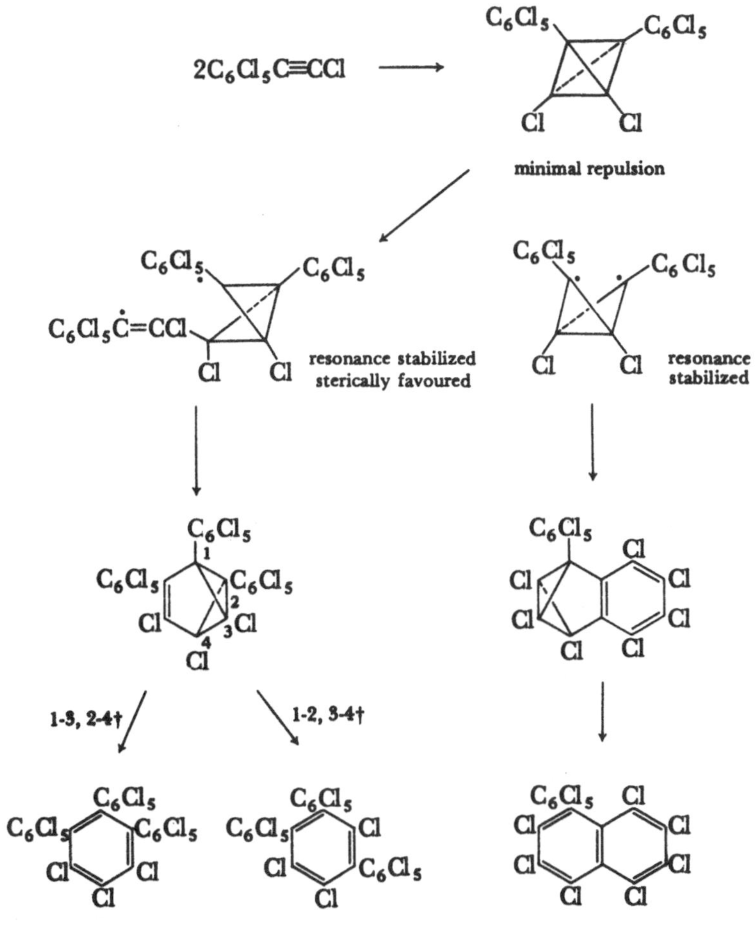

$$2C_6Cl_5C\equiv CCl \longrightarrow$$

C_6Cl_5 ————— C_6Cl_5

Cl   Cl

minimal repulsion

C_6Cl_5 ————— C_6Cl_5

$C_6Cl_5\dot{C}=CCl$

Cl   Cl   *resonance stabilized sterically favoured*

C_6Cl_5 ————— C_6Cl_5

Cl   Cl   *resonance stabilized*

C_6Cl_5
C_6Cl_5 C_6Cl_5
Cl   Cl
Cl

C_6Cl_5 Cl
Cl   Cl
Cl   Cl
Cl   Cl

1-3, 2-4†   1-2, 3-4†

C_6Cl_5
C_6Cl_5 C_6Cl_5
Cl   Cl
Cl

C_6Cl_5
C_6Cl_5 Cl
Cl C_6Cl_5
Cl

C_6Cl_5 Cl
Cl   Cl
Cl   Cl

† Bonds being broken.

1.5.2 Perchlorotolane

Perchlorotolane (perchloro-1,2-diphenylacetylene; CXLVI) was first made by Ballester and co-workers almost quantitatively by dechlorination of *trans*-perchlorostilbene (LII) with zinc in boiling dioxane (Section 1.4.7) [274]. Its identification was based on the mode of synthesis and elemental analysis. Gilman and co-workers have obtained (CXLVI) from pentachlorophenylmagnesium chloride (Section 1.2.3g) in the presence of cobaltous chloride, or penta-

chlorophenylcopper, and tetrabromoethylene and have fully confirmed its structure [331].

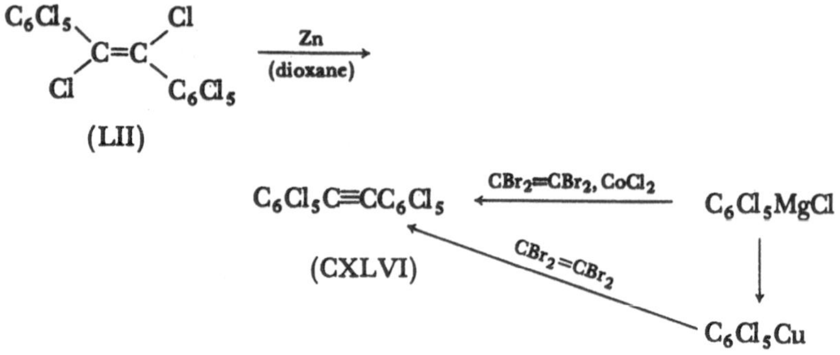

$$
\underset{\text{(LII)}}{\overset{C_6Cl_5}{\underset{Cl}{>}}C=C\overset{Cl}{\underset{C_6Cl_5}{<}}} \xrightarrow[\text{(dioxane)}]{\text{Zn}}
$$

$$
\underset{\text{(CXLVI)}}{C_6Cl_5C\equiv CC_6Cl_5} \xleftarrow{\text{CBr}_2=\text{CBr}_2,\ \text{CoCl}_2} C_6Cl_5MgCl
$$

$$
\xleftarrow{\text{CBr}_2=\text{CBr}_2} \qquad\qquad \downarrow
$$

$$
C_6Cl_5Cu
$$

Perchlorotolane is a white, insoluble chlorocarbon melting at 359°
with known infrared and ultraviolet spectra [331]. Its chemical
properties have not yet been reported.

1.5.3   Perchloro-1,4-diphenylbutadi-ine

Thermolysis of pentachlorophenyliodoacetylene at 165° gives a
67.7% yield of perchloro-1,4-diphenylbutadi-ine (CXXIX) [323].

$$
C_6Cl_5C\equiv CI \xrightarrow{165°} C_6Cl_5C\equiv CC\equiv CC_6Cl_5 + I_2
$$
$$
\text{(CXXIX)}
$$

This probably involves dimerization of perchlorophenylethynyl
radicals.

Perchloro-1,4-diphenylbutadi-ine is an insoluble white solid
melting at 313.5-5.5°. Its infrared and ultraviolet spectra have been
recorded [323]. Its carbon tetrachloride solution reacts with
chlorine in light at room temperature, yielding *trans,trans*-perchloro-
1,4-diphenylbutadiene (CXXVIII; Section 1.4.9) quantitatively.

1.5.4   Other Perchloroacetylenes

A chlorocarbon possessing both acetylenic and free radical character
is described in Section 1.8.4d.

1.6   MACROMOLECULAR CHLOROCARBONS

The first macromolecular chlorocarbons were reported by Ballester
and co-workers [249, 302]. They are synthesized by reductive
polycondensation of bistrichloromethyl chlorocarbons.

It has previously been observed that aromatic perchloroethylenes, such as perchlorostyrene (Section 1.4.1) and the perchlorostilbenes (Section 1.4.7) possess thermal stability and chemical inertness (Section 1.4). These properties prompted the attempt to synthesize macromolecular chlorocarbons possessing only $C(sp^2)$-Cl bonds. As in the self-condensation of perchlorotoluene (Section 1.3.3b) and related chlorocarbons (Sections 1.3.9c, 1.4.2 and 1.4.6) the poly-condensations were carried out with reducing agents (e.g., alkaline iodides, ferrous chloride, stannous chloride), starting from bistri-chloromethyl chlorocarbons such as perchloro-p-xylene (Section 1.3.9), perchlorobi-p-tolyl (Section 1.3.16), and the perchloro-1,4-dipropenylbenzenes (Section 1.4.5).

Preliminary studies on some of the properties of these macro-molecular chlorocarbons (lamination, water repellency, extreme-pressure properties, and biological activity) have been reported [12]. Applications based on their high thermal stability and chemical inertness are envisaged.

Doorenbos confirmed the results obtained by Ballester and co-workers and also found a new reducing agent for the poly-condensations, namely a solution of a dialkyl (diethyl or dibutyl) hydrogen phosphonate in boiling ethyl ether, in the presence of cuprous chloride [12, 332].

1.6.1   PP-xynene

Perchloro-p-xylene, when treated with iodide ion, gives an excellent yield of perchloro-p-xylylene (LXI; Section 1.3.9c). However, with reducing metal cations in dioxane, the main (67-72% yield) product is the expected perchloropoly-p-xylenediylidene (PP-xynene; CXLVII) [249, 302] and some starting material (4-17%). Similar results have shown at top of page 116; (see Sections 1.3.3b, 1.3.9c, and 1.7.3) [302].

Ballester and co-workers suggested substituted perchlorobenzyl radicals as reaction intermediates for which e.p.r. evidence has now been forthcoming [12, 332], proposing the following mechanism (shown at top of page 116; see Sections 1.3.3b, 1.3.9c, and 1.7.3) [302].

The fact that a perchlorobenzyl radical gives αH-heptachloro-toluene under similar conditions (Sections 1.3.3b and 1.7.1) suggests that the dichloromethyl moiety is an end-group. However, since the thermal stability of PP-xynene is improved by performing the polycondensation in the presence of a hydrogen donor such as chloroform, trichloromethyl groups might still be present in some polymer fractions. Oxygen interferes with the polycondensation so that low-polymers with chlorocarbonyl end-groups are obtained, presumably by reaction with the intermediate chlorocarbon radicals

$p\text{-}Cl_3C\text{-}C_6Cl_4\text{-}CCl_3$

$$\overset{+e}{\underset{(-Cl^-)}{\searrow}}$$

$p\text{-}Cl_3C\text{-}C_6Cl_4\text{-}\overset{\cdot}{C}Cl_2 \quad \xrightarrow[\;(-Cl^-)\;]{+e} \quad p\text{-}Cl_2C\text{=}C_6Cl_4\text{=}CCl_2$

$$\|$$

$p\text{-}Cl_3CC_6Cl_4CCl_2CCl_2C_6Cl_4CCl_3\text{-}p$

$$\overset{+2e}{\underset{(-2Cl^-)}{\swarrow}}$$

$p\text{-}Cl_3CC_6Cl_4CCl\text{=}CClC_6Cl_4CCl_3\text{-}p$

cis and trans

$$\underset{+e}{\overset{(-Cl^-)}{\searrow}}$$

$p\text{-}Cl_3CC_6Cl_4CCl\text{=}CClC_6Cl_4\text{-}\overset{\cdot}{C}Cl_2\text{-}p$

$$\swarrow$$

$p\text{-}Cl_3CC_6Cl_4\text{-}(CCl\text{=}CClC_6Cl_4)_m\text{-}\overset{\cdot}{C}Cl_2\text{-}p$

$$\downarrow$$

$(-CCl\text{=}CClC_6Cl_4)_n$
(CXLVII)

[302]. Doorenbos found that when the reaction is carried out in air the product is essentially tetrachloroterephthaloyl dichloride [12], formed from perchloro-p-xylyl radicals and oxygen.

$$\begin{array}{c} \overset{HS}{\nearrow} \quad \text{---}C_6Cl_4\text{-}CHCl_2 \\ \text{---}C_6Cl_4\text{-}\overset{\cdot}{C}Cl_2 \\ \underset{O_2}{\searrow} \quad \text{---}C_6Cl_4\text{-}COCl \end{array}$$

With dialkyl hydrogen phosphonate and cuprous chloride the initiation step occurs probably as follows [12, 332]:

$$p\text{-}Cl_3C\text{-}C_6Cl_4\text{-}CCl_3 + Cu_2^{+} \longrightarrow p\text{-}Cl_3C\text{-}C_6Cl_4\text{-}\overset{\cdot}{C}Cl_2 + Cu^{++}$$

$$Cu^{++} \xrightarrow{HPO(OR)_2} Cu_2^{+}$$

It is, however, noteworthy that cuprous chloride (in pyridine) *does not* generate radicals [12, 332] in the absence of the phosphonate. Similar reductive condensations of non-di-*ortho*-substituted bistrichloromethylbenzenes yield polymers containing exclusively bibenzyl structural units [333, 334].

$$p\text{-}Cl_3C-C_6H_5-CCl_3 \rightarrow (-C_6H_5-CCl_2CCl_2-)_n$$

This is due to the lack of steric assistance to vicinal chlorine elimination giving the corresponding stilbene structural units. By contrast, as in the reductive self-condensation of perchlorotoluene (Sections 1.3.3b and 1.7.1), no bibenzyl units have been detected in the synthesis of PP-xynene.

Kinetic measurements indicate that the activation energy for the reductive polycondensation is about 6 kcal.mole^{-1}. The polarographic behaviour of perchloro-*p*-xylylene has been studied using different solvents and electrolytes [12]. The average half-wave reduction potential is -1.0 volt approximately.

PP-Xynene is a whitish, infusible powder containing fractions of different solubility, one of which (26-62% yield) is completely insoluble. The soluble fractions obtained by Ballester's procedure have molecular weights up to 4000, while those from the Doorenbos procedure have higher molecular weights (up to 12,700). The ultraviolet spectrum indicates the presence of both *cis*- and *trans* structural units [302], *trans*-forms decreasing with solubility as expected (see Section 1.4.7). The infrared spectrum of the insoluble fraction shows a higher *trans* character [302].

PP-Xynene is stable *in vacuo* up to 500° [302] and thermogravimetric studies indicate an even higher stability [12]. Ultraviolet light or heat (450-500°) causes *cis*-to-*trans* stereomutation and, consequently, a further decrease of solubility [302]. Accordingly, differential thermal analysis has shown a gradual exotherm up to 400° attributed partly to stereomutation, as well as a large and sharp endotherm at 625° corresponding to the decomposition of PP-xynene [12]. PP-Xynene is inert towards highly reactive chemicals such as oleum, red fuming nitric acid, chlorine, or bases [302].

1.6.2   PP-Bitylene

Perchloropolybi-*p*-tolyldiylidene (PP-bitylene; CXLVIII) has been first synthesized by Ballester and co-workers from perchlorobi-*p*-tolyl (LXXXIX) by the reaction with stannous chloride in dioxane at 100° in a 67% yield [302]. Doorenbos repeated this reductive polycondensation and also performed it with his phosphonate-cuprous chloride procedure with nearly quantitative yields [12, 332]. The mechanism is believed to be essentially similar to the formation of PP-xynene (see preceding Section).

$$(-C_6Cl_4-C_6Cl_4-CCl=CCl-)_n$$

(CXLVIII)

↑

$$p\text{-}Cl_3C-C_6Cl_4-C_6Cl_4-CCl_3\text{-}p$$

(LXXXIX)

Because of the extreme twisting in the central ethylene bond caused by repulsions of the four proximal *ortho* chlorines in a planar conformation, the chlorocarbon (CIL) is not formed (compare with preceding Section). Consequently, polycondensation with alkaline iodides is possible here.

(CIL)

PP-Bitylene is a white, infusible powder, soluble in carbon tetrachloride, methylene chloride, or benzene [12, 302]. Its molecular weight is practically independent of the synthetic procedure (~9000). Its infrared [12, 302] and ultraviolet [302] spectra have been reported. PP-Bitylene withstands temperatures as high as 500° for short periods in the air without significant decomposition [302]. Thermogravimetric analysis shows stability beyond that temperature (~600°), particularly when air is excluded [12]. The ultraviolet- or heat-treated (500°) polymer has a much greater *trans* character than the original material, part of the product becoming insoluble in carbon tetrachloride [302]. The occurrence of some cross-linking in such treatments is not ruled out [12].

1.6.3   PP-Diprobene

Perchloropoly-*p*-dipropenylbenzenediylidene (PP-diprobene; CL) has been obtained from either *cis,syn,cis*- (CXXIIIa; Section 1.4.5) or *cis,anti,cis*-perchloro-*p*-dipropenylbenzene (CXXIIIb; Section 1.4.5) with stannous chloride in dioxane at 100° in about 90% yield [249, 302]. The products from either isomer are spectroscopically indistinguishable.

PP-Diprobene is an off-white powder, soluble in carbon tetrachloride and stable at 400° in air for short periods. Its mean molecular weight ranges from 15,000 to 20,000. Its infrared and ultraviolet spectra have been reported [302]. Although there are no steric reasons for vicinal dechlorination in the intermediate dimers of type (CLI) the product is still essentially an all sp²-carbon chlorocarbon.

$$Cl_3CCCl=CCl-C_6Cl_4-CCl=CClCCl_3\text{-}p$$

(CXXIII)

$$(Cl_3CCCl=CCl-C_6Cl_4-CCl=CClCCl_2-)_2$$

(CLI)

$$(-C_6Cl_4-CCl=CClCCl=CClCCl=CCl-)_n$$

(CL)

Doorenbos has also prepared a relatively low m.wt. ($\sim$4000) PP-diprobene from a crude mixture containing undoubtedly some isomeric perchloro-1,4-dipropenylbenzenes [12] as a black material. Its infrared spectrum indicates it to be a heterogeneous mixture.

PP-Diprobene does not change in ultraviolet light, or by heating at 400° for short periods except that the latter treatment renders it insoluble in carbon tetrachloride, an indication of extensive cross-linking [302].

1.6.4 Perchloropoly-4,4'-dipropenylbiphenyldiylidene

According to Doorenbos, perchloropoly-4,4'-dipropenylbiphenyl-diylidene (PP-Diprobiphene; CLII) is prepared from a mixture containing isomeric perchloro-1,4-dipropenylbiphenyls (CXXV; Section 1.4.6) by his phosphonate–cuprous chloride method [12]. The purple, infusible product obtained appears to be stable at 480°. No analytical data other than the infrared spectrum have, however, been reported.

1.6.5 P-Copolymers

A presumably 1:1 perchloro-copolymer from perchloro-p-xylene (LX; Section 1.3.8) and perchlorobi-p-tolyl (LXXXIX; Section 1.3.16) has been reported by Doorenbos [12]. A nearly quantitative yield of an off-white chlorocarbon is obtained which is soluble in carbon tetrachloride. It softens at 320°, flows at 395° and is stable at 480°. Unfortunately, no analytical data other than its infrared spectrum are available [12], the latter indicating high purity.

Assuming alternating structural units, this copolymer can be depicted as follows:

$$(-C_6Cl_4-CCl=CCl-C_6Cl_4-C_6Cl_4-CCl=CCl-)_n$$

Another 1:1 low perchloro-copolymer has been obtained by Doorenbos from perchloro-bi-p-tolyl (LXXXIX; Section 1.3.16) together with a mixture of isomeric perchloro-1,4-dipropenylbenzenes (Section 1.4.5) as a dark-brown material. Its composition has not been ascertained but its infrared spectrum has been recorded [12].

1.7 REACTIVE FREE RADICALS

This chapter describes polychlorinated radicals which are unstable because of their high reactivity. They are usually generated from polychloro-aromatic structures previously discussed.

1.7.1 Perchlorobenzyl radical

Perchlorobenzyl (PB) radical (LIII) is the simplest alkaromatic chlorocarbon free radical. It has been postulated as an intermediate in various reactions of perchlorotoluene (XLIX), such as thermolysis (Section 1.3.3d), photolysis in carbon tetrachloride (Section 1.3.3e), and reductive condensations with alkaline iodides, copper, stannous chloride, ferrous chloride, etc. (Section 1.3.3b), which are summarized below:

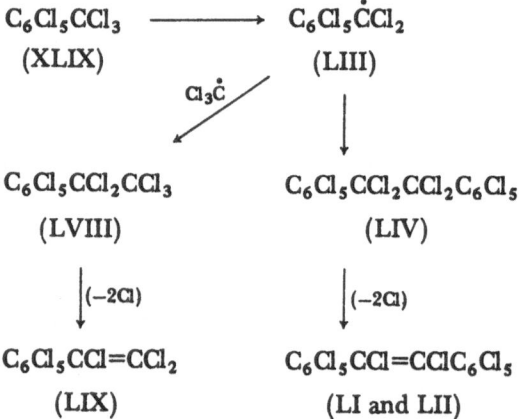

Generation of PB radicals has been performed by irradiating a solution of perchlorotoluene in carbon tetrachloride with ultraviolet light followed by treatment with copper turnings to prevent liberated chlorine from recombining with PB to perchlorotoluene [273].†

$$C_6Cl_5CCl_3 \xrightarrow{h\nu} C_6Cl_5\overset{\cdot}{C}Cl_2 + Cl$$

$$2\,Cl \longrightarrow Cl_2$$

$$Cl_2 + Cu \longrightarrow Cu_2Cl_2$$

$$2\,C_6Cl_5\overset{\cdot}{C}Cl_2 \rightleftharpoons C_6Cl_5CCl_2CCl_2C_6Cl_5$$

† Copper itself reacts slowly with perchlorotoluene in carbon tetrachloride.

Kinetic data indicate an initial second-order decay of the radical, approaching gradually to an equilibrium with a dimer, the predominant species at room temperature [273]. The radical concentration increases with temperature, and is restored to its original value by cooling. This observation together with the formation of perchlorostilbenes in dechlorinating media (Section 1.3.3b) is evidence for the dimer being perchlorobibenzyl (LIV) [273]. It appears to be the sole known benzyl radical which is *only* partly associated at room temperature, undoubtedly due to the repulsions among the four *ortho* and the four *alpha* chlorines of dimer (LIV) (frontal strain).

The ease with which the PB radical is formed from perchlorotoluene is mainly due to steric assistance (cf. 1.3 and 1.3.3c). Space filling models show clearly the steric compression in the planar conformation of the PB radical due to the repulsions among the *ortho* and the *alpha* chlorine atoms. Rotation of the dichloromethylene group about its bond with the ring occurs until a balance between resonance forces and steric repulsions—minimal in the perpendicular benzyl—is reached. Quantum calculations based on the hyperfine coupling in the 4H-hexachlorobenzyl radical (CLIII; Section 1.7.2) show that the angle between the planes of those groups is about 60°.

Perchlorobenzyl radical abstracts a hydrogen from donors such as hydrogen iodide, toluene, etc., giving αH-heptachlorotoluene (1.3.3b).

$$C_6Cl_5\dot{C}Cl_2 + HD \rightarrow C_6Cl_5CHCl_2 + D\cdot$$

1.7.2   4H-Hexachlorobenzyl Radical

4H-hexachlorobenzyl radical (CLIII) is an intermediate in the reaction of 4H-heptachlorotoluene with potassium iodide in acetic acid at 100° giving a mixture of cis- and trans-4H,4'H-decahlorostilbene [287]. It is also generated from 4H-heptachlorotoluene with

(CLIII)

cis and trans

stannous chloride in dioxane or by ultraviolet light in carbon tetrachloride [273].

Its chemical behaviour is similar to that of perchlorobenzyl (PB) radical described in the preceding Section. It possesses the same substituent constellation in the vicinity of the trivalent carbon atom and, therefore, the same angle of twist with the benzene ring.

1.7.3   p-Substituted Perchlorobenzyl Radicals

p-Substituted perchlorobenzyl radicals $p\text{-RC}_6\text{Cl}_4\overset{\cdot}{\text{C}}\text{Cl}_2$ have been detected by Doorenbos [12, 332] and Ballester et al. [273]. They occur in the reductive polycondensation of perchloro-p-xylene (LX; Sections 1.3.8c and 1.6.1), where R = $p\text{-Cl}_3\text{C}-(-\text{C}_6\text{Cl}_4\text{CCl}=\text{CCl}-)_n$, perchloro-p-xylenyl (LXII) being the radical initially found [12, 332]. Related perchlorobenzyls, particularly (LXII), are generated by ultraviolet light [273].

$$p\text{-R}-\text{C}_6\text{Cl}_4-\text{CCl}_3 \xrightarrow[\text{or } h\nu]{+\text{e}(-\text{Cl}^+)} p\text{-R}-\text{C}_6\text{Cl}_4-\overset{\cdot}{\text{C}}\text{Cl}_2$$

The epr spectrum of these radicals is presumably insensitive to the structural features of the p-substituent. The activation energy for their formation with stannous chloride in dioxane or tetrahydrofurane is approximately six kcal/mole [12]. Oxygen prevents the reductive polycondensation of perchloro-p-xylene to PP-xynene (Section 1.6.1), tetrachloroterephthaloyl dichloride being obtained instead [12]. This result is interpreted as follows [335-337]:

$$p\text{-Cl}_3\text{C}-\text{C}_6\text{Cl}_4-\text{CCl}_3 \xrightarrow[(-\text{Cl}^-)]{+\text{e}} p\text{-Cl}_3\text{C}-\text{C}_6\text{Cl}_4-\overset{\cdot}{\text{C}}\text{Cl}_2 \xrightarrow{\text{O}_2}$$

$$\text{(LXII)}   p\text{-Cl}_3\text{C}-\text{C}_6\text{Cl}_4-\text{CCl}_2\text{O}\overset{\cdot}{\text{O}}$$

$$\Big/$$

$$p\text{-ClCO}-\text{C}_6\text{Cl}_4-\text{COCl} \longleftarrow \text{Cl}_3\text{C}-\text{C}_6\text{Cl}_4-\text{COCl-}p$$

Perchloro-4-p-tolylbenzyl (CLIV) and related radicals are formed from perchlorobi-p-tolyl (LXXXIX) by reductive polycondensation yielding PP-bitylene (Section 1.6.2) and by illumination with ultraviolet light [273].

1.7.4   Perchloro-3-phenylallyl Radicals

Perchloro-3-phenylallyl radicals (CXXXI) are formed in the reductive self-condensation of the perchloropropenylbenzenes (CXI and CXVI) which has been discussed in Section 1.4.2. The formation of intermediate chlorocarbon radicals is supported by epr evidence [273].

$$C_6Cl_5CCl=CClCCl_3 \xrightarrow[(-Cl^-)]{+e} C_6Cl_5CCl=CCl\overset{\bullet}{C}Cl_2$$

(CXI and CXVI)   (CXXXI)

1.7.5 αH-Decachlorodiphenylmethyl Radical and Perchlorodiphenyl Carbene

Potassium perchlorodiphenylmethide (CLV; Section 1.10.2) decomposes slowly (four months at −10°; 12 hours at 35°) in ethyl ether giving αH-undecachlorodiphenylmethane (CLIX), αH,αH-decachlorodiphenylmethane (CLVIII), αH,α'H-eicosachlorotetraphenylethane (CLVII) and perchlorodiphenylmethyl (PDM) radical [316].
These products are accounted for in the following manner:

$$(C_6Cl_5)_2\bar{C}Cl \xrightarrow{(Cl^-)} {}^1\{(C_6Cl_5)_2C\} \longrightarrow {}^3\{(C_6Cl_5)_2C\}$$

(CLV)   (S-XCVI)   (T-XCVI)

$${}^3\{(C_6Cl_5)_2C\} + HS \longrightarrow (C_6Cl_5)_2\overset{\bullet}{C}H + S^{\bullet}$$

(CLVI)

$$2\,(C_6Cl_5)_2\overset{\bullet}{C}H \longrightarrow (C_6Cl_5)_2CHCH(C_6Cl_5)_2$$

(CLVII)

$$(C_6Cl_5)_2\overset{\bullet}{C}H + HS \longrightarrow (C_6Cl_5)_2CH_2 + S^{\bullet}$$

(CLVIII)

$$S^{\bullet} \longrightarrow R^{\bullet} + HD$$

$$(C_6Cl_5)_2\bar{C}Cl + HD \longrightarrow (C_6Cl_5)_2CHCl$$

(CLIX)

$$(C_6Cl_5)_2\bar{C}Cl + R^{\bullet} \longrightarrow (C_6Cl_5)_2\overset{\bullet}{C}Cl + R^-$$

HS = solvent   (PDM)

The mechanism deserves some comment: the carbene resulting from chloride-ion elimination from the anion (CLV) must be a singlet (S-XCVI) [338-340] and, therefore, the two bonds of its central carbon atom must form an angle causing significant repulsions between the two pentachlorophenyl groups. This back strain assists the linear (sp) hybridization of that carbon and, consequently, the conversion into the triplet carbene (T-XCVI) which behaves chemically as a biradical [338-340]. It will thus abstract a hydrogen from the solvent forming αH-decachlorodiphenylmethyl radicals

(CLVI) which dimerizes to (CLVII) or gives (CLVIII). Accordingly, it was found that the higher the reaction temperature, the higher the yield of (CLVIII) (0.0-32.6%) and the lower that of (CLVII) (54.0-8.5%). It is assumed that by interaction with the medium a radical S* results causing the formation of an electrophilic radical R* and a proton donor HD. Both species then react with the anion (CLV) yielding PDM and (CLVII) respectively. This hypothesis is consistent with the formation of PDM and (CLIX) in about the same yield (18.7-22.5 and 16.7-20.8%).

When cyclohexene, a typical carbene trap, is added to the ethereal solution of carbanion (CLV) decomposition takes place rapidly with the formation of compounds (CLVIII) (80.1%) and (CLIX) (9.6%). The failure to give the expected 1,1-bispentachlorophenylcyclopropane derivative is attributed to its high steric strain.

The rapid carbene formation from (CLV) in the presence of cyclohexene is probably due to π-electron assistance in the chloride elimination. The hydrogen donor for the subsequent reactions is probably cyclohexene itself.

Other simple ways to generate αH-decachlorodiphenylmethyl radicals [316] are: (1) dechlorination of αH-undecachlorodiphenylmethane (CLIX) with stannous chloride in boiling dioxane yielding dimer (CLVII) and (CLVIII) in 84.5 and 10.6% yields, respectively; (2) conversion of (CLVIII) into αH-decachlorodiphenylcarbanion with sodium hydroxide in DMSO, and then oxidation with iodine producing a 40.1% yield of dimer (CLVII); (3) heating dimer (CLVII) around 160° in diglyme which leads to a 59.0% yield of (CLVIII); (4) photolysis of αH-compound (CLIX) in carbon tetrachloride with ultraviolet light which allows the high decay rate to be followed at room temperature.

$$(C_6Cl_5)_2CHCl$$

$$(CLIX)$$

$$h\nu \Bigg| \begin{matrix} \text{or SnCl}_2 \\ \text{(dioxane)} \end{matrix}$$

$$(C_6Cl_5)_2\bar{C}H \xrightarrow[\text{(DMSO)}]{I_2} (C_6Cl_5)_2\overset{\cdot}{C}H \xleftarrow{160°} (C_6Cl_5)_2CHCH(C_6Cl_5)_2$$

$$(CLVI) \qquad\qquad\qquad (CLVII)$$

1.7.6 αH-Octadecachloro-4,4'-diphenyldiphenylmethyl Radical and Perchloro-4,4'-diphenyldiphenylcarbene

Perchloro-4,4'-diphenyldiphenylcarbene (CLXI) and αH-octadecachloro-4,4'-diphenylmethyl radical (CLXII) are presumably intermediates in the very slow (10 weeks) decomposition of potassium perchloro-4,4'-diphenyldiphenylmethide (CLX; Section 1.10.2) in ethyl ether at room temperature. The product is $\alpha H,\alpha' H$-hexatria-contachloro-4,4′,4″,4‴-tetraphenyltetraphenylethane (CLXIII) (43% yield) [316]. The latter is also formed in the reaction of αH-nona-decachloro-4,4′-diphenyldiphenylmethane (CLXIV) with stannous chloride in boiling dioxane.

$$(C_6Cl_5-C_6Cl_4-)_2\bar{C}Cl\text{-}p \xrightarrow{(-Cl^-)} (C_6Cl_5-C_6Cl_4-)_2\overset{+}{C}: \quad (C_6Cl_5-C_6Cl_4-)_2CHCl$$

$$\text{(CLX)} \qquad\qquad \text{(CLXI)} \qquad\qquad \text{(CLXIV)}$$

$$\diagdown\text{HS} \qquad\qquad \Big/\begin{smallmatrix}\text{SnCl}_2\\ \text{(dioxane)}\end{smallmatrix}$$

$$(C_6Cl_5-C_6Cl_4)_2CHCH(-C_6Cl_4-C_6Cl_5)_2 \longleftarrow (C_6Cl_5-C_6Cl_4-)_2\overset{\bullet}{C}H$$

$$\text{(CLXIII)} \qquad\qquad\qquad \text{(CLXII)}$$

For a more detailed account of the relevant reaction mechanisms, see preceding Section.

1.7.7 $\alpha H,4,4'$-Dimethoxyoctachlorodiphenylmethyl

The reaction of $\alpha H,4,4'$-dimethoxynonachlorodiphenylmethane (CLXV) with stannous chloride in boiling (100°) dioxane gives $\alpha H,\alpha' H,4,4',4'',4'''$-tetramethoxyhexadecachlorotetraphenylethane (CLXVII; 76% yield) and some $\alpha H,\alpha H,4,4'$-dimethoxyoctachlorodi-phenylmethane (CLXVIII; 7.1%) [316]. This reaction takes place through the $\alpha H,4,4'$-dimethoxyoctachlorodiphenylmethyl radical (CLXVI). It can also be generated thermally [316] since treatment of (CLXV) with boiling (161°) diglyme gives (CLXVIII). Also, when the dimer (CLXVII) is treated under the same conditions (CLXVIII) results (64.8% yield).

$$(MeOC_6Cl_4)_2CHCl\text{-}p \xrightarrow[\substack{\text{or SnCl}_2,\, 100°\\ \text{(dioxane)}}]{161°\ \text{(diglyme)}} (MeOC_6Cl_4)_2\overset{\bullet}{C}H\text{-}p$$

$$\text{(CLXV)} \qquad\qquad\qquad\qquad \text{(CLXVI)}$$

$$161°,\ \text{diglyme} \qquad\diagup\qquad\overset{100°}{\diagup}\qquad\diagdown\, 161°,\ \text{diglyme}$$
$$\diagdown\,\text{dioxane}$$

$$(MeOC_6Cl_4)_2CHCH(C_6Cl_4OCH_3)_2 \qquad (MeOC_6Cl_4)_2CH_2\text{-}p$$

$$\text{(CLXVII)} \qquad\qquad\qquad\qquad \text{(CLXVIII)}$$

These results, as well as those reported in Sections 1.7.5 and 1.7.6 concerning the reactivity of αH-quasiperchlorodiphenylmethyl radicals, show the importance of the shielding by the *alpha* chlorine atoms in radicals of the PDM series (Section 1.8.1).

1.7.8 $\alpha H,\alpha'H$-Tetradecahloro-α,α'-diphenyl-p-xylylene

The generation of $\alpha H,\alpha'H$-tetradecachloro-α,α'-diphenyl-p-xylylene radicals (CLXX) for polymerization purposes was attempted from $\alpha H,\alpha'H$-hexadecachloro-α,α'-diphenyl-p-xylene with triphenylphosphine in boiling benzene or with stannous chloride in tetrachloroethylene at 180°. Products were obtained in a high yield [67]. The physical and chemical properties of the supposed biradical indicate, however, that it possesses a p-quinonedimethane (singlet) structure (CLXXI) [67] (see Section 1.8.4a).

$$C_6Cl_5CHCl-C_6Cl_4-CHClC_6Cl_5 \xrightarrow{\text{Ph}_3\text{P}} C_6Cl_5\overset{-}{C}H-C_6Cl_4-CHClC_6Cl_5$$

(CLXIX)

$$\text{SnCl}_2 \downarrow$$
$$(\text{C}_2\text{Cl}_4)$$

$$\downarrow (-\text{Cl}^-)$$

$$C_6Cl_5\overset{\cdot}{C}HC_6Cl_4\overset{\cdot}{C}HC_6Cl_5 \longleftarrow\!\!\!\times\!\!\!\longleftarrow C_6Cl_5CH=C_6Cl_4=CHC_6Cl_5$$

(CLXX)   (CLXXI)

1.8 INERT CARBON FREE RADICALS†

The inert free radicals are a new species of isolable, completely disassociated radicals characterized by their excellent thermal stability and chemical inertness not found in any other carbon radical known. Their stability in a broad sense can even be greater than that of many 'normal' tetravalent carbon compounds, and is caused by extensive shielding of the tricovalent carbon atom due to 'chlorine overcrowding'. Moreover, the chlorine substituents are inert and strongly bonded to sp^2-hybridized carbon (see Sections 1.1 and 1.4).

Some of these trivalent-carbon compounds, particularly the radicals of the so-called PTM series (Section 1.8.2), withstand temperatures as high as 300°, do not combine with oxygen, and are inert towards typical radical reagents (hydroquinone, p-quinone, nitric oxide, toluene) as well as reagents such as chlorine, bromine, concentrated nitric and sulphuric acids.

The first radical of this type was reported in 1964 by Ballester and Riera [314]. While reagent BMC (Sections 1.1 and 1.3.1) has been important for the development of the chemistry described so far,

† Because of the exceptional properties these radicals are described as 'inert' rather than 'stable'.

reagent BCR (named after the initials of its authors) is of significance here.†
The state of dissociation of these radicals was ascertained by magnetic susceptibility measurements at several temperatures in the solid state and, in some cases, in solution by osmometry and their ultraviolet-visible spectra.

1.8.1   Perchlorodiphenylmethyl Radicals

These diphenylmethyl radicals are very stable. Their *alpha* chlorine atom is as important as the *ortho* chlorines for shielding, since its substitution by a hydrogen atom causes dimerization in solution at room temperature (cf. Sections 1.7.5, 1.7.6 and 1.7.7).

(a)   *Perchlorodiphenylmethyl Radical*
The synthesis of the perchlorodiphenylmethyl (PDM) radical by dechlorination of perchlorodiphenylmethane (XCII; Section 1.3.17) [314] has been published [314, 315]. Stannous chloride in ethyl ether (48% yield), ferrous chloride in ethyl ether (95%) or in chloroform-DMSO (85%), and mercury and ultrasonics in ethyl ether-DMSO (46%) were used as reducing agents.
However, (XCII) is not an accessible precursor (Section 1.3.17) and a better way to synthesize PDM is from diphenylchloromethane which by chlorination with reagent BMC gives an 81.6% yield of αH-undecachlorodiphenylmethane (CLIX; 1.9) [315].‡ With reagent BCR it gives a blue solution of the perchlorodiphenylcarbanion (CLV; Sections 1.7.5 and 1.10.2) which is then oxidized with iodine to PDM in a 74% yield (based on the αH-compound).

$$\text{Ph}_2\text{CHCl} \xrightarrow{\text{BMC}} (\text{C}_6\text{Cl}_5)_2\text{CHCl} \xrightarrow{\text{BCR}} (\text{C}_6\text{Cl}_5)_2\bar{\text{C}}\text{Cl}$$
$$\text{(CLIX)} \qquad\qquad\qquad \text{(CLV)}$$
$$\swarrow \text{I}_2$$
$$(\text{C}_6\text{Cl}_5)_2\text{CCl}_2 \xrightarrow[\text{+e}]{(-\text{Cl}^-)} (\text{C}_6\text{Cl}_5)_2\overset{\cdot}{\text{C}}\text{Cl}$$
$$\text{(XCII)} \qquad\qquad \text{(PDM)}$$

PDM has been detected when (XCII) or its isomer, the perchloro-benzylidenecyclohexadiene (XCV; Section 1.4.14), are illuminated with ultraviolet or sunlight in carbon tetrachloride [238]. It is also formed in the thermolysis of these chlorocarbons at 250-300°

---

† A mixture of an ether, DMSO and pulverized alkaline hydroxide possessing strongly basic as well as strongly reducing properties.
‡ The high yield of this preparation is undoubtedly due to the lack of significant strain of the αH-quasiperchloro compound.

(Sections 1.3.17 and 1.4.14) or of (CLIX) at 350° (Section 1.9) and also by oxidation of an ether solution of carbanion (CLV) (Section 1.8.1a) (70.5% yield) and by its spontaneous decomposition (Section 1.7.5).

PDM radicals are orange-red crystals melting at 190° with decomposition and their infrared [315], ultraviolet [315], epr [314, 315, 341], and mass [301b] spectra have been reported. Magnetic susceptibility measurements indicate that PDM is completely dis-associated as a solid [315]. This is confirmed by osmometric and electronic absorption data [315].

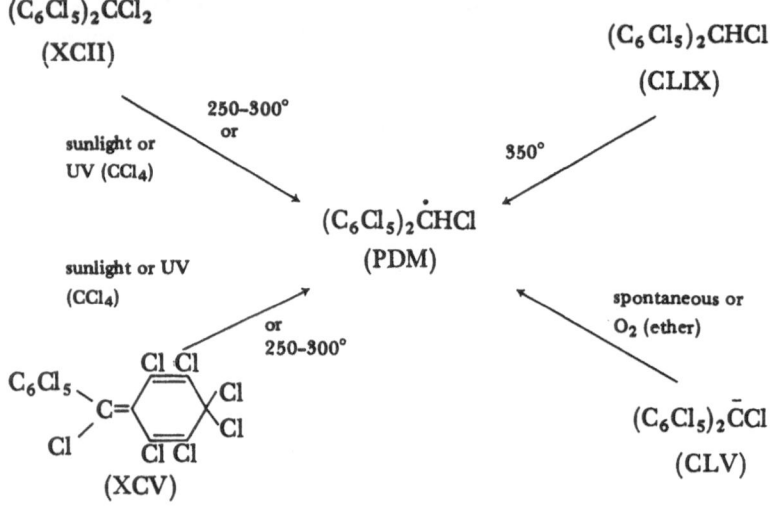

An X-ray structural study on a disordered mixed crystal of PDM indicates twisting of the two pentachlorophenyl groups with respect to the plane of the central carbon bonds of about 45° [342]. The value consistent with the observed hyperfine ^{13}C-splitting is, however, about 60° (Section 1.14.2) [273].

PDM as a solid can be exposed for months to air without significant decomposition. In solution, in air and at room temperature, its half-life is about 2-3 days, perchlorobenzophenone being formed along with other compounds. PDM is stable at room temperature to water, cyclohexane, carbon tetrachloride, chloroform, hydroquinone, p-quinone, nitric oxide, powdered sodium hydroxide in ethyl ether or carbon tetrachloride, chlorine in carbon tetrachloride, and concentrated sulphuric acid. It reacts slowly with bromine at room temperature giving, presumably, perchlorodiphenylmethyl bromide [342], and with boiling dioxane to give (CLIX) [344]. In concentrated nitric acid only 57.6% of PDM is destroyed after eleven days with formation of perchlorobenzophenone. With sodium or potassium in ethyl ether the carbanion (CLV) is formed as ascertained by hydrolysis to (CLIX) (84.4%) [291, 294, 315, 316] and to (6%) $\alpha H,\alpha H$-decachlorodiphenylmethane (see Section 1.7.5).

$$(C_6Cl_5)_2CO$$

$$\nearrow HNO_3$$

$$(C_6Cl_5)_2\dot{C}Cl$$

(PDM)

$$\searrow Na \text{ or } K \text{ (ether)}$$

$$(C_6Cl_5)_2\bar{C}Cl \xrightarrow{H_3O^+} (C_6Cl_5)_2CHCl$$

$$(CLV) \qquad\qquad (CLIX)$$

PDM in boiling toluene gives (CLIX) in 56% yield and 1,1-bis-pentachlorophenyl-2-phenylethylene (CLXXII; 24.5%). The first step of this reaction consists of hydrogen abstraction from toluene with formation of a benzyl radical. This radical is then captured by PDM itself giving a strained ethane (not isolated) which spontaneously eliminates hydrogen chloride [315].

$$(C_6Cl_5)_2\dot{C}Cl + PhMe \rightarrow (C_6Cl_5)_2CHCl + Ph\dot{C}H_2$$

(PDM) $\qquad\qquad$ (CLIX)

$$Ph\dot{C}H_2 + (C_6Cl_5)_2\dot{C}Cl \rightarrow (C_6Cl_5)_2CClCH_2Ph$$

$$\swarrow (-HCl)$$

$$(C_6Cl_5)_2C=CHPh$$

An interesting reaction of PDM occurs at room temperature with reagent BCR giving, unexpectedly, the carbanion (CLV) as ascertained by spectrum and hydrolysis to (CLIX) (71% yield) [315]:

$$(C_6Cl_5)_2\dot{C}Cl \xrightarrow[\text{(ether)}]{HO^-, DMSO} (C_6Cl_5)_2\bar{C}Cl$$

(PDM) $\qquad\qquad\qquad$ (CLV)

It is assumed that the anion $(CH_2SO_2CH_3)^-$ is the reducing species, the electron-transfer transition state being stabilized by the solvating power of DMSO.

The reaction with reagent BCR has been used extensively for elucidation of the structure of the inert radicals, since the resulting carbanions give the corresponding αH-quasiperchloro-compounds with acids (Section 1.9). Polarographic oxidation and reduction of PDM to its corresponding ions in acetonitrile and other solvents, using the platinum spinning-electrode technique and lithium or tetraethylammonium perchlorate as a conducting salt, occurs at a half-wave potential between +1.17 and +1.024, and −0.63 and −0.85 V/ferrocene, respectively [343].

PDM reacts slowly at room temperature with sulphuryl chloride giving perchlorodiphenylmethane (65% yield). When aluminium chloride is added, a deep-green colour, due to perchlorodiphenyl-carbonium ion, develops (Sections 1.3.17, 1.4.14 and 1.11.3) and after treatment with water, perchlorodiphenylmethane (81%) is obtained (Section 1.3.17) [317]. If the preceding reactions are carried out at low (<−15°) temperature, the major (52.8%) product is perchlorobenzylidenecyclohexa-1,4-diene (XCV; Section 1.4.14) [317]. This, and other evidence (Section 1.11.3), indicates the existence of two salt-like complexes where the tetrachloroaluminate ion is still bonded, although weakly, to the carbonium ion through a chlorine bridge.

PDM reacts also with chlorine in carbon tetrachloride in the presence of iodine to give perchlorodiphenylmethane (XCII) in 87% yield [291, 294], presumably due to the formation of iodine chlorides.

PDM thermolyses at ca. 300° [238] to give products similar to those obtained from perchlorodiphenylmethane (XCII; Section 1.3.17) or perchlorobenzylidenecyclohexa-1,4-diene (XCV; Section 1.4.14), i.e., perchlorofluorene (XCIII; Section 1.3.21) (61.3%), perchlorobenzene (22.5%), perchlorodiphenylmethane (9.0%), and perchlorobiphenyl (3.5%).

(b)   Perchloro-4-phenyldiphenylmethyl Radical

The PPDM radical is obtained (75.7%) from α*H*-pentadecachloro-4-phenyldiphenylmethane (CLXXIII; Section 1.9) via the anion (CLXXIV) by oxidation with iodine (see preceding Section) [294, 315].

$$p\text{-}C_6Cl_5\text{-}C_6Cl_4\text{-}CHClC_6Cl_5 \xrightarrow{\text{BCR}} p\text{-}C_6Cl_5\text{-}C_6Cl_4\text{-}\bar{C}ClC_6Cl_5$$
$$\text{(CLXXIII)} \qquad\qquad\qquad \text{(CLXXIV)}$$

$$I_2 \Big/$$

$$p\text{-}C_6Cl_5\text{-}C_6Cl_4\text{-}\dot{C}ClC_6Cl_5$$
$$\text{(PPDM)}$$

PPDM is an orange-yellow crystalline solid melting at 250-60° (dec.). Its infrared, electron absorption and epr spectra have been reported [315]. It is chemically more stable than PDM and its half-life time in solution in air at room temperature is about 13 days.

PPDM reacts slowly with concentrated nitric acid at room temperature, to give a 77% yield of perchloro-4-phenylbenzo-phenone. With reagent BCR (see preceding Section) it is reconverted into its carbanion (CLXXIV) which, with acids, gives (CLXXIII) (56.7%) (see preceding Section). It reacts with toluene in the same manner as PDM giving (CLXXIII) (64%). With aluminium chloride in sulphuryl chloride, it gives a carbonium ion solution which, on hydrolysis, yields perchloro-4-phenyldiphenylmethane (XCVII; Section 1.3.18) (71.7%) or perchloro-4-biphenylylmethylenecyclo-hexa-1,4-diene (CXXXVII; Section 1.4.15) (34.8%), depending upon the reaction temperature (see preceding Section) [317].

Alternative quinonoid structures for the latter chlorocarbon are ruled out on steric grounds (see Sections 1.4.16 and 1.8.1d).

$$p\text{-}C_6Cl_5\text{-}C_6Cl_4\text{-}COC_6Cl_5$$

$$\uparrow \text{HNO}_3$$

$$C_6Cl_5\text{-}C_6Cl_4\text{-}\bar{C}ClC_6Cl_5 \xleftarrow{\text{BCR}} \text{(PPDM)} \searrow \text{(a) AlCl}_3, SO_2Cl_2$$

$$\Big\downarrow \begin{array}{c}\text{PhMe} \\ 110°\end{array} \quad \text{(b) H}_2\text{O} \searrow$$

$$\qquad\qquad\qquad\qquad\qquad \text{(XCVII)}$$
$$p\text{-}C_6Cl_5\text{-}C_6Cl_4\text{-}CHClC_6Cl_5 \qquad \text{or}$$
$$\text{(CLXXIII)} \qquad\qquad \text{(CXXXVII)}$$

(c) *Perchloro-2-phenyldiphenylmethyl Radical*

Chlorination of 9-phenylfluorene with reagent BMC gives a 61.7% yield of αH-pentadecachloro-2-phenyldiphenylmethane (CLXXV; Section 1.9). From this compound perchloro-2-phenyldiphenyl-methyl (PODM) radical is obtained in a 55% yield, in the way described for the preparation of PPDM (preceding Section) [318].

$$C_6Cl_5 \overset{\displaystyle Cl\ Cl}{\underset{\displaystyle C_6Cl_5CHCl}{\diagdown}} Cl \quad \xrightarrow[\text{(b) I}_2]{\text{(a) BCR}} \quad C_6Cl_5 \overset{\displaystyle Cl\ Cl}{\underset{\displaystyle C_6Cl_5\overset{\displaystyle \cdot}{C}Cl}{\diagdown}} Cl$$

(CLXXV)   (PODM)

It is also formed by dechlorination of perchloro-2-phenyldiphenyl-methane with ferrous chloride in boiling ethyl ether (Section 1.3.19).

PODM is a bright orange-red crystalline radical melting at 244-6° (dec.). It is strongly fluorescent in ultraviolet light. Its infrared [318], ultraviolet-visible [318], mass [301b], and epr [318] spectra have been measured.

It is the most stable radical of the PDM series so far owing to the extra shielding of the trivalent carbon by the pentachlorophenyl group. Its half-life in solution at room/ temperature is about two months. It is converted into (CLXXV) by means of reagent BCR (Section 1.8.1a) in 83.8% yield, or with potassium in ethyl ether, at room temperature (Section 1.8.1a), or in boiling toluene (Section 1.8.1a) in 88.6 and 65.7% yields, respectively. With a boiling solution of hydriodic acid in acetic acid, an 86.3% yield of $\alpha H,\alpha H$-tetradeca-chloro-2-phenyldiphenylmethane (CLXXVI) is obtained. As far as the mechanism is concerned, this reaction is probably analogous to that of perchlorotoluene under similar conditions (Section 1.3.2b). Although PODM is stable towards chlorine in carbon tetrachloride, when iodine is present it is readily converted into perchloro-2-phenyldiphenylmethane (XCVIII; Section 1.3.19) (see Section 1.8.1a). The latter is also formed from the perchloro-2-phenyl-diphenylcarbonium ion with aluminium chloride–sulphuryl chloride (see Sections 1.3.19 and 1.8.1a).

PODM reacts with fuming nitric acid at room temperature to give 72% perchloro-4-o-biphenylylmethylenecyclohexa-2,5-dienone (CLXXVII). Oleum, at room temperature, also yields the perchloro-ketone (CLXXVII) (6.1%) and perchloro-3-phenyl-4-benzylidene-cyclohexa-2,5-dienone (IC; Section 1.3.19). The fact that the non-quinonoid ketone (CLXXVIII) is not formed (compare with the hydrolysis of (XCVIII); Section 1.3.17) is due to shielding of the *alpha* carbon atom (Section 1.3.17), which causes the oxidation to occur exclusively on the *para* ring-carbons. Alternatively, the per-chloroketone (CLXXVII) could also be formed by the direct oxi-dation of the radicals as for nitric acid.

PODM undergoes thermolysis at about 275° giving perchloro-9-phenylfluorenyl (PPF) radical (see Section 1.8.3a). The formation of PPF from (XCVIII) takes place most probably through PODM (Sections 1.3.19 and 1.8.3a).

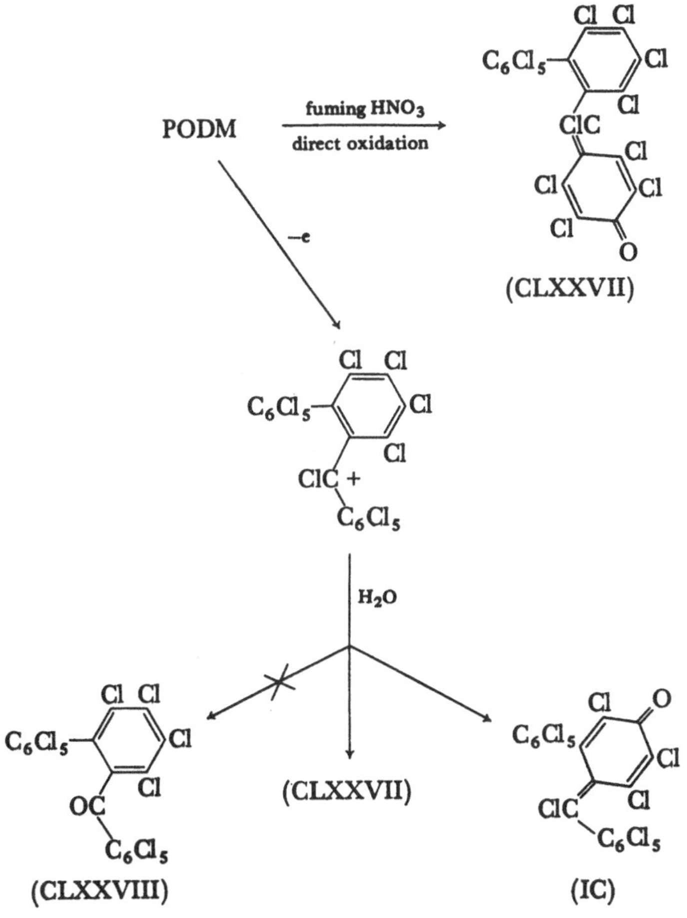

(CLXXVII)

(CLXXVIII)   (CLXXVII)   (IC)

(d) *Perchloro-4,4'-diphenyldiphenylmethyl Radical*

Perchloro-4,4'-diphenyldiphenylmethyl (PDDM) radical is prepared in an analogous manner to the PDM series, i.e., by oxidation with iodine of the corresponding carbanion (CLXXIX) in 72.6% yield [294, 315].

$$(C_6Cl_5-C_6Cl_4-)_2CHCl \xrightarrow{\text{BCR}} (C_6Cl_5-C_6Cl_4-)_2\bar{C}Cl$$

(CLXIV)   (CLXXIX)

$$\Big/ I_2$$

$$(C_6Cl_5-C_6Cl_4-)_2\dot{C}Cl$$

(PDDM)

PPDM is an orange-red radical melting at 280-300° with decomposition. Its infrared, electronic absorption, and epr spectra have been reported [315]. Its chemical stability is similar to that of PPDM. Its half-life time in solution, in air at room temperature, is about 14 days [315].

PPDM reacts slowly with concentrated nitric acid at room temperature giving perchloro-4,4'-diphenylbenzophenone (52%) and with reagent BCR (Section 1.8.1a) yielding the carbanion (CLXXIX) which, on hydrolysis, gives (CLXIV) in 78% yield. Boiling toluene also gives (CLXIV) (53.5%) [315]. Although it is inert to chlorine in the dark, in white, incandescent light a 94% yield of perchloro-4,4'-diphenyldiphenylmethane is obtained (C; Section 1.3.20) [317]. This chlorocarbon is also produced in a 53% yield by treatment with aluminium chloride in sulphuryl chloride (Section 1.8.1a) [317]. No quinonoid species at reaction temperatures ranging from −15 to 80° is formed. Its absence is related to its inherent steric strains (cf. Section 1.4.16).

The reaction of PDDM with potassium in ethyl ether, at room temperature, gives the carbanion (CLXXIX) whose decomposition is discussed in Section 1.7.6 (see also Section 1.8.1a) [291, 316]. Under certain conditions and at moderately high temperature, PDDM

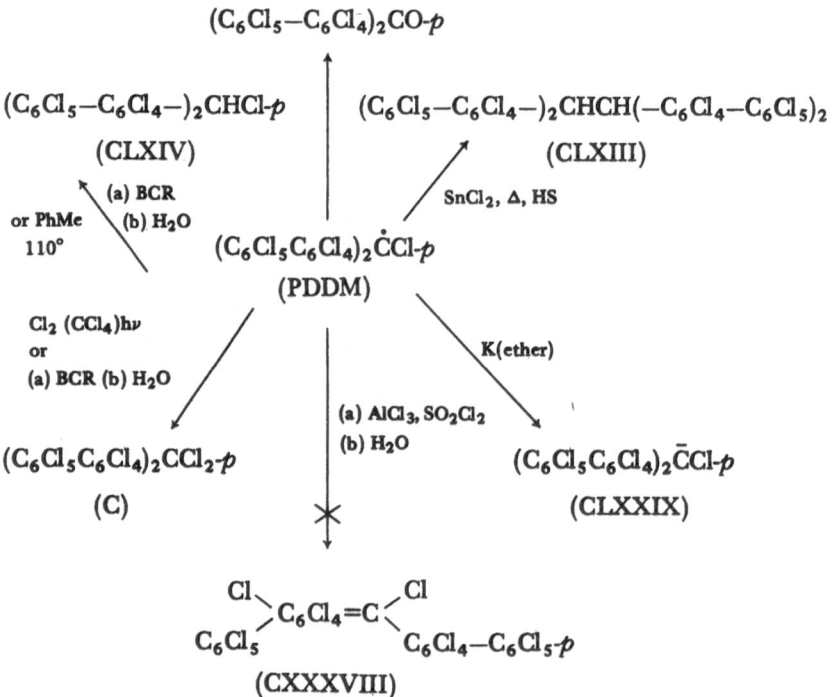

reacts also with stannous chloride giving $\alpha H,\alpha'H$-hexatriacontachloro-tetrakis(4-biphenylyl)ethane (CLXIII; Section 1.7.6) [291, 316].

(e) *4,4'-Dimethoxynonachlorodiphenylmethyl Radical*

αH,4,4'-Dimethoxynonachlorodiphenylmethane (CLXXX; Section 1.9) is converted in the usual way (Section 1.8.1a) into 4,4'-dimethoxynonachlorodiphenylmethyl radical (CLXXXI; 65.0%) [344]. However, the intermediate carbanion is exceptionally unstable. The radical is a bright orange-red substance melting around 130°; its infrared, ultraviolet-visible and epr spectra have been measured [344].

It oxidizes rapidly when exposed to air and light in cyclohexane and slowly in the dark, giving di-4-methoxytetrachlorophenylketone (40% yield). This ketone is also obtained by oxidation with concentrated nitric acid at room temperature.

$$(MeO-C_6Cl_4-)_2CHCl\text{-}p \xrightarrow[\text{(b) I}_2]{\text{(a) BCR}} (MeO-C_6Cl_4-)_2\overset{\cdot}{C}Cl\text{-}p$$

(CLXXX)   (CLXXXI)

$$\Big/ \begin{array}{c} O_2 \text{ (hexane)} \\ or \\ HNO_3 \end{array}$$

$$(MeOC_6Cl_4)_2 > CO\text{-}p$$

(f) *3H,5H,3'H,5'H-Heptachlorodiphenylmethyl and 4,4'-dimethyl-nonachlorodiphenylmethyl Radicals*

3H,5H,3'H,5'H-heptachlorodiphenylmethyl (CLXXXII) and 4,4'-dimethylnonachlorodiphenylmethyl (CLXXXIII) radicals are formed by oxidation of the corresponding carbanions (Sections 1.8.1 and 1.10.2) [316]. They are orange coloured solids with characteristic epr spectra [316]. The stability of (CLXXXII) is lower than that of other related radicals owing to greater electron density at the *alpha* carbon atom and the lack of buttressing of the *ortho* chlorines by *meta* chlorines.

$$sym-C_6H_2Cl_3-\overset{\cdot}{\underset{Cl}{C}}-C_6H_2Cl_3\text{-}sym \qquad p\text{-}MeC_6Cl_4-\overset{\cdot}{\underset{Cl}{C}}-C_6Cl_4Me\text{-}p$$

(CLXXXII)   (CLXXXIII)

1.8.2 Triphenylmethyl Radicals

The members of the so-called PTM series are by far the most stable radicals known. They are inert in reactions which would increase the number of their bonds but active in certain electron-transfer processes, as the radicals of the PDM series previously described. Their thermal stability is also much greater than that of the PDMs. Their inertness is undoubtedly due to almost perfect steric shielding of the central (trivalent) *alpha* carbon atom, much greater than that of the PDM radicals from which they are derived by substitution of the *alpha* chlorine by a chlorinated phenyl group.

(a)   The Perchlorotriphenylmethyl Radical

Treatment of α*H*-pentadecachlorotriphenylmethane (CLXXXIV; Section 1.9) with reagent BRC (Section 1.8.1a) at room temperature gives the carbanion (CLXXXV) (Section 1.10.3) which, by oxidation with iodine, is converted into the perchlorotriphenylmethyl (PTM) radical in a 93% yield [249, 294, 315].

PTM is formed by other methods which have, however, no preparative value.

(1) The reaction of perchloro-3-diphenylmethylenecyclohexene (CXLII; Section 1.4.19) with stannous chloride in ethyl ether [317, 318].   (2)   Perchlorodiphenylmethylenecyclohexa-1,4-diene (CXXXIX; Section 1.4.17) is converted into PTM in a high yield by thermolysis, by photolysis, or by dechlorination with stannous chloride (Section 1.4.17) [317, 318]. (3) PTM is also formed in a high yield from the carbonium hexachloroantimonate (CLXXXVI) with alkaline iodide or bromide (Section 1.11.4), or treatment with cycloheptatriene or other hydrocarbons (Section 1.11.6) [345, 346]. (4) A 92% yield of PTM is obtained by oxidation of tetraethylammonium perchlorotriphenylmethide (CLXXXVII) with iodine (Section 1.10.3) [345, 346]. (5) 4-Bromo- (CLXXXIII; Section 1.8.2i)   and   4-iodotetradecachlorotriphenylmethyl (CLXXXIX; Section 1.8.2h) radicals react with chlorine in carbon tetrachloride in the dark to give PTM, the yields being 91 and 98.5%, respectively [347]. These reactions are summarized in the diagram on page 137.

PTM is a deep-red radical melting with decomposition at 305°. Its infrared [294, 315], electronic absorption [294, 315], epr [273, 315, 341] and mass [301b] spectra have been reported. It is completely disassociated, even in solid form. It is accepted that PTM belongs to the D_3 point group symmetry and, therefore, possesses a large-angle ($\sim$60°), propeller-like conformation. Quantum calculations based on observed hyperfine splittings have been performed.

A solution of PTM exposed to air, at room temperature, has a half-life time of the order of decades [315]. It is stable to water, cyclohexane, carbon tetrachloride, boiling toluene, hydroquinone, *p*-quinone, nitric oxide, certain solutions of sodium hydroxide, concentrated sulphuric acid, concentrated nitric acid, bromine and chlorine (compare with PDM; Section 1.8.1a). PTM does not scavenge short-lived free radicals [348]. It withstands temperatures

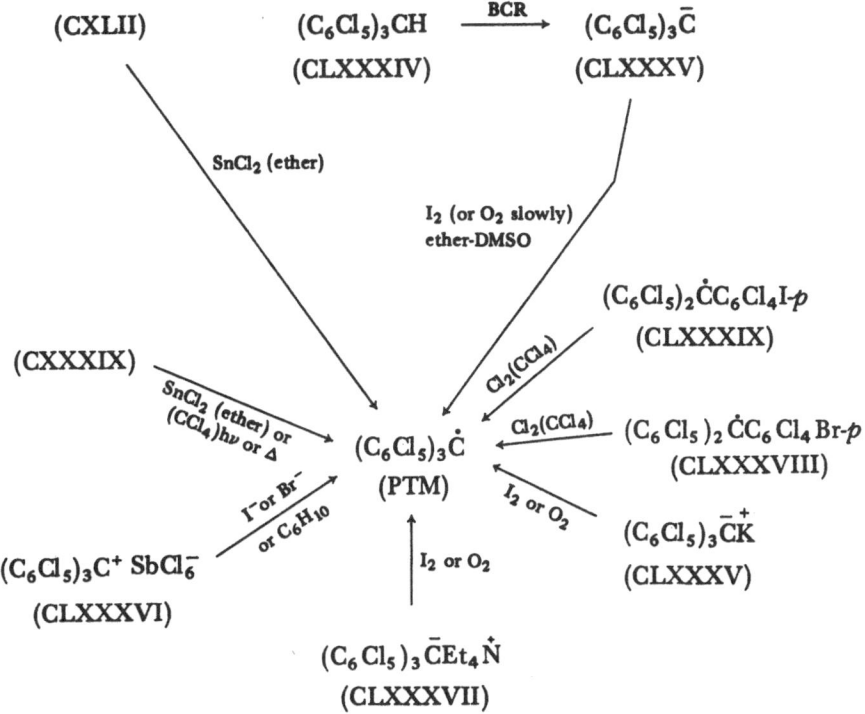

near 300° without significant decomposition [315]. This inertness is due to shielding of the *alpha* carbon which is buried by the substituents, particularly the six *ortho* chlorines. Hence, an attacking species is so hindered that the intermolecular distance where effective bond-formation begins cannot be reached. The inertness of the pentachlorophenyl rings could feasibly be diminished on account of the radical character. However, because of the tilting of those rings with respect to the plane of the *alpha*-carbon sp^2-bonds spin-delocalization is less extensive, which preserves the inertness of the rings. Polar effects apparently play a very minor role in causing such stability since the salts of the corresponding carbanion (CLXXXVII; Section 1.10.3) and carbonium ion (CLXXXVI; Section 1.1.4) are remarkably stable.

Nevertheless, the PTM radical is active under certain electron-transfer conditions. It can pick up an electron or release it forming the corresponding ion. Such processes do not require actual bond formation and may take place either at the *alpha* carbon atom or at the pentachlorophenyl rings, since shielding becomes ineffectual.

Electron-transfers occur in the following processes: (1) The reaction of PTM with potassium or sodium in ethyl ether, at room temperature, gives the carbanion (CLXXXV; Section 1.10.3) (79%) as ascertained by spectra and quenching with aqueous acid [315]. (2) The reaction with BCR (Section 1.8.1a) gives the same carbanion (94%). (3) The reaction with sulphuryl chloride and aluminium chloride yields a perchlorotriphenylcarbonium ion (CXC; Section 1.11.4). (4) Antimony pentachloride in sulphuryl chloride, or carbon tetrachloride, at room temperature gives perchlorotriphenyl-carbonium hexachloroantimonate (CLXXXVI; Section 1.11.4). These reactions are described in some detail in the corresponding sections.

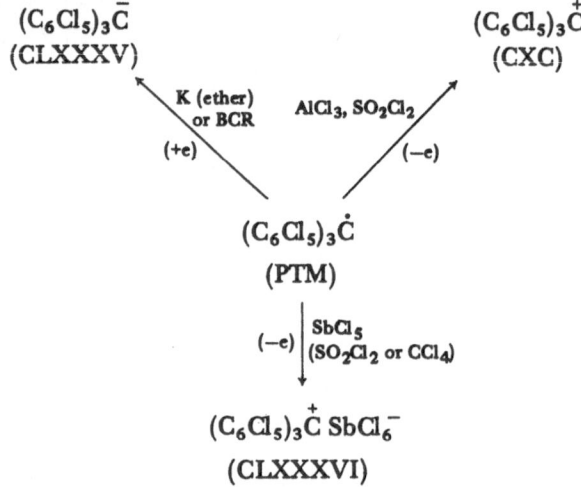

PTM reacts with a boiling solution of hydrogen iodide in acetic acid furnishing the αH-compound (CLXXXIV) [318, 347]. This is probably an electron transfer concerted with, or followed by, a proton transfer rather than a direct hydrogen atom extraction from hydrogen iodide (cf. Sections 1.11.6 and 1.3.3b).

$$(C_6Cl_5)_3\dot{C} + HI \rightarrow (C_6Cl_5)_3\bar{C} + (HI)^{+\cdot} \rightarrow (C_6Cl_5)_3CH + I$$
$$\text{(PTM)} \qquad\qquad \text{(CLXXXV)} \qquad\qquad \text{(CLXXXIV)}$$

Oleum, at 100°, converts PTM into the 2,5-dienone (CXL) [346] by way of an electron-transfer process forming perchlorotriphenyl-carbonium ion (CXC; Section 1.11.4).

$$(C_6Cl_5)_3\dot{C} \xrightarrow{-e} (C_6Cl_5)_3\overset{+}{C} \longrightarrow (C_6Cl_5)_2C=C_6Cl_4=O$$
$$\text{(PTM)} \qquad\qquad \text{(CXC)} \qquad\qquad \text{(CXL)}$$

On heating above its melting point, PTM yields perchloro-9-phenylfluorenyl (PPF) radicals (Section 1.8.3a) [318, 349].

A solution of PTM is light-sensitive [294, 315], being transformed into a mixture containing PPF radicals, apparently as a primary product [349, 350]. When this photolysis is performed in carbon tetrachloride containing molecular chlorine, a variety of chlorocarbons results, such as perchlorodiphenylmethylenecyclohexa-1,4-diene (CXXXIX; Section 1.4.17) [317, 318]. Longer reaction times lead to complex mixtures from which perchloro-3-diphenylmethylenecyclohexene (CXLII; Section 1.4.19) and perchloro-9-phenylfluorene (CII; Section 1.3.21) are isolated [317, 318]. These products are accounted for as follows:

(b)   *The Perchloro-4-phenyltriphenylmethyl Radical*

As PTM (preceding section), the perchloro-4-phenyltriphenylmethyl (PPTM) radical is prepared by oxidation of the corresponding carbanion (CXCII; Section 1.10.3) with iodine in 77% yield [249, 294, 315]. This carbanion is obtained from αH-nonadecachloro-4-phenyltriphenylmethane (CXCI; Section 1.9).

PPTM is a crimson crystalline radical melting at 255-60° (dec.) with recorded infrared, ultraviolet-visible and epr spectra [315]. It is light-sensitive and its thermal and chemical stabilities are similar to those of PTM.

$$p\text{-}C_6Cl_5\text{-}C_6Cl_4\text{-}CH(C_6Cl_5)_2 \xrightarrow{\text{BCR}} p\text{-}C_6Cl_5\text{-}C_6Cl_4\text{-}\bar{C}(C_6Cl_5)_2$$
$$\text{(CXCI)} \qquad\qquad\qquad \text{(CXCII)}$$

$$\Big\downarrow I_2$$

$$p\text{-}C_6Cl_5\text{-}C_6Cl_4\text{-}\dot{C}(C_6Cl_5)_2$$
$$\text{(PPTM)}$$

PPTM is converted into the carbanion (CXCII) with reagent BCR (Section 1.8.1a) at room temperature in good yield (86%) as ascertained by hydrolysis with aqueous acid to (CXCI) [315]. It is oxidized to the carbonium (CXCIII)-chloroaluminate salt with aluminium chloride in sulphuryl chloride (Section 1.11.4). Thermolysis of PPTM at 305° gives the perchloro-9-p-biphenylylfluorenyl (PPBF) radical (Section 1.8.3b).

$$p\text{-}C_6Cl_5\text{-}C_6Cl_4\text{-}\bar{C}(C_6Cl_5)_2 \qquad\qquad p\text{-}C_6Cl_5C_6Cl_4\overset{+}{C}(C_6Cl_5)_2$$
$$\text{(CXCII)} \qquad\qquad\qquad\qquad\qquad \text{(CXCIII)}$$

$$\underset{(+e)}{\nwarrow}\text{BCR} \qquad \underset{(-e)}{\nearrow}\overset{SO_2Cl_2}{\underset{AlCl_3}{\diagup}}$$

$$p\text{-}C_6Cl_5\text{-}C_6Cl_4\text{-}\dot{C}(C_6Cl_5)_2$$
$$\text{(PTM)}$$

$$\Big\downarrow 305°$$

$$\text{(PPBF)}$$

(c)   *The Perchloro-4,4'-diphenyltriphenylmethyl Radical*
This (PDTM) radical is prepared similarly to PTM (Section 1.8.2a); i.e., by oxidation of its carbanion (CXCV; Section 1.10.3) [249, 294, 315]. The reported yield, referring to the precursor αH-tricosa-chloro-4,4'-diphenyltriphenylmethane (CXCIV; Section 1.9), is 73%.

PDTM is a crimson, crystalline radical melting with decomposition at 280-300° with stabilities (thermal, light and chemical) similar to those of PTM (Section 1.8.2a). Its spectra are recorded [315]. Reagent BCR converts it into the carbanion (CXCV) at room temperature and subsequent hydrolysis gives a 59% yield of (CXCIV) [315]. As with PTM (Section 1.8.2a), it forms a blue-green solution containing, presumably, the perchloro-4,4'-diphenyltriphenylcarbonium (CXCVI)-chloroaluminate ion pair (Section 1.11.4) [317].

$$(p\text{-}C_6Cl_5C_6Cl_4)_2CHC_6Cl_5 \underset{H_3O^+}{\overset{BCR}{\rightleftharpoons}} (p\text{-}C_6Cl_5C_6Cl_4-)_2\bar{C}C_6Cl_5$$
$$\text{(CXCIV)} \hspace{5cm} \text{(CXCV)}$$

$$I_2 \left\| BCR \right.$$

$$(p\text{-}C_6Cl_5C_6Cl_4)_2\overset{+}{C}C_6Cl_5 \underset{SO_2Cl_2}{\overset{AlCl_3}{\longleftarrow}} (p\text{-}C_6Cl_5C_6Cl_4)_2\overset{\cdot}{C}C_6Cl_5$$
$$\text{(CXCVI)} \hspace{5cm} \text{(PDTM)}$$

(d) *The Perchloro-4,4',4''-triphenyltriphenylmethyl Radical*

Perchloro-4,4',4''-triphenyltriphenylmethyl (PTTM) radical is obtained as the preceding radicals of the PTM series [249, 294, 315]. The yield referring to the starting material (CXCVII) (Section 1.9), is 65%.

$$(p\text{-}C_6Cl_5-C_6Cl_4-)_3CH \underset{H_3O^+}{\overset{BCR}{\rightleftharpoons}} (p\text{-}C_6Cl_5-C_6Cl_4-)_3\bar{C}$$
$$\text{(CXCVII)} \hspace{5cm} \text{(CXCVIII)}$$

$$I_2 \left\| BCR \right.$$

$$\text{(PTPF, cf. 1.8.3)} \overset{>300°}{\longleftarrow} (p\text{-}C_6Cl_5-C_6Cl_4-)_3\overset{\cdot}{C}$$
$$\text{(PTTM)}$$

PTTM is a purple, microcrystalline radical melting at 316-22° (dec.). Its infrared, ultraviolet-visible, and epr spectra have been reported and studied [315]. Its thermal and chemical stabilities are similar to those of the preceding PTMs. It is also photo-sensitive.

PTTM reacts with BCR (Section 1.8.1a) at room temperature giving perchloro-4,4',4''-triphenyltriphenylcarbanion (CXCVIII; Section 1.10.3), and then, with aqueous acid to yield (CXCVII) (84% yield) [315].

With a solution of aluminium chloride in sulphuryl chloride it gives a blue-green colour due to the presence of a perchloro-4,4',4''-triphenyltriphenylcarbonium (CIC)-chloroaluminate ion pair. The resulting solution when treated with water at −15° fails to give any new chlorocarbon (see Section 1.8.1a) giving back PTTM quantitatively (Section 1.11.4) [317]. As in the PDDM radical (Sections 1.8.1d and 1.8.2a), this abnormal result is due to the prohibitive strain of structures A and B.

Heating PTTM above 300° yields the perchloro-9-(p-biphenylyl)-3,6-diphenylfluorenyl radical (Section 1.8.3d) [334].

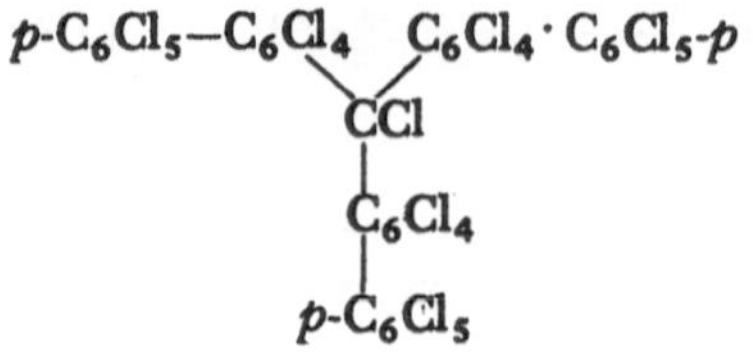
p-C6Cl5—C6Cl4 C6Cl4·C6Cl5-p
CCl
C6Cl4
p-C6Cl5

(A)

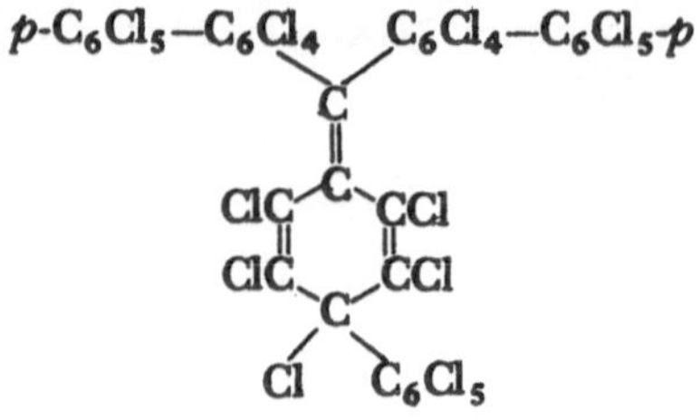
p-C6Cl5—C6Cl4 C6Cl4—C6Cl5-p
C
C
ClC CCl
ClC CCl
C
Cl C6Cl5

(B)

(*e*) *The 4-Carbomethoxytetradecachlorotriphenylmethyl Radical*

Treatment of the methyl ester (CC) (Section 1.9) with reagent BCR (Section 1.8.1a) at room temperature, followed by oxidation with iodine gives the title radical (CCI) in 85.4% yield [347]. Methylation of the acid radical (CCII) (Section 1.8.2h) with diazomethane in ethyl ether [347] gives a better yield (94.7%).

The carbomethoxy-radical (CCI) is a garnet-red crystalline compound and melts at 304-6° (dec.). It is soluble in chlorinated solvents and its infrared, ultraviolet-visible, and epr spectra have been taken [347]. In stability tests it behaves much like PTM. With reagent BCR the corresponding carbanion is obtained which, on acid hydrolysis, yields the *αH*-compound (CC) (85.7%) while hydriodic acid in boiling acetic acid gives *αH*-tetradecachlorotriphenylmethane-4-carboxylic acid (CCIII; 82.8%) (see Section 1.9). It is noteworthy that the carbomethoxy group is not hydrolyzed by reagent BCR.

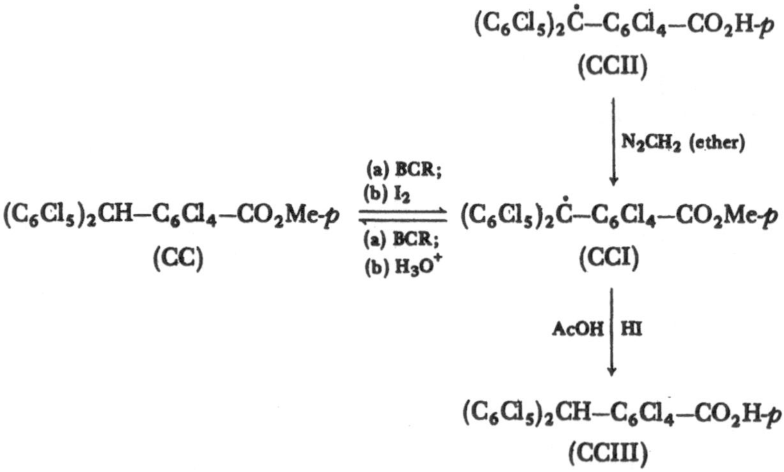
(C6Cl5)2Ċ—C6Cl4—CO2H-p
(CCII)
N2CH2 (ether)
(a) BCR;
(b) I2
(C6Cl5)2CH—C6Cl4—CO2Me-p
(CC)
(a) BCR;
(b) H3O+
(C6Cl5)2Ċ—C6Cl4—CO2Me-p
(CCI)
AcOH HI
(C6Cl5)2CH—C6Cl4—CO2H-p
(CCIII)

(*f*) *The 4,4'-Dicarbomethoxytridecachlorotriphenylmethyl Radical*

The 4,4'-dicarbomethoxytridecachlorotriphenylmethyl radical (CCV) has been synthesized in the usual fashion (Section 1.8.2a) from the di-ester (CCIV) (Section 1.9) via its carbanion [278] in 61% yield.

The radical (CCV) is a red compound melting at 238-9°. Its infrared, ultraviolet-visible, and epr spectra have been measured [278]. It is soluble in organic solvents and its inertness is comparable to that of PTM.

$$C_6Cl_5CH(-C_6Cl_4-CO_2Me\text{-}p)_2 \xrightarrow[\text{(b) } I_2]{\text{(a) BCR}} C_6Cl_5\dot{C}(-C_6Cl_4-CO_2Me\text{-}p)$$
$$\text{(CCIV)} \qquad\qquad\qquad \text{(CCV)}$$

(g)   The 4H-Tetradecachlorotriphenylmethyl Radical

This radical (CCVII) is obtained from αH,4H-tetradecachlorotriphenylmethane (CCVI) (Section 1.9) via its carbanion [347] in 73.2% yield. It is an intense-red crystalline radical of considerable chemical stability melting without decomposition at 262-4° and is very soluble in chlorinated solvents. Its infrared, electronic absorption, and epr spectra have been studied [347].

Just as with PTM, the corresponding carbanion is produced with reagent BCR which follows from conversion into (CCVI) (68.1%) and the latter is also obtained from (CCVII) with hydriodic acid in boiling acetic acid (89.7%).

$$(C_6Cl_5)_2CH-C_6HCl_4 \underset{\substack{\text{(a) BCR; (b) } H_3O^+ \\ \text{HI or AcOH}}}{\overset{\text{(a) BCR; (b) } I_2}{\underset{\longrightarrow}{\longleftarrow}}} (C_6Cl_5)_2\dot{C}-C_6HCl_4$$
$$\text{(CCVI)} \qquad\qquad\qquad\qquad \text{(CCVII)}$$

(h)   The 4-Carboxy-, and 4-Iodotetradecachlorotriphenylmethyl Radicals

When the acid (CCVIII) (Section 1.9) is treated with reagent BCR and then with iodine it gives not only the expected 4-carboxytetradecachlorotriphenylmethyl radical (CCIX) in moderate yields (43.6-55.7%) but also the 4-iodotetradecachlorotriphenylmethyl radical (CLXXXIX; 32.4-19.8%) [347]. This reaction is interpreted as follows:

$$(C_6Cl_5)_2CH-C_6Cl_4-CO_2H\text{-}p \xrightarrow{\text{BCR}} (C_6Cl_5)_2\bar{C}-C_6Cl_4-\bar{C}O_2\text{-}p$$
$$\text{(CCVIII)} \qquad\qquad\qquad\qquad \Big\downarrow I_2$$

$$(C_6Cl_5)_2\dot{C}-C_6Cl_4-CO_2H\text{-}p \xleftarrow{H_3O^+} (C_6Cl_5)_2\dot{C}-C_6Cl_4-\bar{C}O_2\text{-}p$$
$$\text{(CCIX)} \qquad\qquad\qquad\qquad \Big\downarrow I_2$$

$$(C_6Cl_5)_2\dot{C}-C_6Cl_4-I\text{-}p \xleftarrow[(-CO_2)]{I_2} (C_6Cl_5)_2\dot{C}-C_6Cl_4-\dot{C}O_2\text{-}p$$
$$\text{(CLXXXIX)}$$

The carboxy-radical (CCIX) is a red compound melting at 331-2° (dec.), soluble in ethanol. The iodo-radical (CLXXXIX) is garnet-red and melts at 317-8° (dec.). Infrared, ultraviolet-visible, and epr spectra of these radicals have been taken [347]. The stability of these radicals is remarkable, approaching that of PTM (Section 1.8.2a).

While the radical (CCIX) behaves like PTM towards BCR or hydriodic acid in acetic acid reforming (CCVIII) as expected (80 and 76% yields, respectively), radical (CLXXXIX) yields $\alpha H,4H$-tetradecachlorotriphenylmethane (92.6 and 91.6%) (CCVI; Sections 1.8.2g and 1.9).

Radicals (CLXXXIX) and (CCIX) react slowly at room temperature with concentrated nitric acid to give the 2,5-dienone (CXL) (Section 1.4.17) in 98.9 and 80.3% yields respectively. The following mechanisms are suggested:

$$(\text{CCIX}) \xrightarrow{\text{HNO}_3} (C_6Cl_5)_2\overset{+}{C}-C_6Cl_4-CO_2H\text{-}p$$

$$\downarrow \text{HNO}_3$$

$$(C_6Cl_5)_2\overset{\cdot}{C}-C_6Cl_4-I\text{-}p \qquad (C_6Cl_5)_2C=C_6Cl_4\overset{OH}{\underset{CO_2H}{<}}$$

$$\text{(CLXXXIX)}$$

$$\downarrow \text{HNO}_3 \qquad\qquad\qquad \downarrow (-CO, -H_2O)$$

$$(C_6Cl_5)_2\overset{+}{C}-C_6Cl_4-I\text{-}p \qquad\qquad (C_6Cl_5)_2C=C\overset{CCl=CCl}{\underset{CCl=CCl}{<}}\!\!\!>C=O$$

$$\text{HNO}_3 \searrow \qquad\qquad\qquad\qquad \text{(CXL)}$$

$$(C_6Cl_5)_2C=C_6Cl_4\overset{OH}{\underset{I}{<}} \quad \nearrow \quad (-\text{HI})$$

While the carboxy-radical (CCIX) is stable towards chlorine or bromine in carbon tetrachloride in the dark at room temperature, the iodo-radical (CLXXXIX) yields PTM (Section 1.8.2a) and 4-bromo-tetradecachlorotriphenylmethyl (CCXI; Section 1.8.2i) radicals, respectively under these conditions. These are probably four-centre reactions.

$$(C_6Cl_5)_2\overset{\cdot}{C}-C_6Cl_4-I\text{-}p \xrightarrow{X_2} (C_6Cl_5)_2\overset{\cdot}{C}-C_6Cl_4-X\text{-}p + IX$$

$$\text{(CLXXXIX)} \qquad\qquad\qquad\qquad X = Cl \text{ or } Br$$

(i)   *The 4-Bromotetrachlorotriphenylmethyl Radical*
The only known synthesis of this radical (CLXXXVIII) is from the
4-iodotetradecachlorotriphenylmethyl radical (CLXXXIX; preceding
Section). The title compound is a red radical similar to PTM and
melting at 314-6° (dec.). Its spectra have been measured [347]. It
reacts with chlorine in carbon tetrachloride in the dark to give PTM
(Section 1.8.2a).

$$(C_6Cl_5)_2\dot{C}-C_6Cl_4-Br\text{-}p$$

$$(CLXXXVIII)$$

(j)   *The Sodium and Ammonium 4-Carboxytetradecachlorotri-phenylmethyl Radicals*
The formation of the sodium radical (CCX) allows the easy
separation of carboxy-radical (CCIX) from the iodo-radical
(CLXXXIX), both described in Section 1.8.2h. However, although
insoluble in water, (CCX) has not been isolated [347].
The ammonium radical (CCXI) is obtained in a 99.0% yield from
(CCIX) with ammonia in ethyl ether [347]. It is a red compound
melting at 330-1° (dec.) and very soluble in ethanol. Its spectra have
been studied [347].

$$(C_6Cl_5)_2\dot{C}-C_6Cl_4-CO_2Na\text{-}p \qquad (C_6Cl_5)_2\dot{C}-C_6Cl_4-CO_2NH_4\text{-}p$$

$$(CCX) \qquad\qquad\qquad (CCXI)$$

(k)   *The 4-Nitrotetradecachlorotriphenylmethyl Radical*
The preparation of the pure 4-nitrotetradecachlorotriphenyl-
methyl radical (CCXIII) has not been achieved [347]. The best
sample contained only 42% of the radical, the main impurity being
its precursor (CCXII). The formation of a carbanion from the
precursor is possibly accompanied by a nucleophilic substitution of
the nitro group yielding a phenol (Section 1.2.5e).

$$(C_6Cl_5)_2CH-C_6Cl_4-NO_2\text{-}p \xrightarrow[\text{(b) I}_2]{\text{(a) BCR}} (C_6Cl_5)_2\dot{C}-C_6Cl_4-NO_2\text{-}p$$

$$(CCXII) \qquad\qquad\qquad\qquad (CCXIII)$$

(l)   *4-Hydroxy-, 4-Oxide-, 4-Methoxy-, and 4-Acetoxytetradeca-chlorotriphenylmethyl Radicals*
Reduction of the 2,5-dienone (CXL) (Section 1.4.17) by hydriodic
acid-iodine in hexane at room temperature gives an 83% yield of the
4-hydroxy radical (CCXIV) [346]. Its red ethereal solution is
discharged when extracted with an aqueous sodium bicarbonate
solution which becomes green. Presumably, the water-soluble 4-oxy
radical (CCXV) is formed. This phenoxide radical reforms (CCXIV)

with acids, and oxidizes very rapidly in air to the perchloroketone (CXL).

The phenolic radical (CCXIV) reacts with diazomethane in ethyl ether giving the 4-methoxy radical (CCXVI) in a 91% yield and with boiling acetic anhydride the 4-acetoxy radical (CCXVII) (68%).

The methoxy-radical (CCXVI), on treatment with reagent BCR and then with aqueous acid, yields (CCXVIII) (Section 1.8.2b). All these radicals, except (CCXIV) and (CCXV), are inert, and (CCXVI) and (CCXVII) melt at 305-8° (dec.) respectively. Their spectra have been recorded [346].

$$(C_6Cl_5)_2\dot{C}-C_6Cl_4-\bar{O}\text{-}p$$

$$(CCXV)$$

$$O_2 \diagup \qquad \diagdown \begin{array}{c} H_3O^+ \\ HO^- \end{array}$$

$$(CXL) \quad \xrightarrow[\text{(hexane)}]{\text{HI, I}_2} \quad (C_6Cl_5)_2\dot{C}-C_6Cl_4-OH\text{-}p$$

$$(CCXIV)$$

$$\bigg| \begin{array}{c} Ac_2O \\ 140° \end{array}$$

$$(C_6Cl_5)_2CH-C_6Cl_4-OMe\text{-}p \quad \xrightarrow[\text{(ether)}]{CH_2N_2} \quad (C_6Cl_5)_2\dot{C}C_6Cl_4-OAc\text{-}p$$

$$(CCXVIII) \qquad\qquad\qquad (CCXVIII)$$

$$\begin{array}{c} \text{(a) BCR} \\ \text{(b) } H_3O^+ \end{array}$$

$$(C_6Cl_5)_2\dot{C}-C_6Cl_4-OMe\text{-}p$$

$$(CCXVI)$$

(m)   *The 4-Aminotetradecachlorotriphenylmethyl Radical*

The radical (CCXX) is obtained in a 68% yield by reduction of *NH*-tetradecachlorodiphenylmethylenecyclohexa-2,5-dienimine (CCXIX; Section 1.11.4) with diethylhydrophosphonate in ethyl ether at room temperature [346].

$$(C_6Cl_5)_2C{=}C\begin{array}{c} CCl{=}CCl \\ \diagup \qquad \diagdown \\ \diagdown \qquad \diagup \\ CCl{=}CCl \end{array}C{=}NH \quad \xrightarrow[\text{(ether)}]{Et_2HPO_3} \quad (C_6Cl_5)_2\dot{C}-C_6Cl_4-NH_2\text{-}p$$

$$(CCXIX) \qquad\qquad\qquad\qquad (CCXX)$$

Compound (CCXX) is a red radical melting at 282-4° (dec.). Its hexane solution is red while in ethyl ether it turns green. Its infrared, electronic, and epr spectra have been measured [346].

(n) *4,4',4''-Tricarbomethoxy-, and 4,4',4''-Tricarboxydodecachloro-triphenylmethyl Radicals*

The radical (CCXXII) is prepared from the methane (CCXXI) (Section 1.9) via a carbanion (Section 1.8.2a) [278] in 72% yield. When heated in a mixture of sodium hydroxide and aqueous dioxane or in the cold with BCR it reverts to (CCXXI).

The radical (CCXXII) is a red solid melting at 258-60° (dec.), soluble in organic solvents. In concentrated sulphuric acid at 100° it is converted (71% yield) into the 4,4',4''-tricarboxydodecachlorotriphenylmethyl radical (CCXXIII) melting at 312-8° (dec.) which, in aqueous sodium bicarbonate, forms an intense-red solution of the corresponding salt. The radical (CCXXIII) withstands hot concentrated sulphuric acid for several hours.

$$(p\text{-MeO}_2\text{C}-\text{C}_6\text{Cl}_4-)_3\text{CH} \xrightleftharpoons[\substack{\text{HO}^-,\text{ dioxane-H}_2\text{O} \\ \text{or} \\ \text{(a) BCR; (b) H}_3\text{O}^+}]{\text{(a) BCR; (b) I}_2} (p\text{-MeO}_2\text{C}-\text{C}_6\text{Cl}_4-)_3\overset{\cdot}{\text{C}}$$

$$\text{(CCXXI)} \hspace{6cm} \text{(CCXXII)}$$

$$\Big/ \substack{\text{H}_2\text{SO}_4 \\ 100°}$$

$$(p\text{-HO}_2\text{C}-\text{C}_6\text{Cl}_4-)_3\overset{\cdot}{\text{C}}$$

$$\text{(CCXXIII)}$$

The infrared, electronic absorption, and epr spectra of both (CCXXII) and (CCXXIII) have been taken [278].

(o)   *The 4H,4'H,4''H-Dodecachlorotriphenylmethyl Radical*

This radical (CCXXV) is prepared from α*H,4H,4'H,4''H*-dodecachlorotriphenylmethane (CCXXIV; Section 1.9) by carbanion oxidation (Section 1.8.2a) [278] in 77% yield. It reverts to (CCXXIV) (93%) by means of reagent BCR. It is a red, crystalline radical melting at 282-4°, very soluble, even in hexane. Its spectra have been studied [278].

$$(\text{C}_6\text{HCl}_4-)_3\text{CH-}p \xrightleftharpoons[\text{(a) BCR; (b) H}_2\text{O}]{\text{(a) BCR; (b) I}_2} (\text{C}_6\text{HCl}_4-)_3\overset{\cdot}{\text{C-}}p$$

$$\text{(CCXXIV)} \hspace{5cm} \text{(CCXXV)}$$

(p) *4,4',4''-Trimethoxy-, 4,4',4''-Trihydroxy-, and 4,4',4''-Triacetoxy-dodecachlorotriphenylmethyl Radicals*

The radical (CCXXVII) is prepared in a 55.4% yield from the (CCXXVI) (Section 1.9) via its carbanion (Section 1.8.2a) [344]. (CCXXVII) is a red radical melting at 295-7°. Its spectra have been recorded [344]. In chemical stability it resembles that of PTM. Treatment with reagent BCR (Section 1.8.1a) and then with acidified water converts it into (CCXXVI) (55.9%). With oleum at 90° a 72% yield of dark-red dodecachloro-4-di-p-hydroxyphenylmethylenecyclohexa-2,5-dienone (CCXXVIII) is obtained which, by treatment with hydriodic acid–iodine in benzene at room temperature, yields almost pure 4,4',4''-trihydroxydodecachlorotriphenylmethyl (CCXXIX; 69.9%), a radical of relatively low stability which reverts spontaneously to (CCXXVIII). This radical gives, with boiling acetic anhydride, the red 4,4',4''-triacetoxydodecachlorotriphenylmethyl radical (CCXXX; 70.8%), melting at 250-2°. Its spectra have been registered [344].

$$(p\text{-MeOC}_6\text{Cl}_4)_3\text{CH} \underset{\text{BCR/H}_3\text{O}^+}{\overset{\text{BCR/I}_2}{\rightleftharpoons}} (p\text{-MeOC}_6\text{Cl}_4)_3\text{C}^{\cdot}$$

(CCXVI)   (CCXXVII)

$\downarrow$ oleum

$$(p\text{-HO}-\text{C}_6\text{Cl}_4)_2\text{C}=\text{C} \Big\langle {}^{\text{CCl}=\text{CCl}}_{\text{CCl}=\text{CCl}} \Big\rangle \text{CO}$$

$(p\text{-MeCO}_2-\text{C}_6\text{Cl}_4)_3\text{C}^{\cdot}$ (CCXXVIII)

(CCXXX) $\overset{140°}{\underset{(\text{Ac})_2\text{O}}{\nwarrow}}$ $\underset{\text{HI/I}_2\text{/hexane}}{\nearrow}$

$$(p\text{-HO}-\text{C}_6\text{Cl}_4)_3\text{C}^{\cdot}$$

(CCXXIX)

(q) *4-Methyl-, 4,4'-Dimethyl-, and 4,4',4''-Trimethyltriphenylmethyl Radicals*

The radicals (CCXXXI), (CCXXXII) and (CCXXXIII) are synthesized from the corresponding αH-quasiperchloro compounds by oxidation of their carbanions (Section 1.8.1a) [318]. The yields are: CCXXXI, 80.0; CCXXXII, 78.2; and CCXXXIII, 49.6%, and they melt at 303-30° (dec.), 295-315° (dec.) and 315-23° (dec.), respectively.

$$\begin{matrix} p\text{-RC}_6\text{Cl}_4 \\ p\text{-R}'\text{C}_6\text{Cl}_4 \\ p\text{-R}''\text{C}_6\text{Cl}_4 \end{matrix}\!\!\text{CH} \overset{\text{BCR}}{\longrightarrow} \begin{matrix} p\text{-RC}_6\text{Cl}_4 \\ p\text{-R}'\text{C}_6\text{Cl}_4 \\ p\text{-R}''\text{C}_6\text{Cl}_4 \end{matrix}\!\!\text{C}^- \overset{\text{I}_2}{\longrightarrow} \begin{matrix} p\text{-RC}_6\text{Cl}_4 \\ p\text{-R}'\text{C}_6\text{Cl}_4 \\ p\text{-R}''\text{C}_6\text{Cl}_4 \end{matrix}\!\!\text{C}^{\cdot}$$

CCXXXI: R=Me; R'=R'', Cl
CCXXXII: R=R'=Me; R''=Cl
CCXXXIII: R=R'=R''=Me

These three red radicals withstand temperatures near 300° when carbonization starts to occur. They are light-sensitive in solution. Although they possess the chemical inertness of PTM (Section 1.8.2a) in carbon tetrachloride at room temperature, they react slowly with bromine, presumably by attack at the methyl groups. Various spectra of these radicals have been studied [318].

(r) *The α′H-Tetracosachloro-4-diphenylmethyltriphenylmethyl Radical*

αH,α′H-Tetracosachloro-α,α,α′,α′-tetraphenyl-*p*-xylene (CCXXXIV; 1.9) reacts with reagent BCR to give a solution containing the corresponding mono- and dicarbanion which by oxidation with iodine yield a mixture of perchloro-α,α,α′,α′-tetraphenyl-*p*-xylylene (CCXXXV; Section 1.8.4a) and α′H-tetracosachloro-4-diphenyl-methyltriphenylmethyl radical (CCXXXVI), the latter being present in about 37% amount, as ascertained by magnetic susceptibility measurements [351].

$$(C_6Cl_5)_2 CH-C_6Cl_4-CH(C_6Cl_5)_2 \text{-} p$$

$$\text{(CCXXXIV)}$$

$$\left| \begin{array}{l} \text{(a) BCR} \\ \text{(b) I}_2 \end{array} \right.$$

$$(C_6Cl_5)_2C=C_6Cl_4=C(C_6Cl_5)_2 \quad + \quad (C_6Cl_5)_2 CH-C_6Cl_4-\dot{C}(C_6Cl_5)_2 \text{-} p$$

$$\text{(CCXXXV)} \qquad\qquad\qquad \text{(CCXXXVI)}$$

Although the isolation of the pure radical has not been achieved, its visible and epr spectra have been obtained and studied [351].

1.8.3   Fluorenyl Radicals

As far as chemical inertness is concerned, the perchlorofluorenyl radicals occupy an intermediate position between the PDMs (Section 1.8.1) and the PTMs (Section 1.8.2). Thermally, they are more stable than the PTMs and they are completely disassociated.

(a)   *The Perchloro-9-phenylfluorenyl Radical*

This (PPF) radical may be regarded as resulting from the perchlorotriphenylmethyl (PTM) radical (Section 1.8.2a) by loss of two *ortho* chlorines from two benzene rings with formation of a bond between them resulting in two significant changes:

(1) The shielding of the 9-carbon in PPF becomes less effective than that of the corresponding *alpha* carbon of PTM since the number of chlorines in its immediate neighbourhood is smaller.

(2) Because of the quasi planar arrangement of the fluorene system the odd electron is more delocalized than in PTM.

Consequently, it was expected that PPF would be less inert (smaller shielding) and thermally more stable (greater resonance) than PTM. However, an all-planar conformation of PPF is impossible on account of the repulsive forces of the *ortho* chlorines in the pentachlorophenyl group and those of positions 1 and 8 of the fluorenyl system.

The PPF radical is made in a number of ways [318, 349], some of them lacking preparative value since they utilize PPF itself:

(1) Thermolysis of αH-pentadecachloro-2-phenyldiphenylmethane (CLXXV; Section 1.9) at 280-300° (93.1% yield). A likely mechanism is:

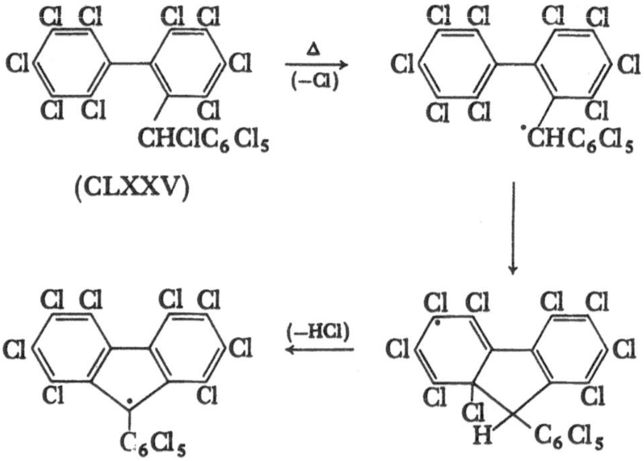

(CLXXV)

(PPF)

(2) Thermolysis of perchloro-2-phenyldiphenylmethane (XCVIII; Section 1.3.19) (84.4%).

$$o\text{-}C_6Cl_5-C_6Cl_4CCl_2C_6Cl_5$$

(XCVIII)

(3) Thermolysis of perchloro-2-phenyldiphenylmethyl (PODM) radical (Section 1.8.1c) at 260-270° (73.0%).

(4) Thermolysis of PTM (Section 1.8.2a) at 310-320° (78.6%) which is a general method for the synthesis of chlorinated fluorenyl radicals. In fact, it has been shown that even the PDM radical cyclizes in an analogous fashion (Section 1.8.1a).

(5) Illumination with white light of a solution of PTM (Section 1.8.2a) [318, 349, 350]. Electron absorption and epr spectra show conversions up to 16%.

(6) Treatment of perchloro-9-phenylfluorene (CII; Section

1.3.22) with ferrous chloride in boiling ethyl ether (84.6% yield) or
(7) Thermolysis of (CII) at 230-240° (80.0%).
(8) From 9H-tridecachloro-9-phenylfluorene (CCXXXVII; cf. Section 1.9) via the corresponding carbanion (31.8%), as in PTM (Section 1.8.2a). However, the only known synthesis of (CCXXXVII) starts from PPF.

PPF is a dark-green radical melting around 300°. Its infrared, electronic absorption, epr [318, 319, 349, 350] and mass [301b] spectra have been recorded. Its inertness towards oxygen in solution at room temperature in the dark is as high as that of PTM (Section 1.8.2a). However, it photo-oxidizes readily, although it is moderately stable to chlorine and white light, in carbon tetrachloride.

PPF is highly inert to p-quinone, hydroquinone, chlorine, bromine, sodium hydroxide, and concentrated sulphuric and nitric acids. It is attacked by hydriodic acid in boiling acetic acid and by oleum and destroyed in boiling toluene, and by nitric oxide in carbon tetrachloride at room temperature (cf. PDM (Section 1.8.1a) and PTM (Section 1.8.2a)). Thus, in some aspects it resembles PTM, while in others it behaves more like PDM. PPF withstands temperatures as high as 320° without significant decomposition and it is recalled that it can be prepared by heating PTM at 310-320°.

Like the members of the PDM and PTM series, PPF reacts with reagent BCR (Section 1.8.1a) giving the corresponding carbanion which with aqueous acids yields 9H-tridecachloro-9-phenylfluorene (CCXXXVII; Section 1.9) (66.5%). This reduction occurs directly with hydriodic acid in boiling acetic acid (49.6% yield).

Chlorination of PPF with chlorine in carbon tetrachloride, in the presence of iodine, gives 80% of perchloro-9-phenylfluorene (CII; Section 1.3.22). This product results also from the chlorination of PTM with chlorine and white light, albeit in small yield [318, 319].

PPF reacts with oleum at room temperature giving a mixture of isomeric perchloro-4-(9-fluorenylidene)cyclohexa-2,5-dienone (CIII; Section 1.3.22) (25.3% yield) and perchloro-9-phenyl-3-fluorenone (CIV; Section 1.3.22) (41.7%). At 120° only the latter perchloroketone is obtained (59.6%). The first step of this process is presumably oxidation to carbonium ion (CV), followed by the sequences suggested in Section 1.3.22. The formation of oxygen-bridged carbonium ion pyrosulphates is not ruled out (see Sections 1.11.3, 1.11.4 and 1.11.5).

A mixture of these perchloroketones is also obtained by photo-oxidation with white light in the air, the yields being 21.3 and 29.6%, respectively.

Like PTM (Section 1.8.2a), PPF reacts with a solution of aluminium chloride, in sulphuryl chloride at room temperature giving, presumably, a chlorine bridged carbonium ion chloroaluminate

(see Sections 1.11.3, 1.11.4 and 1.11.5), which by hydrolysis yields perchloro-9-phenylfluorene (CII) in an overall yield of 64.1%.
The chemistry of PPF has been studied by Ballester and co-workers [318, 319].

(b)   *The Perchloro-9-p-biphenylylfluorenyl Radical*
The perchloro-4-phenyltriphenylmethyl (PPTM) radical when heated at 300-305° gives a 51.0% yield of the title (PPBF) radical (Section 1.8.2b) [334]. It is also obtained by thermolysis of αH-nonadecachloro-2,4'-diphenyldiphenylmethane   (CCXXXVIII) (Section 1.9) at 290-300° (62.7%) [334]. (See PPF; Section 1.8.3a). It appears that no perchloro-3,9-diphenylfluorenyl radical is formed.

$(C_6 Cl_5)_2 \overset{\cdot}{C}C_6 Cl_4 C_6 Cl_5 \text{-} p$ $o \text{-} C_6 Cl_5 C_6 Cl_4 CHClC_6 Cl_4 C_6 Cl_5 \text{-} p$

(PPTM)   (CCXXXVIII)

(PPBF)

PPBF is a deep-green radical melting at about 370° (dec.). Its infrared and epr spectra have been measured [334].

(c)   *The Perchloro-9-o-biphenylylfluorenyl Radical*
αH-Heptadecachloro-9-o-biphenylylfluorene   (CCXXXIX; Section 1.9) reacts with reagent BCR (Section 1.8.1a) to give a solution containing perchloro-o-biphenylen(2-biphenylyl)carbanion (Section 1.10.4) which, on treatment with iodine, yields a green solid containing about 60% of perchloro-9-o-biphenylylfluorenyl (POBF, CVIII; Section 1.3.23) radical [318]. This radical decomposes, even at room temperature, giving perchloro-9,9'-spirobifluorene (CVII; Section 1.3.23) (82% yield) which precludes its isolation in pure form. Its electronic absorption and epr ·spectra have been taken [318].

(d)   *The Perchloro-9-(p-biphenylyl)-3,6-diphenylfluorenyl Radical*
Thermolysis   of   perchloro-4,4',4''-triphenyltriphenylmethyl
(PTTM; Section 1.8.2d) radical at 290-305° gives a 36% yield of this
(PTPF; Section 1.8.2d) radical, a green chlorocarbon decomposing
around 370°. Its infrared, electronic absorption, and epr spectra have
been recorded [334].

1.8.4   Chlorocarbon Biradicals

Ballester and co-workers have synthesized a number of chlorocarbon
biradicals which are also regarded as new members of the PTM series
(Section 1.8.2) of general formula

$$(C_6 Cl_5)_2 \dot{C}-Ar-\dot{C}(C_6 Cl_5)_2$$

(CCXL)

They are completely disassociated and possess high thermal and
chemical stability.

(a) *The Perchloro-α,α,α',α'-tetraphenyl-p-xylylene Biradical*

The perchloro-α,α,α',α'-tetraphenyl-*p*-xylylene (CCXL; Ar = $-C_6Cl_4-p$) has been ascertained by magnetic susceptibility measurements to possess a singlet electron structure and, in spite of the high angle of twist of the two *exo*-ethylene bonds which should favour a triplet state, it is actually the perchloro-α,α,α',α'-tetraphenyl-*p*-quinodimethane (PTPX, CCXXXV; Section 1.8.2r). It is noteworthy that about the same degree of twisting exists in perchlorodiphenylmethylenecyclohexa-1,4-diene (CXXXIX; Section 1.4.17) and perchloro-4-diphenylmethylenecyclohexa-2,5-dienone (CXL; Section 1.4.17). The parent chlorocarbon, the perchloro-*p*-xylylene (LXI; Section 1.3.9c), has also a quinonoid structure with remarkable thermal stability and chemical inertness, in spite of the twisting of its *exo* ethylene bonds.

PTPX is prepared from the tetraphenyl-*p*-xylene (CCXXXIV) (Sections 1.8.2r and 1.9) by the BCR method (Section 1.8.1a) in 42% yield [294, 351]. An important by-product is the radical (CCXXXVI) (see Section 1.8.2r). It is a sparingly soluble, red powder, m.p. 392° (dec.). Its infrared and electronic absorption spectra have been taken [351].

PTPX reacts slowly with reagent BCR (Section 1.8.1a) at room temperature turning quantitatively into the α*H*-compound (CCXXXIV).

PTPX does not react with boiling mesitylene, hydroquinone in carbon tetrachloride, *p*-quinone in carbon tetrachloride, sodium hydroxide in ethyl ether, concentrated sulphuric or nitric acid, bromine, and chlorine (in dark).

$$(C_6Cl_5)_2C=C\underset{CCl=CCl}{\overset{CCl=CCl}{<>}}C=C(C_6Cl_5)_2 \quad \underset{I_2}{\overset{BCR}{\searrow}}$$

(PTPX, CCXXXV) $(C_6Cl_5)_2\bar{C}-C_6Cl_4-\bar{C}(C_6Cl_5)_2$

$$\diagdown HI, AcOH \qquad \overset{BCR}{\diagup\!\diagup}_{H_3O^+}$$

✗

$$(C_6Cl_5)_2CH-C_6Cl_4-CH(C_6Cl_5)_2$$

(CCXXXIV)

(b) *The Perchloro-α,α,α',α'-tetraphenylbi-p-tolyl-α,α'-ylene Biradical*

This (PTBT) biradical corresponds to $-Ar- = -C_6Cl_4-C_6Cl_4-$ in the general formula (CCXL). It is obtained from the bi-*p*-tolyl (CCXLI) (Section 1.9) in the conventional manner (Section 1.8.1a) [294, 315] in 64.7% yield. PTBT is a purple chlorocarbon melting at

250° (dec.) of chemical inertness comparable to that of PTM (Section 1.8.2a) but less inert to concentrated nitric acid.

Its infrared, ultraviolet-visible and epr spectra have been studied [294, 315] and the evidence indicates that it is a true biradical. Its molecular shape is assumed to be that of two twin connected propellers. PTBT reacts with reagent BCR (Section 1.8.1a) giving, as expected, the dicarbanion (CCXLII).

$$(C_6Cl_5)_2CH-C_6Cl_4-C_6Cl_4-CH(C_6Cl_5)_2$$

(CCXLI)

BCR
H$_3$O$^+$

$$(C_6Cl_5)_2\bar{C}-C_6Cl_4-C_6Cl_4-\bar{C}(C_6Cl_5)_2$$

(CCXLII)

I$_2$ / BCR

$$(C_6Cl_5)_2\dot{C}-C_6Cl_4-C_6Cl_4-\dot{C}(C_6Cl_5)_2$$

(PTBT)

(c) *The Perchlorotriptycen-α,α'-ylene Biradical*

The synthesis of this (CCXLIV) biradical from α*H*,α'*H*-dodecachlorotriptycene (CCXLIII; Section 1.9) has been abortive [315]. No detectable carbanion formation takes place, even under drastic conditions (see Section 1.10.5).

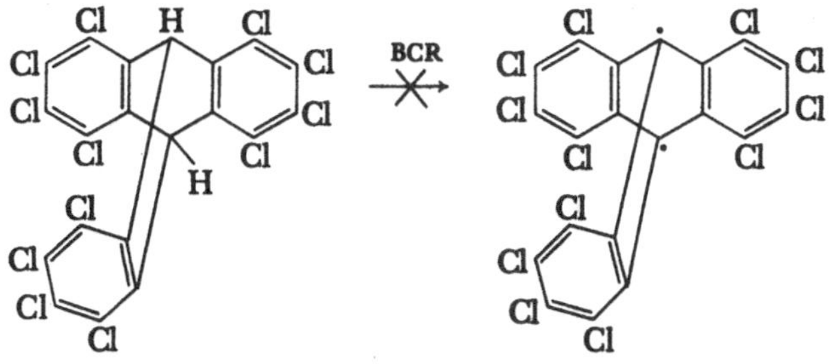

(CCXLIII)   (CCXLIV)

(d) *The Perchloroethynylenebis-p-triphenylmethyl Biradical*

When the bis-*p*-triphenylmethane (*cis* CCXLV) (Section 1.9) is treated with reagent BCR (Section 1.8.1a) it gives a blue carbanion solution which by oxidation with iodine yields the biradical (CCXLVII) in 35.8% yield [347]. This biradical corresponds to −AR ≡ −C$_6$Cl$_4$−C≡C−C$_6$Cl$_4$− in the general formula (CCXL).

It has been found that treatment with aqueous acid of the dicarbanion solution yields at least 62% of the dicarbanion (CCXLVI; Section 1.10.3) by unexpected *cis*-elimination. The following mechanism is suggested:

$$(C_6Cl_5)_2CH-C_6Cl_4-CCl=CCl-C_6Cl_4-CH(C_6Cl_5)_2$$
$$(cis\text{-CCXLV})$$

$$\searrow \text{BCR}$$

$$(C_6Cl_5)_2\bar{C}-C_6Cl_4-CCl=CCl-C_6Cl_4-\bar{C}(C_6Cl_5)_2$$
$$(cis)$$

$$\left\downarrow \begin{array}{l} -2\ Cl, \\ +2e \end{array}\right.$$

$$(C_6Cl_5)_2\bar{C}-C_6Cl_4-C\equiv C-C_6Cl_4-\bar{C}(C_6Cl_5)_2$$
$$H_3O^+ \swarrow \qquad \text{(CCXLVI)}$$

$$(C_6Cl_5)_2CH-C_6Cl_4-C\equiv C-C_6Cl_4-CH(C_6Cl_5)_2 \qquad \bigg\downarrow I_2$$
$$\text{(CCIL)}$$

$$(C_6Cl_5)_2\dot{C}-C_6Cl_4-C\equiv C-C_6Cl_4-\dot{C}(C_6Cl_5)_2$$
$$\text{(CCXLVII)}$$

By the same treatment the vinylenebismethane (*trans*-CCXLV) (Section 1.9) undergoes slow vicinal dechlorination. After oxidation, the product is a mixture of the *trans*-perchlorovinylenebis-*p*-triphenylmethyl biradical (CCXLVIII; Section 1.8.4e) (~1/3) and the acetylene biradical (CCXLVII) (~2/3).

The biradical (CCXLVII) is also obtained in 77.4% yield from α*H*,α′*H*-compound (CCIL) (Section 1.9).

$$(C_6Cl_5)_2CH-C_6Cl_4-C\equiv C-C_6Cl_4-CH(C_6Cl_5)_2$$
$$\text{(CCIL)}$$

$$\begin{array}{cc} \text{(a) BCR} & \text{(a) BCR} \\ \text{(b) } I_2 & \text{(b) } H_3O^+ \end{array}$$

$$\text{(CCXLVII)}$$

It is an intense-green, microcrystalline powder melting at 333-4° (dec.). Because of its abnormal magnetic behaviour spin-counting is not possible but, combined physical and chemical evidence confirms its biradical nature. Its infrared, electronic absorption, and epr data have been studied [347].

The reactivity of this biradical is limited to electron-transfer processes, e.g., reagent BCR gives its dicarbanion (62.9% yield of the $\alpha H, \alpha' H$-compound). Additional chemical activity resides in its unsaturated character. For example, in the presence of iodine it adds chlorine (see next Section).

(e)   The trans-perchlorovinylenebis-p-triphenylmethyl Biradical
The trans-biradical (CCXLVIII) is formed from the trans-ethylene (CCXLV), as pointed out in the preceding Section. Because of the difficulty in separating it from the accompanying acetylene biradical (CCXLVII), the reaction mixture is treated directly with chlorine in carbon tetrachloride at room temperature, in the presence of iodine, giving the trans-biradical (CCXLVIII) exclusively [347].
It is also obtained in 94.4% yield by chlorination of the acetylene biradical (CCXLVII) (Section 1.8.4d) with chlorine in carbon tetrachloride at room temperature, in the presence of iodine.
This biradical is a red, microcrystalline chlorocarbon melting at 354-6° (dec.). Magnetic susceptibility data shows it to be completely disassociated. Its infrared, electronic absorption, and epr spectra have been measured [347]. Reagent BCR converts it into a mixture of $\alpha H, \alpha' H$-quasiperchloro compounds (CCXLV) and (CCIL) (Section 1.8.4d).

$$(C_6Cl_5)_2CH-C_6Cl_4-CCl=CCl-C_6Cl_4-CH(C_6Cl_5)_2$$

(trans-CCXLV)

(a) BCR
(b) I_2

$$(C_6Cl_5)_2\dot{C}-C_6Cl_4-C\equiv C-C_6Cl_4-\dot{C}(C_6Cl_5)_2 \ +$$

(CCXLVII)

$$(C_6Cl_5)_2\dot{C}-C_6Cl_4-CCl=CCl-C_6Cl_4-\dot{C}(C_6Cl_5)_2$$

(trans-CCXLVIII)

Cl_2, I_2
(CCl_4)

1.9 αH-QUASIPERCHLOROAROMATIC COMPOUNDS†

The H-quasiperchloroaromatic compounds include the important ring-perchlorinated precursors for the synthesis of inert and other

† The prefix 'quasi-perchloro' indicates substitution of all hydrogens in the parent hydrocarbon by chlorine, except those indicated by the preceding symbol. See also reference 1.

radicals described in Sections 1.7 and 1.8. Although the *H*-quasiper-chloro compounds possess an interesting chemistry, the study of their general reactions is beyond the scope of the present review.

In this chapter, polychlorinated arylmethanes with one *alpha* hydrogen atom per tetrahedral carbon are described as so-called α*H*-quasiperchloro compounds. Since fluorene is 2,2′-biphenylylene-methane its 9*H*-quasiperchloro derivatives are also α*H*-compounds. Some α*H*,α′*H*-quasiperchloro compounds which are precursors for the biradicals and related compounds reviewed in Section 1.8.4 are also included here.

The most significant general syntheses are:

BMC procedure: An α-chlorodiarylmethane or a triarylmethane is treated with reagent BMC (Section 1.1)

$$Ph_2CHCl \xrightarrow{\ BMC\ } (C_6Cl_5)_2CHCl$$

$$Ph_3CH \xrightarrow{\ BMC\ } (C_6Cl_5)_3CH$$

In most cases the yields are 50-90% of theory and in some chlorinolysis has been observed which may render this method inoperable.

It is remarkable that while ring-perchlorination of 9-*o*-phenyl-fluorene causes chlorinolysis, that of 9-*o*-biphenylylfluorene even leads partly to ring closure. This result is traced to the additional shielding due to the extra *ortho* pentachlorophenyl group. The reactions are set out below:

(CVII)

BCR procedure: This is a synthesis starting from the corresponding free radicals. Although in most cases devoid of preparative value, it has been useful in establishing the structure of these radicals which are first treated with reagent BCR (Section 1.8.1a), and then with acidified water.

$$(C_6Cl_5)_3\dot{C} \xrightarrow{\text{BCR}} (C_6Cl_5)_3\bar{C} \xrightarrow{\text{H}_3\text{O}^+} (C_6Cl_5)_3CH$$

The yields are usually 70-90% and in few cases are structural changes effected.

Hydriodic acid (HI) procedure: It essentially yields the same results as the BCR procedure. The free radical is treated with hydriodic acid in acetic acid giving moderate to excellent yields (50-90%).

Alkaline metal (AM) procedure: The radical is reduced to its carbanion by potassium (or sodium) in ethyl ether at room temperature. Because of the technical difficulties this method is seldom used. The yields are, nevertheless, good to excellent.

Other procedures useful for αH-quasiperchloro compounds with functional substituents are shown in the following examples:

Ring perchlorination (see Section 1.2.1f)

$$(p\text{-HO}_2CC_6H_4)_3CH \xrightarrow[\text{(ClSO}_3\text{H)}]{\text{Cl}_2, \text{I}_2} (p\text{-HO}_2CC_6Cl_4)_3CH$$

Substitution (see Sections 1.2.3h and 1.11.2)
Example 1

$$(C_6Cl_5)_2CH(OH) + AlCl_3 \longrightarrow (C_6Cl_5)_2\overset{+}{C}H \ldots Cl-Al\bar{C}l_3$$

$$\swarrow \text{H}_2\text{O}$$

$$(C_6Cl_5)_2CHCl$$

Example 2

$$(C_6Cl_5)_3CH \xrightarrow[\text{MeO}^-]{} (p\text{-MeOC}_6Cl_4)_3CH$$

Example 3

$$(C_6Cl_5)_2CHCl \xrightarrow{\text{MeO}^-} (p\text{-MeOC}_6Cl_4)_2CHOMe \xrightarrow[\text{(b) H}_2\text{O}]{\text{(a) AlCl}_3}$$

$$(p\text{-MeOC}_6Cl_4)_2CHCl$$

Reductive Condensation (see Sections 1.3.3b and 1.4.7)
Example 1

$$(C_6Cl_5)_2CHC_6Cl_4CCl_3\text{-}p \xrightarrow{Ph_3P} (C_6Cl_5)_2CH-C_6Cl_4-CCl_2\overset{+}{P}(Ph)_3\bar{Cl}$$

$$(C_6Cl_5)_2CH-C_6Cl_4-CCl=CCl-C_6Cl_4-CH(C_6Cl_5)_2$$
cis and *trans*

Example 2

$$p\text{-}(C_6Cl_5)_2CH-C_6Cl_4-CHClC_6Cl_5 \xrightarrow{Sn^{++}} p\text{-}(C_6Cl_5)_2CH-C_6Cl_4-\overset{.}{C}HC_6Cl_5$$

$$(C_6Cl_5)_2CH-C_6Cl_4-CH(C_6Cl_5)-CH(C_6Cl_5)-C_6Cl_4-CH(C_6Cl_5)_2$$
(*meso*) and (*dl*) forms

The syntheses for a number of significant αH-quasiperchloro compounds are given next in a concise form. Some data such as melting points, by-products (in brackets), references and other useful indications, are also included

$(C_6Cl_5)_2CHCl$ (CLIX; Section 1.8.1a); m.p. 231-3° [315]: BMC (81.6%) [$C_6Cl_5CHCl_2$ (10.3%), C_6Cl_6 (7.6%)]; BCR (72%); AM (84.4%) [294, 316]. $(C_6Cl_5)_2CHOH + AlCl_3 (CH_2Cl_2)$ (78%).

$(p\text{-MeOC}_6Cl_4)_2CHCl$ (CLXXX; Section 1.8.1e); m.p. 160-2° [344]: $(C_6Cl_5)_2CHCl$ + [(a) NaOMe (MeOH-dioxane); (b) $AlCl_3 (CH_2Cl_2)$] (30.5%).

$(p\text{-CCl}_3C_6Cl_4)_2CHCl$; m.p. 263.5-7.0° [291]: BMC (58.0%).

$p\text{-}C_6Cl_5C_6Cl_4CHClC_6Cl_5$ (CLXXIII; Section 1.8.1b); m.p. 295-8° [294, 315]: BMC (71.8%); BCR (56.7%).

$o\text{-}C_6Cl_5C_6Cl_4CHClC_6Cl_5$ (CLXXV; Section 1.8.1c); m.p. 242° [318]: BCR (83.8%); AM (88.6%). $(C_6H_4)_2$=CHC_6H_5 + BMC (61.7%) [$o\text{-}(C_6HCl_4)C_6Cl_4CHClC_6Cl_5$ (64.5%)].

$(p\text{-}C_6Cl_5C_6Cl_4)_2CHCl$ (CLXIV; Section 1.8.1d); m.p. 310-6° [294, 315]: BMC (90.6%); BCR (78%); AM (53.6%).

$p\text{-HC}_6Cl_4CH(C_6Cl_5)_2$ (CCVI; Section 1.8.2g); m.p. 238-40° [346]: BCR (68.1%); HI (89.7%). $p\text{-CCl}_3C_6Cl_4CH(C_6Cl_5)_2$ + BCR (53.3%) [$p\text{-HO}_2CC_6Cl_4CH(C_6Cl_5)_2$ (19.9%)]. $p\text{-IC}_6Cl_4\overset{.}{C}(C_6Cl_5)_2$ + BCR (92.6%). *Idem* + HI (91.6%). $p\text{-NH}_4O_2CC_6Cl_4CH(C_6Cl_5)_2$ + $\Delta(330°)$ (57.7%).

$(C_6Cl_5)_3CH$ (CLXXXIV; Section 1.8.2a); m.p. 320°: BMC (92.7%) [315]; BCR (87%)[315]; AM (79%)[315]; HI (74.2%)[318, 346]. $(C_6Cl_5)_2C=C_6Cl_8$ (CXLII; Section 1.4.19) + KI (AcOH) (90%) [318, 319].

$p\text{-}O_2NC_6Cl_4CH(C_6Cl_5)_2$ (CCXII; Section 1.8.2k); m.p. 334-6°[346]: $p\text{-}HC_6Cl_4CH(C_6Cl_5)_2$ + HNO_3 (fum.) (57.3%).

$p\text{-}MeC_6Cl_4CH(C_6Cl_5)_2$(Section 1.8.2q); m.p. 336-9°[318]: BMC (50.6%).

$p\text{-}CCl_3C_6Cl_4CH(C_6Cl_5)_2$; m.p. 272-4°[346]: BMC (57.3%).

$p\text{-}HO_2CC_6Cl_4CH(C_6Cl_5)$ (CCVIII; Section 1.8.2h); m.p. 368-70° [346]: BCR (80%); HI (76.1%). $p\text{-}CCl_3C_6Cl_4CH(C_6Cl_5)_2$ + BCR (19.9%) [$p\text{-}HC_6Cl_4CH(C_6Cl_5)_2$ (53.3%)]. *Idem* + oleum (110-20°) (74.8%). $p\text{-}MeO_2CC_6Cl_4CH(C Cl_5)_2$ + oleum (70-80°) (89.9%). *Idem* + HI (82.8%).

$p\text{-}NH_4O_2CC_6Cl_4CH(C_6Cl_5)_2$ (Section 1.8.2j); m.p. 312-5°[346]: $p\text{-}HO_2CC_6Cl_4CH(C_6Cl_5)_2$ + NH_3 (ether) (96.7%).

$p\text{-}MeO_2CC_6Cl_4CH(C_6Cl_5)_2$ (CC; Section 1.8.2e); m.p. 286-9°[346]: BCR (85.7%). $p\text{-}HO_2CC_6Cl_4CH(C_6Cl_5)_2$ + N_2CH_2 (ether) (94.7%).

$p\text{-}C_6Cl_5C_6Cl_4CH(C_6Cl_5)_2$ (CXCI; Section 1.8.2b); m.p. 410-2° [315]: BMC (75.8%); BCR (86%).

$(p\text{-}MeOC_6Cl_4)_2CHC_6Cl_5$; m.p. 313-4°[343]: $(C_6Cl_5)_3CH$ + NaOMe (MeOH-dioxane) (22.8%) [$(p\text{-}MeOC_6Cl_4)_3CH$ (29.2%)].

$(p\text{-}MeC_6Cl_4)_2CHC_6Cl_5$ (Section 1.8.2q); m.p. 339-42°[318]: BMC (40.5%).

$(p\text{-}CHCl_2C_6Cl_4)_2CHC_6Cl_5$; m.p. 356-8°[346]: $(p\text{-}MeC_6Cl_4)_2CHC_6Cl_5$ + Cl_2 ($h\nu$, CCl_4) (66.9%) [$(C_6Cl_5)_3CH$ (−%)]. $(p\text{-}CCl_3C_6Cl_4)_2CHC_6Cl_5$ + $FeCl_2$ (dioxane) (44.3%).

$(p\text{-}CCl_3C_6Cl_4)_2CHC_6Cl_5$; m.p. 235-40°[278, 244]: BMC (50.5%) [$p\text{-}CCl_3C_6Cl_4CH(C_6Cl_5)_2$ (14%)].

$(p\text{-}HO_2CC_6Cl_4)_2CHC_6Cl_5$; m.p. 314-6°[278, 244]: $(p\text{-}CCl_3C_6Cl_4)_2$-$CHC_6Cl_5$ + oleum (100°) (31.7%).

$(p\text{-}MeO_2CC_6Cl_4)_2CHC_6Cl_5$ (CCIV; Section 1.8.2c); m.p. 230-4°[278]: $(p\text{-}HO_2CC_6Cl_4)_2CHC_6Cl_5$ + N_2CH_2 (ether) (91.1%).

$(p\text{-}C_6Cl_5C_6Cl_4)_2CHC_6Cl_5$ (CXCIV; Section 1.8.2c); m.p. 400° [315]: BMC (54%); BCR (84%).

$(p\text{-}HC_6Cl_4)_3CH$ (CCXXIV; Section 1.8.2o); m.p. 312-4°[278]: $(p\text{-}CHCl_2C_6Cl_4)_3CH$ + H_2SO_4 (200°) (12%). $(p\text{-}NH_4O_2CC_6Cl_4)_3CH$ + Δ(250°) (excellent %).

(p-HOC$_6$Cl$_4$)$_3$CH (Section 1.8.2p); m.p. 293-5°[343] :
(p-MeOC$_6$Cl$_4$)$_3$CH + oleum (room temp.) (67%).

(p-MeOC$_6$Cl$_4$)$_3$CH (CCXXVI; Section 1.8.2p); m.p. 299-300°[343]:
BCR (55.9%). (C$_6$Cl$_5$)$_3$CH + NaOMe (MeOH-dioxane) (40%).

(p-NO$_2$C$_6$Cl$_4$)$_3$CH; m.p. 388-9°[278] : (p-HC$_6$Cl$_4$)$_3$CH + HNO$_3$ (fum.)
(73.4%).

(p-NH$_2$C$_6$Cl$_4$)$_3$CH; m.p. 310-2°[278]: (p-NO$_2$C$_6$Cl$_4$)$_3$CH + SnCl$_2$
(MeOH-HCl-H$_2$O) (91%).

(p-MeC$_6$Cl$_4$)$_3$CH (Section 1.8.2q); m.p. 340-5°[318]: BMC (80.4%).

(p-CH$_2$BrC$_6$Cl$_4$)$_3$CH; m.p. 346-51°[278]: (p-MeC$_6$Cl$_4$)$_3$CH + Br$_2$
(hν, CCl$_4$) (84%).

(p-CHCl$_2$C$_6$Cl$_4$)$_3$CH; m.p. 369-70°[278]: (p-MeC$_6$Cl$_4$)$_3$CH + Cl$_2$
(hν, CCl$_4$) (69%).

(p-CCl$_3$C$_6$Cl$_4$)$_3$CH; m.p. 311-3°[346]: BMC (49%).

(p-HO$_2$CC$_6$Cl$_4$)$_3$CH (Section 1.8.2n); m.p. 300°[278]: (p-MeO$_2$CC$_6$-
Cl$_4$)$_3$CH+ H$_2$SO$_4$ (100°) (72%).

(p-NH$_4$O$_2$CC$_6$Cl$_4$)$_3$CH (Section 1.8.2n); m.p. ~250°[278] :
(p-HO$_2$CC$_6$Cl$_4$)$_3$CH + NH$_3$ (ether) (-%).

(p-MeO$_2$CC$_6$Cl$_4$)$_3$CH (CCXXI; Section 1.8.2n); m.p. 286-8°[278]:
BCR (70.0%). (a) (p-HO$_2$CC$_6$H$_4$)$_3$CH + Cl$_2$(I$_2$, ClSO$_3$H) (80°);
(b) + N$_2$CH$_2$ (ether) (20.4%) (p-MeO$_2$CC$_6$Cl$_4$)$_2$CHC$_6$Cl$_5$ (9.1%),
p-MeO$_2$CC$_6$Cl$_4$CH(C$_6$Cl$_5$)$_2$ (5.6%). (a) (p-CHCl$_2$C$_6$Cl$_4$)$_3$CH +
oleum (140°); (b) + N$_2$CH$_2$ (ether) (32.1%). (a) (p-CH$_2$BrC$_6$Cl$_4$)$_3$CH +
oleum (150°); (b) + N$_2$CH$_2$ (ether) (71%). (a) (p-MeC$_6$Cl$_4$)$_3$CH +
oleum (200°); (b) + N$_2$CH$_2$ (ether) (18%).

(p-C$_6$Cl$_5$C$_6$Cl$_4$)$_3$CH (CXCVII; Section 1.8.2d); m.p. 280-5°[315]:
BMC (64.4%).

(C$_6$Cl$_4$)$_2$=CHC$_6$Cl$_5$ (CCXXXVII; Section 1.8.3a); m.p. 292.0-2°[318]:
BCR (65.5%); HI (49.6%). o-C$_6$Cl$_5$C$_6$Cl$_4$CHClC$_6$Cl$_5$ + BCR (91.6%).
(C$_6$Cl$_4$)$_2$=CClC$_6$Cl$_5$ + SnCl$_2$ or FeCl$_2$ (ether) (-%) [319].
(C$_6$Cl$_5$)$_2$C=C$_6$Cl$_8$ (CXLII; Section 1.4.19) + NaI (AcOH) (100%) or
SnCl$_2$ (ether) (-%) [319].

(C$_6$Cl$_4$)$_2$=CHC$_6$Cl$_4$C$_6$Cl$_5$-o (CCXXXIX; Section 1.8.3c); m.p.
300.0-3.5° [318]: BMC (58.5%) [(C$_6$Cl$_4$)$_4$C; Section 1.3.23]; HI (50%).

p-(C$_6$Cl$_5$CHCl)$_2$C$_6$Cl$_4$; m.p. 349°[350]: BMC (63%). p-(C$_6$Cl$_5$-
CH(OH))$_2$C$_6$Cl$_4$ + AlCl$_3$, SO$_2$Cl$_2$ (81.7%). αH,$\alpha' H$, α,α'-diphenyl-p-
xylylene + AlCl$_3$, SO$_2$Cl$_2$ (85.6%).

4,4'-(C$_6$Cl$_5$CHCl)$_2$(C$_6$Cl$_4$-C$_6$Cl$_4$); m.p. 329-32°[316]: BMC (38%).

p-$(C_6Cl_5)_2CH-C_6Cl_4-CHClC_6Cl_5$; m.p. 391-3°[350]: BMC (62%).

p-$((C_6Cl_5)_2CH)_2C_6Cl_4$ (CCXXXIV; Section 1.8.4a); m.p. > 450°[350]: BMC (26.4%); BCR (100%).

$4,4'$-$((C_6Cl_5)_2CH)_2(C_6Cl_4-C_6Cl_4)$ (CCXLI; Section 1.8.4b); m.p. ~400°[315]: BMC (49.3%).

cis-$4,4'$-$((C_6Cl_5)_2CH)_2(C_6Cl_4CCl=CClC_6Cl_4)$ (cis-CCXLV; Section 1.8.4d); m.p. 416-8° [346]: p-$CCl_3C_6Cl_4CH(C_6Cl_5)_2$ +$Ph_3P(C_6H_6)$ (50.2%) [$trans$-CCXLV (13.2%)].

$trans$-$4,4'$-$((C_6Cl_5)_2CH)_2(C_6Cl_4CCl=CClC_6Cl_4)$ ($trans$-CCXLV; Section 1.8.4d); m.p. 452-5°[346] : p-$CCl_3C_6Cl_4CH(C_6Cl_5)_2$ + $Ph_3P(C_6H_6)$ (13.2%) [cis-CCXLV (50.2%)]. $4,4'$-$((C_6Cl_5)_2CH)_2$-$(C_6Cl_4C\equiv CC_6Cl_4)$+ $Cl_2(I_2, CCl_4)$ (94.4%). $Idem$ + Cl_2 ($h\nu$, CCl_4) (98.4%).

$4,4'$-$((C_6Cl_5)_2CH)_2(C_6Cl_4C\equiv CC_6Cl_4)$ (CCIL; Section 1.8.4d); m.p. 392-5°[346] : BCR (62.9%). cis-$4,4'$-$((C_6Cl_5)_2CH)_2(C_6Cl_4CCl=CClC_6Cl_4)$ + BCR (61.9%).

dl-$(p$-$(C_6Cl_5)_2CH-C_6Cl_4-CH(C_6Cl_5)$-$)_2$; m.p. 331° [350] : p-$(C_6Cl_5)_2CH-C_6Cl_4-CHClC_6Cl_5$ + $SnCl_2(C_2Cl_4)$ (10%).

$meso$-$(p$-$(C_6Cl_5)_2CH-C_6Cl_4-CH(C_6Cl_5)-)_2$; m.p. 400°[350]: p-$(C_6Cl_5)_2CH-C_6Cl_4-CHClC_6Cl_5$ + $SnCl_2(C_2Cl_4)$ (58%).

$(C_6Cl_4)_3(CHCH)$ (CCXLIII; Section 1.8.4c); m.p. 360°[315]: BMC (88.2%).

In spite of high overcrowding by chlorine, the αH-quasiperchloro-compounds possess an ultraviolet spectrum indicating that they are non-strained compounds. Their infrared spectrum is very similar to that of the benzenoid chlorocarbons, except for the presence of carbon-hydrogen stretching peaks about or below 3000 cm^{-1}. The n.m.r. spectrum affords valuable information. It usually consists of a singlet at $\tau 2.65$-3.00.

1.10   AROMATIC CHLOROCARBON ANIONS

It was shown in Section 1.8 that poly- and perchloro carbanions are of importance in the synthesis of stable radicals. Only stable and related carbanions are discussed here with emphasis on the di- and triarylcarbanions.

Since perchlorocyclopentadiene is a carbanion with six π-electrons it is regarded as an aromatic chlorocarbon, according to Hückel's (4n + 2) rule.

1.10.1   Perchlorocyclopentadiene Anion

McBee and Smith showed in 1955 that perchlorocyclopentadiene (CCL), perchlorocyclopentene (CCLI), and 5H-pentachlorocyclopentadiene (CCLII) react with lithium aluminium hydride in ethyl ether at 50° yielding solutions containing lithium perchlorocyclopentadienide ($C_5Cl_5^-$, CCLIII) [352]. This lithium derivative is also obtained from (CCL) and lithium metal in tetrahydrofurane at −25°, with or without the presence of zinc chloride [353], or from (CCLII) and alkyl lithium compounds [354].

(CCL) reacts with alkyl phosphites to produce high yields of the corresponding 5-alkylpentachlorocyclopentadienes (CCLI; H=alkyl). Mark has assumed that a chlorophosphonium perchlorocyclopentadienide is reversibly formed and decomposes giving those products [355].

The anion $C_5Cl_5^-$ is generated by polarographic reduction of (CCL) or (CCLI) in aqueous ethanol at the mercury dropping electrode in the presence of sodium ions [356] and is hence of low basicity.

In this connection, it is of interest that perchlorocyclopentadienyl-magnesium bromide is obtained from (CCL) and methylmagnesium bromide in ethyl ether [357]. Recently, West and co-workers have prepared a number of solid perchlorocyclopentadienides, such as the thallium, trialkyl- and tetra-alkylammonium, tetra-alkylphosphonium, and pyridinium salts, from (CCLI) [358, 359].

The $C_5Cl_5^-$ salts are generally colourless, which may be unexpected since the perchlorocarbanions considered next are deeply coloured. However, perchlorobenzene, which also possesses six π electrons, is colourless as well, and in fact, the ultraviolet spectrum of $C_5Cl_5^-$ resembles that of any neutral benzenoid chlorocarbon [359].

The infrared spectrum of $C_5Cl_5^-$ is, on account of the high symmetry of this compound, rather simple [358, 359]. Its quadrupole nuclear resonance spectrum has also been studied [358, 359]. All $C_5Cl_5^-$ salts are thermally unstable. The lithium salt decomposes rapidly above zero, while the other salts—which are prepared at about $-80°$—are unstable at room temperature. It is believed that the primary decomposition product is the perchlorocyclopentadienyl radical (CCLIV) since perchlorobiscyclopentadienyl (CCLV), the 5H-compound (CCLI), and (CCL), along with tarry materials, are formed. Furthermore, a transient epr signal is detected.

With iodine in tetrahydrofurane, 1,2-di-iodoethane in ethyl ether, or hexamminecobalt(III) in aqueous methanol, it yields the dimer (CCLV), presumably by means of the radical (CCLIV). However, with bromine it gives readily bromopentachlorocyclopentadiene [359]. The anion $C_5Cl_5^-$ gives (CCLI) with acids in an excellent (84%) yield [353].

Some nucleophilic substitutions have been performed below zero, particularly between the lithium salt and methyl, allyl, and benzyl halides in *ca.* 70% yield. These reactions are successful only in the presence of Lewis acids (aluminium or zinc salts). It has accordingly been suggested that the Lewis acids coordinate with the halogen (bromine or iodine) of the halides weakening the carbon–halogen bond and so facilitating its displacement [353].

C_5Cl_5Li reacts also with methyl sulphate and with ethylene oxide giving good yields of the expected products [353] but not with carbonyl compounds [353].

The poor nucleophilicity of $C_5Cl_5^-$ is assumed to be due partly to the relatively low negative charge on each carbon atom since they are equivalent and bonded to chlorine and partly to its electronic structure of its ground state being similar to that of benzene. Calculations indicate that the π-electron density per carbon atom is 1.2083 [360].

Hedberg and Rosenberg have recently synthesized iron and ruthenium perchlorometallocenes (CCLIV) by multi-step chlorination of lithium metallocenes [361, 362]. Efforts to obtain them

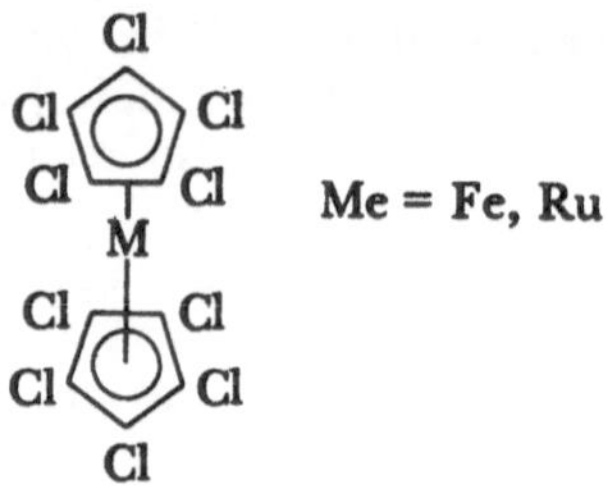

Me = Fe, Ru

(CCLIV)

from perchlorocyclopentadienide salts have been unsuccessful [357, 362]. These perchlorometallocenes are remarkably resistant to concentrated sulphuric or nitric acids (cf. Chapter 3).

1.10.2   Perchlorodiphenylmethides

There are two main ways to obtain perchlorodiphenylcarbanions (or perchlorodiphenylmethides) in an excellent yield: (1) The reaction of an αH-quasiperchlorodiphenylmethane (Section 1.9) with reagent BCR (Section 1.8.1a) and (2) The electron transfer reaction between a perchlorodiphenylmethyl radical with alkaline metal in ethyl ether or with reagent BCR (Section 1.8.1). They are also formed from the corresponding chlorocarbons or quinonoid isomers with reagent BCR.

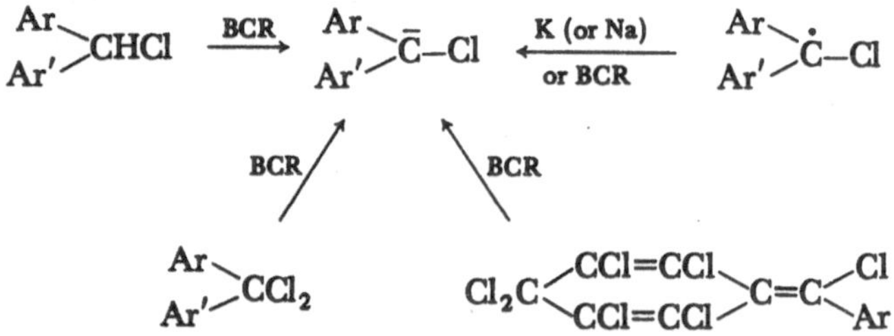

In general, these carbanions are soluble in the reaction media to which they impart a deep-blue colour (strong maximum at about 600 nm). Salts of these carbanions have not been isolated but their solutions in media containing dimethyl sulphoxide are stable at room temperature and even above. They are the most stable α-chloro-carbanions known.† However, in the absence of this polar solvent, they decompose slowly (weeks and even months at room temperature) to give radicals, presumably through a perchlorodiphenyl-carbene, as indicated in Sections 1.7.5, 1.7.6, 1.8.1a and 1.8.1d.

An unexpected result is observed with the blue perchloro-2-

† The lithium trichloromethide is reported to be stable at −100° (see reference 294).

phenyldiphenylcarbanion (CCLVIa) which, even in solvents containing DMSO, forms slowly, at room temperature, a pink solution of the perchloro-9-phenylfluorenyl anion (CCLVIb) in a nearly quantitative yield. This is due to an intramolecular nucleophilic substitution of aromatic chlorine (cf. Section 1.8.1c).

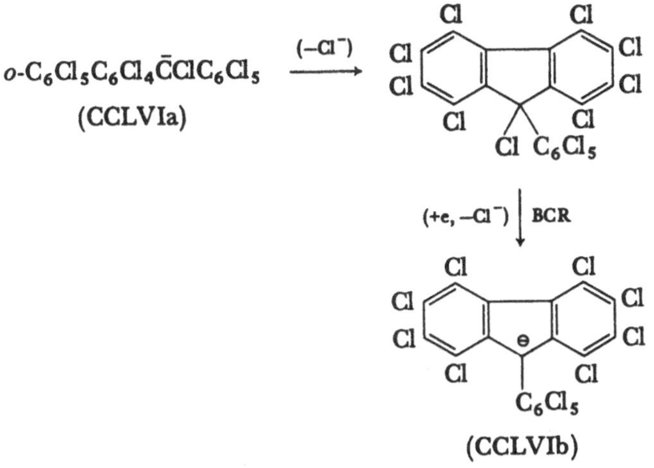

$$o\text{-}C_6Cl_5C_6Cl_4\bar{C}ClC_6Cl_5 \xrightarrow{(-Cl^-)}$$

(CCLVIa)

$$(+e, -Cl^-) \downarrow BCR$$

(CCLVIb)

The perchlorodiphenylcarbanions react with water to give the corresponding αH-quasiperchloro compounds (Section 1.9).

$$\frac{Ar}{Ar'}\bar{C}\text{-}Cl \xrightarrow{H_2O} \frac{Ar}{Ar'}CHCl$$

They react readily at room temperature with oxygen in solution. For instance, perchlorodiphenylcarbanion (CCLVII) gives a mixture of PDM radical (42%; Section 1.8.1a) and perchlorobenzophenone (CCLVIII; 21.9%) [291].

$$(C_6Cl_5)_2\bar{C}Cl \xrightarrow{O_2} (C_6Cl_5)_2\dot{C} + (C_6Cl_5)_2CO$$
(CCLVII)   (PDM)   (CCLVIII)

Perchloro-2-phenyldiphenylcarbanion (CCLVIa) gives slowly a mixture of perchloro-3-phenyl-4-benzylidenecyclohexa-2,5-dienone (IC; 10.7%) and 9H-tridecachloro-9-phenylfluorene (CCXXXVII;

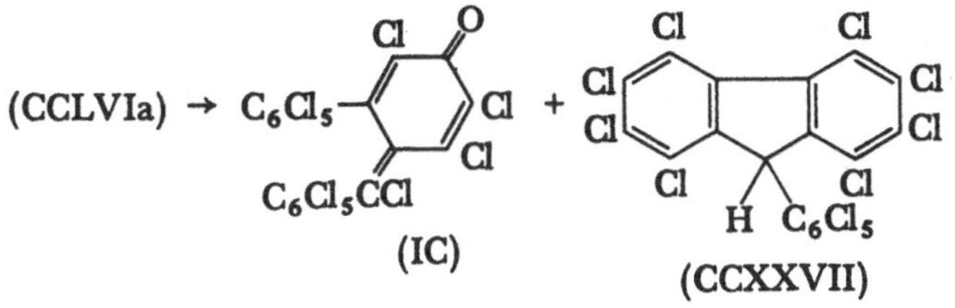

(CCLVIa) →

(IC)   (CCXXVII)

55.5%) [318]. The attack of oxygen occurs at the *para*, rather than at the α- position because of the shielding of the latter by the *ortho*-pentachlorophenyl ring which also hinders the formation of PODM radical (Section 1.8.1c). It is noteworthy that these α-chlorocarbanions are more reactive towards oxygen than the corresponding α-chlororadicals (the PDMs) since oxidation proceeds more readily in molecules with greater electron availability (in this case the negative charge). The oxidation of the perchlorodiphenylcarbanions with iodine is of practical value in the synthesis of α-chlorodiphenylmethyl radicals described in Sections 1.7 and 1.8.

$$\begin{matrix} Ar \\ {}^{\diagdown}\bar{C}{-}Cl \\ Ar' {}^{\diagup} \end{matrix} \quad \xrightarrow{\;I_2\;} \quad \begin{matrix} Ar \\ {}^{\diagdown}\dot{C}{-}Cl \\ Ar' {}^{\diagup} \end{matrix}$$

The α*H*-quasiperchlorodiphenylcarbanions, such as (CCLIX) (CCLX) and (CLX) are prepared by similar methods. They form red solutions. The influence of an *alpha* chlorine on the electronic absorption spectrum of the PDMs is therefore pronounced, possibly because of a significant participation of the chlorine 3d-orbitals.

$$(C_6Cl_5)_2\bar{C}H \qquad\qquad o\text{-}C_6Cl_5C_6Cl_4\bar{C}HC_6Cl_4C_6Cl_5\text{-}o$$

(CCLIX)   (CCLX)

$$(p\text{-}C_6Cl_5C_6Cl_4)_2\bar{C}H$$

(CLX)

1.10.3   Perchlorotriphenylmethides

The perchlorotriphenylcarbanions are formed in excellent yields, either by proton abstraction by base, or by an electron transfer in ether-DMSO [294, 315].

$$ArAr'Ar''CH \xrightarrow{\;BCR\;} ArAr'Ar''\bar{C} \underset{or\ BCR}{\overset{K\ (or\ Na)}{\longleftarrow}} ArAr'Ar''\dot{C}$$

The corresponding alkali metal salts are soluble in ether giving wine-red solutions (strong absorption maximum at about 510 nm) [315]. These carbanions are remarkably passive towards oxygen, particularly in the presence of DMSO, do not decompose at room temperature, and react slower with iodine than the perchlorodiphenylcarbanions (Section 1.10.2) [315]. This stability is due to shielding of the *alpha* trivalent carbon atom and to resonance stabilization which is greater than in the corresponding radical (Section 1.8.2a).

Although no structural data are, as yet, available, the trivalent carbon is assumed to be flat (sp²-hybridized) [315]. The reason that these carbanions are not as inert as the corresponding radicals

(Section 1.8.2) lies in the inherent chemical affinity arising from their negative charge.

Attempts to isolate the alkaline salts from their solutions have failed as they decompose to the corresponding free radicals (the PTMs). However, Ballester and de la Fuente have recently obtained tetraethylammonium perchlorotriphenylmethide (CCLXI) in the form of a stable, garnet-coloured, crystalline powder [363].

$$(C_6 Cl_5)_3 \overset{-}{C} \overset{+}{N} Et_4$$

(CCLXI)

The stability of the salt (CCLXI) is probably due to the inability of the quaternary ammonium counter-ion to act as an electron-acceptor.

As indicated in the foreword of this chapter, the oxidation of these anions with iodine is an all-important process in the synthesis of the PTM radicals (Section 1.8.2).

$$Ar Ar' Ar'' \overset{-}{C} \xrightarrow{\ I_2\ } Ar Ar' Ar'' \overset{\cdot}{C}$$

Although the perchlorotriphenylcarbanions are converted into the corresponding αH-quasiperchlorotriphenylmethanes (Section 1.9) with aqueous acids, they are stable towards water or ethanol [363, 364], presumably due to shielding. It is relevant that the αH-quasi-perchlorotriphenylmethane (CLXXXIV) does not exchange deuterium in the presence of base [364]. With deuterium chloride in deuterium oxide (CCLXI) gives αD-pentadecachlorotriphenyl-methane [363].

$$\underset{\text{(CLXXXIV)}}{(C_6 Cl_5)_3 CH} \underset{H_3 O^+}{\overset{\text{H}_2\text{O or C}_2\text{CH}_5\text{OH}}{\rightleftharpoons\!\!\!\!\!\times}} \underset{\text{(CLXXXV)}}{(C_6 Cl_5)_3 \overset{-}{C}} \xrightarrow[D_3 O^+]{} \underset{\text{(CCLXII)}}{(C_6 Cl_5)_3 CD}$$

1.10.4   Perchloro-9-phenylfluorenides

These anions are formed by analogy with the perchlorinated diphenylcarbanions (Section 1.10.2) from the 9H-quasiperchloro-9-phenylfluorenes or the perchloro-9-phenylfluorenyl radicals with reagent BCR (Section 1.8.1a) [318]. They also play an important role in the chemistry of these radicals (see Section 1.8.3). They give highly stable (red) solutions. The perchloro-o-biphenylen-(2-biphenylyl)-carbanion, however, decomposes slowly in solution giving probably cyclization products [318]. They behave normally towards aqueous acids or iodine.

$$R = Cl, C_6Cl_5$$

1.10.5   Other Carbanions

The $\alpha H,\alpha'H$-dodecachlorotriptycene (CCXLIII; Section 1.8.4c) cannot yield the corresponding biradical because of its lack of acidity [315]. This is caused by: (1) The complete steric inhibition of resonance in the dicarbanion (CCLXIII), since the atomic orbital of the *alpha* carbon atoms is rigidly fixed at the intersection of the three nodal planes of the benzene π-orbitals and is necessarily prevented from overlap with those π-orbitals; (2) the three perchlorobenzenoid rings cause maximal shielding of the two *alpha* carbon atoms against attacking bases.

(CCLXIII)

1.11   AROMATIC CHLOROCARBON CATIONS

In the preceding chapters aromatic perchlorocarbonium ions have frequently been mentioned as transient intermediates. Those described in this chapter possess sufficient stability to be formed in high concentrations and even to be isolated as salts. They obey Hückel's $(4n+2)$ π-electron rule, when the positive charge resides essentially in the aromatic ring.

1.11.1   Trichlorocyclopropenium Ion

The preparation of salts of the perchlorocyclopropenium ion (CCLXIV) has been effected by West and co-workers from perchlorocyclopropene, by the action of Lewis acids such as aluminium, ferric, gallium or antimony chlorides [354, 365-367].

(CCLXIV)

Perchlorocyclopropenium tetrachloroaluminate is a white solid melting at 157-9° (dec.). Its infrared and Raman spectra have been reported [366]. The carbonium ion (CCLXIV) is not appreciably bonded to its counter-ion and, therefore, belongs to the D_{3h} symmetry species, i.e., its three carbon atoms are at the corners of an equilateral triangle.

The C–C stretching force constant $(K_{CC} = 6.32$ mdyne/Å$)$ is the highest for any known aromatic species (perchlorobenzene: $K_{CC} = 4.81$) [369]. This is assumed to be due to the existence of weak bent bonds which causes the C–C bonds to be shorter and consequently stronger than in any known aromatic compound. The C–Cl stretching force constant is also much higher than that in perchlorobenzene (2.99 as against 1.9) [113, 368] which is attributed to an increased contribution of the p-electron pairs of the chlorines to the ring π-system.

The stability of the ion (CCLXIV) is also higher than that of any other non-aromatic perchlorocarbonium ion. For example, perchloroallyl tetrachloroaluminate reacts with perchlorocyclopropene to give rapidly and quantitatively the perchlorocyclopropenium salt.

(CCLXIV)

Quenching perchlorocyclopropenium tetrachloroaluminate or hexachloroantimonate with water regenerates perchlorocyclopropene ($\sim$60%) instead of the expected perchlorocyclopropenone (CCLXV).

It seems, therefore, that in spite of the infrared data some sort of weak chlorine bridging between the ions occurs, as in perchloro-diphenylcarbonium tetrachloroaluminate (Section 1.11.3). Nevertheless, on treating this salt with wet methylene chloride, at low temperature, gives the perchloroketone perhaps because of the existence of an equilibrium between chlorine bonded and non-bonded ionic pairs.

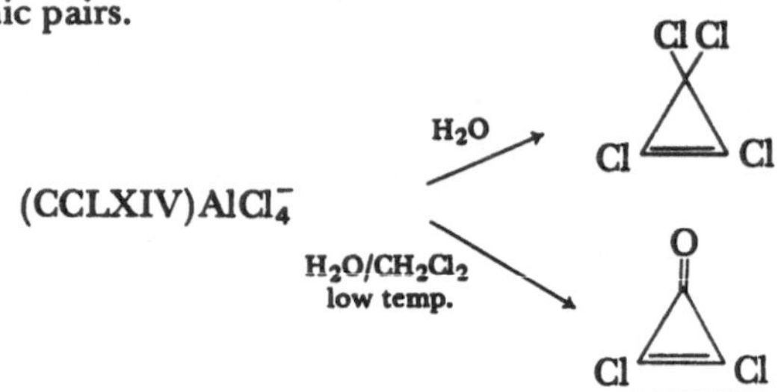

(CCLXIV)AlCl$_4^-$

(CCLXV)

Halogen exchange occurs when the tetrachlorocyclopropene is made to react with boron tribromide to give the tetrabromo-compound, probably involving the cation (CCLXIV) as an intermediate [354]. Exchange with fluorine by means of $SbF_3/SbCl_5$ is similar [354, 370].

Some Friedel–Crafts condensations with cation (CCLXIV) have been studied, including benzenes [371], and phenols and phenol derivatives [272, 273]. Depending upon the reagent and the reaction conditions mono-, di-, or triarylsubstituted derivatives are obtained.

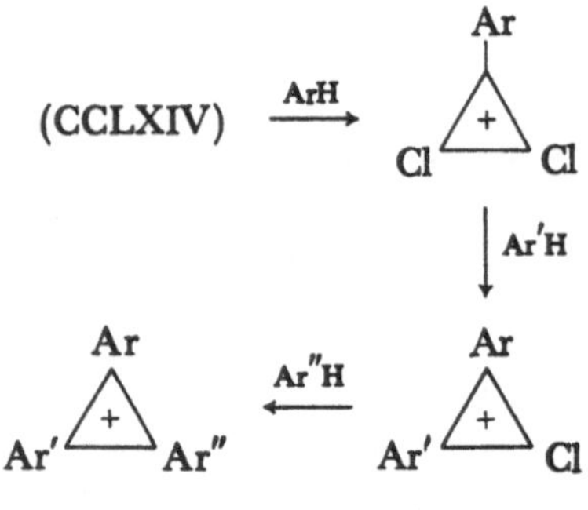

Although tricondensation takes place with phenols, it does not with fluorobenzene. The nucleophilicity of the aromatic reagent rather than the electrophilicity of the cation (CCLXIV) appears to be the major driving force of these Friedel–Crafts condensations.

1.11.2 Perchlorotropylium Ion

Perchlorotropylium (CCLXVII) tetrachloroaluminate has recently been synthesized by West and Kusuda from perchlorobicyclo(3.2.0)-

hepta-2,6-diene by heating between 150-180° with aluminium chloride, followed by treatment with water of an intermediate di-aluminate. The hexachloroantimonate and the pentachlorostannate salts have also been made [280, 354, 367].

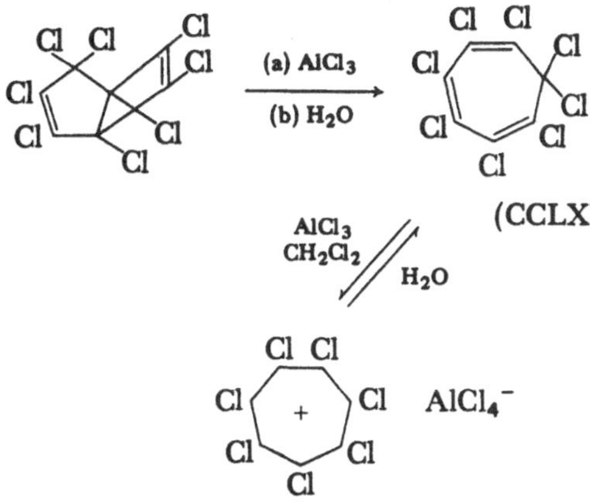

(CCLXVI)

(CCLXVII)

Perchlorotropylium tetrachloroaluminate is a pale-yellow solid of which the infrared spectrum has been reported [280].

In the hydrolysis of (CCLXVI) with concentrated sulphuric acid giving perchlorotropone (CCLXVIII) perchlorotropylium ion is most probably formed [280] as an intermediate.

$$(CCLXVI) \xrightarrow{H_2SO_4} (CCLXVII) \xrightarrow{H_2O}$$

(CCLXVIII)

The tropylium ion reacts with phenols to give stable quinocyclo-heptatrienes or 'quinotropones' [354, 367].

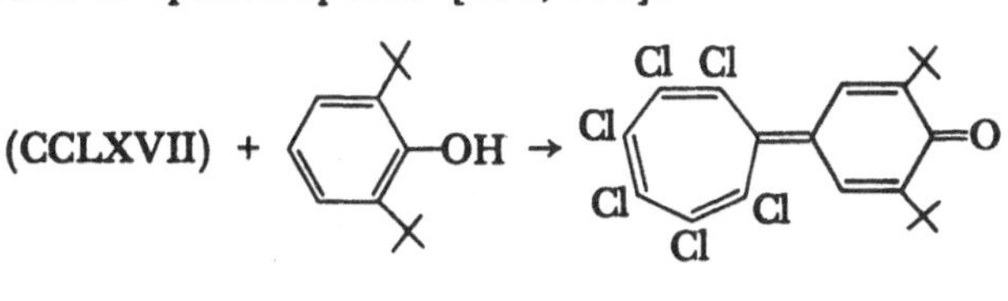

1.11.3   Perchlorodiphenylcarbonium Ions

Perchlorodiphenylcarbonium ions can be formed in three main ways: (1) From the corresponding chlorides (the perchlorodiphenylmethanes or their quinonoid isomers) by strong acids, as in the case of the perchlorocyclopropenium (Section 1.11.1) and the perchloro-cyclotropylium (Section 1.11.2) ions. (2) By oxidation of the corresponding free radicals (the PDMs) in strong acidic media. (3) Electrochemically, from the PDM radicals.

Their incidental formation has been referred to (Sections 1.3.17 to 1.3.22, and 1.8.1a to 1.8.1d) but systematic studies have been carried out on the perchlorodiphenylcarbonium ion (CCLXIX). Ballester and co-workers have recently isolated the perchloro-diphenylcarbonium hexachloroantimonate by the reaction of antimony pentachloride with perchlorodiphenylmethyl (PDM) radical, at room temperature [345], in 94.8% yield. This salt can also be prepared from perchlorodiphenylmethane (XCII; Section 1.3.17) or from the isomeric perchlorobenzylidenecyclohexa-1,4-diene (XCV; Section 1.4.14) [345]. It forms green crystals reacting readily

$$(C_6Cl_5)_2CCl_2$$
$$(XCII)$$
$$\text{or}$$
$$(XCV)$$

$$\xrightarrow{\text{SbCl}_5(\text{CCl}_4)} (C_6Cl_5)_2\overset{+}{C}Cl + SbCl_6^-$$
$$(CCLXIX)$$

$$\Big\uparrow \text{SbCl}_5,\ SO_2Cl_2$$

$$(C_6Cl_5)_2\overset{\cdot}{C}Cl$$
$$(PDM)$$

with moist air or in other media. Its electronic absorption spectrum has been taken [345]. The relative stability of this salt is partly due to the shielding of the *alpha* carbon atom by the four *ortho* chlorines and the two benzene rings, as in PDM radical (Section 1.8.1a). Resonance involving the benzene rings and charge delocalization are additional stabilizing factors. As with the carbanions (Section 1.10.2), the charge is the main reason for its reactivity relative to that of the PDM radical.

Perchlorodiphenylcarbonium tetrachloroaluminate is obtained from the PDM radical with a solution of aluminium chloride in sulphuryl chloride [317]. The oxidizing species is presumably the SO_2Cl^+ present. Interaction of perchlorodiphenylmethane (XCII) or perchlorobenzylidenecyclohexa-1,4-diene (XCV) with aluminium chloride in methylene chloride [317] will also produce it. The tetrachloroaluminate has not been isolated and its chemical behaviour has been studied in solution.

The hexachloroantimonate of (CCLXIX) in methylene chloride reacts with water at room temperature to give a mixture of

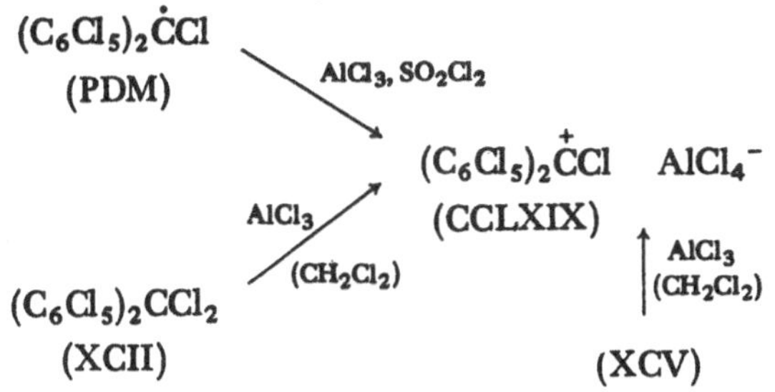

$$(C_6Cl_5)_2\overset{\cdot}{C}Cl$$
(PDM) $\searrow$ $AlCl_3, SO_2Cl_2$

$$(C_6Cl_5)_2\overset{+}{C}Cl \quad AlCl_4^-$$
$AlCl_3$ $\nearrow$ (CCLXIX)
$\diagup$ (CH_2Cl_2) $\uparrow$ $AlCl_3$
                          (CH_2Cl_2)
$$(C_6Cl_5)_2CCl_2$$
(XCII)   (XCV)

perchlorobenzophenone and the perchlorodienone (CCLXX), in 37 and 46%, respectively [345]. Shielding of the *alpha* carbon mentioned above prevents it from being attacked.

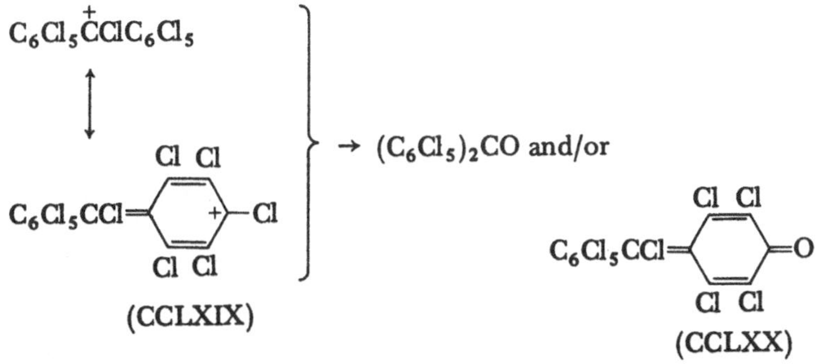

$$C_6Cl_5\overset{+}{C}ClC_6Cl_5$$

$\downarrow$

$$C_6Cl_5CCl = \text{ring} + Cl$$ $\}$ $\rightarrow (C_6Cl_5)_2CO$ and/or

(CCLXIX)

$$C_6Cl_5CCl = \text{ring} = O$$

(CCLXX)

The tetrachloroaluminate of (CCLXIX), on hydrolysis, behaves differently. At low (−12°) temperature, the chlorocarbons (XCII) and (XCV) are formed [317] while at room temperature and above it (XCV) is the sole product. These results indicate that there are probably two types of chlorine-bridged ion pairs. Polarographic measurements have confirmed this assumption. While the hexachloroantimonate is a true salt (reduction potential independent of temperature) the tetrachloroaluminate is not (strong temperature dependence) [343]. The remarkable reactions of cation (CCLXIX)

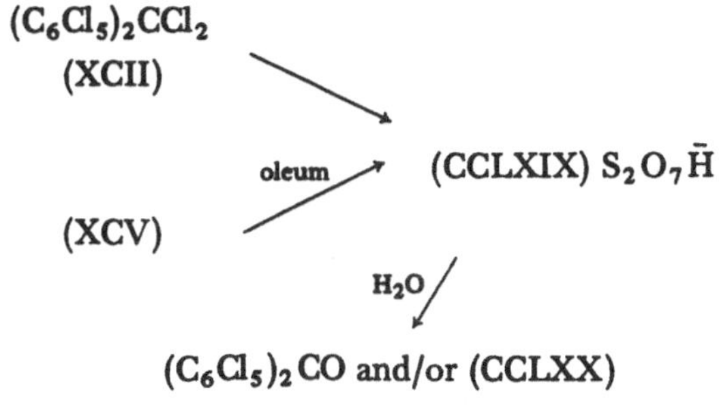

$$(C_6Cl_5)_2CCl_2$$
(XCII) $\searrow$

$oleum$ $\nearrow$ (CCLXIX) $S_2O_7\bar{H}$

(XCV) $\nearrow$

$H_2O \diagup$

$$(C_6Cl_5)_2CO \text{ and/or (CCLXX)}$$

with cycloheptatriene and with 9,10-diphenylanthracene are described in Section 1.11.6.

Solutions of (CCLXIX)-pyrosulphate are obtained when chlorocarbons (XCII) or (XCV) are treated with oleum (see Sections 1.3.17 and 1.4.14). Addition of water yields mixtures of perchlorobenzophenone and perchloroketone (CCLXX), or the former only, depending upon the reaction conditions.

1.11.4   Perchlorotriphenylcarbonium Ions

These ions are usually formed by oxidation of the corresponding perchlorotriphenylmethyl radicals (the PTMs; Section 1.8.2) in the presence of Lewis acids, or in oleum, forming blue-green or blue solutions due to an intense absorption maximum around 700 nm (see Sections 1.8.2a to 1.8.2d).

$$ArAr'Ar''\overset{\cdot}{C} \quad \xrightarrow{\;-e\;} \quad ArAr'Ar''\overset{+}{C}$$

The chemistry of these cations is best exemplified by the perchlorotriphenylcarbonium (CXC); tetrachloroaluminate is prepared by treating perchlorotriphenylmethyl (PTM) radical (Section 1.8.2a) with aluminium chloride in sulphuryl chloride [238, 317]. Addition of water at $-15°$ gives a high yield of perchlorodiphenylmethylenecyclohexa-1,4-diene (Section 1.4.17). This indicates the existence of a chlorine bridge between the counter ions, as in perchlorodiphenylcarbonium tetrachloroaluminate (Section 1.11.3). Because of the much greater shielding of the *alpha* carbon, the tetrachloroaluminate ion is probably bonded to the *para* position.

$$(C_6Cl_5)_3\overset{\cdot}{C} \quad \xrightarrow{AlCl_3/SO_2Cl_2} \quad (C_6Cl_5)_2C=$$
(PTM)

$$\downarrow H_2O$$

$$(C_6Cl_5)_2C=$$

(CXXXIX)

The hexachloroantimonate (CLXXXVI) has been prepared by Ballester and co-workers from the PTM radical (97% yield) or from perchlorodiphenylmethylenecyclohexa-1,4-diene (CXXXIX; ~75%)

[345, 346]. As a solvent, sulphuryl chloride is more advantageous than carbon tetrachloride. It is a crystalline, deep-green salt, decomposing at about 200°. Its infrared and ultraviolet spectra (λ, 690 nm) have been taken [345, 346]. Although stable in water, the

$$(C_6Cl_5)_3\overset{\cdot}{C} \xrightarrow[(-e)]{SbCl_5} (C_6Cl_5)_3\overset{+}{C}\ SbCl_6^- \xleftarrow{SbCl_5} \text{(CXXXIX)}$$
$$\text{(PTM)} \qquad\qquad \text{(CLXXXVI)}$$

hexachloroantimonate salt is readily converted with wet methylene chloride into perchloro-α,α-diphenyl-p-quinomethane (CXL; Section 1.4.17) [345]. The failure to give the carbinol is undoubtedly due to

$$(C_6Cl_5)_3COH$$

$$(C_6Cl_5)_3\overset{+}{C}\ SbCl_6^-$$
$$\text{(CLXXXVI)}$$

$$(C_6Cl_5)_2C=C\!\!\!\overset{CCl=CCl}{\underset{CCl=CCl}{\diagdown}}\!\!\!C=O$$

$$\text{(CXL)}$$

steric shielding of the central carbon atom. The hexachloro-antimonate salt reacts vigorously with reducing agents—even solid potassium bromide or potassium iodide—to give an almost quantitative yield of PTM radical [345, 346]. Action of methanol or ethanol results in the immediate formation of the corresponding 3-alkoxy-diene (CCLXXI) in 95 and 97% yield, respectively [346]. This reaction proceeds further in chloroform giving the 3,3-dialkoxy compound (CCLXXII) (78% yield). (CCLXXI) decomposes slowly at room temperature (rapidly at 100°), giving the perchloroketone (CXL) quantitatively.

$$(C_6Cl_5)_3C\cdot \xleftarrow{+e} (C_6Cl_5)_3\overset{+}{C} \xrightarrow{ROH} (C_6Cl_5)_2C=\!\!\!\!\!\overset{Cl\ Cl}{\underset{Cl\ Cl}{\diagup\!\!\!\diagup}}\!\!\!\overset{Cl}{\underset{OR}{\diagdown}}$$
$$\text{(PTM)} \qquad\qquad \text{(CXC)} \qquad\qquad\qquad \text{(CCLXXI)}$$

$$(C_6Cl_5)_2C=\!\!\!\!\!\overset{Cl\ Cl}{\underset{Cl\ Cl}{\diagup\!\!\!\diagup}}\!\!\!=NH \quad (C_6Cl_5)_2C=\!\!\!\!\!\overset{Cl\ Cl}{\underset{Cl\ Cl}{\diagup\!\!\!\diagup}}\!\!\!=O \quad (C_6Cl_5)_2C=\!\!\!\!\!\overset{Cl\ Cl}{\underset{Cl\ Cl}{\diagup\!\!\!\diagup}}\!\!\!\overset{OR}{\underset{OR}{\diagdown}}$$
$$\text{(CCLXXIII)} \qquad\qquad\quad \text{(CXL)} \qquad\qquad\quad \text{(CCLXXII)}$$

Ammonia in methylene chloride reacts with (CXC)-hexachloro-antimonate, giving about a 70% yield of the 2,5-dienimine (CCLXXIII). This product is reduced to the 4-aminotetradecachloro-triphenylmethyl radical with diethyl hydrophosphonate (Section 1.8.2m).

1.11.5   The Perchloro-9-phenylfluorenyl Cation

As with the corresponding radicals (see Sections 1.8.2a and 1.8.3a) the green perchloro-9-phenylfluorenyl cation (CV) is less shielded at the α-carbon atom (9-carbon) than the perchlorotriphenylcarbonium ion. Consequently, the perchloro-9-phenylfluorenyl tetrachloro-aluminate, synthesized from its (PPF) radical (Section 1.8.3a) by the reaction with aluminium chloride and sulphuryl chloride (Sections 1.3.22 and 1.8.3a) reacts with water yielding perchloro-9-phenyl-fluorene (CII), indicating the existence of a chlorine bridge between its counter-ions.

Nevertheless, the perchloro-9-phenylfluorenyl pyrosulphate generated from (CII) with oleum gives the ketones (CIII) and (CIV) (see Section 1.3.22).

(CV)

The fact that no 9-hydroxytridecachloro-9-phenylfluorene (CVI; Section 1.3.22) is formed indicates either the absence of oxygen bridging between cation (CV) and pyrosulphate ion or irreversible solvolysis.

1.11.6   Hydride Shifts

The reaction of perchlorotriphenylcarbonium hexachloroantimonate (CLXXXVI; Section 1.11.4) with cycloheptatriene (HD) is remarkable. It occurs rapidly at room temperature, with or without solvent, giving an almost quantitative yield of perchlorotriphenylmethyl (PTM) radical (Section 1.8.2a), and a substantial amount of tropylium hexachloroantimonate [345, 346] as set out:

$$4 \, (C_6Cl_5)_3\overset{+}{C} + SbCl_6^- + 3 \, HD = 4 \, (C_6Cl_5)_3\overset{\cdot}{C} + SbCl_3 + 3 \, \overset{+}{D} + 3 \, HCl$$

$$\text{(CLXXXVI)} \qquad\qquad\qquad \text{(PTM)}$$

$\overset{+}{D}$ is the tropylium ion.

It is suggested that some hydride shifts are two-step processes involving first, an electron transfer and second, a hydrogen atom shift.

$$(C_6Cl_5)_2\overset{+}{C}Cl + C_7H_8 \xrightarrow{\text{step 1}} (C_6Cl_5)_2\overset{\cdot}{C}Cl \quad C_7H_8^{+\cdot}$$

$$\text{(CCLXIX)} \qquad\qquad\qquad\qquad \text{(PDM)}$$

$$\text{step 2} \Big/$$

$$(C_6Cl_5)_2CHCl + C_7H_7^+$$

$$\text{(CLIX)}$$

It is reasonable to assume that the two steps occur while the constituents of the original reaction components are still in contact. Moderate shielding of the *alpha* carbon atom as in perchloro-diphenylcarbonium (CCLXIX; Section 1.11.3) hexachloro-antimonate, although ineffectual in the electron-transfer (step 1), may hinder the hydrogen shift of step 2 and thus allow the radical-cation $C_7H_8^{+\cdot}$ to escape before releasing its hydrogen atom to the radical. This is borne out by the excellent yields of (CLIX) and tropylium hexachloroantimonate and the detection of the PDM radical (Section 1.8.1a) (epr technique). In the case of the perchlorotriphenylcarbonium ion (CXC; Section 1.11.4), the steric hindrance in step 2 is very effective and, consequently, PTM radical is the product.

The readiness with which those two carbonium ions accept electrons has been proved both chemically [345, 346] and by voltametry [343]. It is noteworthy that 9,10-diphenylanthracene (a purely aromatic hydrocarbon with no hydride-releasing capacity) reacts instantaneously with (CCLXIX) as well as with (CXC) giving a quantitative yield of PDM and PTM radicals, respectively, and 9,10-diphenylanthracene radical-cation hexachloroantimonate [345, 346].

This result evidently corresponds with the electron-transfer process of step 1.

$$
\left.\begin{array}{c}
(C_6Cl_5)_2\overset{+}{C}Cl \\
(CCLXIX) \\
\text{or} \\
(C_6Cl_5)_3\overset{+}{C} \\
(CXC)
\end{array}\right\} + 9,10\text{-diphenylanthracene} \longrightarrow
$$

$$
\left\{\begin{array}{c}
(C_6Cl_5)_2\overset{\cdot}{C}Cl \\
(PDM) \\
\text{or} \\
(C_6Cl_5)_3\overset{\cdot}{C} \\
(PTM)
\end{array}\right\} [9,10\text{-diphenylanthracene}]^{+\cdot}
$$

In the reaction of (CCLXIX), hexachloroantimonate with some alkylaromatic hydrocarbons, such as toluene, PDM and $\alpha H,\alpha H$-decachlorodiphenylmethane (CLVII; Section 1.7.5) are formed. The presence of these compounds is interpreted as follows:

$$
(C_6Cl_5)_2\overset{+}{C}Cl + PhMe \left\langle\begin{array}{l}
(C_6Cl_5)_2CHCl \\
\\
(C_6Cl_5)_2\overset{\cdot}{C}Cl + PhMe^{+\cdot}
\end{array}\right.
$$

$$PhMe^{+\cdot} \, SbCl_6^- \longrightarrow Ph\overset{\cdot}{C}H_2 + SbCl_5 + HCl$$

$$(C_6Cl_5)_2CHCl + SbCl_5 \longrightarrow (C_6Cl_5)_2\overset{+}{C}H + SbCl_6^-$$

$$(C_6Cl_5)_2\overset{+}{C}H + PhMe \longrightarrow (C_6Cl_5)_2CH_2 + Ph\overset{+}{C}H_2$$

1.12   INFRARED SPECTRA OF AROMATIC CHLOROCARBONS†

Fairly detailed theoretical studies of the vibrational spectra of polychlorobenzenes have been carried out by Scherer during the last decade [113, 139, 145-149]. Other work, mostly experimental, has been performed by Ballester and co-workers (infrared) [108, 296, 302, 374], Delorme et al. (infrared) [114, 128], Plyler et al. (infrared and far infrared) [137], Parodi (far infrared) [135], Darmon et al. (far infrared) [141], Young et al. (infrared) [136], Saeki (infrared and Raman) [138], Truchet and Chapron (Raman) [151], Dadieu et al. (Raman) [150], Ziegler (Raman) [153], Sirkar

---

† The references concerning individual chlorobenzenes are given in the description of their chemistry.

and Bishui (Raman) [154], Nonnenmacher and Mecke [144], and Murray *et al.* [142, 152]. Ballester *et al.* have established some general semi-empirical correlations for chlorocarbons [234].

1.13 ELECTRONIC SPECTRA OF AROMATIC CHLOROCARBONS

Electronic (ultraviolet-visible) absorption spectra have contributed substantially to the chemical progress of aromatic and alkaromatic chlorocarbons. They are still regarded as an important tool for structure elucidation in contrast to the decreasing importance of this spectrometry in general organic chemistry. The exceptional usefulness of ultraviolet spectroscopy in this field is due to three main factors: (1) the lack of degradability of the chlorocarbons; (2) the moderate electronic effects of the chlorine as a substituent, and (3) the almost constantly present steric effects—molecular distortion and steric inhibition—due to repulsions among chlorines.

Ballester and co-workers have given a list of migration moments for a number of substituents useful for calculating the absorptivity of substituted benzenes [133, 316, 376-378]. References to ultraviolet spectra are quoted in the sections dealing with the individual compounds.

1.14 ELECTRON PARAMAGNETIC RESONANCE SPECTRUM OF AROMATIC CHLOROCARBON RADICALS†

The purpose of the present chapter is to present concisely the epr data obtained from the radicals in solution described in Sections 1.7 and 1.8.

This chapter is divided into four sections corresponding with four main types of radicals. They are presented in the order of increasing epr complexity: (1) perchlorotriphenylmethyls; (2) perchlorodiphenylmethyls; (3) perchlorobenzyls, and (4) perchloro-9-phenylfluorenyls.

1.14.1 The Perchlorotriphenylmethyl Radicals

Like most trivalent carbon free radicals, the perchlorotriphenylmethyls present in the epr spectrum, in the absence of spin-active nuclei other than chlorine, a single peak with a Landé's factor g near that of the free electron (2.0025) [315]. No hyperfine splittings due to the chlorines are detected, even in solution. A line broadening traced to them is, however, observed.

Three comparatively weak satellite pairs, symmetrically arranged with respect to the main peak (M), are also found. The distal pair, with the highest coupling constant (about 85 MHz), is the weakest

† The references concerning individual free radicals are given in the corresponding section of Sections 1.7 and 1.8.

TABLE 1.3 Epr data on $p\text{-}R\text{–}C_6Cl_4\text{–}\overset{\cdot}{C}\text{–}C_6Cl_4\text{–}R'\text{-}p$ with $C_6Cl_4\text{–}R''\text{-}p$

R	R'	R''	g-value (± 0.0003)	Number of lines	H Aromatic	α	¹³C Aromatic
H	H	H	2.0023	4	5.4	83.7	29, 36
H	Cl	Cl	2.0027	2	5.3	83.2	28.5, 35.5
Cl	Cl	Cl	2.0026	1	—	82.5	30, 35
Br	Cl	Cl	2.0033	1	—	81.8	25
I	Cl	Cl	2.0038	1	—	80.9	25, 34
HO	Cl	Cl	—	1	—	80.9	30, 34
MeO	Cl	Cl	2.0027	1	—	82.5	30, 35
Me·CO₂−	Cl	Cl	2.0028	1	—	82.8	29.5, 35
NH₂	Cl	Cl	2.0027	1	—	78.4	31
NO₂	Cl	Cl	2.0027	1	—	82.8	30, 34.5
Me	Cl	Cl	2.0027	4	6.0	82.9	29
CO₂H	Cl	Cl	2.0027	1	—	82.3	28.5, 34.5
CO₂NH₄	Cl	Cl	2.0027	1	—	83.2	28.5, 34.5
CO₂Me	Cl	Cl	2.0024	1	—	82.5	29, 35
C₆Cl₅	Cl	Cl	2.0025	1	—	82.5	29.5, 34.0
(C₆Cl₅)₂CH	Cl	Cl	2.0028	2	4.9	82.9	28, 35
(C₆Cl₅)₂ĊC₆Cl₄-p	Cl	Cl	2.0028	1	—	81.5	26.5, 37
(C₆Cl₅)₂Ċ–C₆Cl₄ C≡C–	Cl	Cl	2.0026	1	—	—	27, 34
(C₆Cl₅)₂ĊC₆Cl₄·CCl=CCl– (trans)	Cl	Cl	2.0028	1	—	—	14, 39
MeO	MeO	Cl	2.0025	1	—	81.5	28.5, 35
Me	Me	Cl	2.0024	7	6.0	82.9	—
C₆Cl₅	C₆Cl₅	Cl	2.0027	1	—	83.5	29.5, 34.5
MeO	MeO	MeO	2.0025	1	—	82.1	29, 35.5
MeCO₂	MeCO₂	MeCO₂	2.0025	1	—	82.6	28, 35.5
Me	Me	Me	2.0023	10	5.7	83.3	—
CO₂H	CO₂H	CO₂H	2.0023	1	—	83.6	29, 36
CO₂Na	CO₂Na	CO₂Na	2.0023	1	—	80.8	29, 35.5
CO₂Me	CO₂Me	CO₂Me	2.0023	1	—	84.5	29, 36
C₆Cl₅	C₆Cl₅	C₆Cl₅	2.0026	1	—	83.5	28, 34.5

one, and since the maximum spin density resides necessarily in the *alpha* carbon it is attributed to the *alpha* ^{13}Cs. The remaining proximal pairs P_1 and P_2 have coupling constants of about 30 and 35 MHz respectively and are, therefore, due to the aromatic ^{13}Cs [315].

If the spin densities calculated by Falle *et al.* [341] and by Olivella and Ballester [360] are quantitatively correct, P_2 and P_1 should correspond to the *para* and the *ortho* ^{13}Cs, respectively.† Furthermore, the strong influence of substitution of the *para* chlorines in PTM by other atoms devoid of nuclear spin upon the line-widths shows the existence of a very high spin density on the *para* carbon: while the line width for PTM is 4.3 MHz, the average values (±0.5 MHz) for 4-mono-, 4,4'-di-, and 4,4',4"-trisubstituted PTMs are 4.1, 3.3 and 2.2 MHz, respectively.

Table 1.3 lists some epr data of radicals in the PTM series with various substituents in the *para* positions. Some of them have strongly interacting protons. It is noted that the coupling constant for a *para* hydrogen (5.3 MHz) is close to that of a *para* methyl group (6.0). This fact is regarded as being due to hyperconjugation of the methyl group.

The *para* bromo and *para* iodo radicals have a Landé's factor somewhat higher than that of PTM (2.0033 and 2.0038). This is probably due to the greater spin-orbit coupling in *these* halogens also observed in the PDMs (1.14.2) and the PBs (1.14.3).

1.14.2   The Perchlorodiphenylmethyl Radicals

In these radicals a chlorine is directly attached to the *alpha* carbon where most of the spin density resides [358]. It is reasonable to assume that in spite of the low magnetic moment of chlorine the hyperfine splittings are in this case noticeable. Accordingly, these radicals present four lines of equal intensity in the epr spectrum [315]. The relevant coupling constant is 5.9−7.0 MHz (Table 1.4). The couplings with the *alpha* and the aromatic ^{13}Cs are easily detected and measured. Their values are 100-112 and 31.0−35.5 MHz, respectively. The *alpha* ^{13}C coupling is greater than in the PTMs because of the higher, less delocalized spin density at that site [315]. As with the PTMs the values of the coupling constant do not change significantly with the substituents in the *para* position (Table 1.4). The complexity of the epr spectrum precludes in some cases precise measurement of the ^{13}C splittings.

The greater (∼2.005) g values of the PDMs, compared with the PTMs, are due to spin-orbit coupling in chlorine [315].

---

† An alternative possibility is that P_2 corresponds with the bridgehead ^{13}C. This viewpoint was held by Falle *et al.* [341] and initially by Ballester and co-workers [315].

TABLE 1.4 Epr data on

$$Cl-\overset{R''\ Cl}{\underset{\underset{C_6Cl_4R\text{-}p}{Cl\ Cl}}{\dot{C}}}R'$$

| R | R' | R" | g-value (±0.0003) | Number of lines | Coupling constants | | ^{13}C | |
					α-Cl	Aromatic H	α	Aromatic
Cl	Cl	Cl	2.0055	4	6.1	—	102.5	36.5
C$_6$Cl$_5$	Cl	Cl	2.0055	4	6.2	—	104	32
Cl	Cl	C$_6$Cl$_5$	2.0055	4	7.0	—	104	34
MeO	MeO	Cl	2.0051	4	5.9	—	101	33.5
Me	Me	Cl	—	10	6.1	6.1	—	—
C$_6$Cl$_5$	C$_6$Cl$_5$	Cl	2.0050	4	6.1	—	100	31

1.14.3   The Perchlorobenzyl Radicals

These radicals possess two chlorines attached to the *alpha* carbon atom and accordingly, $2nI + 1 = 7$ lines are observed. Low spin delocalization due to the presence of only one benzene ring and the high angle of tilting θ with respect to the CCl_2 causes an exceptionally high value (166 MHz) of the coupling constant with the *alpha* [13]C. The coupling with the chlorines bonded to that carbon is, consequently, significantly higher (8.4 MHz) than that of the PDM radicals (6.1 MHz) dealt with in the preceding Section.

The epr spectra of the 4*H*-hexachlorobenzyl radical has helped much to elucidate the structure of the perchlorobenzyl (PB) radical.

TABLE 1.5   Epr data on $p\text{-}RC_6Cl_4\text{--}\overset{\bullet}{C}\diagdown^{Cl}_{Cl}$

R	g-value (±0.0003)	Number of lines	Coupling constants			
			α-Cl	Aromatic H	[13]C	
					α	Aromatic
H	2.0071	7	8.3	~0	166	–
Cl	2.0072	7	8.4	–	166	–

Table 1.5 contains some epr data on these two benzyl radicals. Notice the high value of Lande's factor (g, 2.007) due to considerable spin-orbit coupling (cf. preceding Section).

1.14.4   The Perchloro-9-phenylfluorenyl Radicals

The epr spectrum of this radical consists of a line about three times as broad (14.1 MHz) as that of PTM (Table 1.3). The Landé's factor (2.0043) lies between those of the PDM and the PTM radicals.

Only the *alpha* [13]C pair is visible since those peaks corresponding to the aromatic ^{13}C are swamped into the tails of the main line. The coupling constant is 58 MHz, considerably lower than that of PTM, indicating a greater spin delocalization due to the almost planar conformation of the fluorene system. Accordingly, there is much broadening of the lines because the couplings with the chlorines are significantly greater.

These characteristics are also found in other fluorenyl radicals such as perchloro-9-(*o*-biphenylyl)-fluorenyl (g, 2.0046; width, 12.9 MHz) and perchloro-9-(*p*-biphenylyl)fluorenyl (2.0038; 12.9).

REFERENCES

1. *Handbook for Chemical Society Authors*, Special publication no. 14 (1960), The Chemical Society, London.
2. M. JULIN, *Ann. Phil*, 1, 216 (1821).
3. M. FARADAY, *Phil. Trans.*, 111, 47 (1820).
4. M. FARADAY, *Ann. Chim. Phys.*, [2] 18, 48 (1821).
5. R. PHILLIPS and M. FARADAY, *Phil. Trans.*, 111, 392 (1821).
6. R. PHILLIPS and M. FARADAY, *Ann. Phil*, 1, 217 (1821).
7. H. MÜLLER, *J. Chem. Soc.*, 15, 41 (1862).
8. H. MÜLLER, *Z. Chem.*, 7, 40 (1864).
9. M. BALLESTER, C. MOLINET, and J. CASTAÑER, *J. Amer. Chem. Soc.*, 82, 4254 (1960).
10. L. F. FIESER and M. FIESER, *Reagents for Organic Synthesis*, Wiley, 1967, p. 1131.
11. O. SILBERRAD, *J. Chem. Soc.*, 121, 1015 (1922).
12. H. E. DOORENBOS, U.S. Govt. Res. Develop. Rep. 70 (10), 55-6 (1970) A.D. 702840 CFSTI (1969). *Chem. Abstr.*, 73, 88515 n (1970).
13. H. E. DOORENBOS, J. C. EVANS, and R. O. KAGEL, *J. Phys. Chem.*, 74, 3385 (1970).
14. M. BALLESTER, J. FERRER, and J. CASTAÑER, unpublished.
15. M. BALLESTER and C. MOLINET, *Chem. Ind. (London)*, 1954, 1290.
16. F. BEILSTEIN and A. KUHLBERG, *Ann. Chem.*, 150, 286 (1869).
17. F. BEILSTEIN and A. KUHLBERG, *Z. Chem.*, 1869, 75.
18. H. MÜLLER, *Z. Chem.*, 5, 99 (1862).
19. E. JUNGFLEISCH, *Ann. Chem.*, [4] 15, 186 (1868).
20. A. G. PAGE, *Ann. Chem.*, 225, 196 (1884).
21. V. THOMAS, *C. R. Acad. Sci., Paris*, 126, 1211 (1898).
22. A. MOUNEYRAT and C. POURET, *C. R. Acad. Sci., Paris*, 127, 1025 (1898).
23. E. CAMPAIGNE and W. THOMPSON, *J. Amer. Chem. Soc.*, 72, 629 (1950).
24. E. CAMPAIGNE and J. R. LEAL, *J. Amer. Chem. Soc.*, 75, 230 (1953).
25. G. W. WHELAND, *Advanced Organic Chemistry*, 3rd ed., 1960, Wiley, pp. 5-8.
26. J. C. TATLOW, *Endeavour*, 22, 89 (1963).
27. M. BALLESTER, *Arbor*, 70, 161 (1968).
28. M. PÉTRICOU, *Bull. Soc. Chim. Fr.*, 3, 189 (1890).
29. F. KRAFFT and V. MERZ, *Chem. Ber.*, 8, 1296 (1876).
30. F. FICHTER and L. GLANTZSTEIN, *Chem. Ber.*, 49, 2473 (1916).
31. H. BRINTZINGER and H. ORTH, *Monatsh. Chem.*, 85, 1015 (1954).
32. M. BERTHELOT and E. JUNGFLEISCH, *Bull. Soc. Chim. Fr.*, [2] 9, 445 (1886).
33. J. B. COHEN and P. HARTLEY, *J. Chem. Soc.*, 87, 1360 (1905).
34. E. GEBAUER-FÜLNEGG and H. FIGDOR, *Monatsh. Chem.*, 48, 627 (1927).
35. E. H. HUNTRESS and F. H. CARTEN, *J. Amer. Chem. Soc.*, 62, 511 (1940).
36. M. KULKA, *J. Org. Chem.*, 24, 235 (1959).
37. T. VAN DER LINDEN, *Rec. Trav. Chim. Pays-Bas*, 55, 315 (1936).
38. T. VAN DER LINDEN, *Rec. Trav. Chim. Pays-Bas*, 55, 421 (1936).
39. T. VAN DER LINDEN, *Rec. Trav. Chim. Pays-Bas*, 55, 569 (1936).
40. R. RIEMSCHNEIDER and R. OSWALD, *Monatsch. Chem.*, 85, 972 (1954).
41. A. LADENBURG, *Ann. Chem.*, 172, 331 (1874).

42. R. RIEMSCHNEIDER, *Chem. Ber.*, 91, 2605 (1958).
43. G. RUOFF, *Chem. Ber.*, 9, 1483 (1876).
44. P. SCHUTZENBERGER, *Bull. Soc. Chim. Fr.*, 4, 102 (1865).
45. W. SANDERMANN, H. STOCKMANN, and R. CASTEN, *Chem. Ber.*, 90, 690 (1957).
46. V. MERZ and W. WEITH, *Chem. Ber.*, 5, 458 (1872).
47. C. GRAEBE, *Ann. Chem.*, 146, 1 (1868).
48. J. POLLAK, E. GEBAUER-FÜLNEGG, and E. BLUMENSTOCK, *Monatsch. Chem.*, 46, 499 (1926).
49. J. POLLAK and E. GEBAUER-FÜLNEGG, *Monatsh. Chem.*, 47, 537 (1927).
50. O. LITVAY, E. RIESZ, and L. LANDAU, *Chem. Ber.*, 62, 1863 (1929).
51. A. LAUBENHEIMER, *Chem. Ber.*, 7, 1765 (1874).
52. F. BEILSTEIN and A. KURBATOW, *Ann. Chem.*, 182, 94 (1876).
53. H. E. FIERZ-DAVID and F. R. STAHELIN, *Helv. Chim. Acta.*, 20, 1458 (1937).
54. T. VAN DER LINDEN, *Rec. Trav. Chim. Pays-Bas*, 57, 342 (1938).
55. M. BATTEGAY and L. DENIVELLE, *Bull. Soc. Chim. Fr.*, [4] 47, 606 (1930).
56. A. T. PETERS, F. M. ROWE, and D. M. STEAD, *J. Chem. Soc.*, 1943, 372.
57. H. GILMAN and S. Y. SIM, *J. Organometal. Chem.*, 1967, 249.
58. P. J. MORRIS, F. W. G. FEARON, and H. GILMAN, *J. Organometal. Chem.*, 1967, 427.
59. T. VAN DER LINDEN, *Rec. Trav. Chim. Pays-Bas*, 57, 217 (1938).
60. F. BECKE and L. WÜRTELE, *Chem. Ber.*, 91, 1011 (1958).
61. J. KRAFT, R. THERMET, and L. PARVI, *Chim. Ind. (Paris)*, 83, 557 (1960).
62. E. BARRAL, *Bull. Soc. Chim. Fr.*, [3] 13, 418 (1895).
63. E. BARRAL, *Bull. Soc. Chim. Fr.*, [3] 17, 744 (1897).
64. M. BALLESTER and J. CASTAÑER, unpublished.
65. W. C. SOLOMON, L. A. DEE, and D. W. SCHULTS, *J. Org. Chem.*, 31, 1551 (1966).
66. C. GRAEBE, *Ann. Chem.*, 263, 16 (1890).
67. M. BALLESTER, J. CASTAÑER, and J. ESPINOSA, unpublished.
68. E. BARRAL, *Bull. Soc. Chim. Fr.*, [3] 13, 340 (1895).
69. P. G. HARVEY, F. SMITH, M. STACEY, and J. C. TATLOW, *J. Appl. Chem. (London)*, 4, 319 (1954).
70. E. T. McBEE, H. B. HASS, P. E. WEIMER, W. E. BURT, Z. D. WELCH, R. M. ROBB, and F. SPEYER, *Ind. Eng. Chem.*, 39, 387 (1947).
70a. S. D. ROSS, M. MARKARIAN, and M. NAZZEWSKI, *J. Amer. Chem. Soc.*, 69, 1914 (1947).
71. G. LOCK, *Chem. Ber.*, 66, 1527 (1933).
72. M. BALLESTER, *Mem. Real Acad. Ciencias Artes (Barcelona)*, 29, 271 (1948).
73. V. MERZ and W. WEITH, *Chem. Ber.*, 16, 2869 (1883).
74. P. G. HARVEY, F. SMITH, M. STACEY, and J. C. TATLOW, *J. Appl. Chem. (London)*, 4, 325 (1954).
75. E. T. McBEE, H. B. HASS, G. M. ROTHROCK, J. S. NEWCOMER, W. V. CLIPP, Z. D. WELCH, and C. I. GOCHENOUR, *Ind. Eng. Chem.*, 39, 384 (1947).
76. S. D. ROSS and M. NAZZEWSKI, *J. Amer. Chem. Soc.*, 69, 3146 (1947).
77. M. BALLESTER, *An. Real Soc. Espan. Fis Quim.*, 50B, 765 (1954).
78. W. SCHWEMBERGER and W. GORDON, *J. Gen. Chem. (USSR)*, 2, 921 (1932); *Chem. Zentr.*, 1934 I, 215.

79. W. I. SCHWEMBERGER, *J. Gen. Chem.* (*USSR*), **8**, 1353 (1938); *Chem. Zentr.*, 1939 II, 3690.
80. T. DIEHL, *Chem. Ber.*, **11**, 173 (1878).
81. G. ZETTER, *Chem. Ber.*, **11**, 164 (1878).
82. R. GNEHM and E. BÄNZIGER, *Ann. Chem.*, **296**, 62 (1897).
83. D. E. PEARSON, H. W. POPE, W. W. HARGROVE, and W. E. STAMPER, *J. Org. Chem.*, **23**, 1412 (1958).
84. K. STEINER, *Monatsh. Chem.*, **36**, 825 (1915).
85. A. ECKERT and K. STEINER, *Monatsh. Chem.*, **36**, 175 (1915).
86. A. ECKERT and K. STEINER, *Chem. Ber.*, **47**, 2628 (1914).
87. H. VOLLMAN in Houben-Weyl, *Methoden der Organische Chemie*, 4th ed., 1962, Vol. 3, p. 700.
88. A. ECKERT and K. STEINER, *Monatsh. Chem.*, **36**, 269 (1915).
89. M. N. DVORNIKOFF, D. G. SHEETS, and F. B. ZIENTY, *J. Amer. Chem. Soc.*, **68**, 142 (1946).
90. A. KIRPAL and H. KUNZE, *Chem. Ber.*, **62**, 2102 (1929).
91. M. BALLESTER, J. CASTAÑER, and S. OLIVELLA, unpublished.
92. F. KRAFFT, *Chem. Ber.*, **9**, 1085 (1876).
93. F. KRAFFT, *Chem. Ber.*, **10**, 801 (1877).
94. E. HARTMANN, *Chem. Ber.*, **24**, 1011 (1891).
95. K. H. BUECHEL and A. CONTE, *Z. Naturforsch.*, **21**, 1111 (1966).
96. J. A. KRYNITSKY and H. W. CARHART, *J. Amer. Chem. Soc.*, **71**, 816 (1949).
97. T. VAN DER LINDEN, *Rec. Trav. Chim. Pays-Bas*, **57**, 401 (1938).
98. A. ROEDIG, G. VOSS, and E. KUCHINKE, *Ann. Chem.*, **580**, 24 (1953).
99. A. ROEDIG and K. KIEPERT, *Ann. Chem.*, **593**, 71 (1955).
100. A. ROEDIG, G. MÄRKL, and H. SCHALLER, *Chem. Ber.*, **103**, 1011 (1970).
101. J. A. KRYNITSKY and R. W. BOST, *J. Amer. Chem. Soc.*, **69**, 1918 (1947).
102. E. T. McBEE, W. L. DILLING, and H. P. BRAENDLIN, *J. Org. Chem.*, **28**, 2255 (1963).
103. A. ROEDIG, *Ann. Chem.*, **569**, 161 (1950).
104. F. KUSUDA, R. WEST, and V. N. M. RAO, *J. Amer. Chem. Soc.*, **93**, 3627 (1971).
105. E. T. McBEE, C. W. ROBERTS, and J. D. IDOL, *J. Amer. Chem. Soc.*, **78**, 996 (1956).
106. H. HOLTSCHMIDT, E. DEGENER, H. G. SCHMELZER, H. TARNOW, and W. ZECHER, *Angew. Chem., Int. Ed.*, **7**, 856 (1968).
107. M. BALLESTER, C. MOLINET, and J. ROSA, *An. Real Soc. Espan. Fis. Quim.*, **57B**, 393 (1961).
108. M. BALLESTER and J. CASTAÑER, *An. Real Soc. Espan. Fis. Quim.*, **62B**, 397 (1966).
109. M. P. CAVA, M. J. MITCHELL, D. C. DeJONGH, and R. Y. VAN FOSSEN, *Tetrahedron Lett.*, 1966, 2947.
110. R. F. C. BROWN, D. V. GARDNER, J. F. W. McOMIE, and R. K. SOLLY, *Aust. J. Chem.*, **20**, 139 (1967).
111. C. A. COULSON and D. STOCKER, *Mol. Phys.*, **2**, 397 (1959).
112. C. A. COULSON in *Theoretical Organic Chemistry. IUPAC. Kekulé Symposium*, Butterworths, 1959, p. 62.
113. J. R. SCHERER, *Spectrochim. Acta*, **20**, 345 (1964).
114. P. DELORME, F. DENISSELLE, and V. LORENZELLI, *J. Chem. Phys.*, **64**, 591 (1967).
115. G. GAFNER and F. H. HERBSTEIN, *Acta Crystallogr.*, **13**, 706 (1960).
116. K. LONSDALE, *Proc. Roy. Soc.*, **A133**, 536 (1931).
117. W. G. PLUMMER, *Phil. Mag.*, [6] **50**, 1214 (1925).

118. H. MARK, *Chem. Ber.*, 57, 1820 (1924).
119. A. L. PATTERSON, *Phys. Rev.*, 46, 372 (1934).
120. R. KAISER, *Physik Z.*, 36, 92 (1935).
121. A. TULINSKY and J. G. WHITE, *Acta Crystallogr.*, 11, 7 (1958).
122. I. N. STRELTSOVA and Y. T. STRUCHKOV, *Zh. Strukt. Khim.*, 2, 312 (1961); *Chem. Abstrs.*, 56, 8112c (1962).
123. L. O. BROCKWAY and K. J. PALMER, *J. Amer. Chem. Soc.*, 59, 2181 (1937).
124. O. BASTIANSEN and O. HASSEL, *Acta Chem. Scand.*, 1, 489 (1947).
125. T. SAITO, *Bull. Chem. Soc. Jap.*, 33, 343 (1960).
126. T. J. STRAND and H. L. COX, *J. Chem. Phys.*, 44, 2426 (1966).
127. R. KOPELMAN and O. SCHNEPP, *J. Chem. Phys.*, 30, 597 (1959).
128. P. DELORME, V. LORENZELLI, and M. FOURNIER, *C. R. Acad. Sci., Paris*, 259, 751 (1964).
129. J. DUCHESNE and A. MONFILS, *J. Chem. Phys.*, 22, 562 (1954).
130. R. KOPELMAN, *J. Chem. Phys.*, 30, 868 (1959).
131. H. CONRAD-BILLROTH, *Z. Phys. Chem.*, B19, 76 (1932).
132. M. BALLESTER and J. CASTAÑER, *J. Amer. Chem. Soc.*, 82, 4259 (1960).
133. M. BALLESTER, J. RIERA, and L. SPIALTER, *J. Amer. Chem. Soc.*, 86, 4276 (1964).
134. P. R. HAMMOND, *J. Chem. Soc.*, A 1968, 145.
135. M. PARODI, *C. R. Acad. Sci., Paris*, 206, 337 (1938).
136. C. W. YOUNG, R. B. DuVALL, and N. WRIGHT, *Anal. Chem.*, 23, 709 (1951).
137. E. K. PLYLER, H. C. ALLEN, and E. D. TIDWELL, *J. Res. Nat. Bur. Stand.*, 58, 255 (1957).
138. S. SAEKI, *Bull. Chem. Soc. Jap.*, 35, 322 (1962).
139. J. R. SCHERER and J. C. EVANS, *Spectrochim. Acta*, 19, 1739 (1963).
140. R. D. CHAMBERS, J. HEYES, and W. K. R. MUSGRAVE, *Tetrahedron*, 19, 891 (1963).
141. I. DARMON, C. BROT, G. W. CHANTRY, and H. A. GEBBIE, *Spectrochim. Acta*, 24A, 1517 (1968).
142. J. W. MURRAY, V. DEITZ, and D. H. ANDREWS, *J. Chem. Phys.*, 3, 180 (1935).
143. J. LECOMPTE, *C. R. Acad. Sci., Paris*, 248, 1491 (1959).
144. G. NONNENMACHER and R. MECKE, *Spectrochim. Acta*, 17, 1049 (1961).
145. J. R. SCHERER, *Spectrochim. Acta*, 19, 601 (1963).
146. J. R. SCHERER, *Spectrochim. Acta*, 21, 321 (1965).
147. J. R. SCHERER, *Spectrochim. Acta*, 23A, 1489 (1967).
148. J. R. SCHERER, *Spectrochim. Acta*, 24A, 747 (1968).
149. J. R. SCHERER, *Planar Vibrations of Chlorinated Benzenes*, The Dow Chemical Co., 1963, Midland, Michigan.
150. A. DADIEU, A. PONGRATZ, and K. W. F. KOHLRAUSCH, *Monatsh. Chem.*, 61, 426 (1932).
151. R. TRUCHET and J. CHAPRON, *C. R. Acad. Sci., Paris*, 198, 1934 (1934).
152. J. W. MURRAY and D. H. ANDREWS, *J. Chem. Phys.*, 2, 119 (1934).
153. E. ZIEGLER, *Spectrochim. Acta*, 22, 357 (1966).
154. S. C. SIRKAR and P. K. BISHUI, *Indian J. Phys.*, 1968, 1.
155. T. L. WEATHERLY, E. H. DAVIDSON, and Q. WILLIAMS, *J. Chem. Phys.*, 21, 761 (1953).
156. P. J. BRAY, R. G. BARNES, and R. BERSOHN, *J. Chem. Phys.*, 25, 813 (1956).
157. P. J. BRAY and R. G. BARNES, *J. Chem. Phys.*, 27, 551 (1957).

158. C. B. RICHARDSON, *Acta Crystallogr.*, 16, 1063 (1963).
159. R. W. DIXON and N. BLOEMBERGEN, *J. Chem. Phys.*, 41, 1720 (1964).
160. K. S. KRISHNAN and S. BANERJEE, *Phil. Trans. Roy. Soc., London*, A 234, 265 (1935).
161. M. A. LASHEEN, *Acta Crystallogr.*, A 24, 289 (1968).
162. H. BASSET, *J. Chem. Soc.*, 20, 443 (1867).
163. I. B. JOHNS, E. A. McELHILL, and J. O. SMITH, *Ind. Eng. Chem., Prod. Res. Develop.*, 1, 2 (1962).
164. I. B. JOHNS, E. A. McELHILL, and J. O. SMITH, *J. Chem. Eng. Data.*, 7, 277 (1962).
165. M. ISTRATI, *Bull. Soc. Chim. Fr.*, [3] 3, 184 (1890).
166. S. MEYERSON and E. K. FIELDS, *J. Chem. Soc.*, B 1966, 1001.
167. L. SCHAEFER, *Chem. Commun.*, 1968, 1622.
167a. J. HITZKE, F. PETER, and J. GUION, *Org. Mass Spectrom.*, 6, 349 (1972).
168. W. D. BANCROFT and S. F. WHEARTY, *Proc. Nat. Acad. Sci. U.S.*, 17, 183 (1931).
169. S. F. WHEARTY, *J. Phys. Chem.*, 35, 3121 (1931).
170. L. A. BIGELOW and J. H. PEARSON, *J. Amer. Chem. Soc.*, 56, 2273 (1934).
171. G. M. BROOKE, R. D. CHAMBERS, J. HEYES and W. K. R. MUSGRAVE, *J. Chem. Soc.*, 1964, 729.
172. N. FUKUHARA and L. A. BIGELOW, *J. Amer. Chem. Soc.*, 60, 427 (1938).
173. E. T. McBEE, P. A. WISEMAN, and G. B. BACHMAN, *Ind Eng. Chem.*, 39, 415 (1947).
174. A. J. LEFFLER, *J. Org. Chem.*, 24, 1132 (1959).
175. A. J. LEFFLER, *J. Org. Chem.*, 24, 2074 (1959).
176. K. O. CHRISTE and A. E. PAVLATH, *J. Chem. Soc.*, 1963, 5549.
177. E. T. McBEE, V. V. LINDGREN, and W. B. LIGETT, *Ind. Eng. Chem.*, 39, 379 (1947).
178. R. E. FLORIN, W. J. PLUMMER, and L. A. WALL, *J. Res. Nat. Bur. Stand.*, 62, 107 (1959).
179. P. JOHNCOCK, W. K. R. MUSGRAVE, J. FEENEY, and L. H. SUTCLIFFE, *Chem. Ind. (London)*, 1959, 1314.
180. G. C. FINGER, C. W. KRUSE, R. H. SHILEY, R. H. WHITE, and H. A. WHALEY, Abstracts, Organic Chemistry Division, XVIth. IUPAC Congress, Paris, July 1957, p. 303.
181. J. T. MAYNARD, *J. Org. Chem.*, 28, 112 (1963).
182. G. W. HOLBROOK, L. A. LOREE, and O. R. PIERCE, *J. Org. Chem.*, 31, 1259 (1966).
183. G. FULLER, *J. Chem. Soc.*, 1965, 6264.
184. N. N. VOROZHTSOV, V. E. PLATONOV, and G. G. YAKOBSON, *Izv. Akad. Nauk USSR, Ser. Khim.*, 1963, 1524.
185. J. F. BUNNETT and R. E. ZAHLER, *Chem. Rev.*, 49, 315 (1951).
186. A. L. ROCKLIN, *J. Org. Chem.*, 21, 1478 (1956).
187. P. J. BRAY and R. G. BARNES, *J. Chem. Phys.*, 27, 551 (1957).
188. G. HUETT and S. I. MILLER, *J. Amer. Chem. Soc.*, 83, 408 (1961).
189. M. BERTHELOT, *Bull. Soc. Chim. Fr.*, [2] 9, 17 (1868).
190. M. BERTHELOT and E. JUNGFLEISCH, *Am. Chim. Phys.*, [4] 15, 330 (1868).
191. M. BERTHELOT, *An. Chim. Phys.*, [4] 20, 488 (1870).
192. J. D. BROOKS, P. J. COLLIN, H. SILBERMAN, and G. H. TAYLOR, *Carbon (Oxford)*, 4, 375 (1966).
193. J. GIBSON, M. HOLOHAN, and H. C. RILEY, *J. Chem. Soc.*, 1946, 456.

194. B. A. BOLTO, D. E. WEISS, and D. WILLIS, *Aust. J. Chem.*, 18, 487 (1965).
195. M. BALLESTER and J. RIERA, *An. Real Soc. Espan. Fis. Quim.*, 55B, 785 (1959).
196. H. WICHELHAUS, *Chem. Ber.*, 38, 1725 (1905).
197. H. NORMANT, *C. R. Acad. Sci., Paris*, 239, 1510 (1954).
198. H. E. RAMSDEN, A. E. BALINT, W. R. WHITFORD, J. J. WALBURN, and R. CSERR, *J. Org. Chem.*, 22, 1202 (1957).
199. S. D. ROSENBERG, J. J. WALBURN, and H. E. RAMSDEN, *J. Org. Chem.*, 22, 1606 (1957).
200. J. CHATT and B. L. SHAW, *J. Chem. Soc.*, 1961, 285.
201. F. E. PAULIK, S. I. E. GREEN, and R. E. DESSY, *J. Organometal. Chem.*, 1965, 229.
202. J. R. MOSS and B. L. SHAW, *J. Chem. Soc. (A)*, 1966, 1793.
203. I. HAIDUC and H. GILMAN, *J. Organometal. Chem.*, 1968, 73.
204. M. D. RAUSCH and F. E. TIBBETTS, *J. Organometal. Chem.*, 1970, 487.
205. D. E. PEARSON, D. COWAN, and J. D. BECKLER, *J. Org. Chem.*, 24, 504 (1959).
206. D. E. PEARSON and D. COWAN, *Org. Synth.*, 44, 78 (1964).
207. J. F. DURAND and L. WAI-HSUN, *C. R. Acad. Sci., Paris*, 191, 1460 (1930).
208. M. D. RAUSCH, F. E. TIBBETTS, and H. B. GORDON, *J. Organometal. Chem.*, 1966, 493.
209. S. C. COHEN, D. E. FENTON, A. J. TOMLINSON, and A. G. MASSEY, *J. Organometal. Chem.*, 1966, 301.
210. H. HEANEY and J. M. JABLONSKI, *J. Chem. Soc. (C)*, 1968, 1895.
211. K. SHIINA, T. BRENNAN, and H. GILMAN, *J. Organometal. Chem.*, 1968, 471.
212. M. D. RAUSCH, Y. F. CHANG, and H. B. GORDON, *Inorg. Chem.*, 8, 1355 (1968).
213. A. E. JUKES, S. S. DUA, and H. GILMAN, *J. Organometal. Chem.*, 1970, 791.
214. G. A. MOSER, F. E. TIBBETTS, and M. D. RAUSCH, *Organometal. Chem. Syn.*, 1, 99 (1970/1971).
215. G. A. MOSER, E. D. FISCHER, and M. D. RAUSCH, *J. Organometal. Chem.*, 1971, 379.
216. I. HAIDUC and H. GILMAN, *Rev. Roum. Chim.*, 16, 907 (1971).
217. F. W. G. FEARON and H. GILMAN, *J. Organometal. Chem.*, 1968, 73.
218. H. GILMAN and K. SHIINA, *J. Organometal. Chem.*, 1967, 369.
219. A. E. JUKES, S. S. DUA, and H. GILMAN, *J. Organometal. Chem.*, 1970, 241.
220. K. SHIINA and H. GILMAN, *J. Amer. Chem. Soc.*, 88, 5367 (1966).
221. D. BALLARD, T. BRENNAN, F. W. G. FEARON, K. SHIINA, I. HAIDUC, and H. GILMAN, *Pure Appl. Chem.*, 19, 449 (1969).
222. A. WEBER and C. SÖLLSCHER, *Chem. Ber.*, 16, 882 (1883).
223. A. STEPANOW, *J. Russ. Phys. Chem. Soc.*, 37, 12 (1904); *Chem. Zentr.*, 1905 I, 1273.
224. R. H. CLARK and R. N. CROZIER, *Trans Roy. Soc. Canada*, [3] 19, 153 (1925).
225. A. WEBER and N. WOLF, *Chem. Ber.*, 18, 335 (1885).
226. A. F. HOLLEMAN, *Rec. Trav. Chim. Pays-Bas*, 39, 736 (1920).
227. A. HANTZSCH and K. SCHOLTZE, *Chem. Ber.*, 40, 4875 (1907).
228. N. OHTA and K. KAGAMI, *Repts. Govt. Chem. Ind. Research Inst. Tokyo*, 47, 327 (1952); *Chem. Abstr.*, 48, 9941 (1954).
229. S. IMAI, *Noyaku Seisan Gijutsu*, 1, 17 (1961); *Chem. Abstr.*, 56, 5869a (1962).

192   MANUEL BALLESTER AND SANTIAGO OLIVELLA

230. B. B. SCHAEFFER, M. C. BLAICH, and J. W. CHURCHILL, *J. Org. Chem.*, 19, 1646 (1954).
231. M. BALLESTER, J. RIERA, and L. JULIÁ, unpublished.
232. N. OHTA and K. KAGAMI, *J. Soc. Org. Syn. Chem. (Japan)*, 10, 297 (1952); *Chem. Abstr.*, 47, 5375c (1953).
233. N. N. VOROZHTSOV, G. G. YAKOBSON, and T. D. RUBINA, *Dokl. Akad. Nauk. USSR*, 134, 821 (1960); *Chem. Abstr.*, 55, 6415b (1961).
234. M. BALLESTER, J. CASTAÑER, and J. RIERA, forthcoming publication.
234a. O. HUTZINGER, S. SAFE, and V. ZITKO, *Intern. J. Environ., Anal. Chem.*, 1972, 1.
235. K. ADRIANOV, *Ind. Org. Chem. USSR*, 2, 196 (1936); *Chem. Zentr.*, 1937 I, 2058.
236. J. P. WIBAUT, J. OVERHOFF, and K. GRATAMA, *Rec. Trav. Chim. Pays-Bas*, 59, 298 (1940).
237. F. L. W. Van Roosmalen, *Rec. Trav. Chem. Pays-Bas*, 53, 359 (1934).
238. M. BALLESTER and G. de la FUENTE, unpublished.
239. J. RIERA, Doctoral Thesis, University of Barcelona, 1962.
240. M. BALLESTER, J. PALAU, and J. RIERA, *J. Quant. Spectrosc. Radiat. Transfer*, 4, 819 (1964).
241. S. SAFE and O. HUTZINGER, *J. Chem. Soc., Perkin I*, 1972, 686.
242. F. BINNS and H. SUSCHITZKY, *J. Chem. Soc. (C)*, 1971, 1913.
243. M. BALLESTER, J. RIERA, and M. RIERA-MONTESINOS, unpublished.
244. M. BALLESTER and J. CASTAÑER, unpublished.
245. R. F. C. BROWN, D. V. GARDNER, J. F. W. McOMIE, and R. K. SOLLY, *Chem. Commun.*, 1966, 407.
246. W. SCHWEMBERGER, *J. Gen. Chem. USSR*, 8, 1353 (1938); *Chem. Zentr.*, 1939 II, 3690.
247. J. GOUBEAU, H. LUTHER, K. FIELDMANN, and G. BRANDES, *Chem. Ber.*, 86, 214 (1953).
248. M. BALLESTER, U.S. Govt. Res. Develop. Rept. 40 (4), 14 (1965). A. D. 609569, CFSTI (1964); *Chem. Abstr.*, 63, 16733e (1965).
249. M. BALLESTER, *Bull. Soc. Chim. Fr.*, 1966, 7.
250. M. BALLESTER, J. CASTAÑER, J. RIERA and J. PARÉS, forthcoming publication.
251. A. CLAUS and C. WENZLIK, *Chem. Ber.*, 19, 1165 (1886).
252. A. CLAUS and P. MIELCKE, *Chem. Ber.*, 19, 1182 (1886).
253. J. POLLAK, E. GEBAUER-FÜLNEGG, and E. BLUMENSTOCK-HALWARD, *Montash. Chem.*, 49, 187 (1928).
254. W. SCHWEMBERGER and W. GORDON, *J. Gen. Chem. USSR, Ser. A.*, 4, 695 (1934); *Chem. Zentr.*, 1935 II, 514.
255. G. L. FARRAR and P. W. STORMS, *J. Chem. Eng. Data.*, 13, 248 (1968).
256. A. E. GINSBERG, R. PAATZS, and F. KORTE, *Tetrahedron Lett.*, 1962, 779.
257. W. L. MOSBY, *J. Amer. Chem. Soc.*, 77, 758 (1955).
258. H. LUTHER, G. BRANDES, H. GÜNZLER, and B. HAMPEL, *Z. Elektrochem.*, 59, 1112 (1955).
259. G. GAFNER and F. H. HERBSTEIN, *Acta Crystallogr.*, 13, 702 (1960).
260. G. GAFNER and F. H. HERBSTEIN, *Nature*, 200, 130 (1963).
261. M. BERTHELOT, *Bull. Soc. Chim. Fr.*, [2] 9, 295 (1868).
262. M. BALLESTER, J. CASTAÑER and J. M. GÓMEZ-FRAGA, unpublished.
263. R. GRINBAUM and L. MARCHLEWSKI, *Bull. Acad. Pol. Sci.*, [3] 1937 A, 171 (1937).
264. J. D. BROOKS, P. J. COLLIN and H. S. SILBERMAN, *Aust. J. Chem.*, 19, 2401 (1966).
265. M. ISHIMORI, R. WEST, B. K. TEO, and L. F. DAHL, *J. Amer. Chem. Soc.*, 93, 7101 (1971).

266. H. VOLLMANN, H. BECKER, M. CORELL and H. STREECK, *Ann. Chem.*, 531, 1 (1937).
267. H. REIMLINGER and G. KING, *Chem. Ber.*, 95, 1043 (1962).
268. R. MECKE and W. E. KLEE, *Z. Elektrochem.*, 65, 327 (1961).
269. H. C. BROWN, D. GINTIS, and L. DOMASH, *J. Amer. Chem. Soc.*, 78, 5387 (1953).
270. H. C. BROWN and M. GRAYSON, *J. Amer. Chem. Soc.*, 75, 20 (1953).
271. O. SILBERRAD, *J. Chem. Soc.*, 127, 2677 (1925).
271a. J. SILVERMAN, A. P. KRUKONIS, and N. F. YANNONI, *Cryst. Struct. Commun.*, 2, 37 (1973).
271b. V. MARK and V. A. PATTISON, *Chem. Commun.*, 1971, 553.
272. G. LOCK, *Chem. Ber.*, 72, 300 (1939).
273. S. OLIVELLA, M. BALLESTER, and J. CASTAÑER, forthcoming publications.
274. M. BALLESTER, C. MOLINET, and J. ROSA, *Tetrahedron*, 6, 109 (1959).
275. H. C. BRIMELOW, R. L. JONES, and T. P. METCALFE *J. Chem. Soc.*, 1951, 1208.
276. O. NICODEMUS, *J. Prakt. Chem.*, [2] 83, 312 (1911).
277. S. D. ROSS and M. MARKARIAN, *J. Amer. Chem. Soc.*, 71, 2756 (1949).
278. M. BALLESTER, J. RIERA, and M. RIERA-MONTESINOS, unpublished.
279. E. T. McBEE, J. R. CHIRAKAIKARAN *et al.*, private communication.
280. R. WEST and K. KUSUDA, *J. Amer. Chem. Soc.*, 90, 7354 (1968).
281. M. BALLESTER, J. CASTAÑER, and F. DÍAZ-ALZAMORA, unpublished.
282. E. D. HUGHES, *Trans. Faraday Soc.*, 38, 625 (1941).
283. A. STREITWIESER, *Chem. Rev.*, 56, 622 (1956).
284. C. KIRPAL, A. GALUSCHKA, and E. LASSAC, *Chem. Ber.*, 68, 1330 (1935).
285. P. KARRER, W. WEHRLI, E. BIEDERMANN, and M. DALLA VEDOVA, *Chem. Zentr.*, 1928 I, 1394.
286. L. V. JOHNSON, F. SMITH, M. STACEY, and J. C. TATLOW, *J. Chem. Soc.*, 1952, 4710.
287. M. BALLESTER and J. ROSA, *An. Real Soc. Espan. Fis. Quim.*, 56B, 203 (1960).
288. H. HOFFMAN and A. J. DIEHR, *Angew. Chem., Int. Ed.*, 3, 737 1964).
289. R. RABINOWITZ and R. MARCUS, *J. Amer. Chem. Soc.*, 84, 1312 (1962).
290. A. J. BURN and J. I. G. CADOGAN, *J. Chem. Soc.*, 1963, 5788.
291. M. BALLESTER, J. CASTAÑER, and J. RIERA, U.S. Govt. Res. Develop. Rept., 68 (18), 48 (1968). A.D. 672319, CFSTI; *Chem. Abst.*, 70, 46980t (1969).
292. M. BALLESTER, J. ROSA, and E. GUARDIOLA, unpublished.
293. M. BALLESTER and C. MOLINET, *An. Real Soc. Espan. Fis. Quim.*, 54B, 151 (1958).
294. M. BALLESTER, *Pure Appl. Chem.*, 15, 123 (1967).
295. S. D. ROSS, M. MARKARIAN, and M. NAZZEWSKI, *J. Amer. Chem. Soc.*, 69, 2468 (1947).
296. M. BALLESTER and J. RIERA, *An. Real Soc. Espan. Fis. Quim.*, 56B, 897 (1960).
297. E. T. McBEE and T. HODGINS, Final Report Contract No. DA-49-193-MD-3003, US Army Medical Research and Development Command, May 1969.
298. E. T. McBee *et al.*, unpublished.
299. M. BALLESTER and J. CASTAÑER, *An. Real. Soc. Espan. Fis. Quim.*, 56B, 207 (1960).
300. F. SMITH and L. M. TURTON, *J. Chem. Soc.*, 1955, 1350.

194   MANUEL BALLESTER AND SANTIAGO OLIVELLA

301. M. BALLESTER, J. CASTAÑER, and E. GUARDIOLA, *An. Real. Soc.*, *Espan. Fis. Quim.*, 66B, 723 (1960).
301a. A. MARTINEZ, A. F. Institute of Technology, Wright-Patterson AFB (Ohio), Thesis, June 1971.
301b. M. E. FREEBURGER, B. M. HUGHES, L. SPIALTER, and T. O. TIERNAN, *Org. Mass. Spectr.*, 5, 885 (1971).
302. M. BALLESTER, J. CASTAÑER, and J. RIERA, *J. Amer. Chem. Soc.*, 88, 957 (1966).
303. V. W. GASH, *J. Org. Chem.*, 32, 2007 (1967).
304. E. T. McBEE and R. E. LEECH, *Ind. Eng. Chem.*, 39, 393 (1947).
305. A. ROEDIG and R. KOHLHAUPT, *Tetrahedron Lett.*, 1964, 1107.
306. A. ROEDIG, G. BONSE, R. HELM, and R. KOHLHAUP, *Ber.*, 104, 3378 (1971).
307. A. ROEDIG, *Angew. Chem.*, *Int. Ed.*, 8, 150 (1969).
308. A. ROEDIG, R. HELM, R. WEST, and R. M. SMITH, *Tetrahedron Lett.*, 1969, 2137.
309. T. ZINCKE and K. H. MEYER, *Ann. Chem.*, 367, 1 (1909).
310. A. ROEDIG, *Chem. Ber.*, 80, 206 (1947).
311. J. BERNIMOLIN, *Chem. Ber.*, 87, 640 (1954).
312. H. VOLLMANN, *Chem. Abstr.*, 50, 4227 (1956); *ibid.* 50, 10782 (1956), U.S.P. 2,734,927 (1956), G.P. 870,997 (1953), G.P. 844,143 (1952), and G.P. 927,149 (1955).
313. M. BALLESTER and J. CASTAÑER, *An. Real Soc. Espan. Fis. Quim.*, 66B, 487 (1970).
314. M. BALLESTER and J. RIERA, *J. Amer. Chem. Soc.*, 86, 4505 (1964).
315. M. BALLESTER, J. RIERA, J. CASTAÑER, C. BADÍA, and J. M. MONSÓ, *J. Amer. Chem. Soc.*, 93, 2215 (1971).
316. M. BALLESTER, J. RIERA, and J. CASTAÑER, unpublished.
317. M. BALLESTER, J. RIERA, C. BADÍA, J. M. MONSÓ, and I. ALFARO, unpublished.
318. M. BALLESTER, J. CASTAÑER, and J. PUJADAS, unpublished.
319. M. BALLESTER, J. RIERA, and C. BADÍA, unpublished.
320. J. M. M. DASHEVSKII and G. P. PETRENKO, *Sbornik Statei Obshchei Khim.*, *Akad. Nauk USSR*, 1, 630 (1953); *Chem. Abstr.*, 49, 989e (1953).
321. W. MACK, *Tetrahedron Lett.*, 1966, 2875.
322. M. BALLESTER, J. CASTAÑER, J. M. CODINA, and F. LLUCH, *An. Real Soc. Espan. Fis Quim.*, 56B, 197 (1960).
323. M. BALLESTER, J. CASTAÑER, and J. I. TABERNERO, unpublished.
324. J. N. SEIBER, *J. Org. Chem.*, 36, 2000 (1971).
325. M. BALLESTER, J. CASTAÑER, and J. RIERA, unpublished.
326. M. BALLESTER and J. ROSA, *Tetrahedron*, 9, 156 (1960).
327. T. W. TAYLOR and A. R. MURRAY, *J. Chem. Soc.*, 1938, 2078.
328. T. ZINCKE and H. GÜNTHER, *Ann. Chem.*, 272, 243 (1892).
329. P. EATON, E. CARLSON, P. LOMBARDO, and P. YATES, *J. Org. Chem.*, 25, 1225 (1960).
330. H. VOLLMANN, *Chem. Abstr.*, 50, 4226 (1956); G. P. 857,351 (1952).
331. A. E. JUKES, S. S. DUA, and H. GILMAN, *J. Organometal. Chem.*, 1968, 44.
332. H. E. DOORENBOS, Polymers Preprints, 158th ACS Nat'l Mtg. (Sept. 1969), N.Y., p. 1351.
333. V. W. GASH, *J. Org. Chem.*, 32, 2007 (1967).
334. M. BALLESTER, J. CASTAÑER, and J. M. RIO, unpublished.
335. C. WALLING, *Free Radicals in Solution*, Wiley, 1957, pp. 447-8.
336. E. W. R. STEACIE, *Free Radical Mechanisms*, Reinhold, 1946, pp. 192 and 233.

337. E. S. HUYSER, *Free Radical Chain Reactions*, Wiley-Interscience, 1970, .pp. 312-3.
338. J. HINE, *Bivalent Carbon*, Ronald, 1964, Chapter 3.
339. W. KIRMSE, *Carbene Chemistry*, Academic Press, 1964, Chapters 5 and 8.
340. A. LEDWITH, *The Chemistry of Carbenes*, Lecture Series No. 5, The Royal Institute of Chemistry, London.
341. H. R. FALLE, G. R. LUCKHURST, A. HORSEFIELD, and M. BALLESTER, *J. Chem. Phys.*, **50**, 258 (1969).
342. J. SILBERMAN, L. J. SOLTZBERG, N. F. YANNONI, and A. P. KRUKONIS, *J. Phys. Chem.*, **75**, 1246 (1971).
343. G. de la FUENTE and P. FEDERLIN, *Tetrahedron Lett.*, 1972, 1497.
344. M. BALLESTER, J. RIERA, and M. de BARRIOS, unpublished.
345. M. BALLESTER, J. RIERA, and A. RODRÍGUEZ-SIURANA, *Tetrahedron Lett.*, 1970, 3615.
346. M. BALLESTER, J. RIERA, and A. RODRÍGUEZ-SIURANA, unpublished.
347. M. BALLESTER, J. CASTAÑER, and A. IBÁÑEZ, unpublished.
348. P. D. BARTLETT, private communication.
349. M. BALLESTER, J. CASTAÑER, and J. PUJADAS, *Tetrahedron Lett.*, 1971, 1699.
350. G. R. LUCKHURST and J. N. OCKWELL, *Tetrahedron Lett.*, 1968, 4123.
351. M. BALLESTER, J. CASTAÑER, J. M. MONSÓ, and J. ESPINOSA, unpublished.
352. E. T. McBEE and D. K. SMITH, *J. Amer. Chem. Soc.*, **77**, 389 (1955).
353. E. T. McBEE, R. A. HALLING, and C. J. MORTON, unpublished; Doctoral Thesis of R. A. Halling, Purdue University (1965).
354. R. WEST, *Accnts. Chem. Res.*, **3**, 130 (1970).
355. V. MARK, *Tetrahedron Lett.*, 1961, 295.
356. L. G. FEOKTISTOV and A. S. SOLONAR, *J. Gen. Chem. (USSR)*, **37**, 931 (1967); *Chem. Abstr.*, **68**, 26320 (1968).
357. R. RIEMSCHNEIDER and R. NEHRING, *Monatsh.*, **94**, 74 (1963).
358. G. WULFSBERG and R. WEST, *J. Amer. Chem. Soc.*, **93**, 4085 (1971).
359. G. WULFSBERG and R. WEST, *J. Amer. Chem. Soc.*, **94**, 6069 (1972).
360. S. OLIVELLA and M. BALLESTER, unpublished.
361. F. L. HEDBERG and H. ROSENBERG, *J. Amer. Chem. Soc.*, **92**, 3239 (1970).
362. H. ROSENBERG and F. L. HEDBERG, *V International Conference on Organometallic Chemistry* (Moscow, August 1971), Abstracts, Vol. I, pp. 236-7.
363. M. BALLESTER and G. de la FUENTE, *Tetrahedron Lett.*, 1970, 4509.
364. M. BALLESTER, G. de la FUENTE, and J. MAÑÉ, unpublished.
365. S. W. TOBEY and R. WEST, *J. Amer. Chem. Soc.*, **86**, 1459 (1964).
366. R. WEST, A. SADÔ, and S. W. TOBEY, *J. Amer. Chem. Soc.*, **88**, 2488 (1966).
367. R. WEST in *Aromaticity, Pseudo-Aromaticity, Anti-Aromaticity*, The Jerusalem Symposium on Quantum Chemistry and Biochemistry, III (1971), p. 363.
368. T. HIRAISHI, J. MAKAGAWA, and G. T. SHIMANOUCHI, *Spectrochim. Acta*, **20**, 819 (1964).
369. R. WEST, J. CHICKOS, and E. OSAWA, *J. Amer. Chem. Soc.*, **90**, 3885 (1968).
370. S. W. TOBEY and R. WEST, *Tetrahedron Lett.*, 1963, 1179.
371. S. W. TOBEY and R. WEST, *J. Amer. Chem. Soc.*, **86**, 4215 (1964).
372. R. WEST and D. C. ZECHER, *J. Amer. Chem. Soc.*, **89**, 152 (1967).

373. R. WEST, D. C. ZECHER, and W. GOYERT, *J. Amer. Chem. Soc.*, 92, 149 (1970).
374. M. BALLESTER, J. CASTAÑER, and J. RIERA, *Miscellanea Barcinonensia*, 4, 19 (1965).
375. H. WITTEK, *Z. Phys. Chem.*, B 48, 1 (1941).
376. M. BALLESTER and J. RIERA, *Spectrochim. Acta*, 23A, 1533 (1967).
377. M. BALLESTER and J. RIERA, *Tetrahedron*, 20, 2217 (1964).
378. M. BALLESTER and J. RIERA, *Tetrahedron*, 21, 686 (1965).

CHAPTER 2

Polychloroheteroaromatic Compounds

B. Iddon and H. Suschitzky

The Ramage Laboratories, Department of Chemistry and
Applied Chemistry, University of Salford,
Salford, M5 4WT, Lancashire, England

2.1   INTRODUCTION

In writing this chapter we have attempted to show that polychloro-heteroaromatic compounds are more readily available than most heterocyclic textbooks lead one to believe. Some coverage of individual systems is to be found in the usual specialized reviews of heterocyclic chemistry and reference will be made to the existing reviews where appropriate. Research in this field is not a recent fashion. The chemistry of 2,3,4,5-tetrachloropyrrole, for example, dates back to 1837 and that of pentachloropyridine probably to Kekulé's time. However, the industrial importance of cyanuric chloride (trichloro-s-triazine) and the discovery that many poly-chloroheteroaromatic compounds exhibit a broad spectrum of bio-logical activity, and are useful, therefore, as herbicides, pesticides, etc., restimulated interest in these compounds. This has been encouraged also by the development of newer synthetic methods which have made compounds such as pentachloropyridine potentially available on an industrial scale; the use of polychloro-heterocycles in the synthesis of the corresponding polyfluoro-compounds has also played an important role.

This review is not intended to be comprehensive. It does not cover compounds, such as cyanuric chloride, which are adequately reviewed elsewhere, nor have we made extensive reference to papers or patents concerned solely with biological activity (see Chapter 4). Our coverage of the literature is up to and including 1972. We make no apology for giving a detailed account of the chemistry of pentachloropyridine, since this compound has been studied more than any other, with the possible exception of cyanuric chloride, and its reactions are considered typical of those expected for analogous systems. Occasionally, we have discussed mixed halogeno-compounds and made comparisons with the corresponding polyfluoro-derivatives. The latter systems are frequently better known than the polychloro-heterocycles from which they may often be derived. This can be attributed to the stimulus given to fluorine chemistry by the American war-time effort (The Manhattan project)[†] and, in more recent years, to the availability of ^{19}F NMR techniques and the lack

[†] See *Ind. Eng. Chem.*, **39**, 235–434 (1947).

of a comparable technique for the study of the polychloro-systems. The use of ^{35}Cl NQR (nuclear quadrupole resonance) spectroscopy and other more recent analytical techniques, for example ESCA (electron spectroscopy for chemical analysis—high energy photo-electron spectroscopy) [1, 2], are likely to solve the often difficult problems of structure determination of polychloroheterocycles. Studies of polychloroheterocycles have proved more interesting in many ways than comparable studies of polyfluoroheterocycles because of the interplay between steric and electronic effects in the former compounds. Also, the polychloro-compounds are more readily accessible, and therefore cheaper to prepare, and show potentially useful industrial properties (Chapter 4).

The lay-out of our review is that followed by most textbooks of heterocyclic chemistry. Five-membered ring systems are discussed before six-membered ring systems and compounds containing one hetero-atom are discussed before those containing two or more hetero-atoms.

Usually, we have discussed the chemistry of the perchloro-compound together with the chemistry of compounds containing one less chlorine atom. With heterocycles such as thiazole, however, which can be substituted by a few chlorine atoms only, our discussion is usually confined to the perchloro-compound.

A number of polychloroheterocycles are vesicants and others irritate mucous membranes. Irritant activity is usually observed in compounds containing more than one hetero-atom (particularly nitrogen atoms) in which the chlorine atoms are highly activated. It is advisable, therefore, that all polychloroheterocycles be handled with care. Reference to a specific irritant action is made at the appropriate place.

2.2   FURANS

Halogenation of furan is difficult to accomplish because of its sensitivity to acid; polymerization is sometimes violent [3]. Nevertheless, a chlorination process has been developed [4, 5] in which a solution of furan in methylene chloride is treated with chlorine at −40° and then allowed to warm up slowly in order to remove hydrogen chloride (presumably blowing nitrogen through the reaction mixture would assist this process). With 1.6 moles of chlorine per mole of furan, this yields a mixture of 2-chloro-, 2,5-dichloro-, and 2,3,5-trichloro-furan (ratio 9:4:1) and with 3.0 moles of chlorine per mole of furan a mixture of these three compounds, tetrachloro-furan (1; R = Cl), and perchloro-2,5-dihydrofuran (2) (as a mixture of two isomers) is obtained. Chlorination of furan probably occurs via an addition-elimination mechanism (see below) [6, 7].

Tetrachlorofuran is most conveniently prepared by dehalogenation

(1)   (2)

of perchloro-2,5-dihydrofuran (2) with iron or zinc [8, 9]. Compound (2) has become readily available recently through successive rearrangement and cyclization of perchlorocrotonyl chloride [10, 11]. It is also the major product of chlorination of 2,3-dichlorotetrahydrofuran with chlorine at 180° in the presence of a radical initiator [12].

The action of chlorine on furan-2-carboxylic acid (pyromucic acid) was studied as long ago as 1837 by Malaguti [13, 14], who obtained a tetrachloro-addition compound. Later studies by Denaro [15] and Hill *et al.* [16, 17] showed that chlorination of ethyl furan-2-carboxylate (ethyl furoate) at 0° followed by distillation of the product under reduced pressure and treatment of the various fractions with base yields 5-chlorofuran-2-carboxylic acid, 3,4-dichlorofuran-2-carboxylic acid, and a small amount of trichloro-furan-2-carboxylic acid (1; R = CO$_2$H). The trichloro-acid (1; R = CO$_2$H) is formed in greater amounts when ethyl furan-2-carboxylate is treated with chlorine at 0° and the resulting tetra-chloro-addition compound is decomposed immediately with base at 0° [16, 17]. If the tetrachloro-adduct is decomposed thermally by distillation at atmospheric pressure and the separated furan esters (fractional distillation) are hydrolyzed with base, 3,5- and 4,5-dichlorofuran-2-carboxylic acid are produced also [16-21]. These acids were assigned the wrong structures by Hill and his co-workers [6, 21]. Hill and Jackson [16, 18] also prepared trichlorofuran-2-carboxylic acid (1; R = CO$_2$H) by successive chlorination of ethyl 5-chlorofuran-2-carboxylate at 120-145° and treatment of the resulting tetrachloro-addition compound with base at 0°. In a similar manner, 2-chlorofuran may be converted into a trichlorofuran (unidentified) [22] and diethyl furan-2,5-dicarboxylate into 3,4-dichlorofuran-2,5-dicarboxylic acid (3) [23].

(3)   (4)

Trichlorofuran-2-carboxylic acid (1; R = CO$_2$H) is most conveniently prepared from tetrachloro-2*H*-pyran-2-one (4), which gives trichloro-2-furoyl chloride (1; R = COCl) on being heated [24, 25]. This acid chloride can be converted into the corresponding acid, ethyl ester, nitrile, and various amides by using conventional procedures (see also refs. 16 and 18). Trichlorofuran-2-carbonamide is reduced by sodium amalgam in ethanol to 3,4-dichlorofuran-2-carbonamide [25].

Partial decarboxylation of the dicarboxylic acid (3) yields 3,4-dichlorofuran-2-carboxylic acid [23]. At 210-300°, trichlorofuran-2-carboxylic acid (1; R = CO_2H) similarly gives 2,3,4-trichlorofuran (1; R = H) [26].

There appear to be no reports of nucleophilic substitution reactions of tetrachlorofuran or of the trichlorofurans, but the tetrachloro-compound has been shown to undergo a number of Diels–Alder reactions. For example, with maleic anhydride it gives the adduct (5) [27], with acrylonitrile it gives compound (6) [27], and with 2,5-dihydrofuran it gives (7) as a separable mixture of *exo*- and

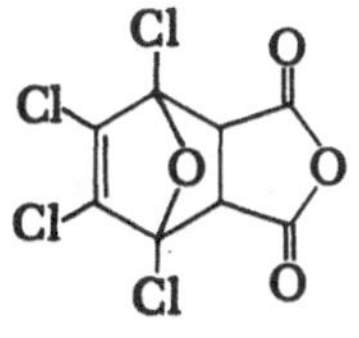

(5) 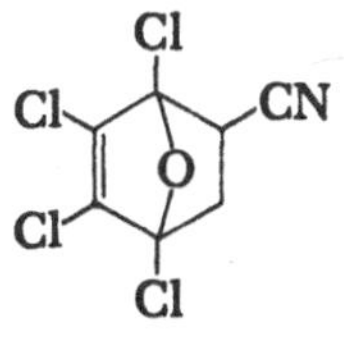   (6)

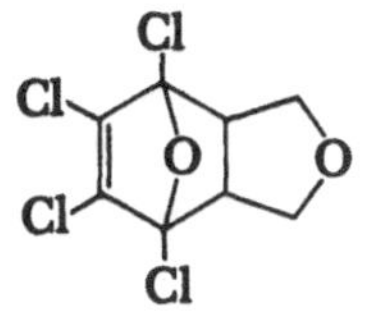

(7) 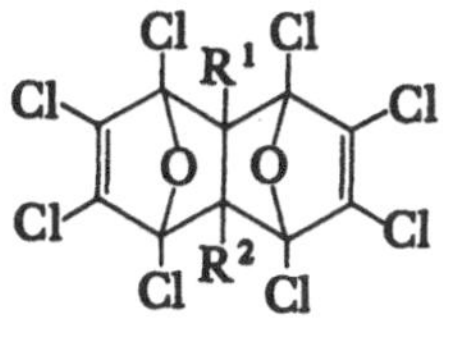   (8)

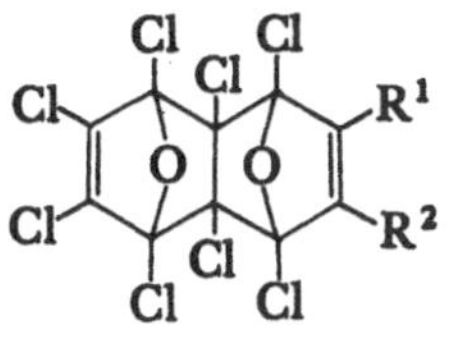

(9) 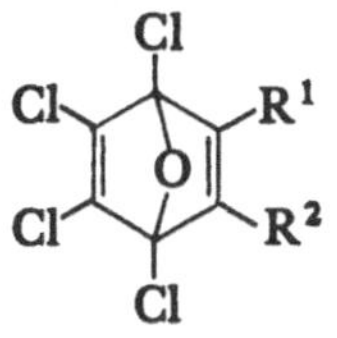   (10)

endo-isomers [28]. It also reacts with various alkynes in the ratio 2:1 [29, 30]; two types of bis-adduct, (8) or (9), are formed, depending on the alkyne used [30]. A mono-adduct (10) is formed initially, although a considerable amount of the bis-adduct is present, even when a 1:1 mole ratio of reactants is used.

With bromine water, trichlorofuran-2-carboxylic acid (1; R = CO_2H) yields 2-bromotrichlorofuran (1; R = Br) together with dichloromaleic acid [6, 16, 18] and it gives the lactone (11) with

(11)

concentrated hydrochloric acid [20]. The last compound also arises when 3,4-dichlorofuran-2-carboxylic acid is treated with bromine water [20]. The structures of some of the lactones prepared in this way by Hill and his co-workers have been questioned [21].

Dichlorofuran-2-carboxylic acids undergo electrophilic substitution in the vacant position; for example, the 3,4-dichloro-isomer is brominated or sulphonated in the 5-position [16, 18, 19]. The resulting sulphonic acid undergoes displacement of the sulphonic acid group by a nitro-group on treatment with fuming nitric acid [16, 18, 19].

3,4-Dichloro- [16, 17] and trichloro-furan-2-carboxylic acid [16, 18] each yield dichloromaleic acid on oxidation with moderately concentrated nitric acid. This method was used to prove the structure of the dichloro-compound (see also ref. 23).

2.3   BENZOFURANS

Dormal et al. [31] prepared a mixture of a tetra- and a tri-chloro-benzo[b]furan by chlorination of the parent heterocycle with chlorine, but no attempt was made to separate or prove the structures of these compounds.

In the presence of zinc chloride, ethyl acetoacetate condenses with 2,3-dichloro-p-benzoquinone to give the benzo[b]furan (12; R = Me)

(12)

[32]. A similar condensation yields compound (12; R = Ph). Both esters are hydrolyzed by hot aqueous sodium hydroxide, and the methyl compound (12; R = Me) is chlorinated in the 4-position with chlorine in acetic acid [32].

Reduction of the quinone (13) with stannous chloride in acetic acid gives 4,6,7-trichloro-2,5-dihydroxy-2,3-dihydrobenzo[b]furan (14) (cf. ref. 33) [34]. These reactions proceed via formation of the aldehyde (15; R = CHO). The corresponding acid (15; R = CO$_2$H) can be prepared by hydrolysis of the quinone (13) with concentrated hydrochloric acid and made to cyclize at about 210° to give 4,6,7-trichloro-5-hydroxybenzo[b] furan-2(3H)-one (16) [35, 36]. Reduction of compound (16) gives the dihydroxy compound (14) [34]. The quinone (13) gives the epoxide (17) on treatment with

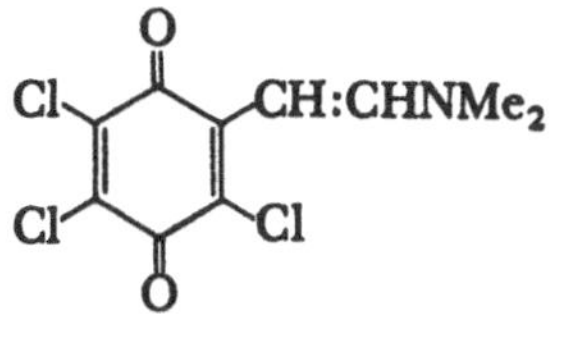

(13)

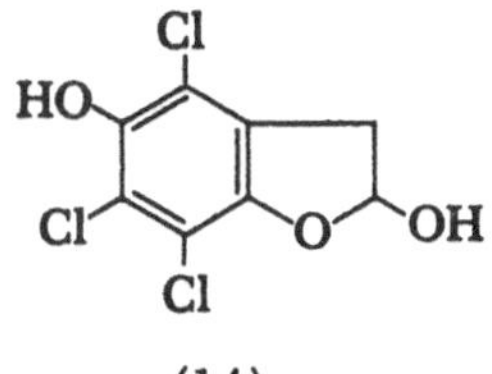

(14)

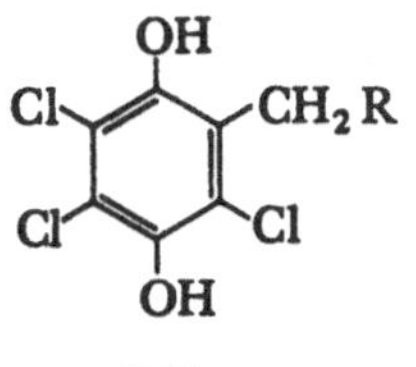

(15)

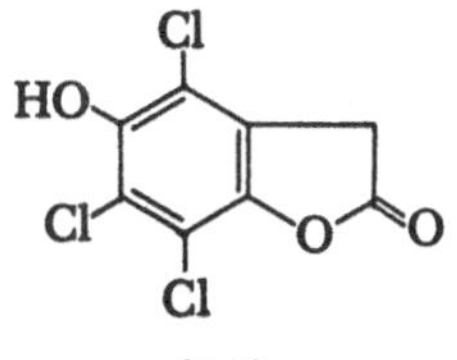

(16)

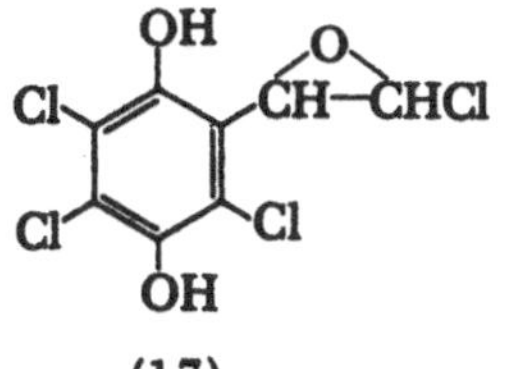

(17)

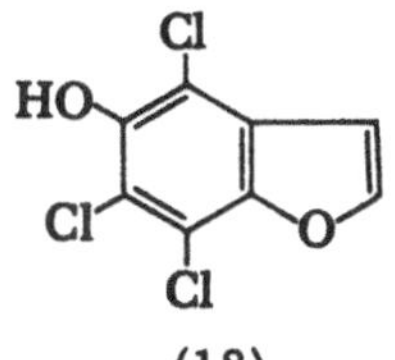

(18)

hydrochloric acid in dioxan, and both the epoxide (17) and the parent quinone (13) yield 4,6,7-trichloro-5-hydroxybenzo[b] furan (18) on reduction with zinc and acid [34, 36]. The same compound (18) can be prepared also by dehydration of the dihydroxy compound (14) [34] and can be methylated with methyl iodide or acetylated using standard procedures. 4,6,7-Trichloro-5-hydroxybenzo[b] furan-2(3H)-one (16) is also formed when the epoxide (17) is heated at 150-160° or treated with boron trifluoride [36].

Huisgen et al. [37-39] have shown that tetrachloro-o-benzoquinone diazide loses nitrogen on photolysis or thermolysis to give the dipolar species (19) which undergoes 1,3-dipolar cycloaddition reactions with various alkynes to give the 4,5,6,7-tetrachlorobenzo[b] furans (20-24) [37, 38] and with various alkenes, to give compounds

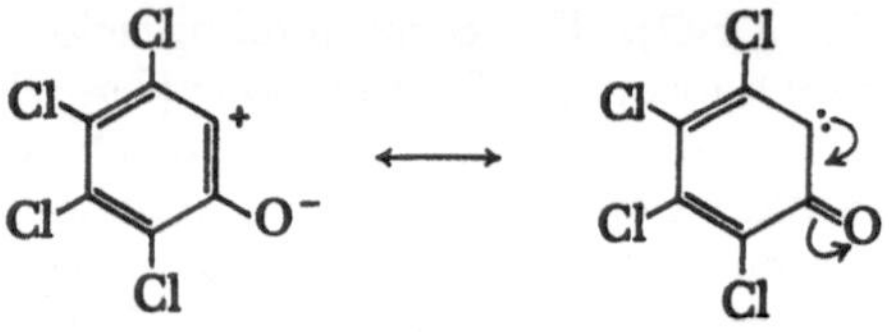

(19)

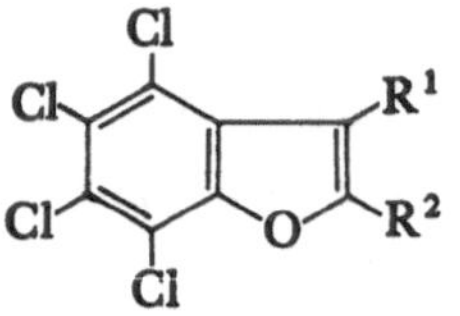

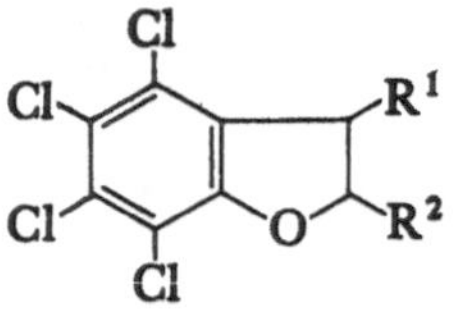

(20) R^1 = H, R^2 = Ph
(21) R^1 = R^2 = Ph
(22) R^1 = R^2 = CO_2Me
(23) R^1 = CO_2Et, R^2 = Ph
(24) R^1 = COPh, R^2 = Ph

(25) R^1 = H, R^2 = Ph
(26) R^1 = R^2 = Ph
(27) R^1 = R^2 = CO_2Me
(*trans*-isomer)
(28) R^1 = CO_2Et, R^2 = Ph

(25-28) [37, 39]. The 2,3-dihydro-compounds (25-28) are converted into the corresponding tetrachlorobenzo[*b*]furan (20-23) on dehydrogenation with sulphur [37, 39]. The structure of the 2-phenyl derivative (20) was proved by catalytic hydrodehalogenation, which gave a mixture of 2-phenylbenzo[*b*]furan (23%) and the ring-opened product, 1-phenyl-2-(2-hydroxyphenyl)ethane (43%) [37, 38].

Tetrachlorobenzyne gives the 4-acetyl-2,2,4-trialkylbenzo-1,3-dioxans (29-31) with acetone, diethyl ketone, and ethyl methyl

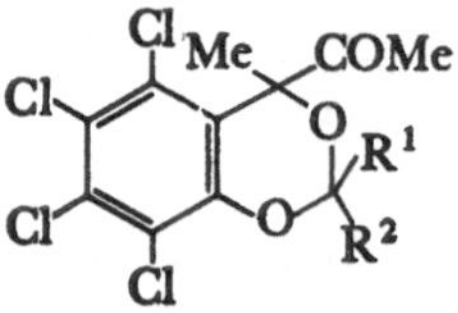

(29) R^1 = R^2 = Me
(30) R^1 = R^2 = Et
(31) R^1 = Me, R^2 = Et

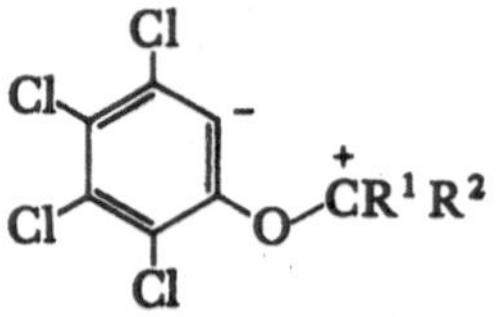

(32)

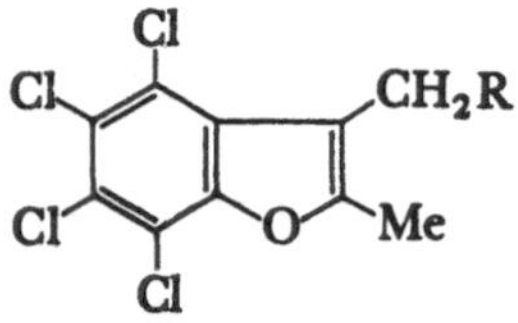

(33)

ketone, respectively, in the presence of buta-2,3-dione [40]. The dipolar species (32) have been postulated as intermediates. Each of the adducts (29-31) yields the same product (33; R = OAc) on treatment with sulphuric acid in acetic anhydride and an alternative product (33; R = Br) with hydrobromic acid in acetic acid. Each adduct is reduced at the carbonyl group by sodium borohydride and treatment of each of the resulting carbinols with sulphuric acid in acetic anhydride gives 4,5,6,7-tetrachloro-2,3-dimethylbenzo[b]-furan (33; R = H).

In the presence of chlorosulphónic acid, 2,4,5-trichlorophenol condenses with glyoxylic acid monohydrate in methylene chloride to

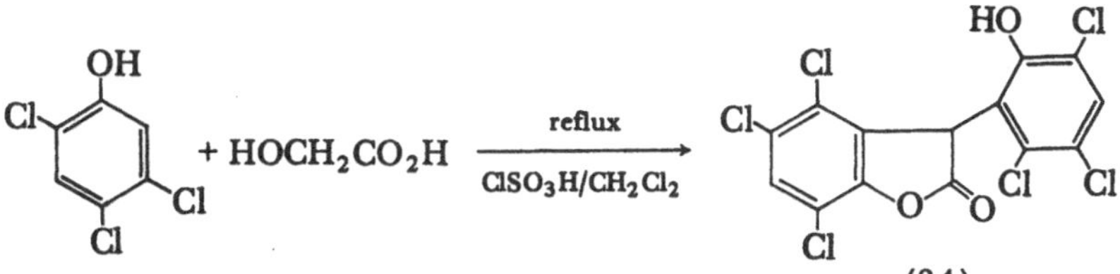

(34)

$$(2.1)$$

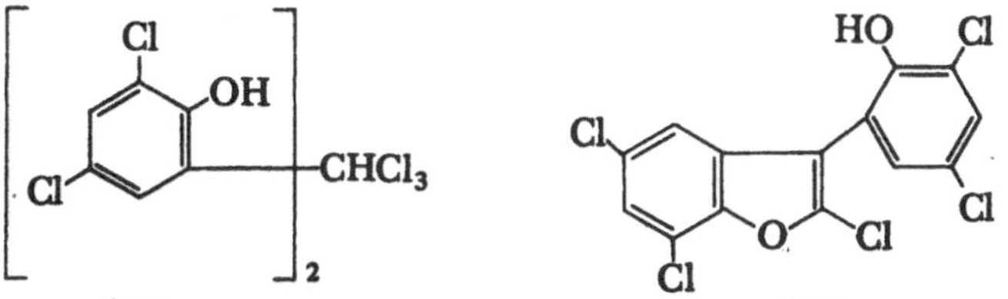

(35)   (36)

give the 3-aryltrichlorobenzo[b]furan-2(3H)-one (34) [equation (2.1)], the lactone of bis-(3,5,6-trichloro-2-hydroxyphenyl)acetic acid [41]. Successive treatment of compound (35) with hot aqueous alkali and acid similarly yields the benzo[b]furan (36) [42].

Hexachlorobenzo[b]furan and pentachlorobenzo[b]furans appear to be unknown.

When compound (37) is subjected to flow pyrolysis at ca. 600° and 3 mmHg under nitrogen, it gives a quantitative yield of 4,5,6,7-tetrachlorobenzo[c]furan (38) [43]. This compound is more

(37)   (38)

stable than the parent heterocycle (which was unknown until recently) and forms a 2:1 mixture of the endo- and exo-adducts with N-phenylmaleimide.

It is noteworthy that the toxicity of polychlorobiphenyls has been shown to be due principally to the presence of impurities such as polychlorodibenzofurans [44] (cf. Chapter 4).

2.4 PYRROLES

Much of the early work on polyhalogenopyrroles was carried out by Ciamician and Silber and by Mazzara and Borgo and is extremely confusing. Hence, we have attempted to give a clarifying summary of their work on the polychloro-derivatives. Work published prior to 1904 has been summarized by Ciamician [45] while Schofield [46] has reviewed halogenopyrroles more recently. Durham *et al.* [47] have studied the chlorination of various pyrrole derivatives, but a discussion of their results is outside the scope of this review.

Halogenation of pyrrole occurs so readily that monohalogenation is difficult. Thus 2,3,4,5-tetrachloropyrrole is formed when a cold ethanolic solution of pyrrole is treated with chlorine [45, 48, 49], or when pyrrole is treated with aqueous sodium hypochlorite [45, 50, 51]. 2,3,4,5-Tetrachloropyrrole may be prepared also by reduction

(39)   (40)

of pentachloro-2H-pyrrole (39) (see later) with sodium amalgam in acetic acid [52], by reduction of heptachloro-Δ^1-pyrrolidine (40) (see later) with zinc dust in acetic acid [53, 54] and by treatment of pyrrole-2-carboxylic acid with an alcoholic solution of chlorine [48].

Sulphuryl chloride has been used extensively for the preparation of chloropyrroles. The reactions, which are normally carried out in dry ether (usually at 0°), may occur partly by an electrophilic process and partly by a radical mechanism [46]. According to the amount of reagent used, pyrrole yields 2-, 2,5-di-, 2,3,5-tri-, or 2,3,4,5-tetra-chloropyrrole [45, 46, 55-58]. Prolonged chlorination yields pentachloro-2H-pyrrole (39) [45, 55-57].

With sulphuryl chloride (ether is the preferred solvent) pyrrole homologues suffer both nuclear and side-chain chlorination. 2,5-Dimethylpyrrole, for example, yields the dialdehyde (41; R = CHO) via formation and hydrolysis of the bis(dichloromethyl) compound (41; R = CHCl$_2$) [59]. Prolonged chlorination followed by hydrolysis [59, 60], or oxidation of the dialdehyde *in situ* with potassium permanganate [61], gives the corresponding dicarboxylic acid (41; R = CO_2H). Compounds (42; R^1 = H, R^2 = Me) [62], (42; R^1 = H, R^2 = CHO) [63], and (42; R^1 = Me, R^2 = CHO) [63] may be prepared similarly. Likewise, with three equivalents of sulphuryl chloride, ethyl 2-methylpyrrole-5-carboxylate gives the chloromethyl compound (43; R = Cl) which, on being heated in boiling water, yields a

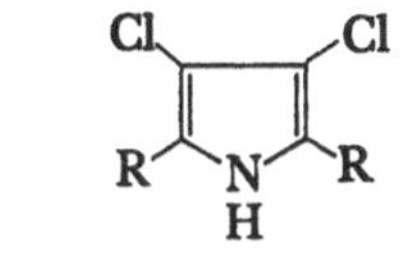

(41) 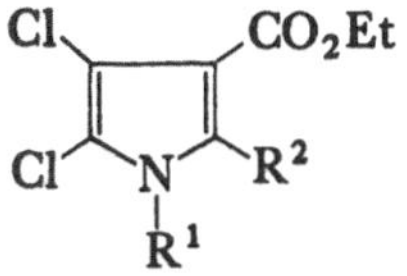   (42)

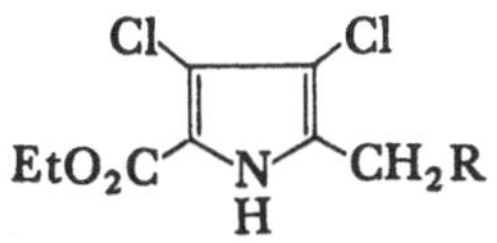

(43) 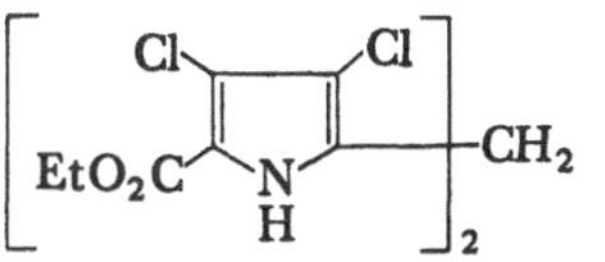   (44)

mixture of formaldehyde, the hydroxymethyl derivative (43; R = OH), and the dimer (44) [64]. With four equivalents of sulphuryl chloride the product is ethyl 3,4-dichloro-2-dichloromethylpyrrole-5-carboxylate, which can be hydrolyzed to the corresponding aldehyde, while chlorination with five equivalents of the reagent followed by treatment of the product with boiling ethanol yields diethyl 3,4-dichloropyrrole-2,5-dicarboxylate [64].

Chlorination of methyl 1-phenylpyrrole-2-carboxylate with two equivalents of chlorine in acetic acid gives a mixture of the 4-chloro-(8%), 5-chloro- (20%), 4,5-dichloro- (40%), and 3,4,5-trichloropyrrole esters (45; R = CO_2Me) (25%), whilst four equivalents of chlorine yield a compound to which structure (46) has been assigned [65]. It is noteworthy that the chlorination of methyl pyrrole-2-carboxylate under various conditions has been studied in some detail by Hodge and Rickards [66]. Only lower chlorinated derivatives were reported by these workers, although Mazzara and Borgo [45, 67] claim to have prepared methyl 3,4,5-trichloropyrrole-2-

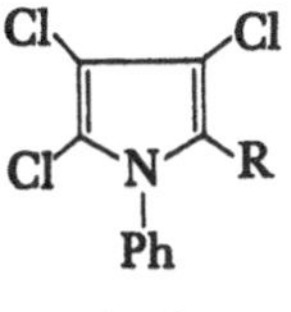

(45) 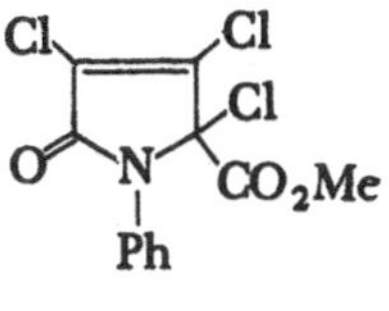   (46)

carboxylate by treatment of methyl pyrrole-2-carboxylate with three equivalents of sulphuryl chloride in ether at 0°. Less sulphuryl chloride results in the formation of lower chlorinated products.

N-Aryltetrachloropyrroles may be prepared by chlorination of a *N*-arylpyrrole with chlorine in carbon tetrachloride in the presence of ferric and aluminium chlorides (see later, however) [67a].

On treatment with chlorine in the dark in the presence of hydrogen chloride, a solution of succinonitrile gives 2,3,4-trichloro-5-iminopyrrole hydrochloride (47; R = Cl) [68]. α-Methylsuccino-nitrile similarly yields (47; R = Me).

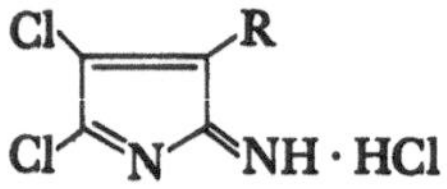

(47)

Prechlorination of *N*-chlorocarbonylpyrrolidine with chlorine at 50-150° followed by further chlorination at high temperatures in the presence of a catalyst is claimed [69] to yield a mixture of pentachloropyrrole [which probably exists as pentachloro-2*H*-

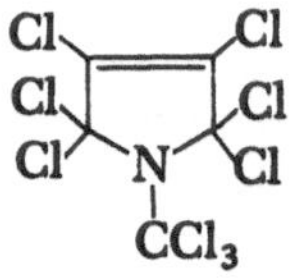

(48)

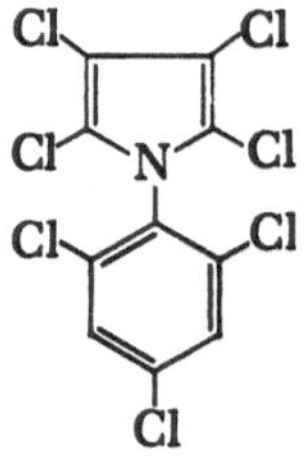

(49)

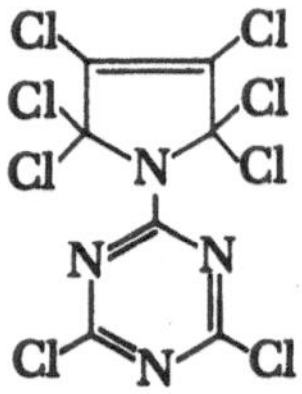

(50)

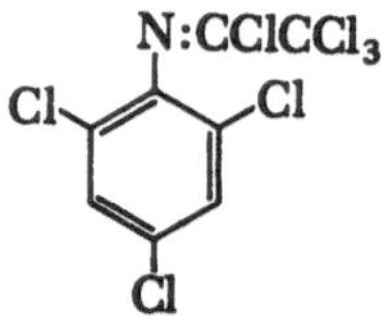

(51)

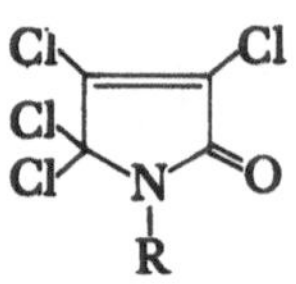

(52)

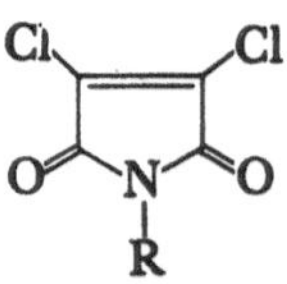

(53)

pyrrole (39)] and heptachloro-Δ^1-pyrrolidine (40). Similar treatment of N-methylpyrrolidine initially gives the polychloropyrroline (48), which loses carbon tetrachloride on prolonged reaction to give pentachloro-2H-pyrrole (39) [70]. N-Phenylpyrrolidine similarly gives (49) (60-70%) [70, 71], whilst 2,4-dichloro-6-pyrrolidino-s-triazine yields compound (50) [70, 71]. Compound (49) is also formed (38%), together with (51), when NN-diethylaniline is treated with chlorine at 200° [70, 71]. Under similar conditions N-methylpyrrolid-2-one yields compound (52; R = CCl$_3$) [70, 71].

On being heated with phosphorus pentachloride under pressure, succinimide gives a small amount of dichloromaleimide (53; R = H), together with the pyrrolidone (52; R = H) and pentachloro-2H-pyrrole (39) [45, 52]. As expected, with phosphorus pentachloride under similar conditions, dichloromaleimide (53; R = H) gives compound (52; R = H) or compound (39), or a mixture of both, depending on the conditions [52-54]. In addition to tetrachloro-N-phenylpyrrole (45; R = Cl), the pyrrolidone (52; R = Ph) (cf. ref. 52) is formed when N-phenylsuccinimide, N-phenylmaleimide, or dichloro-N-phenylmaleimide (53; R = Ph) is treated similarly with phosphorus pentachloride [52, 72]. Compound (52; R = C$_6$H$_4$Me-p) and related compounds may be prepared similarly [52, 73, 74].

Chlorination of N-phenyl- or N-m-tolyl-pyrrole with chlorine in carbon tetrachloride in the presence of a Lewis acid followed by mild hydrolysis of the product with boiling water gives the corresponding maleimide derivative (53; R = Ph or C$_6$H$_4$Me-m), presumably via formation of the N-aryltetrachloropyrrole and adducts analogous to the adduct (50) [67a]. Similar treatment of N-m(or p)-chlorophenylpyrrole gives the corresponding N-aryl-2,2,3,3,4,4,5,5-octachloropyrrolidine; these are hydrolyzed by boiling water to N-aryl-3,3,4,4,5,5-hexachloropyrrolidin-2-ones.

Halogenopyrroles with less than four halogen atoms are unstable; indeed, lower chlorinated derivatives may decompose explosively [45, 46, 57, 58]. Therefore, all chloropyrroles must be handled with caution, particularly those unsubstituted in the 1-position. Stability in the polychloro-series appears to increase with the number of halogen atoms and with N-substitution. Thus, 2,3,4,5-tetrachloropyrrole appears to be more stable than 2,3,5-trichloropyrrole, which is affected by heat and light [56]. Whereas the tetrachloro-compound partly decomposes on attempted steam distillation [55], its N-methyl derivative can be steam-distilled without decomposition. [56].

Polychloropyrroles are markedly acidic and generally dissolve in aqueous alkali [45]. This is the cause of the bathochromic shifts shown by their electronic spectra in alkaline solution [66]. The Raman spectrum has been recorded for 2,3,4,5-tetrachloropyrrole [75].

2,3,4,5-Tetrachloropyrrole reacts with ethanolic potassium iodide to give 2,3,4,5-tetraiodopyrrole [48], which is reduced to pyrrole by zinc and potassium hydroxide [76]. Reduction of the tetrachloropyrrole to pyrrole under the same conditions is reported [76] to be more difficult. 2,3,5-Tri- and 2,3,4,5-tetra-chloropyrrole are N-methylated by methyl iodide in the presence of base [56], and dimethyl sulphate converts ethyl 4,5-dichloro-2-methylpyrrole-3-carboxylate (42; R^1 = H, R^2 = Me) into the N-methyl compound (42; R^1 = R^2 = Me) [62]. Bromination of 2,3,5-trichloropyrrole [77] and its N-methyl derivatives [78] with bromine in acetic acid or water yield the corresponding 4-bromo-derivative (54; R = H or Me).

        (54)   (55)

Polychloropyrroles are oxidized by fuming nitric acid to derivatives of maleimide (55) [45]; thus, 2,3,5-trichloropyrrole gives compound (55; R^1 = R^2 = H) [56], its N-methyl derivative gives (55; R^1 = Me, R^2 = H) [56], 2,3,4,5-tetrachloro-N-methylpyrrole gives (55; R^1 = Me, R^2 = Cl) (this compound is also formed when bromine in acetic acid is used instead of nitric acid) [78], the bromotrichloro-compound (54; R = H) gives (55; R^1 = H, R^2 = Br) [77], and its N-methyl derivative (54; R = Me) gives (55; R^1 = Me, R^2 = Br) [78].
        Treatment of the aldehydes (42; R^1 = H, R^2 = CHO) and (42; R^1 = Me, R^2 = CHO) with ethanolic hydrazine yields compounds (56; R^1 = H or Me), each of which gives the corresponding trichloro-diazaindole (57; R^2 = Cl) on treatment with phosphorus oxychloride [63]. In these compounds the chlorine atom in the diazine ring is

        (56)   (57)

displaced readily by nucleophiles to give for instance (57; R^2 = NHCH$_2$Ph, piperidino, or morpholino) [63].
        Pentachloro-2H-pyrrole (39) is unstable in moist air and in boiling water or ethanol is converted into dichloromaleimide (53; R = H)

[52]. On treatment with silver fluoride it gives 3,4-dichloro-2,2,5-trifluoro-2H-pyrrole (58) [79]. We have already referred to its reduction to 2,3,4,5-tetrachloropyrrole with zinc amalgam and acid (*loc. cit.*).

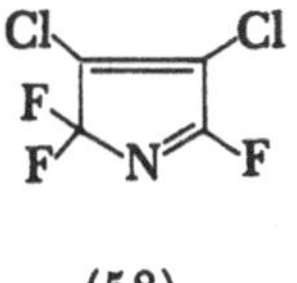

(58)

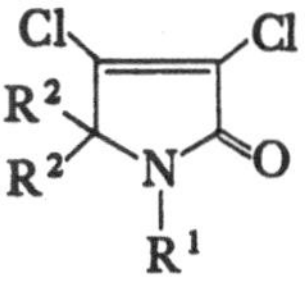

(59)

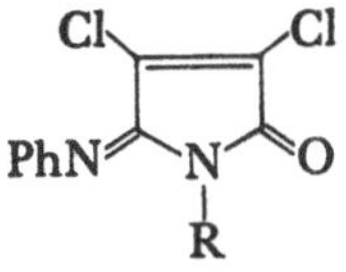

(60)

The pyrrol-2-one (52; R = H) is unstable also in the presence of moisture or ethanol, giving rise to dichloromaleimide (53; R = H) [52]. With phenol, it gives compound (59; R^1 = H, R^2 = OPh), and with aniline it gives the iminopyrrol-2-one (60; R = H) [52]. The N-phenylpyrrol-2-one (52; R = Ph), similarly gives compound (59; R^1 = Ph, R^2 = OMe or OEt) with methanol or ethanol, respectively [52, 72], and with aniline it gives the imine (60; R = Ph) [52, 72]. It is hydrolyzed on boiling with water to dichloro-N-phenylmaleimide (53; R = Ph) [52, 72] and is reduced by zinc and acetic acid to N-phenylsuccinimide [52, 72]. Compound (52; R = C_6H_4Me-p) [73] and related compounds [52, 74] undergo analogous reactions.

Interest in chloropyrroles has been stimulated recently by the discovery that the naturally occurring compound, pyrrolnitrin (61; R^1 = R^2 = H), and a number of related compounds [e.g., 2-chloropyrrolnitrin (61; R^1 = H, R^2 = Cl)], which are produced by a number of *Pseudomonas* strains in fermentation reactions, possess antifungal activity [65, 80-84] and by the fact that pyoluteorin (62), which is produced by certain strains of *Pseudomonas aeruginosa*,

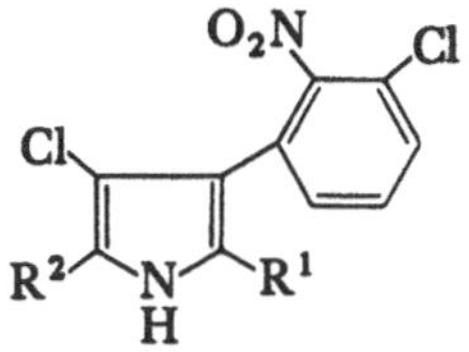

(61)

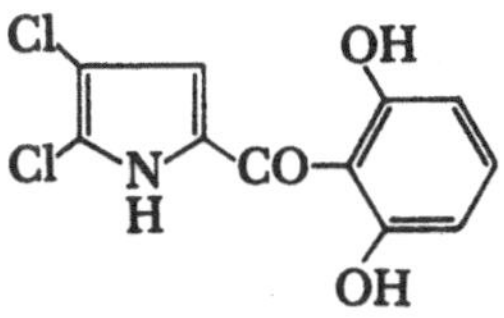

(62)

possesses both bactericidal and fungicidal activity [47, 65, 85, 86].
Compound (62) is active against the fungus causing Dutch elm
disease [65]. Recently, pyrrolnitrin (61; $R^1 = R^2 = H$) has been
chlorinated at -15 to $-20°$ with sulphuryl chloride in chloroform to
give the aryltrichloropyrrole (61; $R^1 = R^2 = Cl$) [87], which is
claimed to possess bactericidal and fungicidal properties.

The recent paper by Motekaitis et al. [88] on 2,5-disubstituted
3,4-dihalogenopyrroles is noteworthy, although a discussion of this
work falls outside the scope of this review.

2.5   INDOLES

Chlorination of indole and its derivatives† can give a variety of
products, depending on the reagent and reaction conditions [89-99].
A mixture of a tetra- (63) and a hexa-chlorocompound (64) is
obtained when methyl 1-methylindole-2-carboxylate is treated with
sulphuryl chloride [91]. The positions of the chlorine atoms in the
benzene ring were not established. Chlorination of sodium 1-acetyl-
indoline-2-sulphonate with chlorine in aqueous solution followed by
hydrolysis of the product and further chlorination of the resulting

(63)   (64)

(65)

mixture of trichloroindoles with phosphorus pentachloride in an
autoclave at 290° yields a heptachloroindole, which is probably
heptachloro-3H-indole (65) [100]. Chlorination of the mixture of
trichloroindoles with chlorine in carbon tetrachloride gives an unidentified pentachloroindole; a hexachloroindole is obtained if a
small amount of iodine is present during the chlorination [101].
These highly chlorinated indoles are reported to be extremely
sensitive to moisture; the nature of the products formed on

† One author [89] was forced to abandon work on the chlorination of
indoles because contact with the products produced a severe skin rash.

hydrolysis has not been established, but they all contain a carbonyl group [100, 101].

Heptachloro-3H-indole (65) may be prepared also by the action of heat on N-pentachlorophenyltrichloroacetimidoyl chloride (66) [70, 102, 103]. 2,4-Dichloro-6-trichloromethylphenyl isocyanide dichloride (67) similarly yields 2,3,3,5,7-pentachloro-3H-indole.

When heated at 200° for 30 min with polyphosphoric acid, acetophenone pentachlorophenylhydrazone gives 4,5,6,7-tetrachloro-2-phenylindole (68) [104]. This cyclization must involve loss of chlorine which, surprisingly, does not attack the 3-position of the

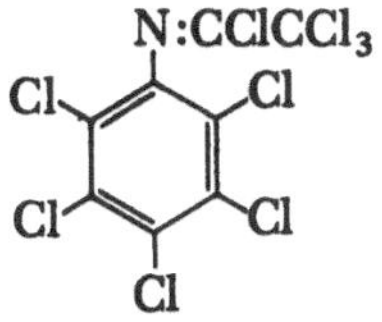

(66) 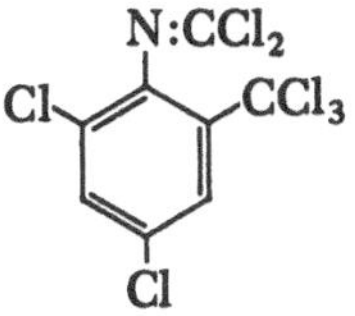   (67)

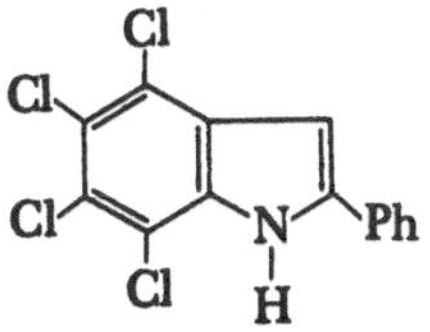

(68) 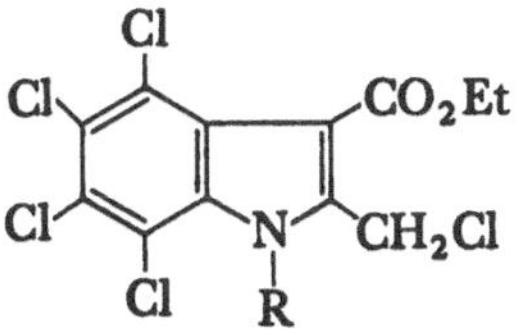   (69)

product. The synthesis is of limited use, however, since the pentachlorophenylhydrazones of acetone, propiophenone, and cyclohexanone do not cyclize under similar conditions.

Tetrachloro-o-benzoquinone condenses with ethyl β-aminocrotonate and its N-methyl derivative in the presence of acid to give unstable 1:1-adducts, which cyclize to give the polychloroindoles (69; R = H or Me) [105].

4,5,7- and 4,6,7-Trichloroisatin may be prepared by treatment of the appropriately chlorinated aniline with a mixture of chloral hydrate, hydroxylamine hydrochloride, and sodium sulphate, followed by cyclization of the resulting isonitrosoacetanilide (70) with concentrated sulphuric acid [106]. When (2-carboxy-3,4,5,6-tetrachlorophenyl)glycine is heated with a mixture of acetic anhydride and sodium acetate, it yields N O-diacetyl-4,5,6,7-tetrachloroindoxyl (71; R = OAc), which gives the monoacetyl compound (71; R = H) following hydrolysis and reacetylation under controlled conditions [106]. O-Acetyl-4,5,7-, -4,6,7-, and -5,6,7-trichloroindoxyl may be prepared similarly.

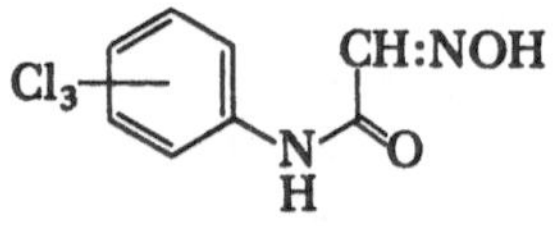

(70) 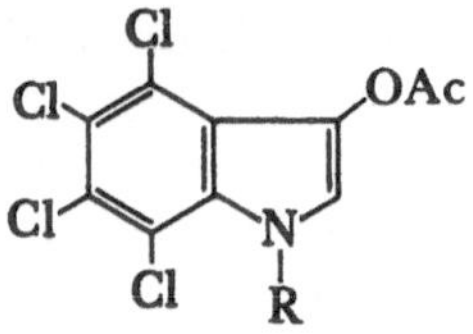   (71)

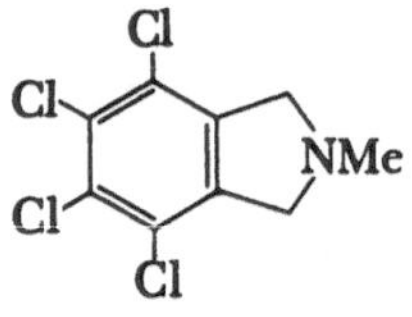

(72) 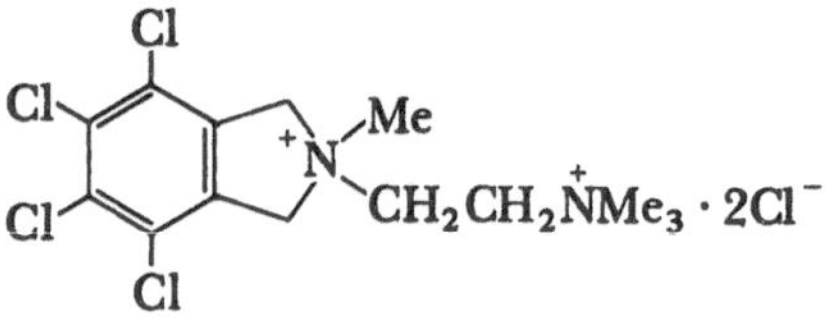   (73)

3,4,5,6,7-Pentachloroindole is reported [107] to possess weak herbicidal activity.

Various alkylation reactions of 4,5,6,7-tetrachloro-N-methyliso-indoline (72) have been reported as a result of the discovery that 4,5,6,7-tetrachloro-2-(2-dimethylaminoethyl)isoindoline   dimetho-chloride (73) is a ganglionic blocking agent [108]. The parent compound, 4,5,6,7-tetrachloroisoindoline, may be prepared by the action of heat on a mixture of tetrachlorophthalic anhydride and ammonium hydroxide followed by reduction of the tetrachloro-phthalimide produced with lithium aluminium hydride [108, 109]. Its N-methyl derivative (72) and other N-alkyl-4,5,6,7-tetrachloroiso-indolines may be prepared similarly.

2.6   THIOPHENS

2.6.1   Synthesis

Tetrachlorothiophen was first prepared in 1884 by Weitz [110, 111], who treated 2,5-dibromothiophen successively with chlorine and alkali. This reaction is improved if the initial mixture of addition compounds is decomposed by distillation [112]. Similar treatment of tribromo-3-methylthiophen yields trichloro-3-methylthiophen [113].

Chlorination of thiophen appears to proceed initially by substi-tution to give mainly 2-chloro- and then 2,5-dichloro-thiophen, together with smaller amounts of 3-chloro- and 2,3-dichloro-thiophen, respectively [111, 113-117]. Addition then occurs to an increasing extent [116]; the principal addition compounds formed are the three chlorinated thiolanes (74; $R^1 = R^2 = H$), (74; $R^1 = H$, $R^2 = Cl$), and (74; $R^1 = R^2 = Cl$) [114]. Thiolanes (74; $R^1 = H$,

Cl   Cl
H─┬──┬─H
R¹─┤ ├─R²
Cl   S   Cl

(74)

Cl   Cl
Cl─┤ ├─Cl
Cl   S   Cl

(75)

R^2 = Cl) and (74; R^1 = R^2 = Cl) are more conveniently prepared by similar treatment of 2-chloro- and 2,5-dichloro-thiophen, respectively. All of the possible chlorothiophens are available by chemical or thermolytic dehydrochlorination of these thiolanes. Thus, pyrolysis of the thiolane (74; R^1 = R^2 = H) gives mainly a 50:50 mixture of 2,3- and 2,4-dichlorothiophen together with a trace of 2,5-dichlorothiophen, whilst its treatment with alcoholic potassium hydroxide yields a mixture of 2,4- (44 mole %), 3,4- (54 mole %), and 2,5-dichlorothiophen (2 mole %). Pyrolysis of the thiolane (74; R^1 = H, R^2 = Cl) or its treatment with alkali gives varying amounts of 2,3,4- and 2,3,5-trichlorothiophen, whilst the thiolane (74; R^1 = R^2 = Cl), which is the principal product of exhaustive chlorination of thiophen at 80–130°, gives tetrachlorothiophen under all conditions. Tetrachlorothiophen is most conveniently prepared (80% yield) by exhaustive chlorination of thiophen at temperatures rising from its boiling point to 200°, at which temperature thermolytic dehydrochlorination of the intermediate thiolane (74; R^1 = R^2 = Cl) occurs [114, 115, 117]. Some 2,3,4-trichlorothiophen (about 14%) is formed also, together with a trace of hexachloro-2,2'-bithienyl (see also ref. 116) [115].

Whereas exhaustive chlorination of thiophen can be directed to give either tetrachlorothiophen at high temperatures (200°) or the thiolane (74; R^1 = R^2 = Cl) at lower temperatures in the absence of a catalyst, in the presence of iodine similar treatment at the higher temperatures yields 2,2,3,4,5,5-hexachloro-3-thiolene (75) [118-120]. It is probable that this product arises by iodine-promoted addition of chlorine to tetrachlorothiophen. The same compound (75) is formed when an irradiated solution of tetrachlorothiophen in carbon tetrachloride is treated with chlorine [120]. The thiolene (75) is converted into tetrachlorothiophen on being heated alone [118, 119, 121] or with sulphur [120, 122], sulphur monochloride [120], zinc dust [119], or copper powder [120].

Exhaustive chlorination of thiophen in solution in chloroform or carbon tetrachloride in the presence of iodine or iodine monochloride [120, 123, 124] or under the influence of light for prolonged periods [120] gives octachlorothiolane. Polychlorothiolanes are converted into chlorothiophens by passage over activated carbon at temperatures as low as 145° [125-127]. Hexachloro-(74; R^1 = R^2 = Cl) and octachloro-thiolane are converted into tetrachlorothiophen on being heated with sulphur [122].

Excellent yields of tetrachlorothiophen are obtained also when perchlorobuta-1,3-diene is heated with sulphur [124, 128, 129]. Tri- and tetra-chloroethylene [130, 131], perchloroethane [130], and several polychloro-butanes [132, 133], -butenes [132], and -buta-dienes [132] may be used also as starting materials.

2,3,5- and 2,3,4-Trichlorothiophen may be prepared by chlori-nation of 2,5- or 3,4-dichlorothiophen, respectively [115]. They can be differentiated by treatment with mercuric chloride which reacts only with thiophens containing an α-hydrogen atom.

Treatment of 2,5-dibromo-4-iodothiophen-3-carboxylic acid or 2,3-dibromothiophen-5-carboxylic acid with chlorine in acetic acid gives trichlorothiophen-3-carboxylic acid and 2,3-dichlorothiophen-5-carboxylic acid, respectively [113]. A trace of tetrachlorothiophen is formed also in the latter reaction.

A twelve-component mixture containing a small amount of 2-chloro-2-(3,4,5-trichloro-2-thienyl)vinyl chloride (76) is obtained when 3-(2-thienyl)acrylic acid is treated with thionyl chloride in

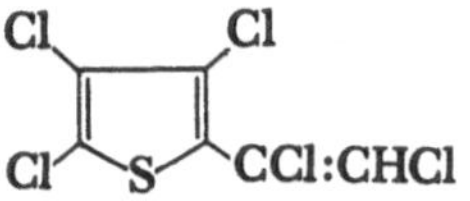

(76) 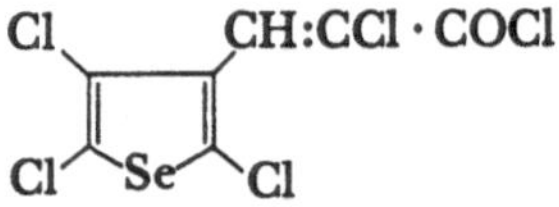   (77)

chlorobenzene in the presence of pyridine [134]. 3-(3-Selenienyl)-acrylic acid similarly gives a mixture containing a moderate (15%) amount of 2-chloro-3-(2,4,5-trichloro-3-selenienyl)acrylic acid chlor-ide (77) [134].

Photochlorination in carbon tetrachloride of compound (78; $R^1 =$ $R^2 =$ Br), which may be prepared by photobromination of 2,5-dimethylthiophen, yields the perchloro-compound (79; $R = CCl_3$) [112]. Compound (78; $R^1 =$ Br, $R^2 =$ Cl) gives compound (78; $R^1 =$ $R^2 =$ Cl) with chlorine in boiling carbon tetrachloride [113]. The

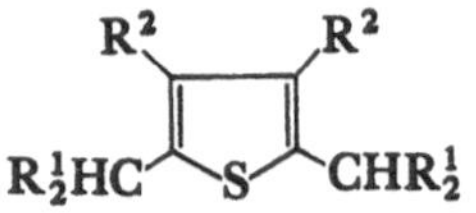

(78) 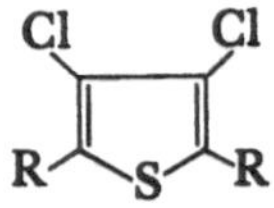   (79)

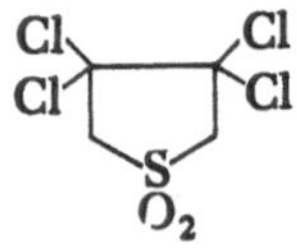

(80)

former compound is hydrolyzed by calcium carbonate to the dialdehyde (79; R = CHO) [113]. With aqueous sodium carbonate, a precipitate of the dialdehyde is formed and acidification of the filtrate yields 3,4-dichloro-2-hydroxymethylthiophen-5-carboxylic acid [113]. With silver monofluoride, compound (78; $R^1 = R^2 = Br$) yields the bromofluoro-compound (78; $R^1 = F$, $R^2 = Br$), which undergoes photochlorination to give 3,4-dichloro-2,5-bis(chlorodifluoromethyl)thiophen (79; R = $CClF_2$) [112].

Buta-1,3-diene-1,4-dicarboxylic acid gives compound (79; R = COCl) with thionyl chloride in pyridine [135].

3,3,4,4-Tetrachlorosulpholane (80) gives 3,4-dichlorothiophen 1,1-dioxide on treatment with methanolic ammonia [136]. 2,4-Dichloro- and 2,3,4-trichloro-thiophen 1,1-dioxide may be prepared similarly.

2.6.2   Reactions

Tetrachlorothiophen is substituted by phenoxide ion or piperidine in the 2-position [137]. Its reduction over 5% palladium-charcoal at 200° gives a mixture of 2,3,4-trichloro- (47 mole %), 2,3-dichloro- (4 mole %), and 3,4-dichloro-thiophen (12 mole %), whilst 2,3,4-trichlorothiophen similarly gives good yields of 3,4-dichlorothiophen [138]. These results are surprising in view of the ease with which thiophens normally poison group VIII noble metal catalysts. Reduction of tetrachlorothiophen with sodium telluride in aqueous methanol gives mainly 2,3,4-trichlorothiophen (74% yield) together with a small amount of the 3,4-dichloro-compound [137]. Similar reduction of 2,3,4-trichloro-, 2,3,4-trichloro-5-phenoxy-, or 2,3,4-trichloro-5-piperidino-thiophen results in loss of the remaining α-chlorine atom.

Oxidation of tetrachlorothiophen with concentrated nitric acid yields dichloromaleic acid thioanhydride as the major product together with a small amount of dichloromaleic acid anhydride (see also ref. 121) [139]. The thioanhydride is best prepared by hydrolysis of the thiolene (75) with concentrated sulphuric acid [120].

Trichloro-2-thienylmagnesium chloride† may be prepared from tetrachlorothiophen by means of the entrainment technique (see also ref. 143) [113, 140-142]. With tetrahydrofuran as the solvent and one equivalent of ethylene bromide as the entrainer, the yield of the Grignard compound is essentially quantitative [142]. It reacts normally with acid [113, 140, 142], carbon dioxide [140, 142], acetaldehyde [141], iodine [142], iodobenzene [142], mercuric chloride [142], cadmium chloride [142], and various chlorosilanes, for example chlorotrimethylsilane [142]. Trichloro-2-thienyl-

† Metal derivatives of polychlorothiophens are discussed also in Chapter 3; however, the approach given there is different and deliberately less comprehensive.

magnesium chloride is less stable in tetrahydrofuran than the corresponding lithium compound. After 6 days at room temperature, for example, subsequent derivatization indicated a 20% decrease in concentration; little change is observed with the lithium compound under comparable conditions [142]. Although tetrachlorothiophen does not form a di-Grignard reagent to any significant extent, a mixture of magnesium, tetrachlorothiophen, and chlorotrimethylsilane in tetrahydrofuran in the presence of a few drops of ethylene bromide yields dichloro-2,5-bis(trimethylsilyl)thiophen [142].

With one equivalent of n-butyl-lithium in ether [141, 143-145] or tetrahydrofuran [146] at −70°, or methyl-, t-butyl-, or phenyl-lithium in tetrahydrofuran at −70° [146], tetrachlorothiophen gives trichloro-2-thienyl-lithium. In the case of t-butyl-lithium, the lithium compound is formed exclusively in tetrahydrofuran but, with methyl- or n-butyl-lithium in the same solvent, considerable ring-fragmentation also occurs [146]. Ring-fragmentation also occurs in the case of phenyl-lithium but to a lesser extent. The ring-opened products have not been identified. With n-butyl-lithium in ether, the lithium compound is formed in essentially quantitative yield [145]. Treatment of tetrachlorothiophen with two equivalents of n-butyl-lithium in ether yields the 2,5-dilithio-derivative, which reacts normally with chlorosilanes and carbon dioxide [147, 148]. When a mixture of equivalent amounts of tetrachlorothiophen and the 2,5-dilithio-compound in ether is kept at 0°, a metal-halogen interconversion reaction occurs to give a mixture in favour of trichloro-2-thienyl-lithium, which suggests that an equilibrium may be involved [147].

Cleavage of the C—Si bond in (trichloro-2-thienyl)dimethylsilane with n-butyl-lithium also yields trichloro-2-thienyl-lithium [149]. This reaction occurs under much milder conditions than those required to effect chlorine-lithium exchange in tetrachlorothiophen.

Trichloro-2-thienyl-lithium reacts normally with acid [143, 145], carbon dioxide [141, 143-146, 148, 149], acetaldehyde [141], acetone [141], benzophenone [143], mercuric chloride [143], various chlorosilanes [145, 146, 149], and chlorodiphenylphosphine [143, 145]. With various transition metal halides, it gives σ-bonded (trichloro-2-thienyl) derivatives, for example compound (81), which show enhanced thermal and oxidative stabilities relative to their hydrocarbon analogues [143].

Attempts to eliminate lithium chloride from trichloro-2-thienyl-lithium have failed [143], presumably because the expected product, 4,5-dichloro-2,3-dehydrothiophen is a highly strained ring-system. An ethereal solution of this lithium compound remains essentially unchanged after 4 days at 0°. Pyrolysis of the mercury derivative (82) (see later) in the presence of tetraphenylcyclopentadienone gives a high yield of the benzo[b]thiophen (83), but this reaction is

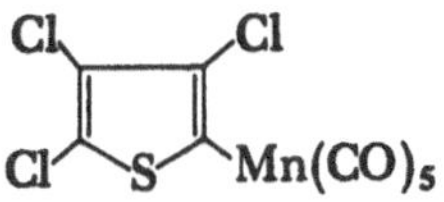

$$\text{(81)}$$

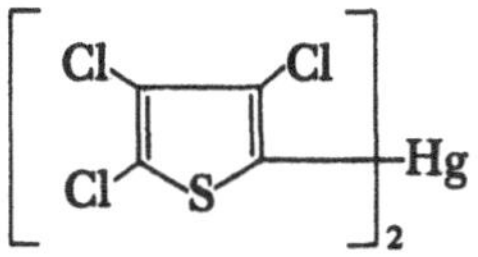

Cl, Cl
Cl S Mn(CO)$_5$

(81)

Cl, Cl
Cl S Hg
                    ]$_2$

(82)

Ph
Ph, Cl
Ph S Cl
Ph

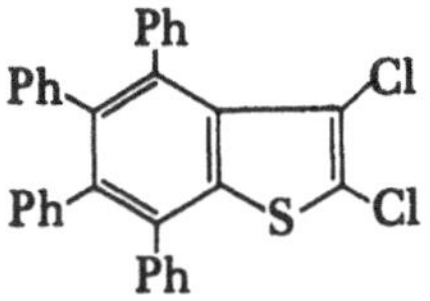

(83)

more likely to proceed via initial addition of the cyclopentadienone to the mercury compound (82) rather than via formation of 4,5-dichloro-2,3-dehydrothiophen [143].

Dichloro-2-dimethylsilyl-5-thienyl-lithium (84; R = SiHMe$_2$) or dichloro-2,5-dilithiothiophen (84; R = Li) may be prepared by treatment of dichloro-2,5-bis(dimethylsilyl)thiophen with one or two equivalents of n-butyl- or phenyl-lithium, respectively [149]. Cleavage of a trimethylsilyl group from a polychlorothiophen nucleus is more difficult. Thus, trichloro-2-trimethylsilylthiophen

Cl, Cl
R S Li

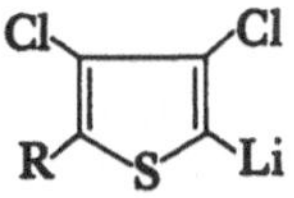

(84)

gives the lithium compound (84; R = SiMe$_3$) exclusively on treatment with n-butyl-lithium [149]. One or both of the trimethylsilyl groups can be cleaved from dichloro-2,5-bis(trimethylsilyl)thiophen under more forcing conditions, but the yields of derived lithium compounds are low. Each of the lithium compounds mentioned here has been characterized by hydrolysis with water or by reaction with carbon dioxide or chlorotrimethylsilane.

Trichloro-2-thienylcopper is best prepared by treatment of tri-chloro-2-thienyl-lithium or -magnesium chloride with a copper(I) halide in various solvents (e.g., ether) [144, 150]. It may be prepared also by decarboxylation of copper(I) trichloro-2-thenoate in boiling pyridine [144], and by treatment of tetrachlorothiophen with lithium dimethylcuprate [150]. It reacts normally with various reagents, for example acid [144], iodine [150], acetyl or benzoyl chloride [150], iodobenzene [150], p-iodoanisole [144], tribromoethene [150], and allyl bromide [150], and is oxidized by 2,4,6-trinitrobenzene to hexachloro-2,2′-bithienyl (cf. ref. 150) [144]. Similarly, 3,4-dichloro-2,5-dicopper-thiophen may be prepared by treatment of the 2,5-dilithio-derivative (see earlier) with a copper(I) halide [147]. It reacts normally with iodine, acetyl chloride, and allyl bromide to give the corresponding 2,5-disubstituted 3,4-dichlorothiophen.

3,4-Dichloro-2,5-dilithiothiophen reacts much less successfully with cadmium chloride, to give a bis(chlorocadmium) derivative [147].

Tetrachlorothiophen does not react with silver monofluoride, even at 220°, but with silver difluoride it yields the dichlorotetrafluoro-3-thiolene (85) [112]. The same product (85) (37%) is formed,

(85)

together with a number of minor products, when tetrachlorothiophen is heated with the recently discovered fluorinating agent, potassium tetrafluorocobaltate [151]. With cobalt trifluoride, tetrachlorothiophen gives mainly ring-opened products (various chlorofluorobutanes) [151].

The perchloro-compound (79; R = CCl$_3$) reacts with silver monofluoride to give compound (79; R = CF$_3$), which is ring-opened on treatment with silver difluoride at 130° to give 3,4-dichloro-1,1,1,2,5,6,6,6-octafluorohexa-2,4-diene [112]. Compound (79; R = CF$_3$) may be prepared also by treatment of compound (79; R = CClF$_2$) (see earlier) with silver monofluoride [112].

2,3,4-Trichlorothiophen, which is most conveniently prepared from tetrachlorothiophen via the 2-magnesium compound or the 2-lithium compound (see before), undergoes nitration [113, 140], bromination [113], acetylation [113], and sulphonation in the vacant α-position [113, 152, 153]. 2,3,5-Trichlorothiophen similarly undergoes nitration, acetylation, and sulphonation in the vacant β-position [113].

Treatment of 2,3,4-trichlorothiophen with mercuric oxide in acetic acid yields the mercury derivative (82), which reacts with mercuric chloride in acetone to give trichloro-2-thienylmercuric chloride [113]. The latter compound, which may be prepared also by reacting trichloro-2-thienyl-lithium with mercuric chloride (see before), reacts with aqueous potassium iodide in the presence of iodine to give trichloro-2-iodothiophen and, with bromine water containing potassium bromide, it gives 2-bromotrichlorothiophen

[113]. 2,3,5-Trichlorothiophen similarly gives trichloro-3-thienyl-mercuric chloride which can be converted into the corresponding bromo- and iodo-trichloro-compounds [113].

Normal procedures allow sulphonamides to be prepared from trichlorothiophen-2-sulphonyl chloride [152, 153], and amides to be prepared from trichlorothiophen-2-carboxylic acid [154-156]. It is noteworthy also that a thiophen analogue of DDT has been prepared by condensation of 2,5-dichlorothiophen with chloral hydrate [157].

The preparations and properties of a large number of mixed tetrahalogenothiophens have been described by Steinkopf and Köhler [113], and the syntheses and a few reactions of several poly-chlorobithienyls have been reported also [111, 158-161].

2.7   BENZO[b]THIOPHENS AND DIBENZOTHIOPHENS

Prolonged treatment of styrene or α,β-dibromostyrene with a mixture of thionyl chloride and sulphuryl chloride in a sealed tube at 270° gives a low yield of hexachlorobenzo[b]thiophen [162, 163]. Exhaustive chlorination of benzo[b]thiophen with chlorine in the presence of iron is reported [164] to give a tetrachloro- and a pentachloro-benzo[b]thiophen, both of unknown identity. We [163] have prepared hexachlorobenzo[b]thiophen similarly; in the later stages of the chlorination the temperature was increased slowly to 200° and small amounts of ferric chloride were added during the terminal stages of the reaction.

Ethyl 4,5,6,7-tetrachlorobenzo[b]thiophen-2-carboxylate may be prepared by treatment of pentachlorobenzaldehyde with ethyl thio-glycolate in the presence of triethylamine [165].

Octachlorodibenzothiophen is readily prepared by photolysis of bis(pentachlorophenyl) sulphide or pentachlorobenzenesulphenyl chloride in carbon tetrachloride [166]. The sulphide may be prepared by exhaustive chlorination of benzene in a mixture of sulphur monochloride and sulphuryl chloride in the presence of aluminium chloride. Exhaustive chlorination of 4,4'-dichlorobiphenyl in carbon tetrachloride in the presence of sulphur and aluminium chloride is also reported to give octachlorodibenzothiophen, together with tar [167].

2.8   PYRIDINES

2.8.1   Introduction

Pentachloropyridine is available commercially; consequently its chemistry has been studied in some detail. The first synthesis of this compound is usually attributed to Sell and Dootson (see later)

[168], but Kekulé may have been the first to prepare it as well as 2,3,5,6-tetrachloropyridine [169].

Pentachloropyridine is a white crystalline solid with a characteristic smell which distils without decomposition at atmospheric pressure, b.p. 279-280° (pentafluoropyridine is a liquid which is more volatile than pyridine). Pentachloropyridine and many of its derivatives are steam volatile and many of them sublime.

Polychloropyridines show a decreasing base strength with increasing chlorine content. Consequently, the highly chlorinated pyridines are reluctant to form salts. However, pentachloropyridine, tetrachloro-2-fluoropyridine and 3,5-dichlorotrifluoropyridine are readily methylated by methyl fluorosulphonate to give the corresponding N-methylpyridinium fluorosulphonate [170]. O-Methyldibenzofuranium fluoroborate similarly gives pentachloro-N-methylpyridinium fluoroborate [171] with pentachloropyridine while triethyloxonium fluoroborate gives the corresponding N-ethylpyridinium salt [170]. Treatment of pentachloropyridine with a mixture of acetic acid and concentrated sulphuric acid gives tetrachloro-2-hydroxypyridine [172]. The latter reaction proceeds via exclusive attack of acetic acid in the α-position of the pentachloropyridinium ion followed by loss of hydrogen chloride and hydrolysis of the product (see also Sections 2.9, 2.11, and 2.14). In addition, the hexafluoroantimonate salts of 3,5-dichlorotrifluoropyridine and various other polyhalogenoheterocycles have been isolated as stable crystalline salts [173]. 4-Aminotrichloropyridine-2-carboxylic acid

(86)

(picloram) (86) is reported [174] to form salts with hexafluoro-arsenic and -phosphonic acid. In this case, however, salt formation probably involves the exocyclic nitrogen atom.

Mixtures of polychloropyridines, which arise from liquid or vapour-phase chlorination procedures (see later), can be separated partially by treatment in solution in dichloromethane with 98% sulphuric acid [175]. Some of the compounds are soluble in the acid as salts and are reprecipitated on dilution with water. For example, 2,3,4,5-tetrachloropyridine (soluble) can be separated from its 2,3,5,6-tetrachloro-isomer.

The position of a ring proton in a polychloropyridine can often be established by reference to its chemical shift [176-179]. For example, the α-, β-, and γ-protons in 2,3-dichloropyridine appear at

τ 1.70, 2.80, and 2.22, respectively, in the NMR spectrum [176]. whilst those in 2,3,4,5-, 2,3,4,6-, and 2,3,5,6-tetrachloropyridine appear at τ 1.65, 2.59, and 2.10, respectively [177]. The structures of polychloropyridines in which ring protons are absent have been determined mainly by classical methods. Thus, pentachloropyridine gives a mixture of the 2- and the 4-monosubstituted product with piperidine. The structures of these follow from the fact that the former gives two disubstituted products with an excess of piperidine whilst the latter gives only one (Scheme 2.1) [180]. Hydrodechlori-

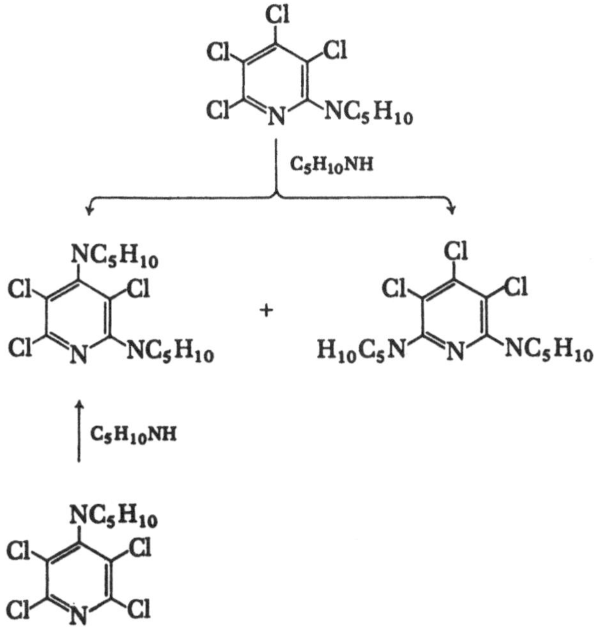

SCHEME 2.1

nation has been used more often to establish structures; examples are given in the following sections. The molecular core binding energies have been measured by means of ESCA for pyridine and its pentachloro- and pentafluoro-derivatives and interpreted in terms of all valence electron CNDO/2 SCF MO calculations [1], and ^{35}Cl NQR spectroscopy has been used also to study polychloropyridines [181]. These more recent techniques may develop further in the future to form the basis for the determination of substitution patterns in these and related (see Sections 2.11-13) molecules.

In contrast to 2- and 4-hydroxypyridine, which exist predominantly as pyridones in solution and in the solid state, comparisons of the UV spectra of tetrachloro-2-hydroxypyridine and tetrachloro-4-hydroxypyridine with the spectra of their *O*- and *N*-alkylated derivatives have shown that they exist exclusively as pyridinols in solution in water or carbon tetrachloride [182, 183]. Tetrafluoro-4-hydroxypyridine also exists as a pyridinol in carbon tetrachloride [184]. Single-crystal X-ray diffraction studies show that tetrachloro-4-hydroxypyridine also exists as a pyridinol in the solid state [185]. In contrast, 3,5-dichloro-4-hydroxy-2,6-dimethyl-

(86a)

pyridine exists exclusively as a pyridone in the solid state [185]. A study of the infrared spectra of 3-, 4-, 5-, and 6-chloro-2-hydroxy-pyridine (pyridones) isolated in an argon matrix shows that the enol/keto ratio decreases in the order $6 > 5 > 4 > 3$ [185a]. This trend can be rationalized on the basis of the inductive effect of the chlorine atom on the stability of the zwitterionic resonance contributor (86a) of the keto-tautomer. A 6-chlorine atom destabilizes such a contributor while a 3-chlorine atom has the opposite effect. A 6-chlorine atom exerts a stronger effect than a 3-chlorine atom. This rationalization also explains why 3,5,6-trichloro-2-hydroxypyridine exists predominantly in the enol form in an argon matrix [185a] and accounts for the contrasting behaviour of tetrachloro-4-hydroxy-pyridine and 3,5-dichloro-4-hydroxy-2,6-dimethylpyridine.

The IR and laser Raman spectra of pentachloropyridine have been measured in both solution and crystalline phases in the region 70-4000 cm^{-1} [186].

The mass spectrum of pentachloropyridine *N*-oxide is noteworthy. At an inlet temperature of 50°, the molecular ion peak is observed at m/e 267 and the base peak due to loss of oxygen is observed at m/e

(87)   (88)

251. At higher inlet temperatures, however, loss of oxygen is not observed and a new base peak appears at m/e 204, corresponding to loss of COCl [180]. This has been attributed to a rearrangement of the *N*-oxide either to the oxaziridine (87) or to the oxazepine (88).

The chemistry of halogenopyridines is discussed in the usual heterocyclic textbooks and has been reviewed elsewhere [46, 187, 188]. However, pentachloropyridine and the tetrachloropyridines have not been given the coverage which they deserve and, therefore, we have discussed much of the early chemistry in this review. Industry has shown a considerable interest in polychloropyridines because they exhibit a broad spectrum of biological activity. Industrial aspects of their chemistry have been reviewed by Ivashchenko and Moschitskii [189] and elsewhere in this volume by Green (Chapter 4).

There are 1624 possible halogenopyridines [190] of which not more than about 100 are reported in the literature. The last of the nineteen possible chloropyridines to be synthesized, namely 2,4,5-trichloropyridine, was reported in 1950 by den Hertog *et al.* [191]. In this paper the authors reviewed much of the earlier work and cleared up a few inconsistencies which existed at that time.

2.8.2   Synthesis of Pentachloropyridine and its Derivatives

By Direct Chlorination Methods
The vapour-phase chlorination of pyridine with chlorine has been discussed in detail elsewhere [46, 187-189, 192, 193]. The product obtained largely depends on the reaction conditions and, in the absence of a suitable catalyst, consists mainly of lower chlorinated pyridines. A small amount of pentachloropyridine was isolated from an experiment carried out at 250° over asbestos as a contact material [192, 194]. More recently, however, vapour-phase chlorination of pyridine has been suitably modified to allow pentachloropyridine to be prepared on a commercial scale [195-197]. The reactants, usually diluted with nitrogen or carbon tetrachloride, are passed through a reactor tube at about 500°. If nitrogen is used as the diluent, the reactor tube is packed with a suitable contact material, for example activated carbon or siliceous earth. 2-Chloropyridine and α-picoline (see later) similarly give pentachloropyridine [195]. At lower temperatures mixtures of lower chlorinated pyridines are produced [195, 196]. The perchloro-derivatives of 2,2'- [196], 2,3'- [198], 2,4'- [198], and 4,4'-bipyridyl [196], tetrachloro-2-, 3-, and 4-cyanopyridine [196, 197], trichloro-2,5-, 2,6-, and 3,5-dicyanopyridine [196, 197], and other polychlorinated heterocycles [196] may be prepared similarly.

A detailed study has been made of the vapour-phase chlorination of pyridine with chlorine diluted with carbon tetrachloride in an unpacked glass tube [199]. This is discussed in Chapter 4 (cf. Fig. 4.17).

Liquid-phase chlorination of pyridine with chlorine gives mainly tars [200] from which a small amount of 3,4,5-trichloropyridine [201, 202] can be isolated. Chlorination of fused pyridine hydrochloride with chlorine for several weeks at 115-120° gives 2,3,4,5-tetrachloropyridine as the major product [203], together with 3,5-dichloro- [204], 2,3,5-trichloro- [205], and pentachloro-pyridine, and a number of other minor products [206, 207]. After 12 h at 165-175°, the major product is 3,5-dichloropyridine, although 3-chloro-, 3,4,5-trichloro-, and pentachloro-pyridine are formed also [193]. In contrast, chlorination of fused 2-chloro-pyridine hydrochloride at 110-125° gives a 70% yield of 2,3,5,6-tetrachloropyridine [208]. 2,6-Dichloropyridine, a by-product in the manufacture of 2-chloropyridine, similarly gives either 2,3,5,6-tetra-chloro- (96%) or pentachloro-pyridine (85%), depending on the amount of chlorine used [209]. A halogen carrier (FeCl₃ or I₂) is necessary in this case.

Chlorination of pyridine with a large excess of phosphorus pentachloride in a sealed tube at 210-220° for 15-20 h [168, 169] gives a mixture containing 2,6-dichloro- [210], 2,3,5- [205], 3,4,5-[201, 202] and 2,3,6-trichloropyridine [191, 211], the isomeric tetrachloropyridines [212], and pentachloropyridine. Increased temperature and time of reaction and the use of a nickel-lined autoclave allow pentachloropyridine to be prepared by this method in nearly quantitative yield [213-216]. However, accompanying pressure and autoclave corrosion problems render the method potentially hazardous on the scales which have been employed (see, however, ref. 217). Tetrachloro-4-cyanopyridine [218] and octachloro-2,2'-bipyridyl [219, 220] may be prepared similarly. 2- or 3-Cyano-pyridine yield only pentachloropyridine under similar conditions [218].

Chlorination of pyridine in refluxing sulphuryl chloride or phosphorus oxychloride is reported to give a mixture of 3,5-dichloro- and 2,3,4,5-tetrachloro-pyridine [221], although some 3,4,5-trichloro-pyridine appears to be formed also [183].

Vapour-phase chlorination of α-picoline [196], chlorination of fused α-picoline hydrochloride [202, 222, 223] with chlorine, or photochlorination of α-picoline at 50-150° in the presence of water [224], give mixtures of polychloro-2-trichloromethylpyridines. Further chlorination of these mixtures can be accomplished in the presence of a Lewis acid and/or on irradiation of the reaction mixture, to give 3,4,5-trichloro- (89; R = H) and tetrachloro-2-tri-chloromethylpyridine (89; R = Cl) as the major products [225-230].

(89)

The latter product (89; R = Cl) is the starting material for the synthesis of the industrially important compound picloram (86) (see Chapter 4). Vapour-phase photochlorination of the mixtures of polychloro-2-trichloromethylpyridines provides a convenient synthesis of the tetrachloropyridines and pentachloropyridine [231]. It is noteworthy here that a trichloromethyl group in these compounds is converted into a trifluoromethyl group on treatment with antimony trifluorodichloride [232, 232a].

Exhaustive vapour-phase chlorination of 2-, 3-, or 4-ethylpyridine similarly gives perchloro(2-, 3-, or 4-vinylpyridine) [196, 233]. Perchloro(3,5-divinylpyridine) [233], perchloro(2-cyano-5-vinyl-pyridine) [234], and tetrachloro-2-trifluoromethylpyridine [196] may be prepared similarly. In the last example some pentachloro-pyridine is formed also.

With sulphuryl chloride at 120° in a sealed tube, pyridine N-oxide hydrochloride gives a mixture (65% yield) of 2- and 4-chloropyridine (57:43), which contains a trace of pentachloropyridine [235]. 4-Nitropyridine-N-oxide similarly gives 2,4-dichloropyridine as the major product, together with 2,3,4,5-tetrachloropyridine [236]. With an excess of the same reagent at 140°, 4-hydroxypyridine-N-oxide gives compound (90), probably via the formation of tetra-chloro-4-hydroxypyridine and the N-chloro-compound (91) [237].

(90)   (91)   (92)

Reduction of compound (90) with various reagents yields tetra-chloro-4-hydroxypyridine. As expected, 3-chloro- and 3,5-dichloro-4-hydroxypyridine-N-oxide give similar yields of the adduct (90), whilst 4-hydroxypyridine and 4-hydroxy-3-nitropyridine-N-oxide also give (90), albeit in low yields [237].

Chlorination of the pyridine nucleus is usually facilitated by the presence of an amino- or hydroxyl group in the starting material, depending on the reaction conditions. One of the first reactions of this kind to be described in the literature was the reaction of glutazine (92) with a mixture of phosphorus pentachloride and phosphorus oxychloride, which yields 4-aminotetrachloropyridine and 4-aminotrichloro-2-hydroxypyridine as the major products, to-gether with 4-amino-2,3,6-trichloropyridine and a fourth unidenti-fied compound [238, 239]. With phosphorus pentachloride, citra-zinic acid (2,6-dihydroxypyridine-4-carboxylic acid) gives a variety

of products, depending on the reaction conditions [168, 240].
Prolonged reaction times yield 2,3,5,6-tetrachloro- and pentachloropyridine. This reaction was the first recorded synthesis of these
compounds [168].

A mixture of phosphorus pentachloride and phosphorus oxychloride converts 3,5,6-trichloro-2-hydroxypyridine into 2,3,5,6-
tetrachloropyridine [241], 2,6-dihydroxy-3-phenylpyridine into
tetrachloro-3-phenylpyridine [242], and 3,4,5-trichloro-2-hydroxy-
pyridine into 2,3,4,5-tetrachloropyridine [206]. Phosphorus oxychloride alone converts 3,5,6-trichloro-2-hydroxy-4-methylpyridine
into   tetrachloro-4-methylpyridine   [243],   3,5-dichloro-2,4-
dihydroxypyridine into 2,3,4,5-tetrachloropyridine [244], 2,3-
dichloro-4-hydroxy-6-pyridone into 2,3,4,6-tetrachloropyridine,†
[244a], 5-bromo-2,4-dihydroxy-3-nitropyridine into 5-bromo-2,4-
dichloro-3-nitropyridine [244], 3-bromo-5-chloro-2,4-dihydroxy-
pyridine into 3-bromo-2,4,5-trichloropyridine [236], 5-bromo-3-
chloro-2,4-dihydroxypyridine into 5-bromo-2,3,4-trichloropyridine
[244, 245], and 3,5-dichloro-2-hydroxy-6-methylpyridine into
2,3,5-trichloro-6-methylpyridine [246].

3,5,6-Trichloro-2-hydroxypyridine may be prepared by chlorination of 6-chloro-2-hydroxypyridine with a mixture of hydrogen
peroxide and hydrochloric acid in acetic acid [241].

A convenient laboratory synthesis of pentachloropyridine starts
with commercially available 2,6-diaminopyridine (Scheme 2.2)

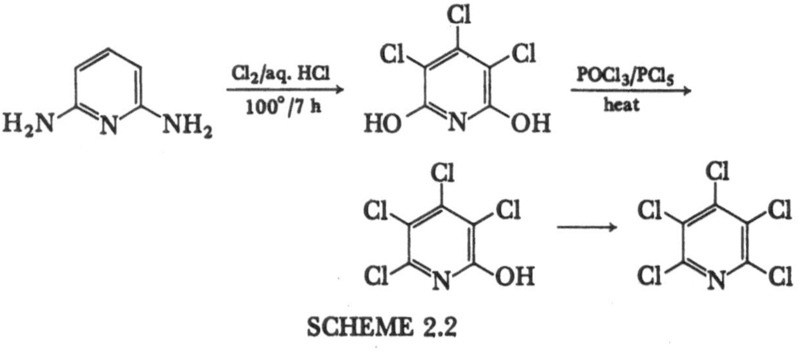

SCHEME 2.2

[247]. On treatment with chlorine in hydrochloric acid, this gives
trichloro-2,6-dihydroxypyridine, which can be further chlorinated
with a mixture of phosphorus pentachloride and phosphorus oxychloride to give either pentachloropyridine or, under milder con-

† This reaction was carried out at 180° under pressure; refluxing phosphorus
oxychloride gave another product, which was assumed to be 2,3,6-trichloro-4-
hydroxypyridine. Recent evidence [302] suggests, however, that the compound
obtained was in fact the 3,4,6-trichloro-2-hydroxy isomer.

ditions, tetrachloro-2-hydroxypyridine. A similar conversion of the latter compound into the perchloro-compound is reported [247, 248] to be difficult.

When 3-aminopyridine is heated at $110°$ with a mixture of hydrochloric acid and hydrogen peroxide, it gives 3-amino-2,6-dichloropyridine (4%) and 3-aminotetrachloropyridine (6%) [249]. The same reagent system (usually used in acetic acid) converts 3-amino-2,4-dichloropyridine into 3-amino-2,4,6-trichloropyridine [191], 4-amino-2,3-(and 2,5)-dichloropyridine into 4-amino-2,3,5-trichloropyridine [191], 4-amino-2,6-dichloropyridine into 4-amino-2,3,6-trichloropyridine [191], and 2-amino-6-methylpyridine into 2-amino-3,5-dichloro-6-methylpyridine [246]. Yields are generally good in these cases.

2-Amino-3,4,5(and 3,5,6)-trichloropyridine [206], 4-amino-2,3,5-trichloropyridine [212], 2-amino-6-chloropyridine-4-carboxylic acid [210], and 2-amino-6-hydroxypyridine-4-carboxylic acid [210] may be converted into 2- or 4-aminotetrachloropyridine, respectively, by further chlorination with phosphorus pentachloride. 2,6-Dichloropyridine-4-carboxylic acid and 2,3,5,6-tetrachloropyridine may be converted similarly into pentachloropyridine [168].

When pyridines containing an α- or γ-bromine atom are heated with 25% hydrochloric acid, the bromine atom is replaced by chlorine; at the same time an ethoxyl group in the starting material is usually converted into a hydroxyl group. For example, 6-bromo-3,5-dichloro-2-ethoxypyridine gives 3,5,6-trichloro-2-hydroxypyridine [equation (2.2)], and 2,3,6-tribromo-5-ethoxypyridine gives

$$
\underset{\substack{\text{Br}}}{\overset{\text{Cl}}{\bigcirc}}\text{OEt} \quad \xrightarrow[\substack{160°/4\text{ h}}]{25\%\text{ aq. HCl}} \quad \underset{\text{Cl}}{\overset{\text{Cl}}{\bigcirc}}\text{OH} \tag{2.2}
$$

3-bromo-2,6-dichloro-5-ethoxypyridine [241]. In the latter case, the ethoxyl group proved particularly resistant to hydrolysis.

When the nitrile (93) (p. 230) is heated and photolyzed in the presence of chlorine, it gives a number of products, depending on the reaction conditions (Scheme 2.3) [250]. At $250°$, the nitrile (94)

$$
\bigcirc \text{NCH}_2\text{CH}_2\text{CN}
$$

(94)

gives tetrachloropyrimidine (Section 2.12), pentachloropyridine, and a number of minor chlorinated products [250].

Prechlorination of piperidine with chlorine at low temperatures

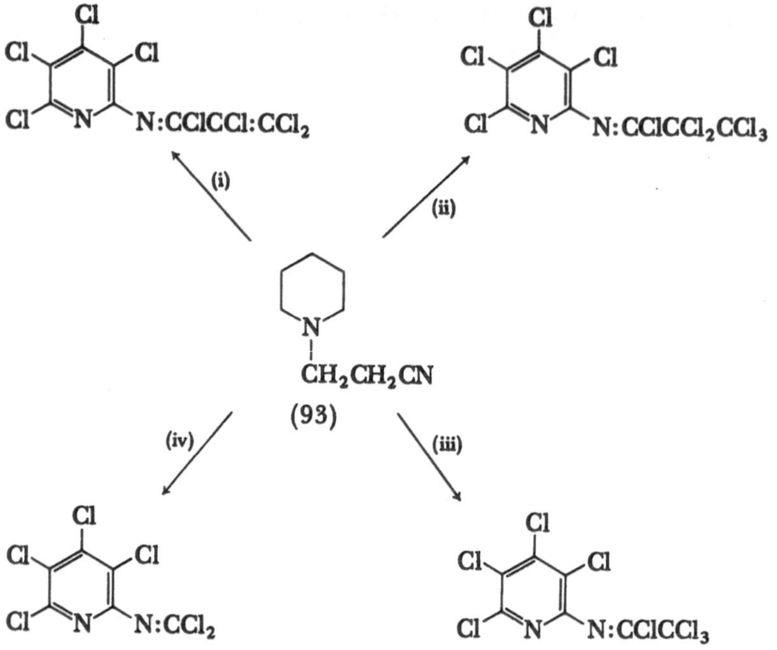

Conditions: (i) Cl_2, $<150°$; (ii) Cl_2, 150–200°; (iii) Cl_2, 250°; (iv) Cl_2, 300°: heat and light in each case.

SCHEME 2.3

(50-150°), initially in a solvent such as chloroform, followed by further chlorination at higher temperatures (150-500°) in the presence of a catalyst (e.g., ferric chloride) and light, yields mainly a mixture of tri- and tetra-chloropyridines [251]. Pentachloropyridine is formed together with other products on chlorination of piperidine with chlorine diluted with carbon tetrachloride at 580° [252]. *N*-Methylpiperidine [253] and piperidine-*N*-carbonyl chloride [69, 71] similarly give pentachloropyridine, together with lower chlorinated pyridines. Photochlorination of 2,6-dicyano-*N*-methylpiperidine in chloroform at 50-60° for 67 h followed by further treatment of the product with chlorine in an autoclave at 195° in the presence of charcoal or further chlorination of the initial product in the vapour phase is claimed [254] to give trichloro-2,6-dicyanopyridine.

An unusual but remarkably simple method for the preparation of pentachloropyridine and mixtures of the isomeric tetrachlorocyanopyridines involves a photochemically initiated self-condensation of acrylonitrile, to give 1,2-dicyanocyclobutane, which is then mixed with chlorine and a diluent (nitrogen, carbon tetrachloride, or chloroform) and passed through a reactor tube at 400° [255]. In a typical experiment the products were pentachloropyridine (6.5 mole%), tetrachloro-2- (44 mole%), 3- (28 mole%), and 4-cyano-pyridine (18.5 mole%). Various lower aliphatic nitriles can be used as starting materials in this process [256]. For example, n-valeronitrile gives mainly pentachloropyridine, whilst adiponitrile and 1,4-dicyanobut-2-ene give mainly a separable mixture of the isomeric tetrachlorocyanopyridines.

A mixture of cyanuric chloride (about 70 mole%), pentachloropyridine (27 mole%), hexachlorobenzene (2 mole%), and two un-identified components (about 0.5 mole% each) is claimed [257] to be the product of a reaction between ammonia and carbon tetrachloride carried out at 450° under pressure.

Of obvious importance, since the starting materials are commercially available on a large scale, is the recent claim [258] that pentachloropyridine is the product of exhaustive chlorination of ε-caprolactam or its industrial precursor, cyclohexanone oxime. The reaction is carried out under conditions similar to those used for high temperature vapour-phase chlorination of pyridine (see before).

By Ring-Closure Methods

β-Hydroxyglutaramide (95) gives 2-aminotetrachloropyridine on treatment with phosphorus pentachloride in a sealed tube [210].

$$\text{OH}$$

(95)

Likewise, glutarimide, α-methyl-, and α,α'-dimethylglutarimide yield 2,3,6-trichloro-, 2,5,6-trichloro-3-methyl-, and 2,6-dichloro-3,5-dimethyl-pyridine [211].

1,2,3,4-Tetrachloro-5,5-dimethoxycyclopentadiene undergoes adduct formation with phenyl cyanoformate to give methyl phenyl trichloro-pyridine-2,6-dicarboxylate (Scheme 2.4) [259], and with various aroyl and sulphonyl cyanides to give compounds (96) and (97), respectively [260, 261].

The aldehyde $CCl_2:CCl\cdot CCl:CCl\cdot CHO$ rearranges on being heated to give the acid chloride $CHCl:CCl\cdot CCl:CCl\cdot COCl$. Hydrazones (98; R^1 = NHAr; R^2 = H) of the aldehyde or amides (99; R^1 = H, Me, Ph; R^2 = H) or hydrazides (99; R^1 = NHAr; R^2 = H) derived from the acid chloride give the corresponding trichloro-2-pyridone (100; R^2 = H)

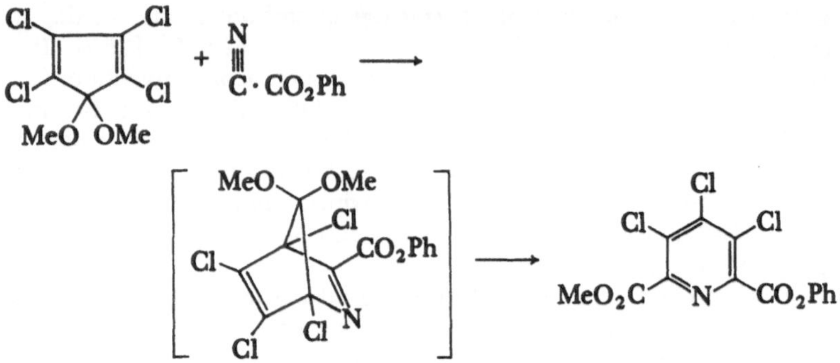

SCHEME 2.4.

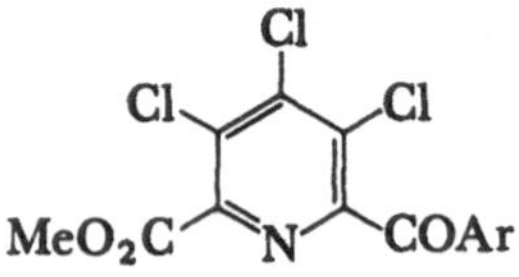

(96)

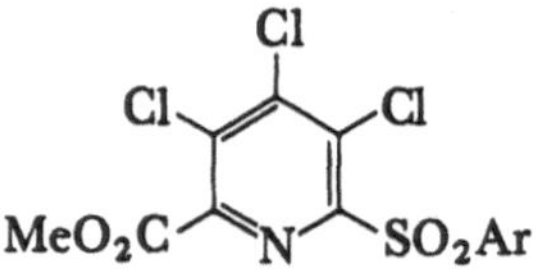

(97)

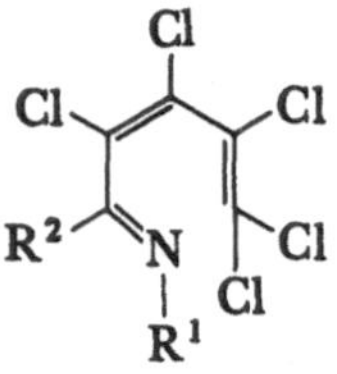

(98)

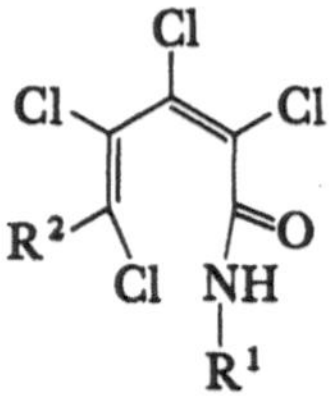

(99)

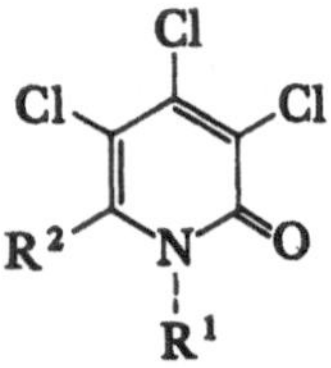

(100)

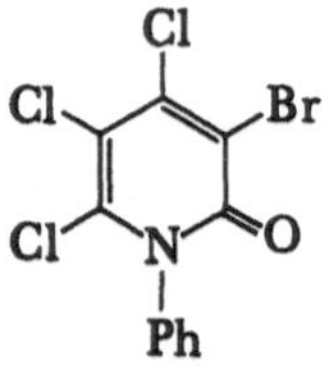

(101)

on being heated in a suitable solvent (e.g., acetic acid) [262, 263]. Similarly, the amides (99: R^1 = H, Me, Ph; R^2 = Cl) give the corresponding tetrachloro-2-pyridone (100; R^2 = Cl) [262, 264], and mixed bromochloro-derivatives [e.g., (101)] may be prepared likewise [265].

This reaction may be extended to the synthesis of pentachloro-pyridine, as shown in Scheme 2.5 [248]. Perchlorocyclopenten-3-one is treated with ammonia and the resulting amide is heated in benzene with phosphorus pentachloride. Although pentachloro-pyridine is the major product, some tetrachloro-2-hydroxypyridine is also formed.

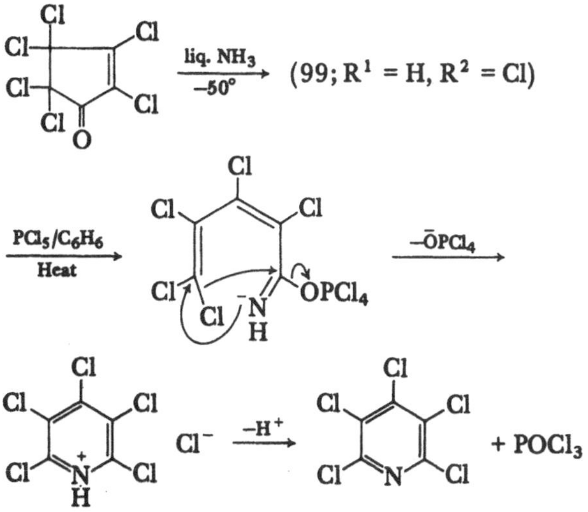

$$\xrightarrow[-50°]{\text{liq. NH}_3} \quad (99; R^1 = H, R^2 = Cl)$$

$$\xrightarrow[\text{Heat}]{\text{PCl}_5/\text{C}_6\text{H}_6}$$

$$\xrightarrow{-\bar{O}PCl_4}$$

$$Cl^- \xrightarrow{-H^+} \quad + POCl_3$$

SCHEME 2.5

Tetrachloro-N-methyl-2-pyridone is converted into pentachloro-pyridine on treatment with a mixture of phosphorus pentachloride and phosphorus oxychloride [264]. Likewise, treatment of compound (100; $R^1 = R^2 = H$) with phosphorus pentachloride yields 2,3,4,5-tetrachloropyridine [265].

The conversions (98) → (100) proceed via the intermediates (102), which are hydrolyzed on work-up. If the oxime (98: $R^1 = OH; R^2 = H$) is employed in this synthesis, the intermediate (103) loses hydrogen chloride spontaneously to give 2,3,4,5-tetrachloropyridine-N-oxide [263]. With acetic anhydride in acetic acid, this N-oxide gives the acetoxy compound (104), which is hydrolyzed to 3,4,5-

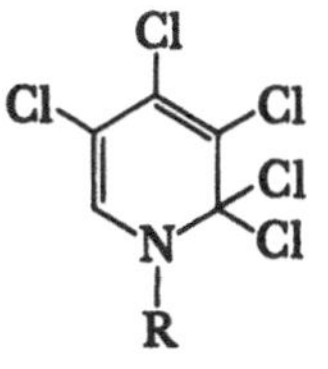

(102)

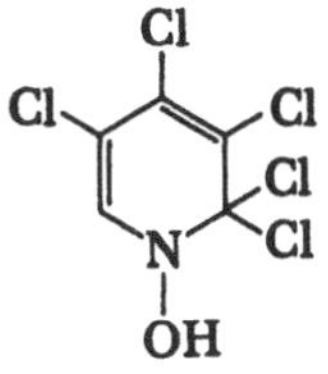

(103)

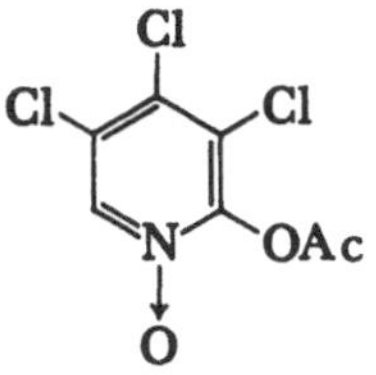

(104)

trichloro-1-hydroxy-2-pyridone on work-up. The latter compound is formed also when the acid chloride mentioned before is treated with hydroxylamine and the intermediate *O*-acylhydroxamic acid is hydrolyzed [263]. 2,3,4,5-Tetrachloropyridine-*N*-oxide gives pentachloropyridine on being heated with sulphuryl chloride [263].

The perchloro-*cis,cis*-heptatrienal (105) reacts with phenyl-

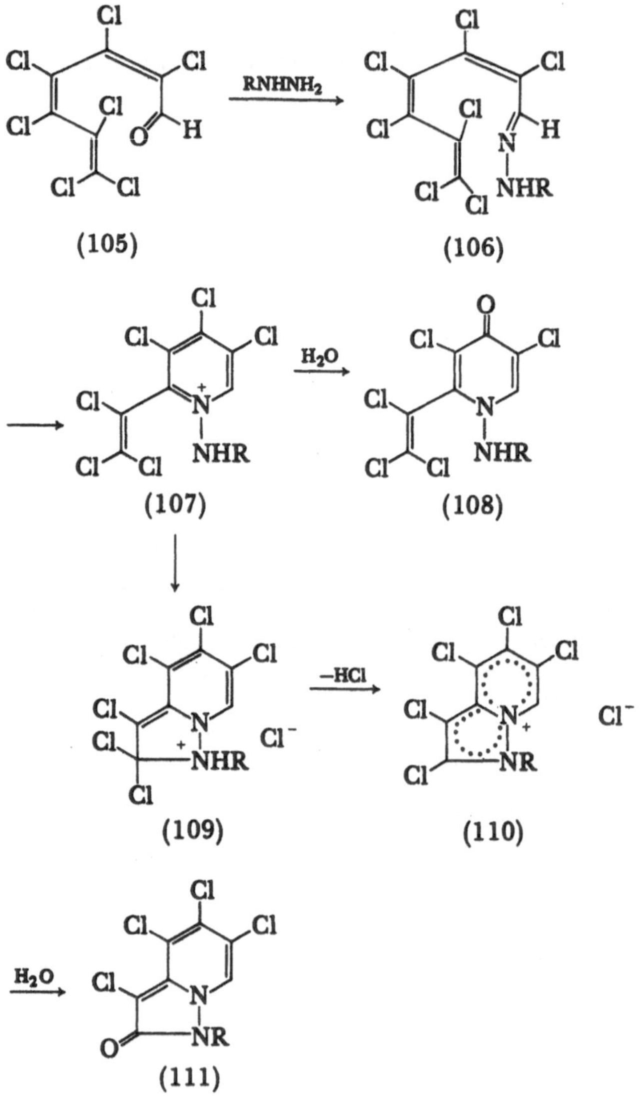

SCHEME 2.6

hydrazine, 2,4-dinitrophenylhydrazine, or semicarbazide to give the carbonyl derivatives [106; R = Ph, $C_6H_3(NO_2)_2$-2,4, or $CONH_2$] (Scheme 2.6) which give the corresponding 3,5-dichloro-2-trichloro-vinyl-4-pyridone (108) on being heated in a refluxing protic solvent containing water [266]. In dry, refluxing aprotic solvents, however, the corresponding 2,3,4,5,6-pentachloropyrazols[1,5a]pyridinium salt (110) is obtained. The latter compounds give quantitative yields of the 2-oxo-3,4,5,6-tetrachloro-1,2-dihydropyrazolo[1,5-a]pyridines (111) on hydrolysis [266]. The reactions proceed via the pyridinium salts (107), which are hydrolyzed in the presence of water or, in its absence, spontaneously cyclize via the intermediates (109) to (110). 3,4,5-Trichloro-1-(2,4-dinitroanilino)-2-phenylpyridinium   chloride (113) which is analogous to the non-isolable intermediates (107), may

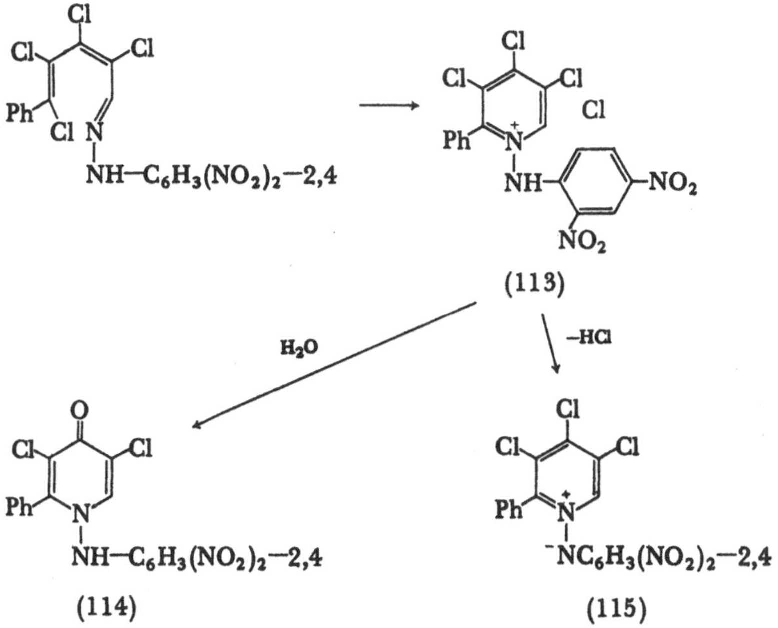

SCHEME 2.7

be prepared similarly from the hydrazone (112) [266]. It is hydrolyzed by water to the 4-pyridone (114), but is deprotonated by non-aqueous bases to give the pyridiniumimin betaine (115) (Scheme 2.7).

The nitrile (116), which may be prepared by treatment of the corresponding amide with phosphorus oxychloride, reacts with alkyl or aryl Grignard reagents to give the corresponding 2-substituted tetrachloropyridine (117; R = Me, Ph, C_6H_4F-p, C_6H_4Me-p,

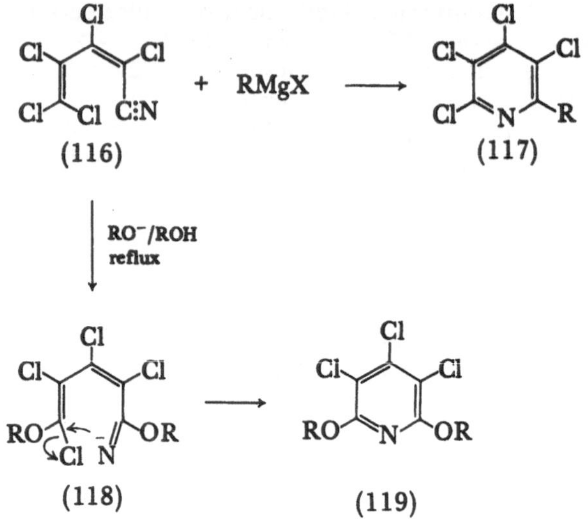

(116)   (117)

(118)   (119)

SCHEME 2.8

$C_6H_4OMe\text{-}p$, 2-thienyl, or α-naphthyl) [267] (Scheme 2.8) and with alkoxide ions in the corresponding alcohol to give a 2,6-disubstituted 3,4,5-trichloropyridine (119; R = Me, Et, CH_2Ph, or cyclohexyl) via the intermediates (118) [268].

By Nucleophilic Substitution of Pentachloropyridine and its Derivatives

As long ago as 1898 Sell and Dootson [206, 239] showed that pentachloropyridine gives mainly 4-aminotetrachloropyridine, together with a small amount of the 2-amino-compound, on treatment with ethanolic ammonia. In Table 2.1 we have listed some of the nucleophiles with which pentachloropyridine has been reacted together with the isomer ratios of the products, where these have been reported. The yields are usually good and in many cases are quantitative. It appears that larger nucleophiles generally react preferentially in the 2-position; some, such as morpholine, react exclusively in this less-hindered position. The report [248] that morpholine reacts exclusively in the 4-position is incorrect. Small nucleophiles (e.g., SH [269, 270]), on the other hand, may react exclusively in the more activated 4-position. With aniline in dimethylformamide, pentachloropyridine yields the expected 4-anilino-derivative. However, when carried out in a mixture of dimethylformamide and pyridine, this reaction gives a mixture of 4-aminotetrachloropyridine (70% yield) and N-phenylpyridinium chloride (82%) [271]. The authors have postulated the quaternary salt (120) as an intermediate without indicating how it reacts further to give the observed products.

(120)

However, the outcome of substitution in pentachloropyridine is not as clear-cut as it appears at first sight. Thus, the reactions are often solvent dependent (Table 2.1). Pentachloropyridine reacts, for example, with dimethylamine, pyrrolidine, or piperidine in benzene to give almost exclusively the 2-substituted product, whilst a mixture of the 2- and the 4-isomer is invariably formed in ethanol. In an aprotic solvent, the amine probably hydrogen bonds to the ring nitrogen atom, thus promoting 2-substitution. In a protic solvent there is competitive hydrogen bonding with the ring nitrogen atom between the solvent and the amine. In contrast, pentafluoropyridine reacts with all nucleophiles in the 4-position.

The chlorine atoms in pentachloropyridine can be replaced stepwise by treatment with sodium or potassium fluoride in a solvent such as dimethylformamide, sulpholane, or N-methyl-2-pyrrolidone [213, 215, 216, 272-275]. It is difficult to replace more than three fluorine atoms in this way, the final product being 3,5-dichlorotrifluoropyridine, because the reflux temperature of the reaction mixture is lowered as the fluorine atoms are successively introduced (certain additives, for example chloroacetic acid, are claimed [275] to promote the reaction). The remaining β-chlorine atoms in 3,5-dichlorotrifluoropyridine may be replaced by continuously leading the collected distillates back into the bottom of the reaction vessel [276]. Attack by fluoride ion occurs initially in both the 2- and the 4-position; the ratio of the two tetrachloromonofluoropyridines is dependent on the reagent, solvent, and time of reaction. With potassium fluoride in sulpholane, for example, the ratio of α:γ-substitution is 1:2 at t_{30} min and 95:5 (i.e., in favour of α-substitution) at t_∞ [277]. Disubstitution yields a mixture of trichloro-2,4 (and 2,6)-difluoropyridine. At t_{30} min the ratio of 2,4:2,6-disubstitution is 2:1 but at t_∞ it becomes 5:95. It is difficult to control these reactions and invariably mixtures of the mono-, di-, and trifluorinated compound are obtained.

Pentafluoropyridine is best prepared by treatment of pentachloropyridine with anhydrous potassium fluoride in an autoclave at 500° [213-216]. Lower temperatures afford mixtures of chlorofluoropyridines. In an attempt to avoid the use of autoclaves in the preparation of pentafluoropyridine the vapour of trichloro-2,6-

difluoropyridine was diluted with nitrogen and led through a fused melt of potassium chloride and potassium fluoride at 750° [278]. This gave a mixture of pentafluoro- (17 mole%), 3-chlorotetrafluoro- (23 mole%), 3,5-dichlorotrifluoro-pyridine (58.5 mole%), and

TABLE 2.1   Nucleophilic Monosubstitution in Pentachloropyridine[†]

Reagent	Solvent	Ratio of 4:2-substitution[a]	Ref.
NaBr[b]	DMF[c]	100:0	284
NH$_3$	EtOH	70:30	206, 237, 239, 247, 248
	DMSO[d]	—	284a
N$_2$H$_4$	EtOH	80:20	104, 248
MeNH$_2$	Dioxan	68:32	285
EtNH$_2$	Dioxan	68:32	285
BuNH$_2$	EtOH	25:75[e]	247
PhCH$_2$NH$_2$	EtOH	71:29	285
	Dioxan	73:27	285
ArNH$_2$[f]	EtOH	100:0	271
	DMF[c]	100:0	271
Me$_2$NH	EtOH	20:80	247
	EtOH	34:66	180, 286
	C$_6$H$_6$	0:100	180, 286
Et$_2$NH	EtOH	1:99	247
Pyrrolidine	EtOH	20:80	180, 286
	C$_6$H$_6$	0:100	180, 286
Piperidine	EtOH	37:63	180, 286
	C$_6$H$_6$	4:96	180, 286
Morpholine	EtOH	0:100	180, 286
	C$_6$H$_6$	0:100	180, 248, 286
NaOH	EtOH—H$_2$O	100:0	237
	Other solvents[g]	—	277
NaOMe	MeOH	85:15	247, 268
NaOEt	EtOH	65:35	247, 243
NaOBu	BuOH	57:43	247
NaO(CH$_2$)$_2$OR		4-O(CH$_2$)$_2$OR	288
CH(CO$_2$Et)$_2$		4-CH$_2$CO$_2$H	248
		4-CH(CO$_2$Et)$_2$	243
CR(CO$_2$Et)$_2$		4-CR(CO$_2$Et)$_2$	243
NaSH	Various	100:0	269, 270, 289, 290, 291
NaSR	Various	100:0	289, 292, 293, 294
Various xanthates	DMF[c] or Me$_2$CO—CH$_2$Cl$_2$	100:0	295
P(OR)$_3$		4-P:O(OR)$_2$	284

[a] First ref. given is the source of the ratio.   [b] The reactions with fluoride ion are discussed in detail in the text.   [c] Dimethylformamide.   [d] Dimethylsulphoxide.   [e] This is probably an error.   [f] Various aromatic primary amines.   [g] Ratio extremely solvent dependent.

† See also ref. 287

starting material. 3,5-Dichlorotrifluoro- and 3-chlorotetrafluoro-pyridine may be used also as starting materials in this process [279]. A mixture of 3,5-dichlorotrifluoro-, trichloro-2,6-difluoro-, and tetra-chloro-2-fluoro-pyridine is obtained when the vapour of pentachloro-pyridine is diluted with nitrogen, mixed with hydrogen fluoride, and passed over carbon carrying a black chromium oxide (Cr_2O_3) catalyst at 500-600° [280, 281]. The reaction with hydrogen fluoride may be carried out also in an autoclave [282]. Under these conditions replacement of the two α-chlorine atoms appears to occur before attack occurs in the γ-position.

Treatment of tetrachloro-4-cyanopyridine with potassium fluoride in dimethylformamide-dimethylsulphoxide gives a mixture of 3,5-dichloro-4-cyanodifluoro- and trichloro-4-cyano-2-fluoro-pyridine [283]. With anhydrous potassium fluoride in an autoclave, this nitrile yields 4-cyanotetrafluoropyridine [218]. Octachloro-2,2'-bi-pyridyl may be converted similarly into a mixture of the perfluoro-compound and lower fluorinated products [220], while tetrachloro-2-trifluoromethylpyridine gives a mixture of 3,5-dichlorodifluoro-(major product), 3-chlorotrifluoro-, and 5-chlorotrifluoro-2-trifluoro-methylpyridine with potassium fluoride in an autoclave at 400° [232a].

On treatment with sodium bromide in dimethylformamide, penta-chloropyridine yields 4-bromotetrachloropyridine in excellent yield [284].

In most monosubstituted tetrachloropyridines the remaining chlorine atoms are readily displaced by further nucleophilic substitu-tion. Some substituents (e.g., −SH [296]), however, strongly de-activate the system and further substitution is not possible, even under forcing conditions. A 4-substituted tetrachloropyridine gives exclusively the 2,4-disubstituted product, whilst its 2-isomer usually gives a mixture of the 2,4- and the 2,6-disubstituted trichloropyrid-ine. 2-sec.-Aminotetrachloropyridines, however, react with potassium or sodium hydrogen sulphide exclusively in the 4-position to give the corresponding 4-mercaptan [296]. The ratio 2,4:2,6-disubstituted product is often solvent dependent (Table 2.2). An excess of the nucleophile and increased temperature of the reaction may lead to breakdown in the case of amine reactants [247], but high yields of

TABLE 2.2   Nucleophilic Disubstitution in Pentachloropyridine

Reagent	Solvent	Ratio of 2,4:2,6-disubstitution	Ref.
NH_3	EtOH	100:0	247
Me_2NH	EtOH	30:70	247
NaOMe	MeOH	100:0	247
NaOEt	EtOH	99:1	247

2,4,6-trisubstituted derivatives are obtained with the anions derived from alcohols or phenols [247]. In other cases, monosubstituted tetrachloropyridines react with nucleophiles by displacement of the group already present (see later).

A number of nucleophilic substitution reactions of 2,3,4,5-tetra-chloropyridine have been reported; these are summarized in Scheme 2.9 [212, 236, 244, 284a, 289, 293, 297-300]. In no case was mention made of substitution having occurred in the 2-position. When this compound is treated with methanolic sodium methoxide at 70-80°, it gives a mixture of 3,5-dichloro-2,4-dimethoxypyridine (major product), 3,5-dichloro-4-hydroxy-2-methoxypyridine, and 2,3,5-trichloro-4-hydroxypyridine (see Scheme 2.9), depending on

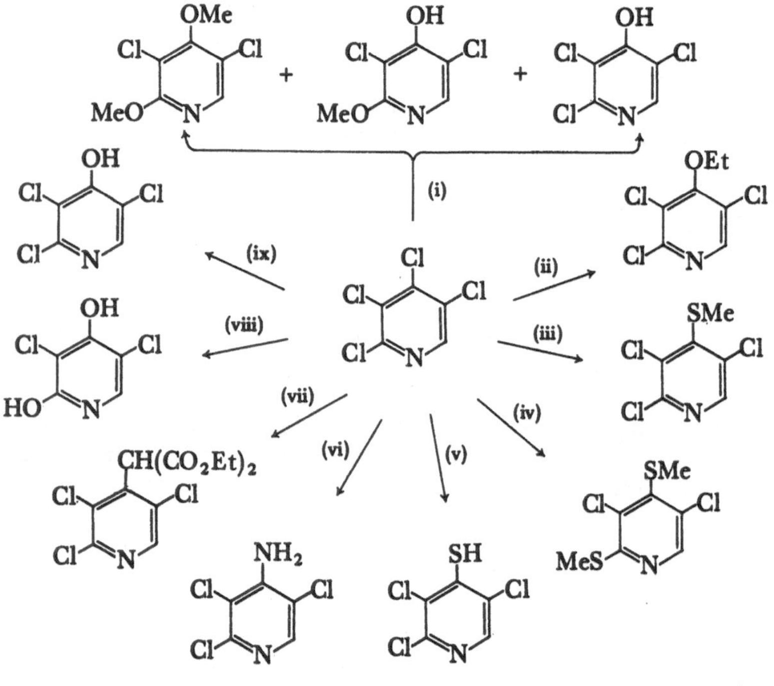

Reagents: (i) NaOMe–MeOH/70-80° [299]; (ii) NaOEt–EtOH/reflux [297]; (iii) NaSMe–MeOCH$_2$CH$_2$OMe/reflux [289]; (iv) KSMe–EtOH/reflux [293]; (v) NaSH–EtOH/65° [289]; (vi) aq. NH$_3$ under pressure/180° [212, 236, 244] or NH$_3$–DMSO [284a]; (vii) NaCH(CO$_2$Et)$_2$–EtOH/reflux [297]; (viii) NaOMe–MeOH under pressure/200-205° [299]; (ix) aq. NaOH/reflux [297].

SCHEME 2.9

the strength of the reagent solution, whilst with a large excess of the reagent at 200° it gives only 3,5-dichloro-2,4-dihydroxypyridine [298, 299]. In the latter reaction dimethyl ether is also formed. Sell [298] has attributed the formation of the hydroxy compounds to demethylation promoted by an excess of the reagent [equation (2.3; $R = C_5Cl_4N$)].

$$ROMe + MeONa \rightarrow RONa + Me_2O \qquad (2.3)$$

2,3,4,6-Tetrachloropyridine may react with nucleophiles in either of the α-positions or in the γ-position [301, 302]. With aqueous sodium hydroxide, for example, it gives a mixture of 2,3,4-trichloro-6-pyridone, 2,3,6-trichloro-4-hydroxypyridine, and 2,4,5-trichloro-6-hydroxypyridine. Amines (dimethylamine or piperidine) similarly give mixtures of three products (the ratio is solvent dependent). Methoxide ion, however, yields only 2,3,6-trichloro-4-methoxypyridine [301, 302], whilst methylthiolate ions are reported [289] to give only 2,3,6-trichloro-4-methylthiopyridine (cf. refs. 269 and 270).

2,3,5,6-Tetrachloropyridine undergoes displacement of an α-chlorine atom with methoxide ion, dimethylamine, or piperidine [300]. With potassium fluoride in dimethyl sulphone it gives 3,5-dichloro-2,6-difluoropyridine [303].

The reactions of 3,4,5-trichloro-2-trichloromethylpyridine (89; R = H) with nucleophiles, particularly amines, have received considerable industrial attention; substitution occurs predominantly, if not exclusively, in the 4-position [226, 230, 304, 305]. Hydrolysis of the 4-amino-compounds with concentrated sulphuric acid give compounds related to picloram (86) [306-308]. It is noteworthy that, whereas tetrachloro-2-trichloromethylpyridine (89; R = Cl) appears to give only the 4-methoxy derivative on treatment with methoxide ion, with ethoxide, isopropoxide, or some other alkoxide ions, it gives a mixture of the 4- and the 6-monosubstituted compound; the 4-isomer predominates in each case [305].

With amines, hydrazines, and hydroxide, alkoxide, phenoxide, thioalkoxide, and thiophenoxide ions, 3,4,5-trichloro-2-trifluoromethyl pyridine gives 4-substituted derivatives [232a]. Tetrachloro-2-trifluoromethylpyridine also gives 4-substituted derivatives with sodium hydroxide, sodium methoxide, sodium sulphide, and isopropylamine.

Tetrachloropyridine-4-carboxylic acid undergoes decarboxylation on being heated with an excess of ammonia at atmospheric pressure to give 2-amino-3,5,6-trichloropyridine (major product) and 2,6-diamino-3,5-dichloropyridine [168, 206]. In an autoclave, decarboxylation is prevented and 2-aminotrichloropyridine-4-carboxylic acid is obtained [309]. 4-Aminotrichloropyridine-2-carboxylic acid may be prepared similarly [309]. 3,4,5-Trichloropyridine-2-carb-oxylic acid gives 4-amino-3,5-dichloropyridine-2-carboxylic acid on being heated with ammonia at atmospheric pressure [201]. Tetra-chloropyridine-2-carboxylic acid and trichloropyridine-2,6-dicarboxylic acid are attacked by hydroxide ion in the 4-position, whilst tetrachloropyridine-4-carboxylic acid gives trichloro-2-hydroxypyridine-4-carboxylic acid [310]. 2,5,6-Trichloropyridine-3-carboxylic acid gives 2,5,6-trichloro-3-hydroxypyridine and 2,3-dichloro-6-hydroxypyridine, respectively, with sodium hydrogen carbonate or 2N-sodium hydroxide (Scheme 2.10) [311, 312]. Like-wise, with alcoholic sodium hydroxide 3,4,5-trichloropyridine-2-carboxylic acid gives 3,5-dichloro-4-hydroxypyridine [201].

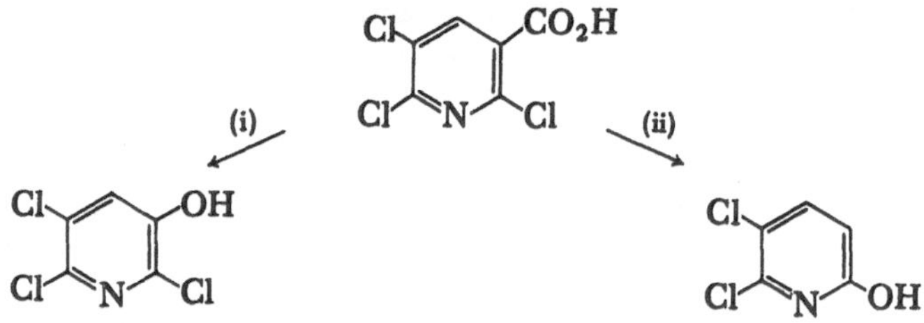

Reagents: (i) NaHCO$_3$; (ii) NaOH

SCHEME 2.10

The nitro-groups in 2- and 4-nitrotetrachloropyridine are more readily displaced than the chlorine atoms [285]. A competition experiment between the 4-nitro-compound and pentachloropyridine for piperidine shows that the nitro-compound is the more reactive. The 2- and the 4-nitro-compound show comparable reactivity towards piperidine in a competition experiment. These displacement reactions can be used advantageously. Thus, the 2-nitro-compound reacts with various nucleophiles by exclusive displacement of the nitro-group; methoxide ion, for example gives a quantitative yield of tetrachloro-2-methoxypyridine [285]. Under the same conditions pentachloropyridine gives a mixture of isomers (Table 2.1). Also, whereas morpholine reacts with pentachloropyridine exclusively in the 2-position, and other bulky nucleophiles react preferentially in this position, the same nucleophiles react with tetrachloro-4-nitro-pyridine to give mixtures of the otherwise elusive 4-substituted tetrachloropyridine (e.g., tetrachloro-4-morphlinopyridine with morpholine) and the corresponding 2-substituted trichloro-4-nitropyridine [285]. The 4-nitro-compound gives a mixture of trichloro-2,4-dimorpholino- and dichloro-2,6-dimorpholino-4-nitro-pyridine with a twofold excess of morpholine [285]. Under the same conditions, pentachloropyridine gives the 2,6-dimorpholino-derivative.

Tetrachloro-4-methylsulphonylpyridine reacts with small nucleophiles (e.g., methylamine, or cyanide, hydroxide, or methoxide ions) by exclusive displacement of the methylsulphonyl group [296]. With dimethylamine, however, attack occurs exclusively in the 2-position, to give trichloro-2-dimethylamino-4-methylsulphonylpyridine, and pyrrolidine gives a mixture of tetrachloro-4-pyrrolidinopyridine and the corresponding 2-aminosulphone [296]. It is claimed [295] that the methylsulphonyl groups in tetrachloro-4-methylsulphonylpyridine and the analogous 2,3,5-trichloro-compound are displaced exclusively by various xanthate salts. Similar reactions on pentachloropyridine yield the same products (see Table 2.1). 2,3,6-Trichloro-4-methylsulphonylpyridine similarly undergoes exclusive displacement of the methylsulphonyl group with methylamine or hydroxide or methoxide ion [302].

The reactivity of tetrachloro-4-methylsulphinylpyridine is comparable to that of the sulphone, whilst the reaction between tetrachloro-4-methylthiopyridine and refluxing aqueous sodium hydroxide, which gives tetrachloro-4-hydroxypyridine, is incomplete after 4 days [296].

Pentachloropyridine-N-oxide (see later) is more susceptible to nucleophilic substitution than pentachloropyridine, and is attacked exclusively in the α-positions. Thus, it reacts with four equivalents of various secondary amines (e.g., pyrrolidine or piperidine) even at room temperature, to give a 2,6-disubstituted trichloropyridine-N-oxide in nearly quantitative yield [180, 286]. Pentachloropyridine gives only monosubstituted products under the same conditions (see before). 2-Monosubstituted tetrachloropyridine-N-oxides are obtained exclusively with two equivalents of the reagent. The resulting N-oxides are readily deoxygenated on treatment with phosphorus trichloride in chloroform [180, 285, 286, 296, 313, 314]. This provides a convenient synthesis of 2-mono- and 2,6-disubstituted polychloropyridines. With certain nucleophiles, substitution and deoxygenation occur concurrently; for example, with hydrazine tetrachloro-2-hydrazinopyridine and 2,3,4,5-tetrachloropyridine (see later) are obtained [104].

Pentachloro-N-methylpyridinium fluorosulphonate (Section 2.8.1) is even more reactive towards nucleophiles than pentachloropyridine-N-oxide [170]. With water at room temperature it gives tetrachloro-N-methyl-2-pyridone (121; R = Cl; X = O) whilst aqueous sodium hydrogen sulphide yields the corresponding thione (121; R = Cl; X = S). Ammonia and methylamine yield the imines (121; R = Cl; X = NH) and (121; R = Cl; X = NMe), respectively. The pyridones are deactivated towards further substitution. With aqueous dimethylamine, however, compound (121; R = NMe$_2$; X = O) is obtained

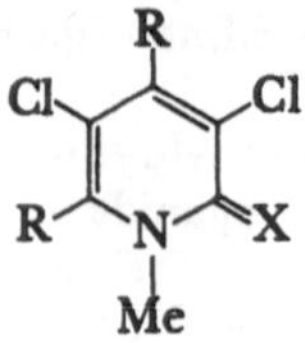

R

Cl, Cl

R   N   X
     |
     Me

(121)

through formation of dichloro-2,4,6-tris(dimethylamino)-N-methyl-pyridinium fluorosulphonate, which undergoes hydrolysis by water. Aqueous sodium azide yields the pyridone (121; R = N$_3$; X = O), presumably by a similar mechanism. Tetrachloro-2-fluoro-N-methylpyridinium fluorosulphonate, 3,5-dichlorotrifluoro-N-methyl-pyridinium fluorosulphonate and pentachloro-N-ethylpyridinium fluorosulphonate (Section 2.8.1) undergo analogous reactions [170].

Pentachloropyridine-N-oxide and 2,3,4,5- and 2,3,5,6-tetrachloro-pyridine-N-oxide each gives a methoxypyridinium methyl sulphate (122, 123, and 124, respectively) on treatment with dimethyl sulphate at 95-100° (cf. the reaction of pyridine-N-oxide with dimethyl sulphate, which is exothermic at ambient temperatures) [315]. Higher yields of these salts are obtained if a large excess of

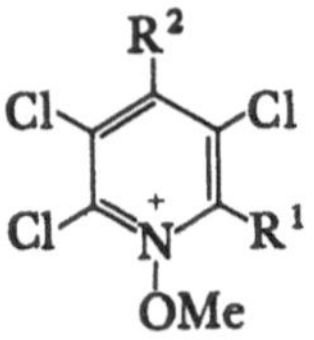

R^2

Cl, Cl

Cl   N   R^1
      |
      OMe

(122) R^1 = R^2 = Cl
(123) R^1 = H, R^2 = Cl
(124) R^1 = Cl, R^2 = H

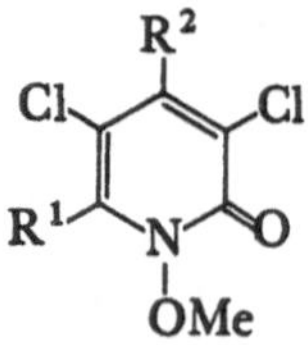

R^2

Cl, Cl

R^1 N   O
         |
         OMe

(125) R^1 = R^2 = Cl
(126) R^1 = H, R^2 = Cl
(127) R^1 = Cl, R^2 = H

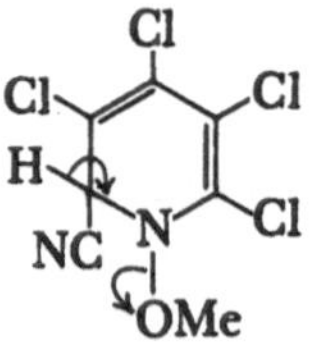

Cl

Cl, Cl

H
      N   Cl
NC   Cl
      |
      OMe

(128)

dimethyl sulphate is used [170]. In the presence of water, the salts yield the corresponding pyridone (125, 126, and 127, respectively) by nucleophilic displacement of an α-chlorine atom, together with pentachloro- or a tetrachloro-pyridine and the corresponding N-oxide

[170, 315]. With aqueous cyanide, the methoxypyridinium salt (123) yields, in addition to the products formed by reaction with water alone, a small amount of tetrachloro-2-cyanopyridine, presumably via the formation of the intermediate (128) [316]. The methoxypyridinium salt (124) similarly gives tetrachloro-4-cyanopyridine, together with 2,3,5,6-tetrachloropyridine and its N-oxide [315].

With methyl fluorosulphonate, pentachloropyridine-N-oxide yields a N-methoxypyridinium fluorosulphonate, which reacts with water to give the pyridone (125) and with aqueous sodium azide to give 2,4,6-triazidodichloropyridine, presumably via attack of water on the intermediate triazido-N-methoxypyridinium fluorosulphonate followed by loss of the methoxyl group as formaldehyde [170].

Unlike tetrachloro-4-methylsulphonylpyridine, which undergoes displacement of the methylsulphonyl group with nucleophiles, the corresponding N-oxide reacts exclusively in the α-positions [296]. With phosphorus pentachloride 2,3,6-trichloro-4-methylsulphonyl-pyridine-N-oxide yields mainly 2,3,4,6-tetrachloropyridine together with 2,3,6-trichloro-4-methylsulphonylpyridine [302].

A number of nucleophilic substitution reactions of 3-chlorotetra-fluoro- (129) and 3,5-dichlorotrifluoro-pyridine (130) have been reported also [104, 184, 273, 274, 290, 291, 317]. As expected, these compounds are intermediate in their reactivity between penta-chloro- and pentafluoro-pyridine; the fluorine atoms are displaced

(129)   (130)

preferentially. Competition experiments have shown that the susceptibility towards nucleophilic substitution of fluorine by ammonia increases in the series $C_5F_5N < C_5ClF_4N < C_5Cl_2F_3N$ in the ratio 1:3.7:12.6, respectively [184]. Substitution occurs in the 4-position in each case.

Hydrazine [104, 184], methoxide ion [184], and other nucleophiles [273, 274, 290, 291] react with 3,5-dichlorotrifluoropyridine (130) by exclusive displacement of the 4-fluorine atom. With morpholine or piperidine in ethanol or dioxan at room temperature, however, 3,5-dichlorotrifluoropyridine (130) gives exclusively the 2-substituted product [317]. Under similar conditions, pentafluoro-pyridine gives only the 4-sec.-aminotetrafluoropyridine. With pyrrolidine or dimethylamine in dioxan, 3,5-dichlorotrifluoropyridine gives exclusively the 2-substituted product, but in ethanol these amines afford mixtures of the 2- and the 4-substituted derivative (ratio 65:35 in both cases) [317]. An excess of piperidine gives only the 2,6-disubstituted product, whilst an excess of dimethylamine yields both the 2,4- and the 2,6-disubstituted product [317]. The interplay between steric and electronic effects is also shown in the reactions of 3,5-dichlorotrifluoropyridine with hydroxide ion. With potassium hydroxide in t-butyl alcohol, attack occurs in the 2- and 4-positions in the ratio 70:30, whilst with aqueous potassium hydroxide the same compounds are produced in the ratio 10:90 [184]. These results suggest involvement of the bulky t-butoxide ion. The highest ratio (100:1) of 3,5-dichloro-2,6-difluoro-4-hydroxypyridine to its 2-hydroxy isomer is obtained when 3,5-dichlorotrifluoropyridine is reacted with potassium or sodium acetate in sulpholane [318] or acetic anhydride [319] and the product is hydrolyzed by acid.

3-Chlorotetrafluoropyridine (129) reacts with piperidine or dimethylamine in ethanol, benzene, or dioxan in both the 4- and 6-positions; the 4-isomer predominates in all cases, but a greater proportion of the 6-isomer is formed in benzene than in ethanol [317]. Hydrazine reacts with this compound in ethanol exclusively in the 4-position [104, 184]. With hydroxide ion, the situation is more complex. In aqueous potassium hydroxide, attack occurs in the 4- and 6-positions in the ratio 90:10 but, with potassium hydroxide in t-butyl alcohol, attack occurs in the 2-, 4-, and 6-positions in the ratio 10:55:35, respectively [184].

As expected, trichloro-2,6-difluoropyridine and tetrachloro-2-fluoropyridine react with nucleophiles mainly by displacement of an α-fluorine atom [273, 274, 320]. With sodium hydrogen sulphide, however, reaction occurs exclusively in the 4-position [320].

With piperidine in benzene, 4-bromotetrachloropyridine gives a mixture of tetrachloro-4-piperidinopyridine and 4-bromotrichloro-2-piperidinopyridine (1:2) (the ratio is 2:3 in ethanol) (cf. ref. 284) [104]. Aromatic amines (e.g., aniline or p-toluidine) are reported [271], however, to attack 4-bromotetrachloropyridine exclusively in the 4-position. Tetrachloro-4-iodopyridine gives a mixture of tetrachloro-4-piperidinopyridine and trichloro-4-iodo-2-piperidinopyridine (1:9) with piperidine in ethanol [104].

Bidentate ligands can be made to react either at one or both functional groups with a polyhalogenopyridine [273, 274, 294].

By the Use of Organometallic Reagents†

Pentachloropyridine can be made to react with magnesium in tetrahydrofuran at $-10°$ using the entrainment procedure [321,

† Metal derivatives of polychloropyridines are discussed also in Chapter 3; however, the approach given there is different and deliberately less comprehensive.

322] or in boiling ether [177, 323] to give tetrachloro-4-pyridyl-magnesium chloride. The reaction of pentachloropyridine with magnesium is more selective than its reactions with most organolithium reagents (see later) and therefore this Grignard compound is probably the preferred organometallic reagent for the synthesis of 4-substituted tetrachloropyridines. Some of its reactions are summarized in Table 2.3.

TABLE 2.3 Products Derived from Tetrachloro-4-pyridylmagnesium Chloride

R	Reagent	Yield (%)	Refs.
H	Acid	46.5, 88	177, 323
CO_2H	CO_2	14, 35–40	177, 323
CHMeOH	MeCHO	30	323
CPh_2OH	Ph_2CO	42	324
Et	EtI	54	325
$C_5Cl_4N^a$	C_5Cl_5N	32	325
$SiMe_3$	$ClSiMe_3$	86	321, 324, 326
$SiPh_3$	$ClSiPh_3$	50	325, 326
$SiMePh_2$	$ClSiMePh_2$	52	326
$SiMe_2Ph$	$ClSiMe_2Ph$	65	321, 326
$SiPh_2(CH:CH_2)$	$ClSiPh_2(CH:CH_2)$	35	326
$SiMe_2H$	$ClSiMe_2H$	75	321, 326
$SiMePhH$	$ClSiMePhH$	55	326
$SiPh_2H$	$ClSiPh_2H$	61	326
$Si(C_5Cl_4N)R_2{}^a$	Cl_2SiR_2	ca. 40	326
$Si(C_5Cl_4N)_2R^a$	Cl_3SiR	40–50	326
Cu	Cu_2I_2	—^{b}	322, 327–329

^{a} C_5Cl_4N = tetrachloro-4-pyridyl. ^{b} Usually converted into other products (Scheme 2.12) (p. 253).

With methylmagnesium iodide in a mixture of ether and tetrahydrofuran at 50-60°, pentachloropyridine yields a mixture of tetrachloro-4-methylpyridine (25%) and octachloro-4,4'-bipyridyl (35%) (cf. ref. 325) [330]. Likewise, with ethyl-, propyl-, n-butyl-, or benzyl-magnesium iodide it gives the corresponding 4-alkyl- or 4-benzyl-tetrachloropyridine, respectively, together with varying amounts of the bipyridyl [330]. Phenylmagnesium bromide is reported not to react with pentachloropyridine, whilst ethylmagnesium bromide also gives the 4-ethyl derivative and benzylmagnesium chloride also yields the 4-benzyl derivative (cf. ref. 321) [325].

Tetrachloro-4-methoxypyridine reacts with sterically non-demanding Grignard reagents, for example n-butylmagnesium chloride but not t-butylmagnesium chloride, by displacement of the methoxyl group to give the corresponding 4-alkyltetrachloropyridine [331].

Whereas pentachloropyridine undergoes attack by Grignard reagents exclusively in the 4-position, pentachloropyridine N-oxide (see later for its preparation) is attacked exclusively in the α-positions. Thus, with a 1:1 mole ratio of methylmagnesium iodide at room temperature, it gives mainly tetrachloro-2-methylpyridine N-oxide [313, 314]. With an excess of the Grignard reagent, it gives the 2-monomethyl derivative (37%) together with trichloro-2,6-dimethylpyridine-N-oxide (29%) [313, 314]. These reactions are believed to proceed via initial 1,3-addition of the Grignard reagent to the N-oxide, to give the intermediate (131) [314]. With a large excess of ethyl- or phenyl-magnesium bromide in tetrahydrofuran (but not ether), the N-oxide gives the corresponding 2,6-disubstituted derivative [314]. In these cases, the use of a 1:1 or 2:1 mole ratio of the Grignard reagent gives a complex mixture of starting material, the mono- and the di-substituted product, and tetrachloro-2-hydroxypyridine. The last compound probably arises by a

(131)   (132)

metal-halogen exchange reaction, to give the Grignard derivative (132), followed by aerial oxidation [314]. In the case of methylmagnesium iodide, either the Grignard compound (132) is not formed or, if it is, it may undergo rapid alkylation with the initially generated methyl halide [314]. The products of these reactions may be deoxygenated with phosphorus trichloride and hence provide useful starting materials for the synthesis of 2-alkyl- and 2-aryl-tetrachloropyridines and the corresponding 2,6-disubstituted compounds.

In contrast to pentafluoropyridine, which is alkylated or arylated with alkyl- or aryl-lithium reagents [332-334], pentachloropyridine undergoes metal-halogen exchange with n-butyl- [177, 321, 325, 328, 335, 336], methyl- [325, 336], or phenyl-lithium [325, 336]. In the case of n-butyl-lithium, the product depends on the solvent used. In hydrocarbon solvents, such as benzene or methylcyclohexane, at room temperature, it gives mainly tetrachloro-2-pyridyl-lithium (68 mole%), together with the isomeric 3- (16 mole%) and 4-lithium compounds (16 mole%) (approx. 43% total yield), whilst in ether at $-70°$ it gives mainly tetrachloro-4-pyridyl-lithium (78 mole%), together with the 3-lithium compound (22 mole%) (approx. 70% total yield) (see also Chapter 3) [177, 335]. With methyl- or phenyl-lithium, tetrachloro-4-pyridyl-lithium is reported to be

TABLE 2.4   Products Derived from Tetrachloro-4-pyridyl-lithium

R	Reagent	Yield (%)	Refs.
H	Acid	–	177
Me	Me_2SO_4	64	178
CO_2H	CO_2	69,[a] 86, 29	148, 149, 180
CPh_2OH	Ph_2CO	65	324
CPh:NH	PhCN	–	337, 338
CPh:NCOPh	PhCN	–	337, 338
C_5Cl_4N[b]	$TiCl_4$	–	339
HgC_5Cl_4N[b]	$HgCl_2$	12	177
HgCl	$HgCl_2$	–	177
$SiMe_3$	$ClSiMe_3$	–	321, 324, 325 336
$SiMe_2H$	$ClSiMe_2H$	–	321
$SiMe_2Ph$	$ClSiMe_2Ph$	–	321
Cu	Cu_2I_2 or MeCu	–[c]	328, 329
PPh_2	$ClPPh_2$	57	340
$PPh(C_5Cl_4N)$[b]	Cl_2PPh	52	340
$P(C_5Cl_4N)_2$[b]	PCl_3	40	340

[a] A mixture of tetrachloropyridine-4-carboxylic acid (85–90%) and tetra-chloropyridine-2-carboxylic acid (10–15%).   [b] C_5Cl_4N = tetrachloro-4-pyridyl.
[c] Usually converted into derivatives (Scheme 2.12) (p. 253).

formed exclusively; optimum reaction conditions have to be used, otherwise polychlorobipyridyls are the major products [325, 335]. Pentachloropyridine is reported to be unreactive towards penta-fluorophenyl-lithium [325]. A solution of isomer-free tetrachloro-4-pyridyl-lithium in ether is probably best prepared by cleavage of (tetrachloro-4-pyridyl)dimethylsilane with n-butyl-lithium [149].

Some reactions of tetrachloro-4-pyridyl-lithium are summarized in Table 2.4. This lithium compound also reacts with an excess of benzonitrile in ether under reflux to give, after addition of water, a mixture of 5,6,8-trichloro-2,4-diphenylpyrido[3,4-d]pyrimidine

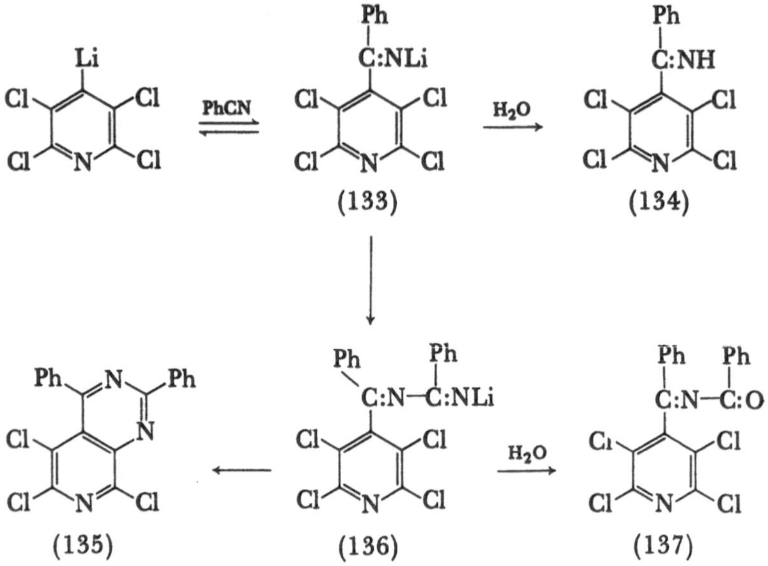

(133)   (134)

SCHEME 2.11

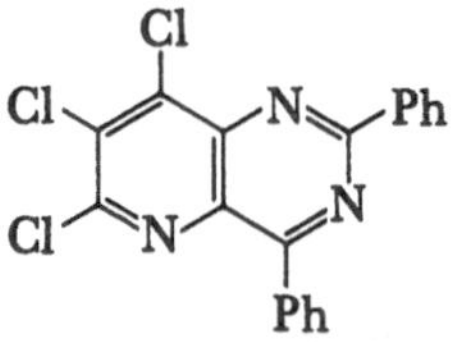

(135)   (136)   (137)

(138)

(135) (11%), 4-benzimidoyltetrachloropyridine (134) (11%), and
4-(N-benzoylbenzimidoyl)-tetrachloropyridine   (137)   (30.5%)
(Scheme 2.11) [337, 338]. The imines are formed by hydrolysis of
the intermediate lithium compounds (133) and (136); the latter is
also the precursor of the pyridopyrimidine. With an excess of
benzonitrile and under more vigorous conditions, tetrachloro-2-
pyridyl-lithium gives a low yield of 6,7,8-trichloro-2,4-diphenyl-
pyrido[3,2-d]pyrimidine (138) [338]. The first stage of these
reactions is reversible [341].

Curiously, when tetrachloro-4-pyridyl-lithium is allowed to
compete for benzophenone and trimethylsilyl chloride in tetrahydrofuran at −70°, it reacts exclusively with the latter, whilst in ether at
−70° the reverse is the case [324].

2,3,5,6-Tetrachloropyridine is metallated by n-butyl-lithium to
give exclusively tetrachloro-4-pyridyl-lithium [178]. Likewise
2,3,4,6-tetrachloropyridine yields tetrachloro-3-pyridyl-lithium,

which is a useful purcursor for 3-substituted tetrachloropyridines [301, 302].

4-Aryl- [179] and 4-*sec.*-amino-tetrachloropyridines [178, 342] react to give the corresponding 3-lithium derivative (139; R = aryl or *sec.*-amino). These may be hydrolyzed to the corresponding 3-hydro-derivative and react with carbon dioxide to give the corresponding 3-carboxylic acid [178, 179, 342]. With *n*-butyl-lithium, tetrachloro-4-methylpyridine is lithiated mainly in the side-chain, to give the lithium compound (140) (80 mole%) [178, 331], but also in the 3-position to give the isomeric lithium compound (139; R = Me) (20 mole%) [331]. In contrast, tetrafluoro-4-methylpyridine is *alkylated* in the 2-position by *n*-butyl-lithium [334]. The lithium compound (140) reacts normally with dimethyl sulphate, to give tetrachloro-4-

(139)

(140)

ethylpyridine [178], and with carbon dioxide, to give (tetrachloro-4-pyridyl)acetic acid [178, 331]. Tetrachloro-4-mercaptopyridine reacts with two equivalents of *n*-butyl-lithium in ether to give a mixture of the 3-lithium derivative (139; R = SLi) (70-75 mole%) and the corresponding 2-lithium derivative (25-30 mole%) [269, 270]. Similar treatment of tetrachloro-4-hydroxypyridine gives, after hydrolysis, only 2,3,5-trichloro-4-hydroxypyridine [343]. 4-Methyl-thio- [269, 270] and 4-phenoxy-tetrachloropyridine [178] react mainly by loss of the 4-substituent to give 4-*n*-butyltetrachloro-pyridine. Likewise, with methyl- [331], ethyl- [331], *n*-butyl- [178, 331], phenyl-, and other aryl-lithium compounds [179, 331], tetrachloro-4-methoxypyridine gives the corresponding 4-alkyl- or 4-aryl-tetrachloropyridine. With two or three equivalents of an aryl-lithium compound, it gives a 2,4-di- or a 2,4,6-tri-substituted derivative, respectively [179]. With *t*-butyl-lithium, however, it gives the 3-lithium compound (139; R = OMe), which may be hydrolyzed to the corresponding 3-hydro-derivative [331].

2-Substituted tetrachloropyridines also undergo metal-halogen exchange reactions with *n*-butyl-lithium to give the corresponding 4-lithium derivative (e.g., 141; R = OMe, NMe₂, pyrrolidino, piperidino) [300]. No displacement of the methoxyl group is observed in the case of the 2-methoxy compound (see before). Trichloro-2,4-dimethoxypyridine reacts with methyl-lithium by displacement of the 4-methoxyl group, to give trichloro-2-methoxy-4-methylpyridine [331]. The lithium compounds (141) give the same trichloro-derivative on hydrolysis that can be prepared by appropriate nucleophilic substitution of 2,3,5,6-tetrachloropyridine.

(141)

(142)

(143)

(144)

Hydrolysis of a solution of tetrachloro-4-pyridyl-lithium in benzene which has been heated at 80° for 45 min gives largely the tetrachloro-compound, indicating that elimination of lithium chloride under these conditions is slow [344, 345]. Trichloro-3-pyridyne (142) may be generated and trapped, however, in a hydrocarbon solvent, such as benzene, in which case the adduct (143) is formed [344, 345]. The pyridyne (142) cannot be generated and trapped in the presence of furan [344, 345], which is reminiscent of a corresponding observation in the fluorine field [346]. It may be formed and trapped, however, in the presence of 1,3-diphenylisobenzofuran [344, 345]. In the absence of a suitable trapping agent, octachloro-4,4'-bipyridyl is formed [325]. The lithium derivatives (141) can be made to lose lithium chloride also to give 3-pyridynes, which may be trapped with furan to give the corresponding adduct (144) [300]. Only one of two possible adducts is formed in these cases. Although it is doubtful whether tetrachloro-2-pyridyl-lithium loses lithium chloride to give trichloro-2-pyridyne [344, 345], 2-pyridynes can be generated from the lithium compounds (139) and trapped with furan or N-methylpyrrole [178, 179, 344, 345].

Recently, Wakefield and his co-workers [339] have synthesized octachloro-4,4'-bipyridyl by treatment of tetrachloro-4-pyridyl-lithium with titanium chloride and studied its reaction with n-butyl-lithium in ether, which gives heptachloro-3-lithio-4,4'-bipyridyl. This lithium compound gives the expected products with water or dimethyl sulphate and, on being heated in p-di-isopropyl-benzene, it loses lithium chloride and gives the adduct (145) [339]. In contrast to the lithium-halogen exchange reaction, treatment of

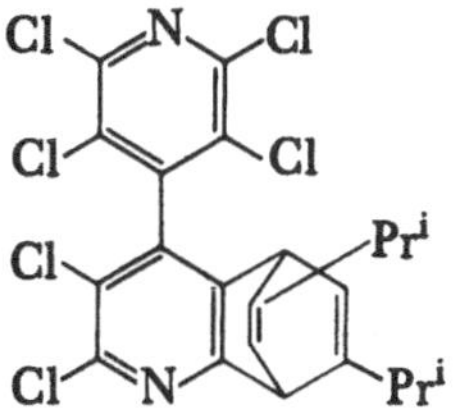

(145)

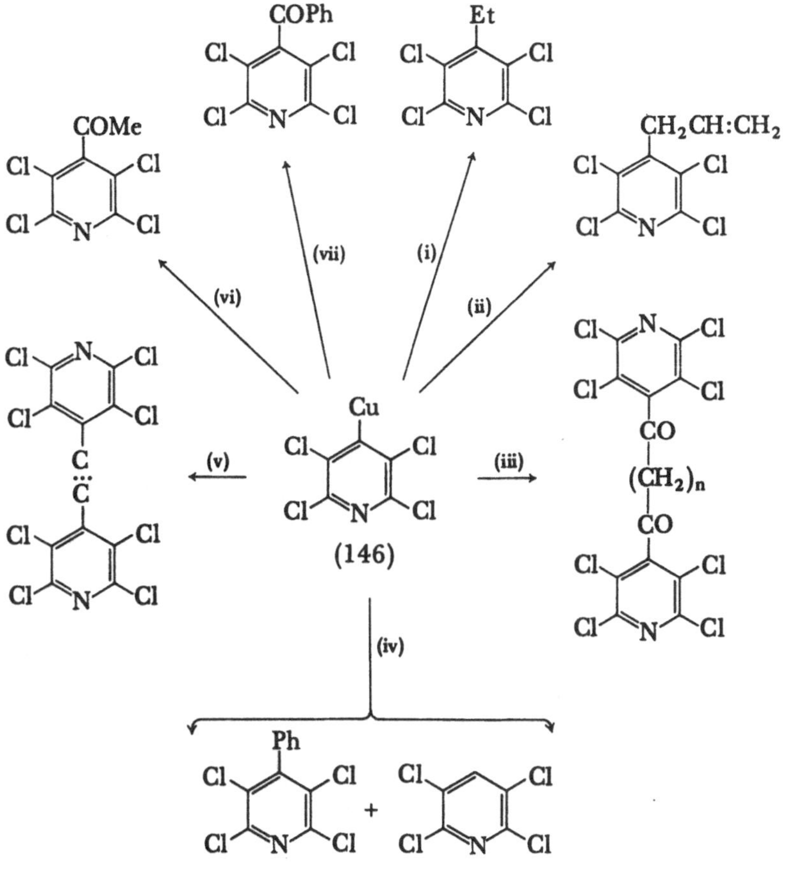

COPh

Et

COMe

CH$_2$CH:CH$_2$

(vii)

(i)

(vi)

(ii)

Cu

(v)

Cl

Cl

(iii)

(146)

CO

(CH$_2$)$_n$

CO

(iv)

Ph

+

Reagents: (i) EtI [329]; (ii) CH$_2$:CHCH$_2$Br [322, 329]; (iii) ClCO(CH$_2$)$_n$COCl (n = 0, 2, 4) [327]; (iv) PhI [329]; (v) Br$_3$CCHBr$_2$, X$_2$CHCHX$_2$, Br$_2$C:CBr$_2$, or XCH:CX$_2$ (X = Br or Cl) [347]; (vi) MeCOBr [328]; (vii) PhCOCl [322, 328].

SCHEME 2.12

octachloro-4,4'-bipyridyl with piperidine yields the 2,2'-dipiperidino-derivative. Proof of the structure of this diamine followed from its unambiguous synthesis by treatment of trichloro-2-piperidino-4-pyridyl-lithium with titanium tetrachloride.

Tetrachloro-4-pyridylcopper (146) (Scheme 2.12) may be prepared by treatment of either tetrachloro-4-pyridyl-lithium or the corresponding Grignard compound with copper(I) chloride or iodide [322, 327-329], by treatment of the lithium compound with methylcopper [328], or by treatment of 2,3,5,6-tetrachloropyridine [327, 328] or tetrachloro-4-iodopyridine [328] with lithium dimethylcopper. Its reactions are summarized in Scheme 2.12. The easiest and least expensive preparation and the one which gives the highest yields of derivatives utilizes the Grignard reagent [322]. The copper compound (146) is very stable, even at elevated temperatures.

2.8.3 Further Reactions of Pentachloropyridine and its Derivatives

Reactions of Pentachloropyridine and its Derivatives with Reducing Agents

Polychloronitro(and nitroso)pyridines may be reduced by stannous chloride [285] or iron and acetic acid [244] to the corresponding amine.

Electrolytic reductions of heptachloro-2(and 3)vinylpyridine give tetrachloro-2(and 3)vinylpyridine, respectively, as the major products [348]. Pentachloro-, tetrachloro-2-cyano-, and trichloro-2,6-dicyano-pyridine are reduced electrolytically in the 4-position [348a].

Reductive hydrodechlorination of a polychloropyridine to a known derivative of pyridine may be used as a means of structure determination. Thus, reductive dehalogenation of tetrachloro-3-phenylpyridine at room temperature (12 h) with hydrogen over 2% palladium on strontium carbonate in ethanol containing 1N-sodium hydroxide surprisingly gives a 79% yield of 3-phenylpyridine [242]. Usually β-chlorine atoms are resistant to reduction. For example, tetrachloro-2-phenylpyridine [267] and the pyridone (100; R^1 = Me; R^2 = H) [262] are reduced by hydriodic acid and zinc and acetic acid, respectively, to 3,5-dichloro-2-phenylpyridine and 3,5-dichloro-N-methyl-2-pyridone. Likewise, 4,5,6-trichloro-2-trichloromethyl-pyridine is reduced by hydriodic acid to 5-chloro-2-trichloromethyl-pyridine [349]. The β-chlorine atoms in 3-chlorotetrafluoro- and 3,5-dichlorotrifluoro-pyridine are replaced, however, on catalytic hydrogenation over a palladium catalyst [350]. This probably reflects the fact that chlorine is more readily reduced than fluorine under these conditions. These compounds behave quite differently with lithium aluminium hydride (see later).

With one equivalent of lithium aluminium hydride in ether, pentafluoropyridine gives 2,3,5,6-tetrafluoropyridine; with two equivalents, it gives 2,3,5-trifluoropyridine [333]. In contrast, pentachloropyridine reacts with an excess of lithium aluminium hydride in ether at 0° to give, after hydrolysis of the products, a mixture of 2,3,4,5-, 2,3,4,6-, and 2,3,5,6-tetrachloropyridine (8:12:80), whilst at room temperature it gives 2,3,6-trichloropyridine as the major product (90 mole%), together with a mixture of the tetrachloropyridines (see also ref. 207) [311, 312]. At 0°, a *trans*-addition of the aluminohydride ion occurs across the 3,4-bond of pentachloropyridine (see Scheme 2.13) to give the intermediate (149), which undergoes hydrolysis by water with concomitant loss of hydrogen chloride in two unusual *cis*-elimination reactions to give 2,3,5,6-tetrachloropyridine and a smaller amount of 2,3,4,6-tetrachloropyridine. *Trans*-addition undoubtedly also occurs across the 2,3-bond of pentachloropyridine, which accounts for the formation of 2,3,4,5-tetrachloropyridine and a further amount of the 2,3,4,6-tetrachloro-compound. At room temperature, the intermediate (149) suffers reduction with inversion of configuration at C-3 to give the intermediates (150) and (147) both of which undergo hydrolysis in the presence of water with concomitant loss of hydrogen chloride to give 2,3,6-trichloropyridine. Successive treatment of pentachloropyridine with lithium aluminium hydride and deuterium oxide in ether at room temperature gives 2,3,6-trichloro-5-deuteriopyridine as the major product (Scheme 2.13; deuterium atoms in parentheses) and its successive treatment with lithium aluminium deuteride and water under similar conditions gives 2,3,6-trichloro-4-deuteriopyridine as the major product. These facts support the proposed mechanism. Additional support for the mechanism is provided by the fact that, if pentachloropyridine is treated with lithium aluminium hydride in ether at room temperature and the resulting mixture is carbonated, the acids (151) and (148) are present in the product [311, 312].

Other polychloropyridines react similarly with lithium aluminium hydride. Thus, at room temperature in ether, 2,3,5,6-tetrachloropyridine gives a mixture of 2,3,6- (75 mole%) and 2,3,5-trichloropyridine (25 mole%) [312]. In this case, addition of the aluminohydride ion occurs across the 2,3-bond to give the intermediate (152) (p. 258), which undergoes subsequent hydrolysis and elimination of hydrogen chloride. Deuteriation studies [312] confirm this mechanism.

At low temperatures (−20° to 0°) in ether, tetrachloro-4-methoxypyridine reacts by successive addition across the 3,4-bond and reduction of the 3-chlorine atom, to give the intermediate (153) [312]. Hydrolysis of the resulting mixture with deuterium oxide gives (154), which subsequently loses hydrogen deuteride to give 2,3,6-trichloro-4-methoxypyridine. At 25-35° in ether, however, addition occurs across the 2,3-bond to give the intermediate (155),

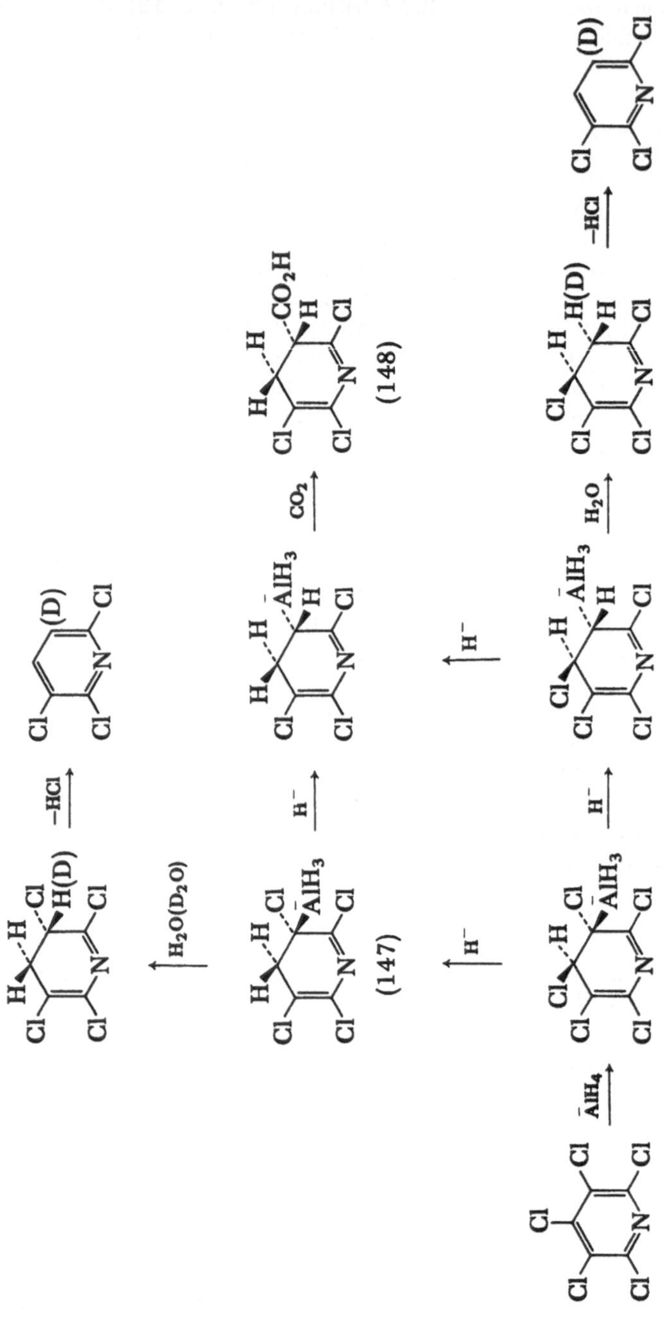
(147)
(148)
(149)
(150)
H₂O(D₂O)
−HCl
CO₂
H₂O
H⁻
AlH₄⁻

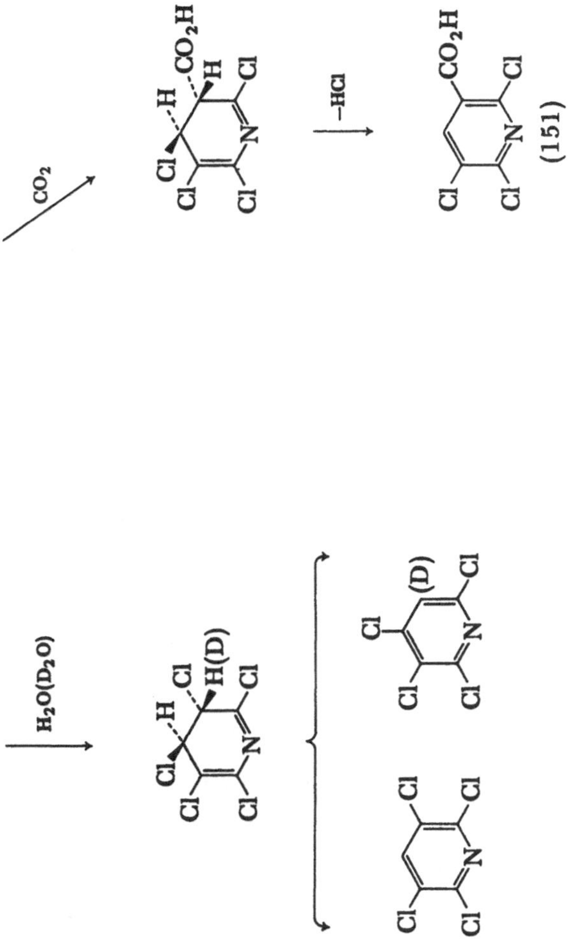

SCHEME 2.13

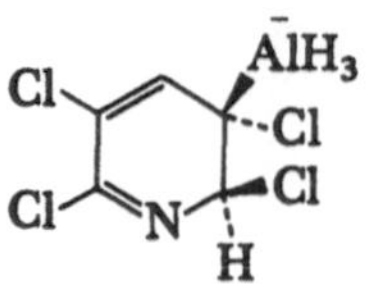

(152)

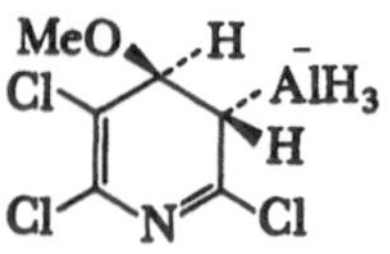

(153)

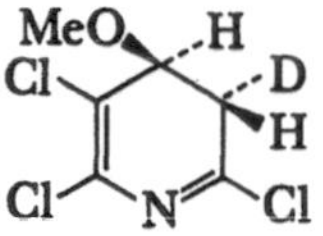

(154)

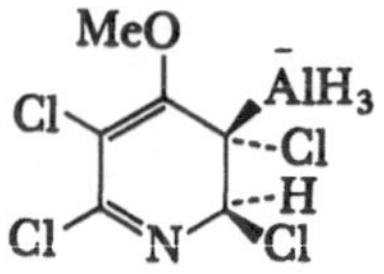

(155)

which undergoes hydrolysis by deuterium oxide with subsequent loss of hydrogen or deuterium chloride, to give a mixture of 2,3,6-trichloro-5-deuterio-4-methoxypyridine (major product) and 2,3,5-trichloro-4-methoxypyridine. At higher temperatures (35-55°) in tetrahydrofuran 2,3,6-trichloro-5-deuteriopyridine is formed exclusively following hydrolysis with deuterium oxide. This suggests replacement of the methoxyl group either by hydride ion or by its elimination as methanol. Tetrachloro-4-ethylamino(and piperidino)pyridine behave similarly [312].

The reaction of tetrachloro-4-methylsulphonylpyridine with lithium aluminium hydride in ether at −25° gives a mixture of 2,3-dichloro-4-methylthiopyridine (the major product), 2,3,6-trichloropyridine, and 2,3,6-trichloro-4-methylthiopyridine (a trace only) [296]. Tetrachloro-4-methylthiopyridine gives the same mixtures of products [296]. In these cases, addition of the aluminohydride ion occurs across both the 2,3- and the 3,4-bond of the starting material; the former mode of addition predominates. It is noteworthy that the substituent is lost in this case under very mild reaction conditions.

With lithium aluminium hydride in boiling ether, tetrachloro-4-nitropyridine gives a high yield of 2,3,6-trichloropyridine, presumably by formation of the intermediate (156) after hydrolysis, with subsequent loss of nitrous acid [312]. In contrast, tetrachloro-

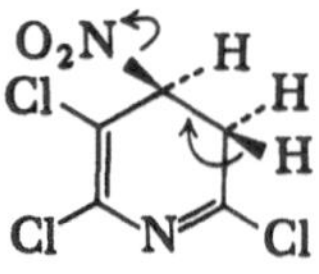

(156)

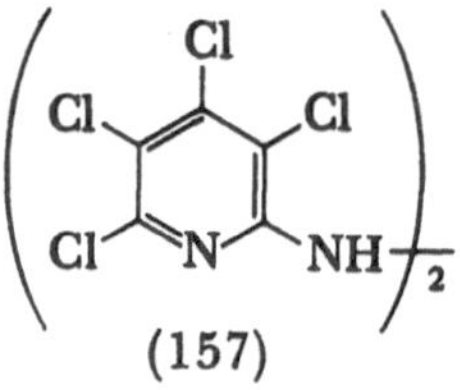

(157)

2-nitropyridine gives the hydrazine derivative (157) under similar conditions [312].

3,5-Dichlorotrifluoropyridine behaves in a similar manner to pentafluoropyridine and gives only 3,5-dichloropyridine with an excess of lithium aluminium hydride [312]. These reactions probably occur through a series of 1,4- and 1,2-additions followed by eliminations as shown, for example, in Scheme 2.14. In pentachloropyridine, the α-chlorine atoms probably hinder approach of the

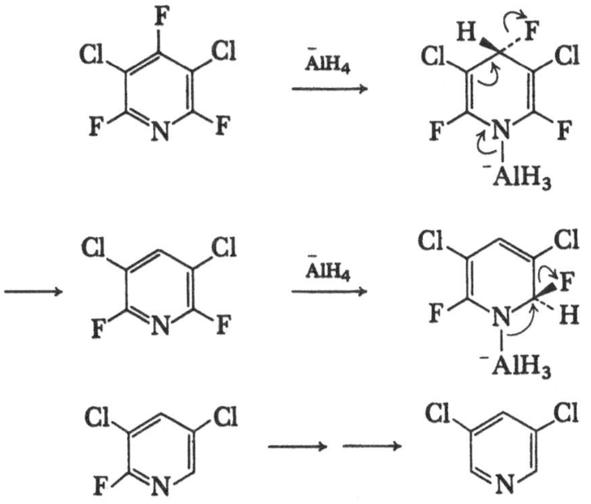

SCHEME 2.14

reagent to the ring nitrogen atom, thereby excluding 1,2- and 1,4-addition.

Pentachloropyridine-N-oxide is reduced by lithium aluminium

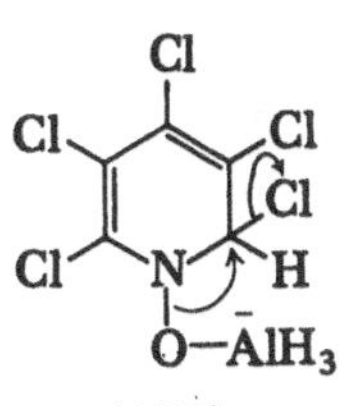

(158)

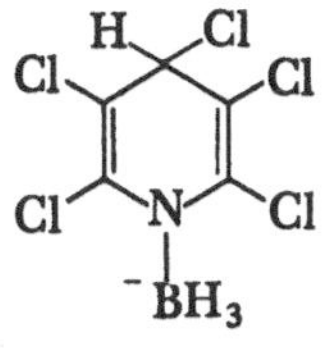

(159)

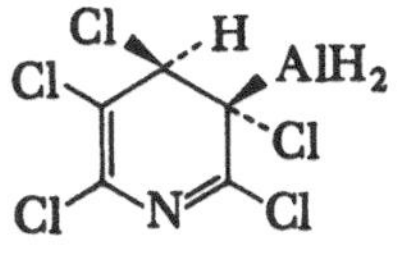

(160)

hydride in ether at 0° to give exclusively 2,3,4,5-tetrachloropyridine [312]. This reaction probably proceeds via the intermediate (158), which undergoes elimination as shown on hydrolysis with water.

Both lithium borohydride and aluminium hydride reduce penta-chloropyridine exclusively to 2,3,5,6-tetrachloropyridine [312]. The former reaction proceeds in boiling tetrahydrofuran probably via 1,4-addition, as shown (159), whilst the latter reaction proceeds in tetrahydrofuran at room temperature probably via 3,4-addition (160) by analogy with lithium aluminium hydride reduction. The smaller borohydride ion is able to co-ordinate with the ring nitrogen atom, unlike the larger aluminohydride ion which cannot. Penta-chloropyridine is also reduced by sodium borohydride, but the results have not been conclusively evaluated [351]. With a large excess of this reagent, 2,3,5,6-tetrachloropyridine is formed together with some 2,3,6-trichloropyridine. Pentachloropyridine does not react with diborane [312, 351].

Reactions of Pentachloropyridine and its Derivatives with Oxidizing Agents

A methyl group in a polychloropyridine is oxidized to a carboxyl group by alkaline potassium permanganate [243, 267, 297, 330].

Pentachloropyridine-*N*-oxide may be prepared by oxidation of pentachloropyridine with hot trifluoroperoxyacetic acid (20% yield) [180, 286, 315, 352] or, preferably, with a mixture of hydrogen peroxide, concentrated sulphuric acid, and acetic acid at ambient temperature (95% yield) [172, 353]. In the former case, thermal deoxygenation of the product is responsible for the low yield.

2,3,4,5- and 2,3,5,6-Tetrachloropyridine are oxidized by trifluoro-peroxyacetic acid to the corresponding *N*-oxide [315].

Various substituted polychloropyridines, for example 2,3,5,6-tetrachloropyridine [172, 353], tetrachloro-4-methyl- [353], tetra-chloro-4-methylsulphonyl- [296], and tetrachloro-4-nitro-pyridine [172, 353], may be oxidized to the corresponding *N*-oxide with a mixture of hydrogen peroxide, concentrated sulphuric acid, and acetic acid. Octachloro-4,4'-bipyridyl gives a mixture of the mono-(30%) and di-*N*-oxide (20%) [172]. The same reagent system has no effect, however, on tetrachloro-4-hydroxypyridine. Under mild conditions, tetrachloro-4-methoxypyridine is not attacked either, but under more forcing conditions it gives a mixture of tetrachloro-4-hydroxypyridine (40%) and its *N*-oxide (45%) [172, 353]. With this reagent system 2,3,6-trichloro-4-methylthiopyridine is oxidized to 2,3,6-trichloro-4-methylsulphonylpyridine-*N*-oxide.

Recently, a mixture of polyphosphoric acid and 90% hydrogen peroxide has been reported to be an effective *N*-oxidizing reagent for weakly basic *N*-heteroaromatic compounds [354]. This reagent system also converts pentachloropyridine into its *N*-oxide, albeit in low yield (20%) [172]. It is advantageous, however, for oxidation of tetrachloro-4-methoxypyridine to its N-oxide, in which case cleavage to the 4-hydroxy compound is avoided [172, 353]. Whereas

(161)

3,5-dichlorotrifluoropyridine can not be converted into its N-oxide with a mixture of peroxyacid and concentrated sulphuric acid, it is converted into 3,5-dichloro-2,4-difluoro-6-hydroxypyridine-N-oxide (161) by the use of the polyphosphoric acid-hydrogen peroxide oxidizing system. This reaction proceeds via the intermediate formation of 3,5-dichlorotrifluoropyridine-N-oxide, which is hydrolyzed in the 2-position on work-up [172].

Alkyl- or aryl-thiopolychloropyridines are oxidized to the corresponding sulphoxide by concentrated nitric acid [293, 355-360], or peroxyacetic acid [270, 361] and to the corresponding sulphone by peroxyacetic acid [232a, 269, 270, 292-294, 296, 355-359, 361], concentrated nitric acid [293, 355-358], potassium dichromate and acid [355-358], alkaline potassium permanganate [355-358], or chlorine water [362].

4-Aminotetrachloropyridine is oxidized by a cold mixture of trifluoroacetic acid and 90% hydrogen peroxide to the corresponding 4-nitro-compound; the 4-nitroso-compound also gives the 4-nitro-compound under these conditions [285]. Likewise, 2-aminotrichloro-6-fluoropyridine gives trichloro-6-fluoro-2-nitropyridine [273, 274]. Oxidation of 2-aminotetrachloropyridine to the 2-nitro-compound occurs under milder conditions, namely with a mixture of trifluoro-acetic acid and 30% hydrogen peroxide at room temperature [285].

The oxidation reactions of 2- and 4-$sec.$-aminotetrachloropyridines are more complex. Thus, oxidation of compounds [162: R = Cl; X = Me$_2$, (CH$_2$)$_{4-6}$, or $-$(CH$_2$)$_2$O(CH$_2$)$_2-$] with peroxyformic acid or trifluoroperoxyacetic acid at room temperature is followed by an $S_N i$ rearrangement (163) of the initially formed N-oxide to give a hydroxylamine derivative (164: R = Cl; X as before) [286]. Similar oxidation of the 2-NN-dialkylamines (162: R = OMe or piperidino; X as before) gives the corresponding hydroxylamine derivative (164), but the N-oxides (163: R = OH; X as before) are the products of oxidation of the 2-NN-dialkylamines (162; R = OH) [286]. An $S_N i$

(162)   (163)

(164)   (165)

rearrangement of the N-oxides derived from these compounds is prevented by resonance interaction (a + M effect) (165) of the hydroxyl group with the pyridine ring. A 4-methoxyl or 4-piperidino-group is sterically prevented from becoming coplanar with the ring and such resonance interaction in these cases cannot prevent the rearrangement from taking place.

At room temperature, peroxyformic acid has no effect on tetrachloro-4-piperidinopyridine. Cold trifluoroperoxyacetic acid, however, yields a small amount of the hydroxylamine derivative (166), but the major product is tetrachloro-4-nitrosopyridine [285]. A small amount of 4-aminotetrachloropyridine is also formed [285]. Cold trifluoroperoxyacetic acid (made from 30% hydrogen peroxide) oxidation of tetrachloro-4-dimethylaminopyridine (167) (Scheme 2.15) gives the corresponding 4-amino- and 4-nitroso-compound, along with tetrachloro-4-methylaminopyridine (169) [285]. The last compound, which arises by demethylation of the initially formed N-oxide (168) in a Polonovski-type reaction, appears to be the precursor of the other compounds (see Scheme 2.15). Thus, the hydroxylamine derivative (170), the expected oxidation product of the secondary amine (169), is oxidized further to the intermediate (171), which fragments to give the pyridylhydroxylamine (172). This may oxidize further to the 4-nitroso-compound or undergo a 'frustrated' Bamberger hydroxylamine rearrangement to give the 4-amine [285]. In keeping with this scheme, other 4-N-alkylamino-tetrachloropyridines are oxidized by a mixture of trifluoroacetic acid and 30% hydrogen peroxide in the cold to give mainly tetrachloro-4-nitrosopyridine [285]. With a mixture of trifluoroacetic acid and 15% hydrogen peroxide, however, 4-aminotetrachloropyridine is the major product. With *hot* peroxyacetic acid, both tetrachloro-4-piperidino-

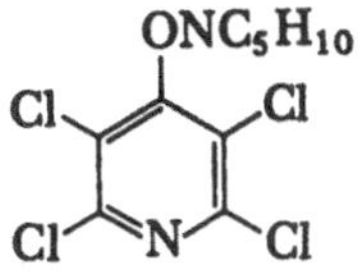
ONC_5H_{10}
Cl
Cl
Cl
N
Cl

(166)

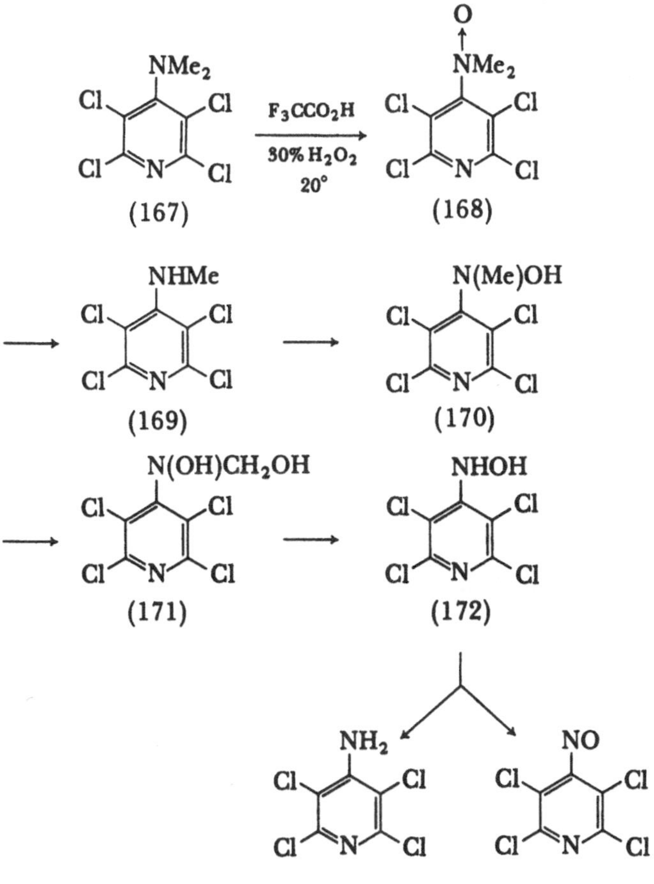
F_3CCO_2H
30% H_2O_2
20°
(167)
(168)
(169)
(170)
(171)
(172)
NMe_2
NHMe
N(Me)OH
$N(OH)CH_2OH$
NHOH
NH_2
NO

SCHEME 2.15

pyridine and the corresponding 4-dimethylamino-compound give tetrachloro-4-nitropyridine [285].

The difference in behaviour between the 2- and the 4-*sec.*-aminotetrachloropyridines on oxidation with peroxyacids at room temperature may be attributed to steric hindrance by the two *ortho*-chlorine atoms in the 4-substituted compounds, which prevent a S_Ni

rearrangement of the initially formed N-oxide (e.g., 168) [317]. In keeping with this suggestion, the 4-*sec.*-aminotetrafluoropyridines (173; X as before), in which such interactions are considerably reduced, undergo oxidation under these conditions with rearrangement of the

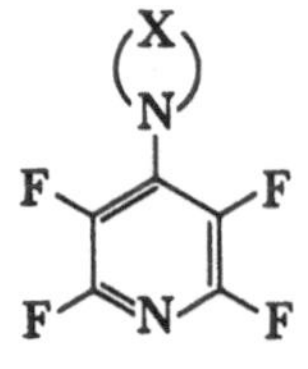

(173)

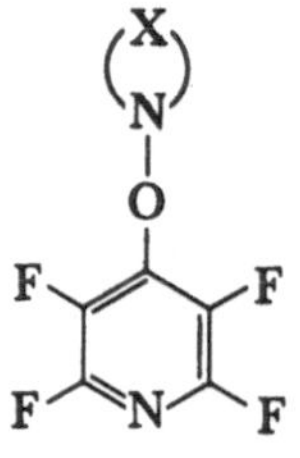

(174)

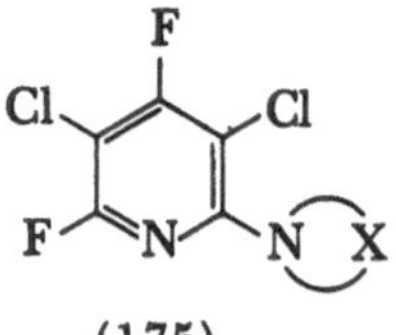

(175)

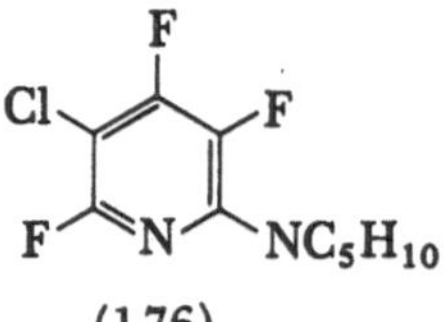

(176)

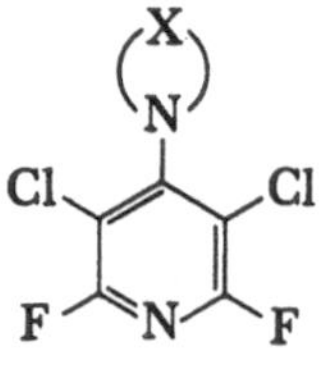

(177)

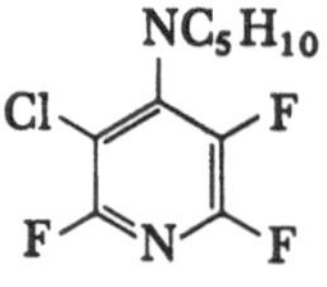

(178)

intermediate N-oxide to give the corresponding hydroxylamine derivative (174) [317]. Likewise, with cold peroxyformic acid, the amines (175; X as before) and (176) may be oxidized with rearrangement to hydroxylamine derivatives, but the amines (177; X as before) and (178) are not affected by this reagent [317]. The amine (178) can be made to undergo oxidation with rearrangement, however, using trifluoroperoxyacetic acid in the cold and is, therefore, intermediate in its behaviour between the amines (173) and the corresponding tetrachloro-compounds.

The 2,6-diamines (179; X as before) and (180) are oxidized by cold peroxyformic acid to the hydroxy compounds (181) and (182), respectively [286, 317]. In each case, the di-N-oxide undergoes one $S_N i$ rearrangement to give a hydroxylamine derivative, which is then cleaved.

Compounds (164: R = Cl, OMe, or piperidino; X as before) and the hydroxylamines derived as described before from compounds

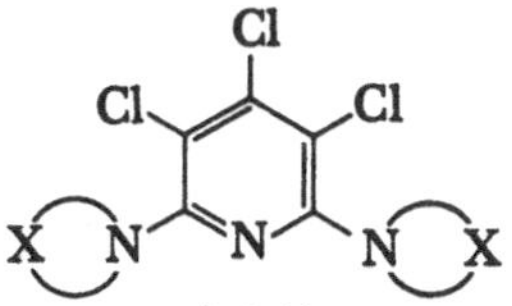

(179)

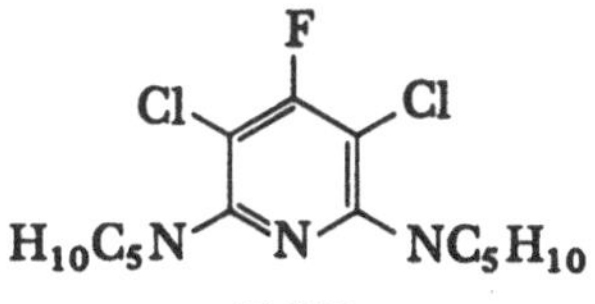

(180)

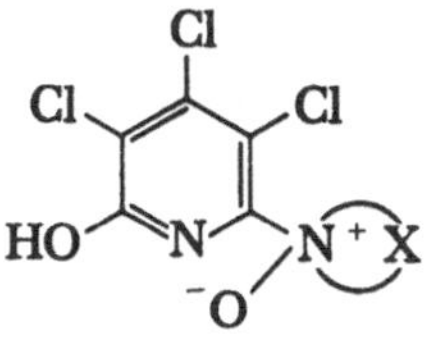

(181)

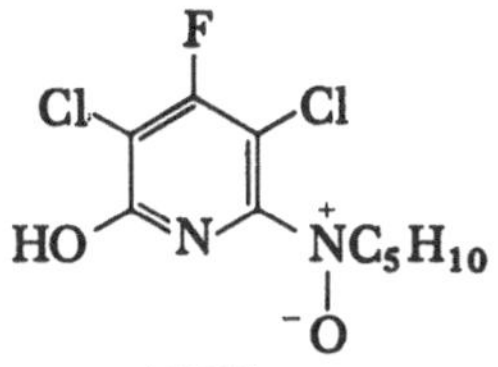

(182)

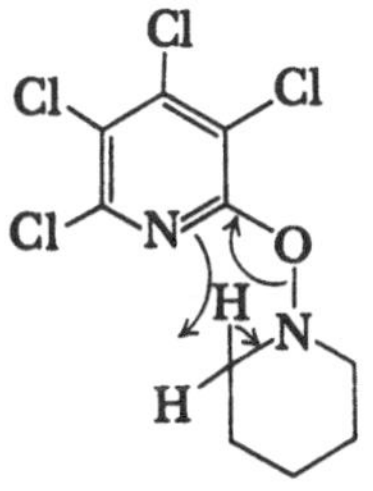

(183)

(175; X as before) and (176) undergo cleavage to the corresponding 2-hydroxypyridine when heated alone or in an inert solvent, or when photolyzed [286, 317]. A concerted mechanism (183) has been proposed [286] for these cleavage reactions.

In keeping with the usual behaviour of hydrazines in the presence of oxidizing agents, both tetrachloro-4-hydrazinopyridine and its 2-isomer react with silver oxide in ethanol to give 2,3,5,6- and 2,3,4,5-tetrachloropyridine, respectively [104]. With silver oxide or manganese dioxide in benzene, the 4-hydrazino-compound gives tetrachloro-4-phenylpyridine contaminated with much 2,3,5,6-tetrachloropyridine. With silver oxide in methyl iodide, it gives tetrachloro-4-iodopyridine together with the tetrachloro-compound (3:1) [104]. Presumably these reactions involve the tetrachloro-4-pyridyl radical. 3,5-Dichlorodifluoro-4-hydrazinopyridine similarly gives 3,5-dichlorodifluoro-4-phenylpyridine on treatment with bleaching powder in benzene [273, 274].

On being treated with bromine in dilute hydrochloric acid tetrachloro-4-hydrazinopyridine gives 4-bromotetrachloropyridine, and its pyrolysis in sand at 160° yields a mixture of 4-aminotetrachloro- and 2,3,5,6-tetrachloro-pyridine [104]. With trifluoro-peroxyacetic acid at 70° it gives tetrachloro-4-hydroxypyridine [104].

When tetrachloro-4-hydrazinopyridine is treated with cuprous oxide in hot water, 2,3,6-trichloropyridine is formed as the major product along with 2,3,5,6-tetrachloropyridine (10:1) [104]. The formation of the trichloro-compound may be attributed to a path involving a prototropic shift catalyzed by the base (cuprous oxide) followed by successive loss of hydrogen chloride and nitrogen (Scheme 2.16). The 4-hydrazino-compound also gives a mixture of

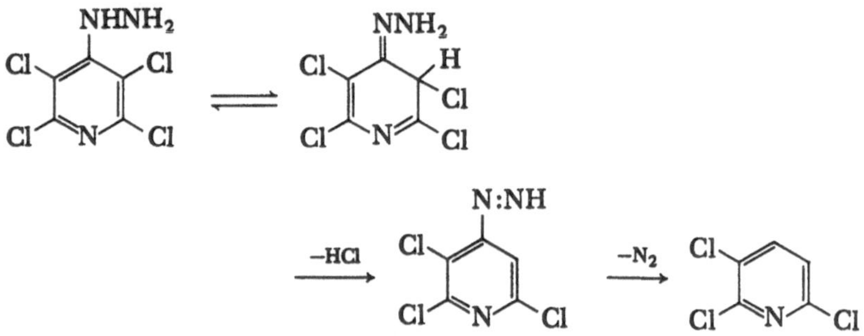

SCHEME 2.16

the tetra- and tri-chloro-compounds (3:7) with hot aqueous cupric sulphate [104]. This reagent acts as an oxidizing agent, and the cuprous oxide produced by its reduction is responsible for the formation of the trichloro-compound. In contrast, tetrachloro-2-hydrazinopyridine reacts with hot aqueous cupric sulphate to give only 2,3,4,5-tetrachloropyridine and no trichloropyridine [104]. This result can be rationalized by assuming that hydrogen bonding occurs between the hydrazino-group and the ring nitrogen atom (184), which facilitates a prototropic shift (185) to the ring nitrogen atom in preference to the *ortho*-carbon atom (cf. Scheme 2.16). The action of hot aqueous sodium hydroxide on the 4-hydrazino-compound gives tetrachloro-4-hydroxypyridine, presumably via the intermediate imino-amine tautomer (186), which is stabilized as its sodium salt [104].

Trichloro-2,4-dihydrazinopyridine reacts with hot aqueous cupric sulphate to give only 2,3-dichloropyridine; this indicates involvement of the intermediate (187) and excludes a stepwise mechanism, which would yield some 2,5-dichloropyridine [104].

It is noteworthy that tetrafluoro-4-hydrazinopyridine gives 2,3-difluoro-6-pyridone with hot aqueous sodium hydroxide [104]. In this case, nucleophilic attack by hydroxide ion occurs prior to

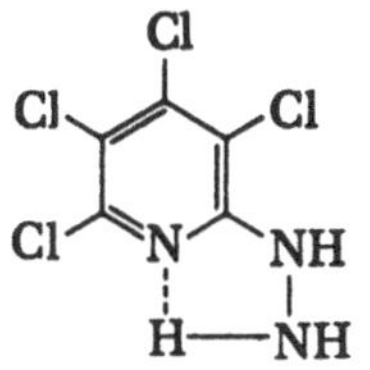

(184)

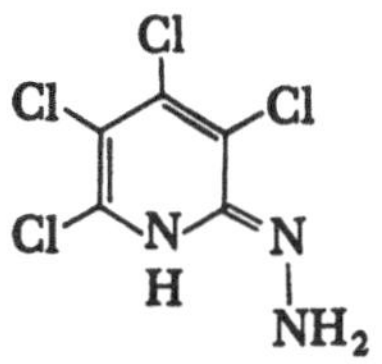

(185)

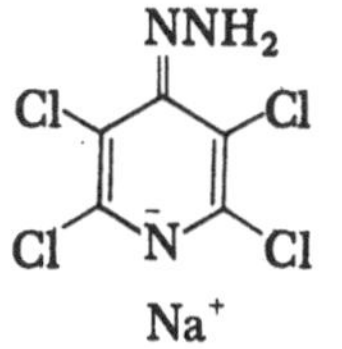

(186)

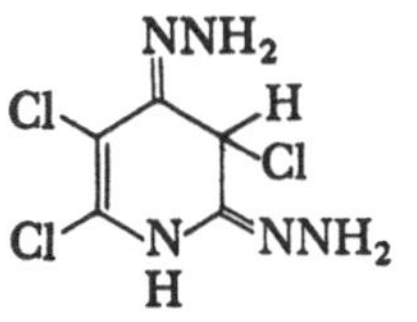

(187)

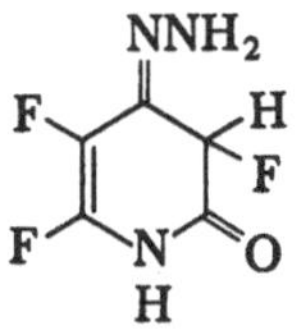

(188)

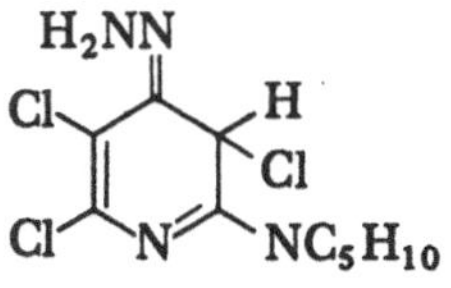

(189)

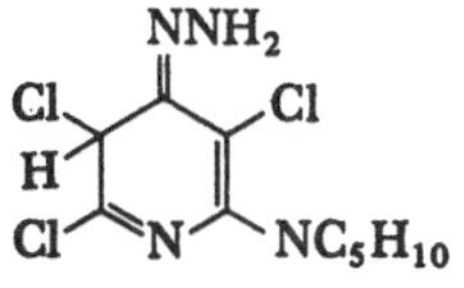

(190)

formation of the intermediate (188), which reacts further as already described to give the difluorohydroxy compound. In a similar manner, 2,3-dichloro-6-piperidinopyridine (major product) and 2,5-dichloro-6-piperidinopyridine are obtained when tetrachloro-4-hydrazinopyridine is treated with piperidine [104]. In this case, nucleophilic attack again precedes the formation of the key intermediates (189) and (190).

With hot aqueous cupric sulphate, 3-chlorotrifluoro-4-hydrazino-pyridine gives a mixture of the 4-hydro-compound and 2,3,6-tri-fluoropyridine, and 3,5-dichlorodifluoro-4-hydrazinopyridine also gives the corresponding 4-hydro-compound together with 3-chloro-2,6-difluoropyridine [104]. Elimination of hydrogen chloride from the key intermediates occurs in preference to elimination of hydrogen fluoride in these cases.

Other Reactions

Photolysis of pentachloropyridine in dioxan or diethyl ether provides a useful route to 2,3,4,6-tetrachloropyridine [301, 302]. In benzene, the product is tetrachloro-3-phenylpyridine.

On the assumption that these reactions involve a 3-pyridyl radical Bratt and Suschitzky [363] photolyzed 2- and 4-phenoxy-, thiophenoxy-, and anilino-tetrachloropyridine in ethanol or tetrahydrofuran. These reactions yielded the heterocycles (191: R = Cl; X = O, S, or NH) and (192: R = Cl; X = O, S, or NH), respectively.

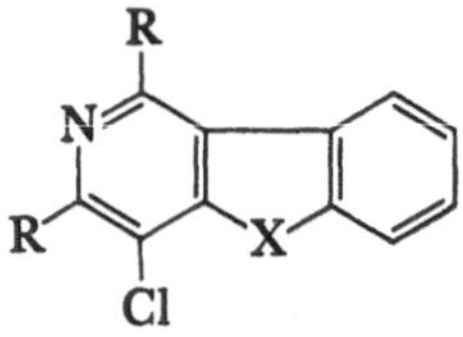

(191) 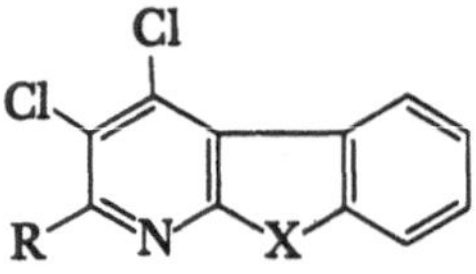   (192)

2,6-Di-(thiophenoxy)trichloropyridine and 2,4,6-tri(thiophenoxy)-dichloropyridine cyclize once only under these conditions to give compounds (192: R = SPh; X = S) and (191: R = SPh; X = S), respectively. Compound (191: R = SPh; X = S) is also the product of treatment of the trichloro-compound (191: R = Cl; X = S) with an excess of thiophenol.

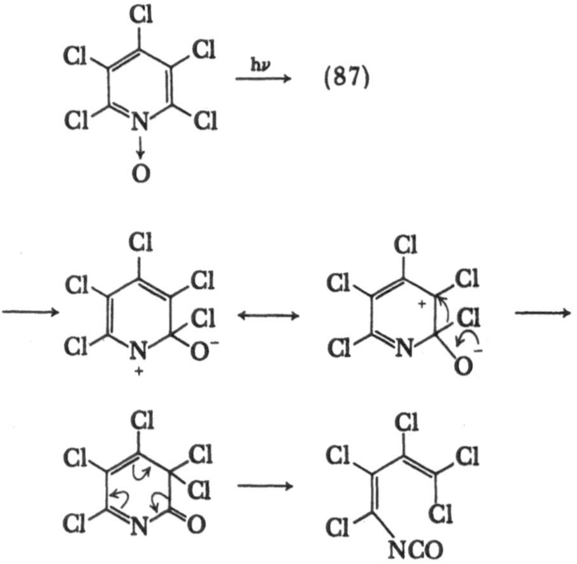

SCHEME 2.17

Photolysis of pentachloropyridine-*N*-oxide in carbon tetrachloride gives a complex mixture of products from which pentachloropyridine and pentachlorobutadienyl-1-isocyanate have been isolated (Scheme 2.17) [301, 302]. Irradiation of tetrachloro-*N*-methyl-2-pyridone at ambient temperatures in light petroleum yields the 'Dewar' pyridone (193), which collapses to starting material in the hot solvent [302].

(193)

We have already discussed the oxidation of aminopolychloropyridines to the corresponding nitro-compounds (Section 2.8.3) and the reduction of nitro-compounds to amines (Section 2.8.3). In addition, aminopolychloropyridines may be diazotized, for example with nitrosylsulphuric acid, and the amine group can thus be replaced by chlorine [191, 244, 287], or a hydroxyl group [201, 206, 237, 239, 246, 297]. 4-Aminotetrachloropyridine reacts with phosphorus pentachloride at room temperature to give (tetrachloro-4-pyridyl)phosphorimidic trichloride (194); compounds (195) and

(194) $R^1 = R^2 = Cl$
(195) $R^1 = CCl_3$, $R^2 = Cl$
(196) $R^1 = CCl_3$, $R^2 = H$

(196) may be prepared similarly [364]. With oxalyl chloride, 2-aminotetrachloropyridine yields the corresponding isocyanate which, unlike its hydrocarbon analogue, is stable as a monomer and withstands distillation [365].

Treatment of 4-amino-3,5-dichloro-2-fluoro-6-methoxypyridine with sodium hydride in dimethylformamide followed by the addition of 3,5-dichlorotrifluoropyridine yields 3,3',5,5'-tetrachloro-2,2'6-trifluoro-6'-methoxy-4,4'-dipyridylamine [366]. A large number of related compounds, including polyhalogenopyridylpyrimidinylamines [367], may be prepared similarly.

Polychloropyridinecarboxylic acids may be prepared by treatment of an organometallic derivative with carbon dioxide (Section 2.8.2) or by exhaustive chlorination of a methylpyridine followed by hydrolysis of the resulting trichloromethyl-substituted polychloropyridine (Section 2.8.2). They may be prepared also by hydrolysis of a polychlorocyanopyridine (Section 2.8.2) [310, 368].

The Hofmann reaction has been used to convert 3,4,5-trichloropyridine-2-carboxylic acid via the corresponding amide into 2-amino-3,4,5-trichloropyridine [202]. Polychloropyridinecarboxylic acids readily decarboxylate, usually on being heated in a suitable solvent [168, 201, 202, 246, 267]. (Tetrachloro-4-pyridyl)- and (2,3,5-trichloro-4-pyridyl)-acetic acid similarly give tetrachloro-4-methyl-[243, 330] and 2,3,5-trichloro-4-methyl-pyridine [297], respectively. Tetrachloropyridine-2-carboxylic acid chloride reacts normally with hydrazine at low temperatures to give the corresponding hydrazide, which gives a hydrazone derivative with acetone [369]. It can be made to react with hydrazine also to give the di-substituted hydrazide (197), which, on treatment with phos-

(197)   (198)

(199)   (200) R = Me or Et
                         (201) R = H

phorus pentachloride, gives the 1,3,4-oxadiazole (198; R = tetrachloro-2-pyridyl) [369]. An attempt to prepare the hydrazide of tetrachloropyridine-2-carboxylic acid by treatment of the corresponding ethyl ester with hydrazine hydrate gave instead the hydrazide (199) of trichloro-4-hydrazinopyridine-2-carboxylic acid, which undergoes ready hydrolysis on treatment with concentrated sulphuric acid to the corresponding acid. It is noteworthy that the esters (200) may be prepared either by alkylation of compound (201) via its potassium salt, or directly from pentachloropyridine by a nucleophilic substitution reaction (Table 2.1) [243].

Considerable attention has been given to tetrachloro-4-mercaptopyridine and its derivatives, many of which are of interest as biological toxicants (see Chapter 4). The parent mercaptan may be prepared in quantitative yield by treatment of pentachloropyridine with potassium or sodium hydrogen sulphide [269, 270, 289-291]. It is oxidized by bromine in acetic acid to the corresponding disulphide [359, 364, 370, 371], but with concentrated nitric acid it gives mainly tetrachloro-4-nitropyridine together with some of the corresponding sulphonic acid [360]. With chlorine in anhydrous carbon tetrachloride it gives tetrachloropyridine-4-sulphenyl chloride [359, 364, 370-372]. This sulphenyl chloride, which may be prepared also by chlorinolysis of 4-chlorobenzyl tetrachloro-4-pyridyl sulphide [360], reacts normally with amines [359, 371], acetone [371], and sodium cyanide [371], and it gives tetrachloro-4-pyridyl phenyl sulphide with phenylmagnesium bromide [359]. With alkenes, it can undergo both Markownikov and anti-Markownikov addition to give a mixture of products, the ratio being solvent dependent [373]. Acrylonitrile is claimed [361] to give exclusively 3-[(tetrachloro-4-pyridyl)thio]propionitrile. 2,3,5-Trichloropyridine-4-sulphenyl chloride, 2,3,5-trichloro-6-cyanopyridine-4-sulphenyl chloride and related compounds may be prepared similarly [372]. Polychloromercaptopyridines are alkylated by dimethyl sulphate [269, 270, 296], ethyl bromide [359], and aralkyl halides [360], and tetrachloro- and 2,3,5-trichloro-4-mercaptopyridine react with chloroacetonitrile to give the corresponding [(polychloro-4-pyridyl)thio]acetonitrile [361].

Treatment of a 2-substituted trichloro-4-mercaptopyridine (where the substituent is CF_3, CN, or SO_2Me) with concentrated nitric acid yields the corresponding 4-nitro-compound [232]. The starting mercaptans may be prepared by treatment of the appropriate 6-substituted tetrachloropyridine with sodium sulphide or sodium hydrogen sulphide.

When applied to tetrachloro-4-mercaptopyridine, the usual procedures for the conversion of a mercaptan into the corresponding sulphonyl chloride (e.g., treatment of the mercaptan with chlorine in aqueous acetic acid) give mainly pentachloropyridine [370, 371]. Important modifications of the usual procedures allow tetrachloropyridine-4-sulphonyl chloride to be prepared from the mercaptan, the corresponding disulphide, or tetrachloropyridine-4-sulphenyl chloride [370-372]. Conversion of tetrachloropyridine-4-sulphonic acid to the sulphonyl chloride with phosphorus pentachloride is only possible if the reaction mixture is saturated initially with sulphur dioxide [360], otherwise pentachloropyridine is the major product [371]. The sulphonic acid may be prepared by oxidation of tetrachloro-4-pyridyl 4-nitrobenzyl sulphide with concentrated nitric acid [360].

Tetrachloropyridine-4-sulphonyl chloride is a remarkably stable compound compared with pyridine-2(or 4)-sulphonyl chloride, which are unstable at room temperature. It is hydrolyzed by base to the corresponding sulphonic acid, is reduced to tetrachloro-4-mercapto-pyridine by zinc dust in water, with amines it gives sulphonamides [360, 371], and with potassium fluoride it gives the corresponding sulphonyl fluoride [372]. With aqueous alkaline sodium sulphite, it is converted into the unstable sulphinic acid [371].

An attempt to prepare tetrachloro-4-pyridylsulphur trichloride by treatment of tetrachloropyridine-4-sulphenyl chloride with chlorine in inert solvents failed, presumably because the equilibrium:

$$C_5 Cl_4 NSCl + Cl_2 \rightleftharpoons C_5 Cl_4 NSCl_3$$

lies far to the left as in the case of trichloromethanesulphenyl chloride, where steric hindrance may account for the failure to react [371].

Amines, mercaptans, alcohols, and phenols react with the sulphone (201a) by addition to the side-chain double bond [373a]. The products of addition of pentachlorothiophenol and tetrachloro-4-mercaptopyridine can be made to undergo an $S_N i$ rearrangement (201b; $R = C_6 Cl_5$ or $C_4 Cl_4 N$) to give tetrachloro-4-pyridyl penta-chlorophenyl sulphide and bis(tetrachloro-4-pyridyl) sulphide, respectively.

(201a)   (201b)

Various other sulphur derivatives are described in ref. 360.

Phosphorus derivatives of polychloropyridines are also of interest as biological toxicants. The phosphines listed in Table 2.4 give the corresponding phosphine oxide on treatment with a mixture of sodium dichromate, concentrated sulphuric acid, and acetic acid [340], and esters with the general formula (202; $R = OR'$) (Table 2.1) are hydrolyzed by alcoholic hydrogen chloride to (tetrachloro-4-pyridyl)phosphonic acid (202; $R = OH$) and they react with phosphorus pentachloride on prolonged heating to give (tetrachloro-4-pyridyl)phosphonic dichloride (202; $R = Cl$) [284]. The last compound reacts with aniline, sodium phenoxide, or anhydrous formic acid to give compounds (202; $R = NHPh$), (202; $R = OPh$), and (202; $R = OH$), respectively [284].

With anhydrous formic acid in dry benzene, compound (194) gives (tetrachloro-4-pyridyl)phosphoramidic dichloride (203) which reacts

O:PR$_2$

Cl, Cl

Cl, N, Cl

(202)

HNPOR$_2$

Cl, Cl

Cl, N, Cl

(203) R = Cl
(204) R = OPh

N:P(OPh)$_3$

Cl, Cl

Cl, N, Cl

(205)

Cl, Cl

Cl, N, O—P(R)(OEt) S

(206)

with sodium phenoxide in dry dioxan to give compound (204) [364]. The same compound is also formed when compound (194) is treated with sodium phenoxide and the resulting product (205) is boiled in aqueous ethanol [364]. Compounds (195) and (196) undergo an identical series of reactions.

When 3,5,6-trichloro-2-hydroxypyridine is treated with methylthiophosphonic acid O-ethyl ester chloride in the presence of potassium carbonate, the ester (206; R = Me) is obtained [374]. The esters (206; R = Et, Ph, and OEt) may be prepared similarly.

2.9   QUINOLINES

Chlorination of quinoline with chlorine gives a number of products depending on the reaction conditions [375-378]. In the presence of aluminium chloride [377, 378] a mixture of 5,6,7,8-tetrachloroquinoline and an unidentified pentachloroquinoline is obtained. The latter compound yields heptachloroquinoline on further chlorination with an excess of phosphorus pentachloride in an autoclave at 315-330° [379]. Vapour-phase chlorination of quinoline [196] or hydroquinolines [252] with chlorine diluted with carbon tetrachloride at high temperatures (ca. 600°) also yields the perchlorocompound. Hexachloro-2-cyanoquinoline may be prepared similarly [196]. Heptachloroquinoline may be prepared also starting from 1,2,3,4-tetrahydroquinoline-N-carbonyl chloride (207) [69, 71], or the N-carbonyl chloride (208) of N-n-propylaniline [69-71, 103, 380, 381]. The carbonyl chlorides are pre-chlorinated at 50-150°, preferably in a solvent such as chloroform to begin with, and then further chlorinated at 150-500°. The temperature is raised slowly and a halogen carrier (e.g., ferric chloride) is used at the higher temperatures [69, 70, 103, 380, 381]. Irradiation of the mixtures is

Structure (207): tetrahydroquinoline with N-COCl

Structure (208): NPr^{n}COCl on benzene

Structure (209): N:CClC$_2$Cl$_5$ with Cl substituents

(207)   (208)   (209)

also advantageous. 1,2,3,4-Tetrahydroquinoline also gives a moderate yield of heptachloroquinoline [70, 251, 381], together with an unidentified pentachloroquinoline [251], when it is chlorinated under similar conditions. Heptachloroquinoline is formed also when compound (209) is heated at 400° [70, 103].

2,4-Dichloro-, and 2,4,5- and 3,4,5-trichloro-aniline may be converted into 3,4,6,8-tetra-, 3,4,5,6,8-penta-, and 3,4,5,6,7-pentachloroquinoline [376], respectively, using the quinoline synthesis developed by Riegel *et al.* [382]. This involves the synthesis of a 4-hydroxyquinoline suitably chlorinated in the benzene ring followed by its successive treatment with chlorine and phosphorus oxychloride, to introduce chlorine in the 3- and 4-positions.

Heptachloroquinoline shows the expected lack of basic properties; for example, it does not form salts with hydrogen chloride or boron trichloride [379]. An attempt to prepare its *N*-oxide using a mixture of peroxyacid and concentrated sulphuric acid (see Sections 2.8.1 and 2.8.3) under forcing conditions probably gave hexachloro-2-quinolone [172]. This reaction proceeds via attack of acetic acid on the intermediate heptachloroquinolinium ion, to give the intermediate (210), which undergoes successive loss of hydrogen chloride

Structure (210): chlorinated quinoline with OAc substituent

(210)

and hydrolysis. Hexachloro-2-quinolone is obtained also by treatment of the perchloro-compound with potassium hydroxide in hot aqueous dimethylformamide or with a refluxing mixture of concentrated sulphuric acid and acetic acid (the acetic acid is essential) [172]. With anhydrous potassium fluoride in an autoclave at 470°, heptachloroquinoline yields the perfluoro-compound [379].

2.10   ISOQUINOLINES

Polychloroisoquinolines have received even less attention than polychloroquinolines. The perchloro-compound may be prepared in good yield by exhaustive chlorination of the complex formed between isoquinoline and aluminium chloride with chlorine followed by treatment of the resulting hexachloroisoquinoline (cf. ref. 377) with phosphorus pentachloride at 270° in an autoclave [379, 383], and by exhaustive chlorination of 1,2,3,4-tetrahydroisoquinoline [251] or 1,2,3,4-tetrahydroisoquinoline N-carbonyl chloride (see also Section 2.9) [71]. Vapour-phase chlorination of isoquinoline with chlorine diluted with carbon tetrachloride at 600° also yields the perchloro-compound [196]. When o-carboxyphenylacetonitriles with α-alkyl substituents or the corresponding methyl esters are treated with an excess of phosphorus pentachloride, 1,3,4-trichloro-isoquinoline is formed via dealkylation of the intermediate 4-alkyl-1,3-dichloroisoquinoline [384, 385].

Conversion of heptachloroisoquinoline into the perfluoro-compound may be accomplished by treatment with anhydrous potassium fluoride in an autoclave [379, 383]. With a mixture of peroxyacid and concentrated sulphuric acid, heptachloroisoquinoline yields a hydroxy derivative, which has not been identified [172].

2.11   PYRIDAZINES

Halogenopyridazines have been reviewed† (see also ref. 386) [387, 388]. Although pyridazines undergo electrophilic substitution with difficulty, tetrachloropyridazine may be prepared in good yield by vapour-phase chlorination of 3-chloropyridazine with chlorine diluted with carbon tetrachloride at 600° [196], or by treatment of 3,6-dichloropyridazine with phosphorus pentachloride in an auto-clave at 280° [389-391]. The 3,6-dichloro-compound reacts exo-thermically with chlorine at 70°; the temperature rises to 130° and further chlorination at 150° yields mainly 3,4,6-trichloropyridazine (see later) together with traces of the perchloro-compound [392]. In the presence of aluminium chloride, the perchloro-compound is obtained in quantitative yield at elevated temperatures [386]. Liquid-phase chlorination of 6-chloro-3-cyanopyridazine with chlorine at 150-200° gives a mixture of 4,6- and 5,6-dichloro-3-cyanopyridazine or trichloro-3-cyanopyridazine, depending on the reaction conditions [393]. 3,6-Dichloro-4,5-dicyanopyridazine may be prepared similarly.

A general synthesis of chloropyridazines starts with maleic hydrazide (211; $R^1 = R^2 = H$) and its mono- (211; $R^1 = Cl$, $R^2 = H$)

† The emphasis is on the lower halogenated derivatives. *Note added in proof*: See, however, ref. 858.

276   B. IDDON AND H. SUSCHITZKY

and di-chloro-derivatives (211; $R^1 = R^2 = Cl$) (these exist as 3-hydroxy-6(1H)pyridazinones), which give 3,6-dichloro- [394, 395], 3,4,6-trichloro- [396-398], and tetrachloro-pyridazine [397], respectively, on treatment with phosphorus oxychloride. A detailed study of the preparation of 3,6-dichloropyridazine from maleic hydrazide showed that the product is usually contaminated with 3-chloro-6(1H)-pyridazinone and compound (212) [399-401]. The latter is formed by a reaction between the 3,6-dichloro-derivative and the pyridazinone. Maleic hydrazides may be prepared by treatment of

(211)   (212)   (213) R^1 = Ph, R^2 = OH
                                      (214) $R^1 = R^2$ = H
                                      (215) R^1 = Me, R^2 = H
                                      (216) R^1 = Ph, R^2 = H
                                      (217) R^1 = Ph, R^2 = Cl

the corresponding maleic anhydride with a hydrazine salt. Similarly, when dichloromaleic anhydride is treated with phenylhydrazine hydrochloride in 5% aqueous hydrochloric acid, it gives 4,5-dichloro-3-hydroxy-1-phenyl-6(1H)-pyridazinone (213) [402, 403].

Dichloromaleic anhydride may be prepared by oxidation of mucochloric acid (218) with nitric acid [403]. Mucochloric acid has

$$HO_2C \cdot CCl{:}CCl \cdot CHO \rightleftharpoons$$

(218)

become readily available recently and is itself an important starting material for the synthesis of polychloropyridazines. Its uses have been reviewed [403]. With hydrazine or a hydrazine salt, for example, it gives 4,5-dichloro-6(1H)-pyridazinone (214) [403-406]. whilst methylhydrazine [407] and phenylhydrazine [405] yield the N-methyl compound (215) and the N-phenyl compound (216), respectively. Many other examples are given in ref. 403.

Treatment of the pyridazinone (214) with phosphorus oxychloride gives 3,4,5-trichloropyridazine [406, 408, 409]. Likewise, a mixture of phosphorus oxychloride and phosphorus pentachloride converts the pyridazinone (213) into 3,4,5-trichloro-1-phenyl-6(1*H*)-pyridazinone (217) [402, 403]. The same compound (217) may be prepared similarly from 3,4-dichloro-1-phenyl-, 3,5-dichloro-1-phenyl-, 3-chloro-5-hydroxy-1-phenyl-, or 3-hydroxy-1-phenyl-6(1*H*)-pyridazinone [402, 410].

Perchloro(vinylacetaldehyde) reacts with arylhydrazines to give 1-aryl-4,5-dichloro-6(1*H*)-pyridazinones as shown in Scheme 2.18 [411]. In some cases the intermediate azo-compounds (219) can be isolated.

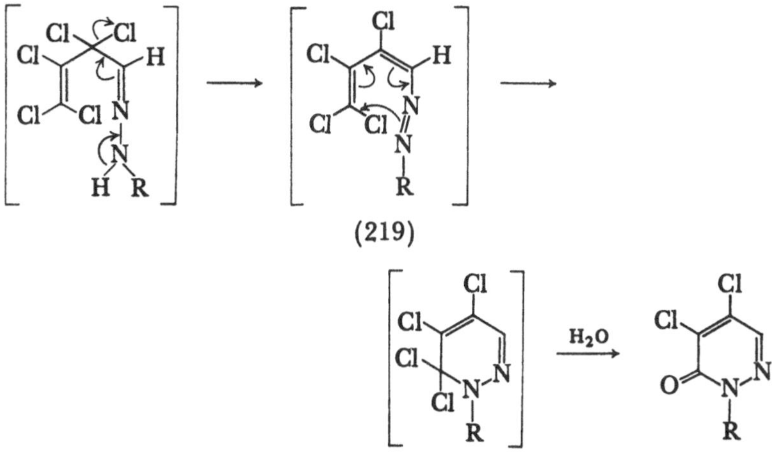

(219)

SCHEME 2.18

The product of the reaction between perchloropropene and diazomethane crystallizes explosively to give a 51% yield of 3,4,5-trichloropyridazine [412]. The explosion has been attributed

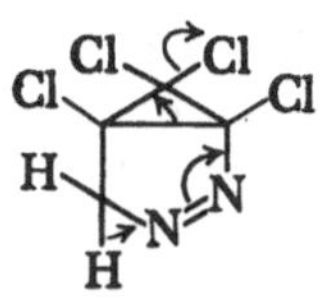

(220)

to the initial formation of the strained ring adduct (220), which then rearranges, as shown.

As expected, tetrachloropyridazine reacts with nucleophiles (e.g., methylamine or hydroxide, methoxide, or ethoxide ions) mainly in the 4(5)-position [386, 413-415]. With trimethylamine in benzene at room temperature, however, it gives 4,5,6-trichloro-3-dimethyl-aminopyridazine and 4,5-dichloro-3,6-bis(dimethylamino)pyridazine (*cf.* the corresponding reactions of trichloropyridazines discussed later) [416, 417]. The formation of these compounds may be attributed to steric hindrance to attack by the trimethylamine in the 4(5)-position; their structures were proved by hydrodechlorination (see also ref. 413). The perchloro-compound gives tetrafluoro-pyridazine on treatment with anhydrous potassium fluoride (see also ref. 418) in an autoclave at 340° [389-391].

With a mixture of peroxyacetic acid and concentrated sulphuric acid [172, 353], or trifluoroperoxyacetic acid [315], tetrachloro-pyridazine gives the mono-*N*-oxide. The *N*-methoxypyridazinium methyl sulphate (221) is obtained on treatment of the *N*-oxide with dimethyl sulphate at 100° (see also Section 2.8.2) [315]. With water this salt gives a mixture of the 1-substituted trichloro-6(1*H*)-pyrid-azinones (222-224) together with the *N*-oxide of compound (223) and

(221)

(222) R = OMe
(223) R = Me
(224) R = CH$_2$Cl

tetrachloropyridazine-*N*-oxide (*cf.* the reaction of pentachloro-1-methoxypyridinium methyl sulphate with water—Section 2.8.2). The last compound is probably residual from its incomplete conversion to the salt (221); the salt was not purified. The formation of the pyridazinone (222) can be attributed to nucleophilic displacement by water of an α-chlorine atom, but the formation of the other products is more difficult to rationalize. The *N*-methylpyridazinone (223) can arise by thermal deoxygenation of its *N*-oxide. The *N*-chloromethyl compound (224) was unambiguously synthesized by treatment of trichloro-6(1*H*)-pyridazinone with formaldehyde followed by treatment of the *N*-hydroxymethyl compound produced with thionyl chloride [315, 386].

Irradiation of tetrachloropyridazine converts it into tetrachloro-pyrazine (see Section 2.13); tetrafluoropyridazine similarly gives tetrafluoropyrazine or, after prolonged irradiation, tetrafluoro-pyrimidine [418].

With ethanolic ammonia under pressure, 3,4,5-trichloropyridazine gives a mixture of two amines, (225) and (226) [386, 406, 419, 420]. These amines give 4-aminopyridazine on hydrodechlorination [406]. Dimethylamine, on the other hand, gives only the amine (227)

$$R^1 \underset{N^{\nwarrow N}}{\overset{R^2}{\bigvee}} Cl$$

(225) R^1 = Cl, R^2 = NH$_2$
(226) R^1 = NH$_2$, R^2 = Cl
(227) R^1 = NMe$_2$, R^2 = Cl

[415], which is also formed when the trichloro-compound is reacted with trimethylamine in benzene at room temperature (cf. the corresponding reaction of the 3,4,6-trichloro-isomer, described later) [416]. The structure of the amine (227) has been proved by hydrodechlorination, which gives a mixture of 4-dimethylamino-pyridazine and its 6-chloro-derivative [415, 416]. The amines (225) and (226) yield the corresponding 1-oxide on treatment with ethereal monoperoxyphthalic acid [421]. Each of these N-oxides gives 3,4,5-trichloropyridazine 1-oxide on diazotization in hydrochloric acid.

With an equimolar amount of methoxide ion, 3,4,5-trichloropyridazine gives 3,4-dichloro-5-methoxypyridazine, but with two equivalents of methoxide ion it yields a mixture of 4-chloro-3,5-dimethoxy- (25%) and 3-chloro-4,5-dimethoxy-pyridazine (41%) [422, 423]. With an excess of the sodium salt of benzyl mercaptan, it gives 3,4,5-tri(benzylthio)pyridazine [408, 409].

When 3,4,5-trichloropyridazine is heated in acetic acid, it is converted into 4,5-dichloro-6(1H)-pyridazinone (214) [424]. A trichloropyridazinium ion is probably responsible for the orientation in this case [416].

It is noteworthy that aqueous solutions of 3,4,5-trichloropyridazine produce severe blisters on contact with the skin [408, 412].

In agreement with theoretical predictions [425], 3,4,6-trichloropyridazine reacts with nucleophiles (e.g., ammonia, methylamine, hydrazine, sulphanilamide, and hydroxide or alkoxide ions) mainly in the 4-position [386, 387, 398, 413, 415, 416, 424-435a]. Where the nucleophile is hydroxide ion, replacement of the 3- or 6-chlorine atom also occurs to the extent of 5% [426]. 3,4,6-Trichloropyridazine gives 4-amino-3,6-dichloropyridazine with methanolic ammonia above 100° [398] but, at room temperature, it gives 3,6-dichloro-4-methoxypyridazine [436]. The former product arises by initial formation of the latter product followed by displacement of the methoxyl group by ammonia. The reaction between the 3,4,6-trichloro-compound and ethanolic ammonia at 120-130° similarly yields the 4-amino-derivative contaminated with the 4-ethoxy derivative [433]. The structures of some of the derived products have been established by hydrodechlorination to known pyridazines [387, 398, 415, 426, 432].

With trimethylamine in benzene at room temperature, 3,4,6-trichloropyridazine gives a mixture of 3,4-dichloropyridazin-6-yl-trimethylammonium chloride (228) and the isomer (229) (*cf.* the

$$(228) \; R^1 = \overset{+}{N}Me_3\overset{-}{Cl}, \; R^2 = Cl$$
$$(229) \; R^1 = Cl, \; R^2 = \overset{+}{N}Me_3\overset{-}{Cl}$$

corresponding reaction of the 3,4,5-trichloro-isomer, already discussed) [416, 417]. When heated in dimethylformamide, these salts give 3,4-dichloro-6-dimethylaminopyridazine and 3,6-dichloro-4-dimethylaminopyridazine, respectively [416]. The trimethylammonium group in compounds (228) and (229) is readily displaced

$$(230) \; R^1 = Cl, \; R^2 = H$$
$$(231) \; R^1 = H, \; R^2 = Cl$$

(232)   (233)

(234)   (235)

by nucleophiles, for example, by hydroxide ion or by sodium 2-ethoxyethoxide [416].

A 3,6-dichloro-4-*sec*.-aminopyridazine is substituted by hydrazine in the 6-position [434].

When 3,4,6-trichloropyridazine is heated in acetic acid, it gives 3,4- or 3,5-dichloro-6(1*H*)-pyridazinone [424]. This reaction is analogous to that of the 3,4,5-trichloro-isomer with acetic acid (see before).

Trifluoroperoxyacetic acid oxidation of 3,4,6-trichloropyridazine yields a mixture of the isomeric *N*-oxides [315]. Successive treatment of this mixture with dimethyl sulphate and aqueous cyanide gives a mixture of the pyridazinones (230) and (231), which arise via nucleophilic displacement by water of an α-chlorine atom in each of the intermediate *N*-methoxypyridazinium methyl sulphates [315]. None of the expected 3,5,6-trichloro-4-cyanopyridazine was isolated (cf. Section 2.8.2).

The reaction between 3,4,6-trichloropyridazine and *o*-aminophenol yields 2-chloro-3,4-diazaphenoxazine (232) [427]. An attempt to prepare the isomer (233) of compound (232) through cyclization of compound (234) resulted in the formation of the diazaphenoxazine (232) via an initial Smiles rearrangement, (234) → (235) [427].

3,4,6-Trichloropyridazine gives the addition compounds (236) and (237) and the ring-opened product, Bu^{t}CH:CBu^{t}CN, together with

(236) R = Cl
(237) R = Bu^t

smaller amounts of unidentified products, on treatment with *t*-butylmagnesium chloride [437].

Several reactions of substituted di- and tri-chloropyridazines have been studied by Landquist and his co-workers [436, 438]. With dimethylamine, for example, both 3,4,5-trichloro-6-dimethylamino- and 3,4,6-trichloro-5-dimethylamino-pyridazine give 3,5-dichloro-4,6-bis(dimethylamino)pyridazine [438]. Complex mixtures are formed when these trichloropyridazines are treated with a sodium alkoxide. In the former case, attack appears to occur predominantly in the 4-position, whilst in the latter case attack occurs predominantly in the 6-position.

Nucleophilic attack by sodium butoxide on 3,6-dichloro-4-dimethylaminopyridazine occurs in the 3-position, as expected, but attack by amines occurs in the 6-position [438]. This difference in orientation has been attributed to steric hindrance to attack by amines. We referred earlier to displacement by ammonia of the methoxyl group in 3,6-dichloro-4-methoxypyridazine. The methoxyl group in this compound is also displaced by hydrazine [436] or acetylsulphanilamide [432]. With secondary aliphatic amines, however, the main reactions are displacement of either or both of the chlorine atoms followed by demethylation of the resulting ethers [436]. Hence, the product arising from this alkoxy compound is governed mainly by the nature of the attacking nucleophile.

3,4-Dichloro-6-(2-ethoxyethoxy)pyridazine (238) undergoes reaction with dimethylamine exclusively by displacement of the 4-chlorine atom, as expected if the most stable transition state is involved in the process [436]. 5-Butoxy-3,4-dichloropyridazine (239), on the other hand, gives the 3-morpholino-compound

(238)   (239)

exclusively with morpholine and 3,5-dibutoxy-4-chloropyridazine exclusively with sodium butoxide [436]. In this case the reactions are controlled by steric hindrance.

With hydrazine, 4-amino-3,6-dichloropyridazines give the corresponding 4-amino-3-chloro-6-hydrazinopyridazine which are claimed [435a] to possess antidepressant activity.

3,6-Dichloropyridazine is reported [439] to give 6-chloro-3-pyridazinyl-lithium on treatment with n-butyl-lithium at $-15°$. It seems surprising that metallation in the 4(5)-position was not observed.

Photoirradiation of 3,6-dichloropyridazine (see also ref. 418) in methanol has been studied [440] and the same authors have described a number of interesting reactions of polychlorobipyridazinyls [441].

Chlorine in a polychloropyridazine may be replaced by a mercapto-group by treatment with phosphorus pentasulphide in pyridine [408]. A large number of nucleophilic substitution reactions of 4,5-dichloro-6(1H)-pyridazinones are described in ref. 403.

The UV spectra have been recorded and discussed for 3,6-dichloro- [442], 3,4,5- [443] and 3,4,6-trichloro- [442, 443], and tetrachloro-pyridazine [443], and the molecular core binding energies have been measured by means of ESCA for pyridazine and its tetrachloro- and tetrafluoro-derivatives (see Section 2.8.1) [1].

2.12   PYRIMIDINES

The chloropyrimidines have been reviewed previously (see footnote on p. 275) [444, 445]. 2,4,6-Trichloro- and tetrachloro-pyrimidine are available commercially and can be prepared in the laboratory from readily available starting materials. Consequently, the chemistry of these compounds has been studied intensively, both in academic and industrial laboratories. The high reactivity of the chlorine atoms in them has resulted in the preparation of a large number of dyestuffs (cf. cyanuric chloride) [446-472]. The 2,4,6-trichloro-compound is also a useful cross-linking agent [473-476]. In addition, the importance of the pyrimidine nucleus in nature has led to the synthesis of a large number of polychloropyrimidines for biological evaluation [477-502]; many of these compounds have been shown to exhibit potentially useful activity.

It is important to note that 2,4,6-trichloropyrimidine irritates eyes and lungs [503, 504] and can cause blisters and lesions when spilled on the skin [504]. All three trichloropyrimidines appear to be lachrymatory [505]. The fact that 2,4- and 4,6-dichloro-5-nitro-pyrimidine and related compounds produce dermatitis in susceptible persons is well known [506].

Considerable interest has been shown in the physical properties of chloropyrimidines, including 2,4,6-trichloro- and tetrachloro-pyrimidine. Thus, the IR spectra of a number of chloropyrimidines have been recorded [507-510]; absorptions in the region 743 to 786 cm^{-1} have been assigned to C—Cl stretching [509]. By comparing the IR and Raman spectra of 2,4,6-trichloropyrimidine with the corresponding spectra of its 5-deuteriated derivative, it has been possible to make a large number of assignments for these compounds [510]. The UV spectra have been recorded for 2,4,6-trichloro- and tetrachloro-pyrimidine, both in the vapour phase [511] and in solution in aqueous methanol [511] and ethanol [512]. The relationship between ^{1}H NMR chemical shifts and electron densities at the carbon atoms to which the protons are bonded (computed by a simple LCAO-MO method) has been studied for a series of chloro- and fluoro-pyrimidines [513], and the molecular core binding energies have been measured by means of ESCA for pyrimidine and its tetrachloro- and tetrafluoro-derivatives (Section 2.8.1) [1]. The [35]Cl NQR spectra have been recorded for each of the trichloro-compounds and the perchloro-compound [514-517]. Two features of these spectra are noteworthy: (i) the observed frequencies indicate appreciable π-bonding between the chlorine and ring carbon atoms; and (ii) the 5-chlorine atoms in 2,4,5-trichloro- and tetrachloro-pyrimidine exhibit particularly high frequency values. Although NQR spectra are necessarily complex and difficult to record, the absence of high frequencies characteristic of a

5-chlorine atom might be used to establish the absence of such a chlorine atom in a polychloropyrimidine. The dipole moments have been recorded for 2,4,6-trichloro- and tetrachloro-pyrimidine [518]. Both 2,4,6-trichloro- and tetrachloro-pyrimidine were known before 1900. The former compound is conveniently prepared by treatment of barbituric acid with phosphorus oxychloride [equation (2.4); R = H] [519-525]. The presence of a tertiary aromatic amine,

(2.4)

(240)

(241)   (242)

such as *NN*-dimethyl- [524, 526-529] or preferably, *NN*-diethyl-aniline [530, 531], promotes the reaction, but a small amount of the corresponding 2-*N*-alkylanilino-4,6-dichloropyrimidine, for example compound (241) (see later) in the former case, is usually to be found in the product if an amine promoter is used [527, 532]. A mixture of phosphorus oxychloride and phosphorus pentachloride converts barbituric acid into tetrachloropyrimidine [533], which may be prepared also by bubbling chlorine through a mixture of the acid, phosphorus oxychloride, and phosphorus trichloride in pyridine while the temperature is raised from 20 to 115° [534]. If barbituric acid is heated with triphenoxyphosphorus dichloride at 100° and the resulting intermediate, assumed to be compound (242), is pyrolyzed at 200°, 2,4,6-trichloropyrimidine is obtained in 80% yield [535]. A related procedure for the synthesis of this trichloro-compound involves the use of phenylphosphonic dichloride (PhPOCl$_2$) [536].

The following are some of the 5-substituted 2,4,6-trichloro-pyrimidines (240; R given) which may be prepared by treatment of the corresponding 5-substituted barbituric acid with phosphorus oxychloride, preferably in the presence of *NN*-diethylaniline [equation (2.4)]: methyl [537, 538], ethyl [539, 540], *n*-butyl [483, 494], MeCHCH$_2$Me [494, 541], *n*-pentyl [494], Me(CH$_2$)$_2$CHMe [494], Me$_2$CHCH$_2$CH$_2$ [494], *n*-hexyl [494], *n*-heptyl [494], *n*-octyl [483, 494], *n*-nonyl [494], *n*-dodecyl [483], allyl (note that the double bond is unaffected) [494], phenyl [479, 532], substituted aryl [542], benzyl [494, 522], methoxyl [543], and nitro [544]. It is not possible to convert 5-nitro-barbituric acid into its trichloro-derivative by treatment with phosphorus oxychloride in the presence of *NN*-dimethylaniline (the reaction proceeds only as far as the 2,4-dichloro-compound) [545], but the reaction proceeds smoothly in the presence of *NN*-diethyl-aniline [544]! Trichloro-5-nitropyrimidine may be prepared also by treatment of 4,6-dichloro-2-methylthio-5-nitropyrimidine with chlorine in ethanol (see also ref. 566) [547]. 5-Chlorobarbituric acid is converted into tetrachloropyrimidine on treatment with phos-phorus oxychloride in the presence of *NN*-dimethyl- [548] or *NN*-diethyl-aniline [548, 549].

The uracils constitute a second major class of starting materials for the synthesis of polychloropyrimidines. Thus, 2,4,5-trichloro-pyrimidine may be prepared by treatment of 5-chlorouracil (243; R^1 = H, R^2 = Cl) with a mixture of phosphorus oxychloride and phosphorus pentachloride [505], or with phosphorus oxychloride alone [492] or in the presence of *NN*-dimethylaniline [499, 550]. Treatment of a 6-substituted uracil (243; R^2 = H) with phosphorus oxychloride yields the corresponding 6-substituted 2,4-dichloro-pyrimidine (244; R^2 = H) [492]. If the 6-substituted uracil (243; R^2 = H) is chlorinated initially in the 5-position using sulphuryl

R^1 ⎤ N ⎤ O
R^2 ⎦ NH
     O

(243)

R^1 ⎤ N ⎤ Cl
R^2 ⎦ N
     Cl

(244)

chloride and the product (243; R^2 = Cl) is treated with phosphorus oxychloride, alone or in the presence of *NN*-diethylaniline [551, 552], the final product is a 6-substituted trichloropyrimidine (244; R^1 = Me [489, 551, 552], Et [492], *n*-Pr or *i*-Pr [492], n-$C_{17}H_{37}$ [492], Ph [492], CO_2Me [553], or CO_2H [553]; R^2 = Cl). The 6-methyl compound (244: R^1 = Me; R^2 = Cl) may be prepared also by treatment of 6-methyluracil (243: R^1 = Me; R^2 = H) with a mixture of phosphorus oxychloride and phosphorus pentachloride [552, 554].

Some groups present in the starting material may be modified on treatment with the reagent. Thus, with phosphorus oxychloride in the presence of *NN*-diethylaniline, the nitro-amide (245) gives trichloro-6-cyanopyrimidine (246) as the major product, together with the nitro-amide (247) and a trace of compound (248) (Scheme 2.19) [555]. Presumably, the nitro-amide (247) is formed initially, since it is converted into the nitrile (246) on further treatment with

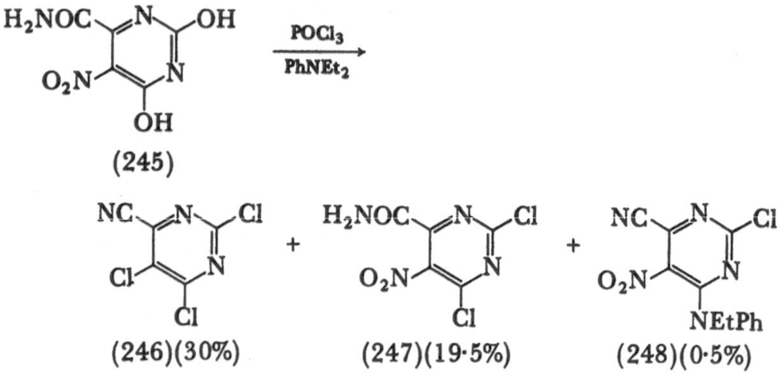

(245)

(246)(30%)   (247)(19·5%)   (248)(0·5%)

SCHEME 2.19

phosphorus oxychloride or phosphorus pentachloride. The nitro-amide (245) gives an improved yield of the nitrile (246) on treatment with a mixture of phosphorus oxychloride and phosphorus pentachloride [555]. Replacement of a nitro-group in this way is unusual [555]. Compound (249; R = OH), for example, yields the dichloro-compound (249; R = Cl) with phosphorus oxychloride and NN-dimethylaniline [556].

When 5-chloro-örotic acid (5-chloro-2,6-dihydroxypyrimidine-4-carboxylic acid) (250) is treated successively with phosphorus oxychloride (in the presence of NN-dimethylaniline) and phosphorus pentachloride, it gives a low yield (4%) of tetrachloropyrimidine [553]. With phosphorus oxychloride in the presence of NN-diethylaniline, however, decarboxylation is prevented and a moderate yield of the trichlorocarboxylic acid (244: R^1 = CO_2H; R^2 = Cl) is obtained.

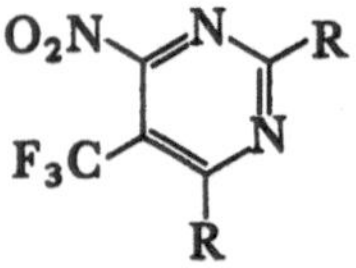

(249) 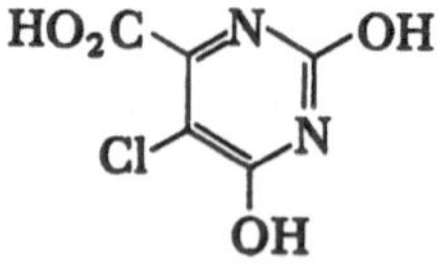   (250) 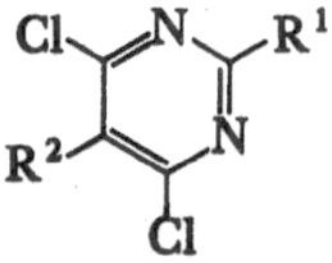   (251)

Phosphorus oxychloride in the presence of NN-dimethylaniline converts 5-chloro-4,6-dihydroxypyrimidine into 4,5,6-trichloropyrimidine [505]. The starting material may be prepared by treatment of 4,6-dihydroxypyrimidine with iodine chloride in acetic acid. Other 4,6-dihydroxypyrimidines similarly yield polychloropyrimidines on treatment with phosphorus oxychloride or a phosphorus oxychloride–phosphorus pentachloride mixture. The following compounds (251) have been prepared in this way (R^1 and R^2 given): Me, Cl [489]; Et, Cl [493]; n-Pr, Cl [493]; i-Pr, Cl [493]; n-C$_{17}$H$_{35}$, Cl [493]; Ph, Cl [493]; various p-ROC$_6$H$_4$CH$_2$, Cl [557]; CH$_2$OMe, Cl [558]; SMe, Cl [546]; Cl, n-Pr [493]; Cl, i-Pr [493]. 2-Amino-4,6-dihydroxypyrimidine gives 2-amino-4,6-dichloropyrimidine with phosphorus oxychloride in the presence of NN-dimethylaniline [524] but, with a mixture of phosphorus oxychloride and phosphorus pentachloride, it gives 2-aminotrichloropyrimidine [533]. 4,6-Dihydroxy-2-hydroxymethyl-5-chloropyrimidine yields trichloro-2-chloromethylpyrimidine (251: R^1 = CH$_2$Cl; R^2 = Cl) on treatment with a mixture of phosphorus oxychloride and phosphorus pentachloride [489], whilst 2-amino-5-ethyl-4,6-dihydroxypyrimidine gives 2-aminodichloro-5-ethylpyrimidine (251: R^1 = NH$_2$; R^2 = Et) with phosphorus oxychloride [539].

Tetrachloropyrimidine may be prepared also by reaction of alloxan with a mixture of phosphorus oxychloride and phosphorus pentachloride at 120° in a sealed tube [559] or by similar treatment of dialuric acid (tetrahydroxypyrimidine) [560]. The former synthesis was reported in 1885. Exhaustive chlorination of 5-ethylpyrimidine with chlorine, initially in the liquid phase and finally in the vapour phase at 300-500° over a catalyst, such as activated charcoal, yields perchloro(5-vinylpyrimidine) [233].

Exhaustive chlorination of 3-dimethylaminopropionitrile (252) leads via the isolable intermediates shown in Scheme 2.20 to

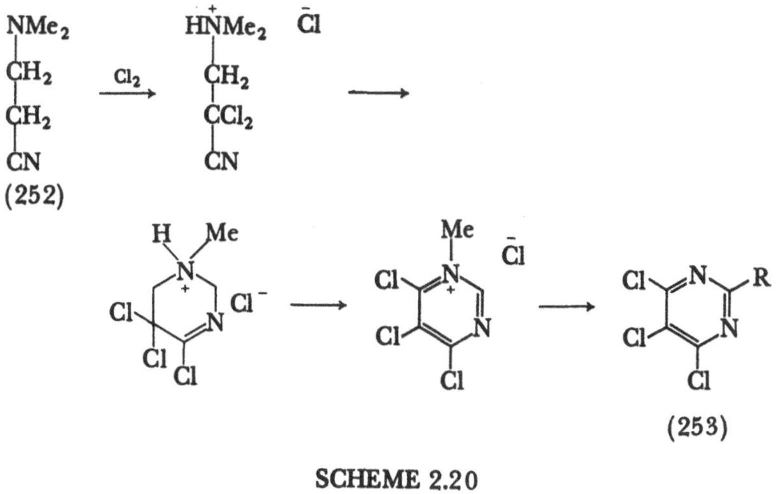

(253)

SCHEME 2.20

4,5,6-trichloropyrimidine (253; R = H); some tetrachloropyrimidine may be formed also, depending on the reaction conditions [70, 561]. At about 200°, 3-pyrrolidinopropionitrile similarly gives a mixture of perchloro(2-methylpyrimidine) (253; R = CCl_3) and perchloro(2-ethylpyrimidine) (253; R = C_2Cl_5) [250]. 3-Perhydro-azepinopropionitrile (94) gives tetrachloropyrimidine, pentachloro-pyridine, and a number of minor chlorinated products at 250° (see also sections 2.8.2 and 2.18.1) [250], whilst exhaustive chlorination of 3-morpholinopropionitrile yields trichloro-2-chlorocarbonyl-pyrimidine (253; R = COCl) under similar conditions [562]. Chlorination of 3-perhydroazepinopropionitrile at 150° gives octa-chloropyrimido[1,2-a]azepine (Section 2.18.1) which rearranges to yield a mixture of isomeric trichloro-2-(heptachlorocyclopentenyl)-pyrimidines [e.g., compound (254)] on further treatment

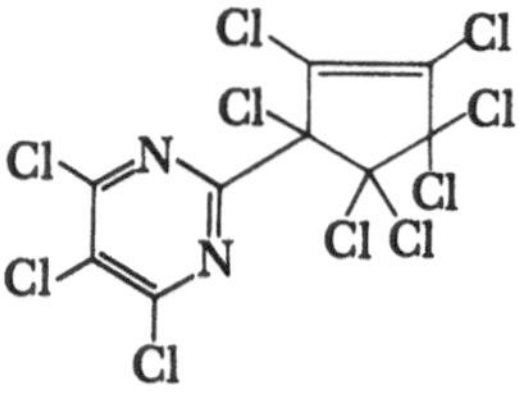

(254)

with chlorine at 310-345° [563]. Nitriles with the general formula $R^1(R^2CH_2)NCH_2CH_2CN$ will undergo the reaction to give a 2-substituted pyrimidine providing that the group R^1, which is normally methyl or benzyl, can be eliminated during the chlorination procedure [564]. Thus, for example, N-chlorocarbonyl-N-methyl-3-aminopropionitrile gives 2,4,5-trichloropyrimidine or tetrachloro-pyrimidine, depending on the reaction conditions (see also Sections 2.18.1 and 2.18.2) [565].

α-Dichloroisocyanide dichlorides with the general formula (255) have become readily accessible recently. They react with nitriles

$$R^1CCl_2N{:}CCl_2 + R^2CH_2CN \xrightarrow[-HCl]{FeCl_3/200°}$$
$$(255) \qquad\qquad (256)$$

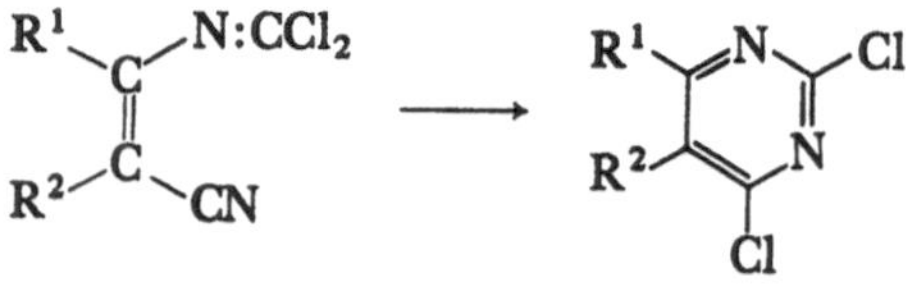

(257)   (244)

SCHEME 2.21

(256) containing two α-hydrogen atoms in the presence of a Lewis acid at elevated temperatures to give polychloropyrimidines (244) (Table 2.5) via the intermediates (257), which can be detected in certain cases (Scheme 2.21) [566, 567]. The reaction proceeds analogously [equation (2.5)] if an isocyanide dichloride of the general formula (258) is reacted with an α,α-dichloronitrile (259) [566, 567]. Of particular interest is the reaction

$$R^1CH_2N=CCl_2 + R^2CCl_2CN \rightarrow (257) \rightarrow (244) \qquad (2.5)$$
$$(258) \qquad\qquad (259)$$

between trichloroacetonitrile (259; R^2 = Cl) and trichloromethyl isocyanide dichloride (255; R^1 = Cl), which yields tetrachloropyrimidine [567].

TABLE 2.5 Polychloropyrimidines (244) Prepared According to Scheme 2.21 [566, 567]

R^1	R^2	Yield (%)
Cl	H	88
Cl	Cl	94
Cl	Me	61
Cl	CH_2Cl	74
Cl	Ph	57
Cl	p-$O_2NC_6H_4$	72
Cl	$3,4$-$Cl_2C_6H_3$	75
Cl	$2,3,4$-$Cl_3C_6H_2$	65
Cl	$NCCH_2$	14
Cl	2-(2,4,6-trichloro-5-pyrimidinyl)ethyl	32
CCl_3	H	5
Ph	H	58

Exhaustive chlorination of 2,4-di(N-ethylamino)-6-chloropyrimidine is reported [251, 568] to give compound (260). Related compounds may be prepared similarly.

$$Cl\text{-pyrimidine ring with } N:CClCCl_3 \text{ substituents}$$

(260)

2,4,6-Trichloropyrimidine may be converted into tetrachloropyrimidine by treatment with an excess of phosphorus pentachloride in an autoclave at 350° [569] as well as by exhaustive chlorination with chlorine at elevated temperatures in the presence of iron or ferric chloride [570] or over activated charcoal [571]. Exhaustive chlorination over activated charcoal at 450-500° of 2,4-dichloro-, 2,4,5-trichloro-, 4,5,6-trichloro-, 2,4-dichloro-6-methyl-, or 4,6-dichloro-2-methyl-pyrimidine similarly yields tetrachloropyrimidine [571]. Chlorination of 4,5,6-trichloropyrimidine with chlorine at 200-210° under UV irradiation gives a low yield of hexachloro-2,2'-bipyrimidyl (261; R = Cl) [572], whilst chlorination of 2,6-dichloro-5-fluoropyrimidine with chlorine in carbon tetrachloride in the

$$
\begin{array}{c}
\text{Cl} \qquad\qquad \text{Cl} \\
\text{Cl} - \underset{R}{\overset{N}{\bigvee}} \underset{N}{\overset{N}{\bigvee}} - \text{Cl} \\
R \qquad\qquad R
\end{array}
$$

(261)

presence of dibenzoyl peroxide gives a low yield of trichloro-5-fluoropyrimidine [546]. It is noteworthy here that the bipyrimidyl (261; R = Cl) gives the disubstituted compound (261; R = $C_{14}H_7O_2NH-$) on treatment with 1-aminoanthraquinone in nitrobenzene in the presence of ferric chloride [572].

Only in one experiment was oxidation of tetrachloropyrimidine to its mono-N-oxide with a mixture of peroxyacid and concentrated sulphuric acid successful; the reaction was found not to be reproducible (see also Sections 2.8.2, 2.11, and 2.13) [172].

Pyrimidine and its derivatives are often prepared by reduction of a more readily accessible polychloropyrimidine. Pyrimidine was first prepared in this way by reduction of its 2,4,6-trichloro-derivative with zinc dust and water [519]. Higher yields (> 90%) are obtained when the 2,4,6-trichloro-compound [530, 545] or the perchloro-compound [530] are hydrodechlorinated over palladium-charcoal in the presence of a suitable base, for example magnesium oxide [545] or sodium hydroxide [530]. In the absence of a suitable base, reduction of the ring also occurs to give 1,4,5,6-tetrahydro-pyrimidine (ca. 95% yield) [530]. Reduction of trichloro-5-methylpyrimidine with zinc and water [537], or catalytically [538], similarly gives 5-methylpyrimidine; with zinc dust in benzene in the presence of 3N-ammonium hydroxide partial reduction occurs to give 2-chloro-5-methylpyrimidine (85%) [573]. Further examples may be found in refs. 444, 445, 522, 530, 532, 539, 545, 574, and 575. In some polychloropyrimidines, other substituents may be reduced without concurrent dehalogenation; for example, catalytic reduction of trichloro-5-nitropyrimidine gives a good yield of 5-aminotrichloropyrimidine [544].

It is claimed [576] that both 2,4,6-trichloro- and tetrachloropyrimidine react with magnesium in tetrahydrofuran to give solvated magnesium compounds, but the structures of these compounds were not given, nor were any of their reactions reported.

With nucleophiles (e.g., ammonia [574], n-butylamine [574], primary [574, 577] or secondary [577] aromatic amines, diethylamine [574], or alkoxide ions [578]) tetrachloropyrimidine appears to react almost exclusively in the 4-position. A detectable amount of a 2-substituted product has been isolated from its reaction with diethylamine [574]. The claim [498, 579] that tetrachloropyrimidine gives a 2-substituted product with triethylamine, tri-n-butylamine, NN-dimethylaniline, N-ethylpiperidine, or N-methylmorpholine is suspect in our opinion (see the discussion of the corresponding reactions of 2,4,6-trichloropyrimidine). With two equivalents of a nucleophile, a 2,4-disubstituted pyrimidine is the usual product [578]. An excess of ammonia is reported [533] to give a mixture of 4,6-diaminodichloropyrimidine and 2,4,6-triamino-5-chloropyrimidine, whilst two equivalents of ethylenimine also yield a 4,6-disubstituted derivative [580]. An excess of ethylenimine is claimed [581, 582] to yield the 2,4,6-trisubstituted derivative. 4,6-Diaminodichloropyrimidine was shown to be different from the 2,4-diamino-isomer by unambiguous synthesis of the latter. 4-Anilinotrichloropyrimidine gives a mixture of 4-anilinodichloro-2-ethoxy- and 4-anilinodichloro-6-ethoxy-pyrimidine on treatment with ethanol containing potassium hydroxide [574]. With an excess of sodium ethoxide or ammonia it is substituted in both the 2- and the 6-position [574]. 4-Anilinotrichloropyrimidine is weakly acidic [574]. The rates of hydrolysis and of quaternization by NN-dimethylhydrazine have been studied for a few water-soluble trichloropyrimidin-4-ylarylamines [577].

Tetrachloropyrimidine may be converted into 5-chlorotrifluoropyrimidine by reaction with silver fluoride [578] or into the perfluoro-compound by reaction with anhydrous caesium or potassium fluoride, preferably at elevated temperatures (410-480°) in an autoclave [214, 569, 583]. When the perchloro-compound is reacted with hydrofluoric acid in a nickel autoclave at 130° for 10 h, it yields a mixture of 5-chlorotrifluoropyrimidine (6%), a dichlorodifluoropyrimidine (39%), which is probably the 5,6-dichloro-2,4-difluoro-compound, and a trichlorofluoropyrimidine (47%), which is probably trichloro-4-fluoropyrimidine [584].

2,4,6-Trichloropyrimidine is converted into 2,4,6-trifluoropyrimidine on treatment with potassium [583, 585-587] or silver fluoride [531, 578, 588]. Lower fluorinated products may be formed also, depending on the reaction conditions. Silver difluoride in the presence of triperfluorobutylamine converts the trifluoropyrimidine into perfluoropyrimidine [578, 588]. When 2,4,6-trichloropyrimidine is treated with sulphur tetrafluoride at 225°, it gives a mixture of 4,6-dichloro-2-fluoro- and 2,6-dichloro-4-fluoro-pyrimidine [589].

A large number of other nucleophilic substitution reactions have been reported for the 2,4,6-trichloro-compound. Alcoholic ammonia at 20-100° gives a separable mixture of 2-amino-4,6-dichloro- and 4-amino-2,6-dichloro-pyrimidine [520, 524, 590]. Each of these compounds reacts further at 160° to give 2,4-diamino-6-chloro-pyrimidine whilst, at 200°, 2,4,6-triaminopyrimidine may be pre-pared from any of the foregoing compounds. Primary amines generally react analogously [502, 575, 591]. Aniline [575] and γ-NN-diethylaminopropylamine [591], however, give the 4-mono-substituted product exclusively, whilst it is claimed [502] that one equivalent of other primary amines gives only a 2-monosubstituted product. Careful product analysis would probably show that most primary amines yield mixtures of products, the larger amines giving mainly the 4-substituted isomer.

Successive treatment of 4-amino-3,5-dichlorodifluoropyridine with sodium hydride in tetrahydrofuran and 2,4,6-trichloropyrimidine yields 3,5-dichlorodifluoro-4-pyridyl-2′,6′-dichloropyrimidin-4′-ylamine (262)

(262)

[367]. Several related compounds may be prepared similarly.

With one equivalent of dimethylamine [524, 591, 592] or diethylamine [593, 594] 2,4,6-trichloropyrimidine is substituted in both the 2- and the 4-position to give a mixture of aminodichloro-isomers. With other amines, for example, ethylenimine (aziridine) [479, 595], N-methylaniline [527], m-aminobenzenesulphonic acid [577], or the acetyl derivative of hydrazine [594], it appears to give exclusively a 4-monosubstituted product. Two equivalents of an amine yield a 2,4-diamino-6-chloropyrimidine [479, 582, 592-596], whilst a large number of 2,4,6-triaminopyrimidines have been prepared by treatment of the trichloro-compound with three equivalents of an amine [523, 581, 582, 593, 594]. The reactions may be carried out stepwise using different amines at each stage (see ref. 593).

Tertiary amines may react with 2,4,6-trichloropyrimidine to give quaternary salts which, in most cases, are thermolabile, but reaction with trimethylamine at room temperature allows isolation of the quaternary salt (263; R = Me) [597]. Similarly, boiling NN-dimethyl-aniline gives 2,4,6-tri-N-methylanilinopyrimidine, presumably by loss of three moles of methyl chloride from the quaternary salt (263; NR_3 = NMe_2Ph) [523]. The reactions of the 2,4,6-trichloro-

(263)

compound with tertiary amines in an excess of boiling toluene or decalin have been studied by Kober and Rätz [498, 579] who showed that tertiary aromatic amines are unreactive under these conditions. Tertiary amines, such as triethylamine, tri-n-butylamine, and NN-dimethylaniline, reacted by displacement of only one chlorine atom. Substitution was tentatively assumed to have occurred in the 2-position. It seems more likely, however, that substitution occurred in the 4-position (see before). N-Ethyl-piperidine was claimed to give a small amount of 6-chloro-2,4-di-(piperidino)pyrimidine (by loss of ethyl chloride) together with 4,6-dichloro-2-piperidinopyrimidine, the major product. The more reactive amine, N-methylmorpholine, gave 2,4,6-tri(morpholino)-pyrimidine. We have already referred to the formation of an analogous monosubstituted product (241) during the preparation of 2,4,6-trichloropyrimidine by the action of phosphorus oxychloride on barbituric acid in the presence of NN-dimethylaniline.

The chlorine atoms of 2,4,6-trichloropyrimidine may be replaced stepwise also by alkoxide ions. Thus, with one equivalent of sodium methoxide at 0°, it is reported [598] to give 2,6-dichloro-4-methoxypyrimidine (a careful analysis would probably show that a mixture of isomers is formed), with two equivalents at room temperature or lower it gives 6-chloro-2,4-dimethoxypyrimidine [520, 525, 529, 598-600], whilst with an excess of the reagent at 70-100° it gives 2,4,6-trimethoxypyrimidine [520, 599]. Sodium ethoxide [575, 601] and sodium benzyloxide [528, 591] are reported to react similarly, whilst 2,4,6-triallyloxypyrimidine may be prepared by treatment of the trichloro-compound with allyl alcohol in the presence of potassium hydroxide [602]. A study has been made of the polymerization of the triallyloxy compound.

6-Chlorouracil is obtained when the 2,4,6-trichloro-compound is hydrolyzed with hot aqueous sodium hydroxide [586, 603]. 6-Chlorouracil is probably best prepared, however, by hydrolysis of 6-chloro-2,4-dimethoxypyrimidine (preceding paragraph) with dilute mineral acid [604]. A large number of uracils have been prepared similarly [444, 445]. For example, treatment of 6-chloro-2,4-dimethoxypyrimidine with sodium hydrogen sulphide followed by hydrolysis of the product yields 6-mercaptouracil [600]. Tetrachloropyrimidine may be used also as a starting material.

With sodium methoxide [499, 505, 550] or hydrazine [505], 2,4,5-trichloropyrimidine gives the 4-substituted compounds (264) and (265), respectively. The former compound (264) gives the quaternary salt (266) on treatment with trimethylamine in benzene

(264) R^1 = Cl, R^2 = OMe
(265) R^1 = Cl, R^2 = NHNH$_2$
(266) R^1 = $\overset{+}{N}$Me$_3\bar{C}$l, R^2 = OMe

[550]. It is noteworthy that a nucleophilic displacement of the methoxyl group occurs in preference to displacement of chlorine when 2,5-dichloro-4-methoxypyrimidine (264) is kept in liquid ammonia overnight; the product is 4-amino-2,5-dichloropyrimidine [550]. Chlorine is a better leaving group than methoxyl but, in this case, the latter is more readily displaced because in the transition state a p-quinonoid structure with a negative charge on the 1-nitrogen atom is most readily accommodated [605].

As far as we are aware, no nucleophilic substitution reactions of the 4,5,6-trichloro-compound have been reported.

A 5-bromine atom in a polyhalopyrimidine is resistant to nucleophilic attack [580, 581, 595]. For example, with ethylenimine (aziridine) 5-bromotrichloropyrimidine [580] and 5-bromo-dichloro-2-methylthiopyrimidine [595] give the 4,6-diethylenimino-compound and the 4(6)-mono-ethylenimino-compound, respectively.

Trichloro-2-methylthiopyrimidine reacts with ethylenimine by displacement of the 4(6)-chlorine atom(s) [606] whilst the corresponding sulphone undergoes successive loss of the 2-methylsulphonyl group and the 4-chlorine atom [580].

Treatment of trichloro-5-methylpyrimidine with ammonia at increasing temperatures gives successively the 4-amino-, the 2,4-diamino-, and the 2,4,6-triamino-derivative [537]. With sodium methoxide, it gives a 2,4-dimethoxy derivative [537] and it reacts with γ-NN-diethylaminopropylamine in the 4-position [591]. Other 5-alkyltrichloropyrimidines react similarly with ammonia and sodium alkoxides [522, 539, 540, 607]. With ammonia, however, 5-benzyl-trichloropyrimidine is reported [522] to give a mixture of the 2- and the 4-amino-derivative. Again, we believe that a careful analysis of the reaction products would reveal that mixtures of monosubstituted derivatives arise from the nucleophilic substitution reactions of most 5-alkyltrichloropyrimidines. Trichloro-5-phenylpyrimidine gives a 2,4-diamino-derivative with ammonia [532] and ethylenimine (aziridine) [479-481]. Other 5-aryltrichloropyrimidines give the corresponding 2,4-diamine on treatment with ammonia [542]. Trichloro-5-methoxypyrimidine gives the 6-chloro-2,4,5-trimethoxy derivative with sodium methoxide [543]. In contrast to the foregoing results, 5-aminotrichloropyrimidine is reported to give 2,5-diaminodichloro-

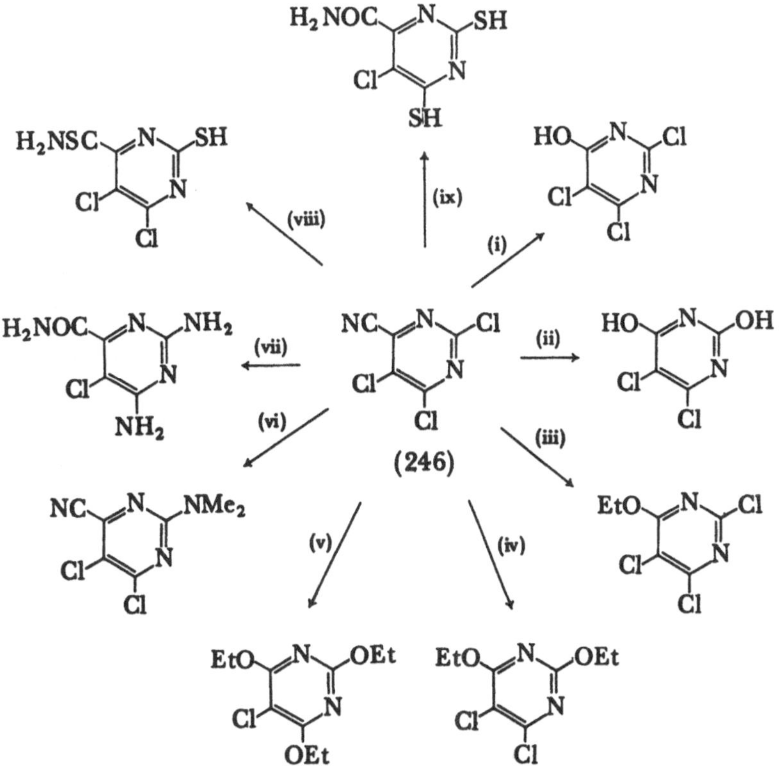

(246)

Reagents: (i) 2N-NaOH/0°; (ii) 4N-NaOH/reflux; (iii) EtOH/NaOH/20°; (iv) EtOH/NaOEt/20°; (v) EtOH/NaOEt/reflux; (vi) Me₂NH; (vii) EtOH/NH₃/120°; (viii) NaSH/20°; (ix) NaSH/reflux.

SCHEME 2.22

pyrimidine, whilst the trichloro-5-nitro-compound gives 4-amino-2,6-dichloro-5-nitropyrimidine with ethereal ammonia at 0° and the 4,6-diamino-compound with ethanolic ammonia at 20° [544]. The reaction between trichloro-5-nitropyrimidine and *p*-bromoaniline to give either 2,4- or 4,6-di(*p*-bromoanilino)-5-nitropyrimidine (probably the latter) is extremely rapid, even at −20°, and a monosubstituted product cannot be isolated [608]. With other nucleophiles the trichloro-5-nitro-compound reacts also by displacement of the 4- and 6-chlorine atoms [580, 582, 609]; with an excess of ethylenimine (aziridine) [581, 582] and other nucleophiles [609] it gives a 2,4,6-trisubstituted derivative. The chlorine atoms may be displaced stepwise by the same or different nucleophiles [609].

Various reactions of trichloro-6-cyanopyrimidine (246) have been reported [555]; these are summarized in Scheme 2.22. With one equivalent of *p*-fluoroaniline, trichloro-6-methylpyrimidine gives the 4-amino-derivative; with two equivalents of aromatic amines, it yields a 2,4-diamino-derivative [552].

The chlorine atoms of 2,6-dichloro-5-chloromethylpyrimidine may be selectively replaced by nucleophiles; their order of reactivity is $CH_2Cl > C_6-Cl > C_2-Cl$ [610].

Trichloro-5-methylpyrimidine [489, 493, 611] and the corresponding 6-methyl compound [489] are each brominated by *N*-bromosuccinimide to give a bromomethyl derivative. A 2-methyl group is unreactive towards this reagent; for example, dichloro-2,5-dimethylpyrimidine gives exclusively 5-bromomethyldichloro-2-methylpyrimidine [489, 493].

Trichloropyrimidine-2-isocyanate, which is stable as a monomer, may be prepared by treatment of the 2-amino-compound with oxalyl chloride [365]. The isocyanates (267; R = Cl, CCl_3, C_2Cl_5, and Ph)

(267)

may be prepared similarly and are stable also as monomers. These isocyanates react normally with alcohols to give urethanes (see, however, ref. 611a) [365].

2.13   PYRAZINES

Halogenopyrazines have been reviewed (see footnote on p. 275) [612, 613].

Tetrachloropyrazine may be prepared by vapour-phase chlorination of pyrazine or chloropyrazine with chlorine at 350-500° in steam (dichloropyrazines and trichloropyrazine are also produced) [614-616], by photochlorination of 2,3-dichloropyrazine at 200° [617], by treatment of various pyrazine derivatives, e.g., piperazine, chloropyrazine, or pyrazine-2,3-dicarboxylic acid, with an excess of phosphorus pentachloride in an autoclave at 250-320° [391, 569, 618], or by exhaustive, high-temperature chlorination (accompanied by aromatization) of piperazine [252] or the NN-disubstituted piperazines (268) and (269) [252, 253, 619] with

(268) R = Me
(269) R = CH_2CH_2OH
(270) R = COCl

(271)

chlorine. When a mixture of NN-di(chlorocarbonyl)piperazine (270) and chlorine is irradiated at 200°, it yields 2,3-dichloropyrazine [70, 71, 620] or a mixture of trichloro- and tetrachloro-pyrazine after prolonged reaction times (cf. ref. 617) [70, 71]. Exhaustive chlorination of 2-chloro-3-N-ethylaminopyrazine with chlorine, initially in a solvent such as chloroform and finally in the absence of a solvent but with addition of ferric chloride, is claimed [251, 568] to give the perchloro-compound (271).

We have already mentioned (Section 2.11) that irradiation of tetrachloropyridazine converts it into tetrachloropyrazine. The mechanism proposed for this and related rearrangements is shown in Scheme (2.23) [621]. Support for this mechanism is provided by the fact that ultraviolet irradiation of 4,5-dichloro-3,6-difluoropyridazine (272; R = F) gives 2,5-dichloro-3,6-difluoropyrazine (273; R = F) [418].

Oxidation of tetrachloropyrazine with a mixture of peroxyacetic acid and concentrated sulphuric acid at 20° for 18 h gives mainly the mono-N-oxide (56% yield) together with a small amount (4-5%) of the di-N-oxide (see also Sections 2.8.3, 2.11, and 2.12) [172, 353]. A mixture of 60% hydrogen peroxide and concentrated sulphuric acid similarly yields either the mono- or di-N-oxide, depending on the ratio of reactant to reagent [622]. Persulphuric acid (Caro's acid)

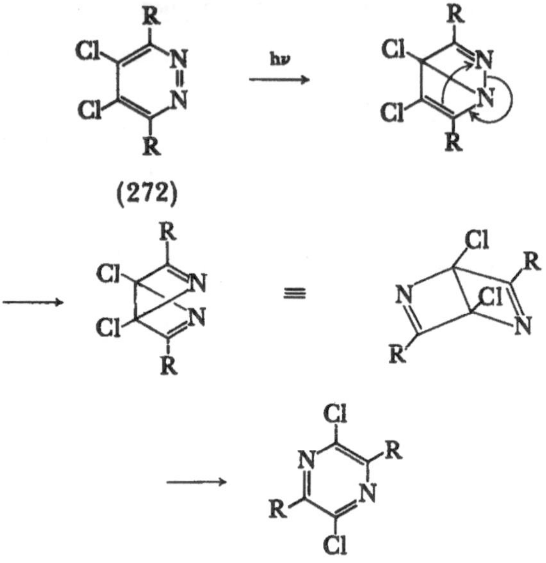

(272)

(273)

SCHEME 2.23

is thought to be the effective reagent in this case. Trifluoroperoxy-acetic acid may be used also to oxidize tetrachloro- as well as trichloro-pyrazine to the corresponding mono-N-oxide (2,3,5-tri-chloropyrazine-1-oxide in the latter case) [622, 623]. It is also claimed [623] that the same reagent system converts 3,5-dichloro-2,6-difluoropyrazine into its 1-oxide (i.e., fluorine adjacent to the N-oxide function). In our hands, however, attempts to N-oxidize nitrogen heterocycles bearing α-fluorine atoms has led to extensive decomposition, probably via nucleophilic displacement by water of the highly activated fluorine atoms followed by ring-cleavage. Hence, we suggest that the compound mentioned above may be the 4-oxide (i.e., chlorine adjacent to the N-oxide function).

Tetrachloropyrazine reacts with nucleophiles in the 2-position [617, 624]. With bidentate nucleophiles, it can react at each functional group; for example, it gives compound (274) with 1,4-butanedithiol [294]. This and related compounds are oxidized to bis-sulphones on treatment with peroxyacids [294]. 2-Aminotri-chloropyrazine (275) gives 2-aminodichloro-3-methoxypyrazine (276) with sodium methoxide, but it yields a mixture of 2,3-di-amino-5,6-dichloropyrazine (277) and the isomeric 2,6-diamino-compound on treatment with ammonia at 120° [617]. The structures of these compounds were proved by removal of the

(274)

(275) R = Cl
(276) R = OMe
(277) R = NH$_2$

(278)

remaining chlorine atoms by hydrogenolysis. With acetic anhydride the 2,3-diamino-compound (277) gives the heterocycle (278) [617]. Trichloropyrazine-2-isocyanate may be prepared by treatment of 2-aminotrichloropyrazine (275) with oxalyl chloride [365]. This compound, which is a stable monomer, reacts normally with allyl alcohol to give a urethane.

On treatment with anhydrous potassium fluoride in an autoclave at 310-320°, tetrachloropyrazine is converted into tetrafluoropyrazine; milder conditions yield mixtures of chlorofluoroderivatives [569, 618, 623]. The tetrachloro-compound is converted into tetrabromopyrazine on treatment with hydrogen bromide in refluxing acetic acid [625].

Whereas tetrachlorothiophen (Section 2.6.2), pentachloropyridine (Section 2.8.2), and other polychloroheterocycles undergo chlorine-lithium exchange reactions with organolithium compounds, surprisingly tetrachloropyrazine is alkylated by methyl- or phenyl-lithium; with n-butyl-lithium no recognizable products could be isolated [338].

Treatment of tetrachloropyrazine N-oxide with sodium methoxide or other alkoxides yields a 2-alkoxytrichloropyrazine-1-oxide [623]. Trichloropyrazine may be prepared by treatment of 2,3,5-trioxopyrazine with phosphorus oxychloride, phosphorus pentachloride, or a mixture of both reagents, preferably in the presence of a tertiary amine [626, 627], by chlorination of piperazine (with aromatization) or pyrazine, initially in chloroform under ultraviolet irradiation and then at higher temperatures in the absence of the solvent but in the presence of ferric chloride [251, 568], by chlorination of 2,3-dichloropyrazine in the presence of ferric chloride [627], and also by treatment of 2,6-dichloropyrazine 4-oxide with phosphorus oxychloride [627]. The last method is probably the most convenient.

2,3,5-Trichloropyrazine reacts with sodium methoxide at 20° to give 2,5-dichloro-3-methoxypyrazine (279) [628]. The structure of this compound has been established beyond doubt by X-ray diffraction analysis. Likewise, treatment of the trichloro-compound with aqueous ammonia at 60-140° under pressure probably gives 3-amino-2,5-dichloropyrazine† (280) [627]. With sodium

(279) R = OMe   (281) R¹ = OMe, R² = Cl
(280) R = NH₂   (282) R¹ = Cl, R² = OMe

methoxide, this amine probably gives a mixture of 3-amino-5-chloro-2-methoxypyrazine† (281) and 3-amino-2-chloro-5-methoxy-pyrazine† (282) (ratio 2:1) [627].

The UV spectra of chloro-, 2,3-dichloro-, and tetrachloropyrazine have been recorded and discussed [629], molecular orbital calculations have been carried out for the $\pi \rightarrow \pi^*$ spectral transition energy for tetrachloropyrazine [630], and the molecular core binding energies have been measured by means of ESCA for pyrazine and its perchloro- and perfluoro-derivatives (see Section 2.8.1) [1].

2.14   CINNOLINES, PHTHALAZINES, QUINAZOLINES, AND QUINOXALINES

The polychloro-derivatives of this group of compounds are reviewed under the same heading because little is known about them.

Treatment of 7,8-dichloro-4-hydroxycinnoline with sulphuryl chloride and acetic anhydride in acetic acid gives 3,7,8-trichloro-4-hydroxycinnoline, which yields 3,4,7,8-tetrachlorocinnoline with a mixture of phosphorus oxychloride and phosphorus pentachloride [631]. Chlorination of the tetrachloro-compound with chlorine in the presence of aluminium chloride gives hexachlorocinnoline (283) [631]. Fluorination of hexachlorocinnoline with anhydrous potassium fluoride converts it into hexafluorocinnoline [631], but an attempt to N-oxidize it with a mixture of concentrated sulphuric acid and peroxyacetic acid required forcing conditions and gave only

† The same authors have claimed [626] that 2,3,5-trichloropyrazine gives 2-amino-3,5-dichloropyrazine with ammonia under slightly different conditions and that reaction of this amine with sodium methoxide gives 2-amino-5-chloro-3-methoxypyrazine. Similarly, sulphanilamide was claimed to react with the trichloro-compound to give 3,5-dichloro-2-sulphanilamidopyrazine which reacts with sodium methoxide by displacement of the 3-chlorine atom.

a hydroxy-derivative of unknown structure [172]. Hexafluoro-cinnoline is attacked by nucleophiles in the 4-position [631]. Octachlorobenzo[c]cinnoline (284) may be prepared by chlori-nation of benzo[c]cinnoline with chlorine in the presence of

(283)

(284)   (2.6)

aluminium chloride [632]. Sublimation of this compound through a silica tube at *ca.* 700°, 0.2 mmHg, gives a high yield of octachloro-biphenylene [equation (2.6)]; presumably steric interaction between the 4- and 5-chlorine atoms (phenanthrene numbering) is relieved as a result.

Hexachlorophthalazine (285) may be prepared similarly from 1,4-dichlorophthalazine [633]. It is converted into the perfluoro-compound on treatment with anhydrous potassium fluoride, while addition of water to its solution in concentrated sulphuric acid yields

(285)   (286)

(287)   (288)

the pentachlorophthalazinone (286) [633]. It is noteworthy that hexafluorophthalazine is attacked by nucleophiles successively in the 1-, 4-, 6-, and 7-positions; the 5- and 8-fluorine atoms are the most resistant to attack.

Hexachloroquinazoline (287) may be prepared by chlorination of 2,4-dichloroquinazoline under autogenous pressure with phosphorus pentachloride at 300° [634]. It is obtained also in a yield of about 40% when the isocyanide dichloride (288) is heated at 300° [70]. Both the perchloro-compound (287) and the perfluoro-compound, which can be prepared from it by treatment with potassium fluoride at 350°, exhibit a vivid, blue fluorescence in UV light [634]. It is noteworthy that hexafluoroquinazoline is attacked by nucleophiles successively in the 4-, 2-, 7-, and 5-positions [634].

When 3,5-dichloroanthranilic acid is heated with urea at 150-200°, it gives the quinazolindione (289). This compound is converted into 2,4,6,8-tetrachloroquinazoline on treatment with a mixture of phosphorus oxychloride and phosphorus pentachloride [635], or

(289)   (290)

with thionyl chloride [636] or phosgene [637] in dimethyl-formamide. Presumably, this general method is capable of extension to the synthesis of other polychloroquinazolines. It is noteworthy that chlorination of o-dimethylaminobenzonitrile leads via ring closure to 4,6,8-trichloroquinazoline (290) [70].

Pentachlorophenyl-lithium reacts with an equimolar amount of various aryl nitriles in ether at room temperature to give the expected adducts (291; Scheme 2.24) [337, 341, 638]. With an excess of the nitrile in ether heated under reflux, however, it also gives the corresponding 2,4-diaryl-5,6,7,8-tetrachloroquinazoline (293) in moderate yield via addition of the N-lithio-imine (291) to a second molecule of the nitrile, to form the intermediate (292), which then undergoes cyclization. The addition of pentachlorophenyl-lithium to aromatic nitriles has been shown to be reversible [341]. Consequently, the method cannot be employed to prepare tetra-chloroquinazolines carrying two different aryl substituents, since inseparable mixtures of quinazolines are usually obtained.

Hexachloroquinoxaline (294) may be prepared by vapour-phase chlorination of quinoxaline [196], or by chlorination of 2,3-di-chloroquinoxaline either with phosphorus pentachloride in an

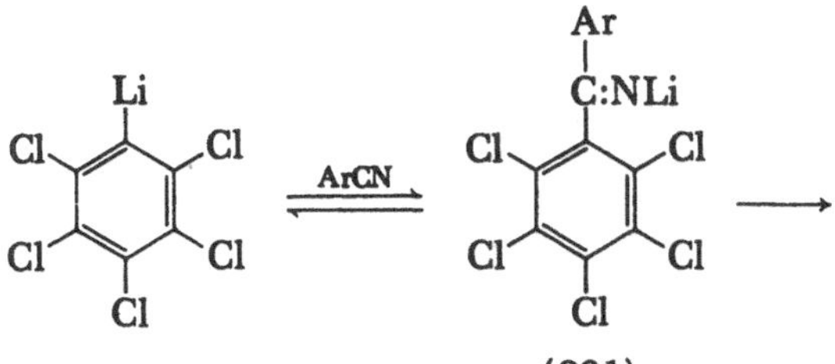

(291)

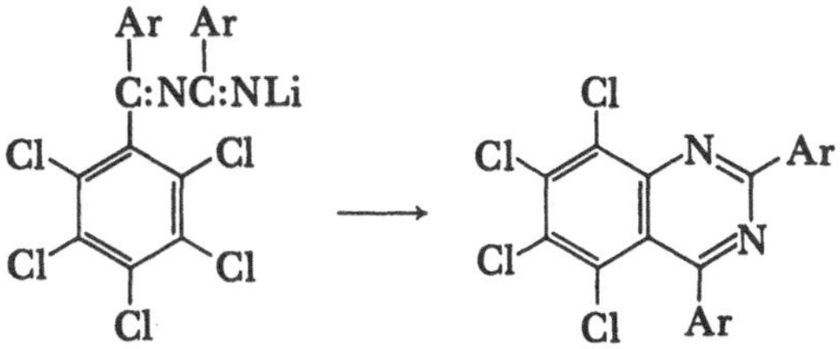

(292)   (293)

Ar = Ph, $C_6H_4Me\text{-}p$, $C_6H_4Cl\text{-}p$, 3-pyridyl, or 4-pyridyl.

SCHEME 2.24

autoclave at 300° [639-641] or with chlorine in the liquid phase in the presence of iron [642]. The liquid phase chlorination yields mixtures of the perchloro-compound and lower chlorinated products [642]. When N-pentachlorophenylcyanoformamidoyl chloride (295) is heated in the gas phase at 600°, it yields pentachlorobenzonitrile, but hexachloroquinoxaline is formed at 400° [70]. High-temperature chlorination of the quinoxaline derivative (296) yields

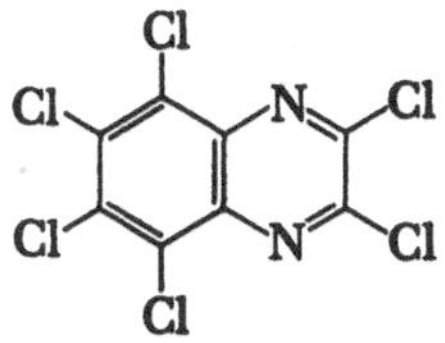

(294) 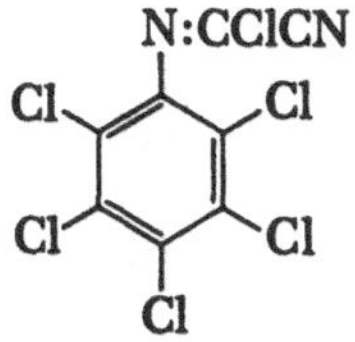   (295)

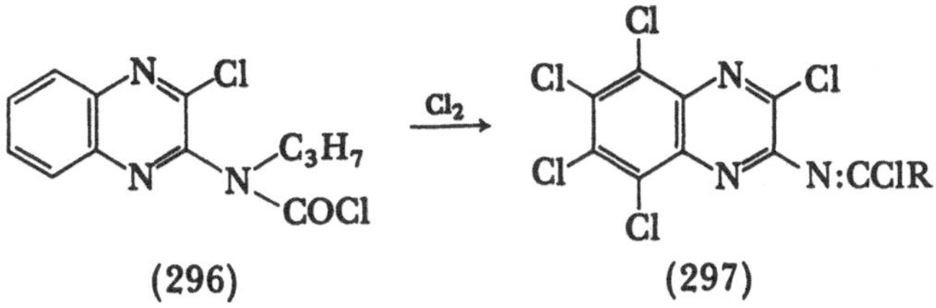

(296)   (297)

the perchlorinated imidoyl chloride (297; R = C_2Cl_5) [70]. The analogous compound (297; R = CCl_3) may be prepared by exhaustive chlorination of 3-chloro-2-N-ethylaminoquinoxaline, initially in chloroform, then in trichlorobenzene, and finally in the absence of a solvent at higher temperatures in the presence of ferric chloride [251, 568]. In the same patents [70, 251, 568] 1,2,3,4-tetrahydroquinoxaline is claimed to give hexachloroquinoxaline under similar conditions.

TABLE 2.6   5,6,7,8-Tetrachloroquinoxalines derived from
Tetrachloro-o-phenylenediamine [643, 644]

(298)

R^1	R^2	Yield (%)	R^1	R^2	Yield (%)
H	H	80	Me	Me	47
Me	H	55	Pr^n	Pr^n	50
Pr^n	H	50	Ph	Ph	88
Ph	H	49	OH	OH^b	45
C_6H_4OH-p	H	66	OH	Me^c	69
OH	H^a	95	OH	CO_2Et^c	90

^{a} Exists as the keto-tautomer (299). ^{b} Exists as the keto-tautomer (301).
^{c} Exists as keto-tautomers analogous to (299).

The classical synthesis of quinoxalines, namely that of condensing an o-phenylenediamine with a 1,2-dicarbonyl compound, can be extended to the synthesis of polychloroquinoxalines. Thus, from tetrachloro-o-phenylenediamine can be prepared the tetrachloro-compounds (298) listed in Table 2.6 [643, 644]. The product (299) from the reaction between tetrachloro-o-phenylenediamine and glyoxylic acid yields 2,5,6,7,8-pentachloroquinoxaline (300) on treatment with a phosphorus oxychloride–phosphorus pentachloride mixture [643, 644]. Likewise, the product (301) from the reaction between the o-phenylenediamine and ethyl oxalate yields hexa-chloroquinoxaline [643], whilst the product from the reaction between 3,5-dichloro-o-phenylenediamine and oxalic acid gives 2,3,5,7-tetrachloroquinoxaline (302) [642]. The 6-substituted compounds (303; R = Me, Br, F, OMe) may be prepared similarly from the appropriately substituted trichloro-o-phenylenediamine [643, 644]. These in turn can be conveniently synthesized by reduction of

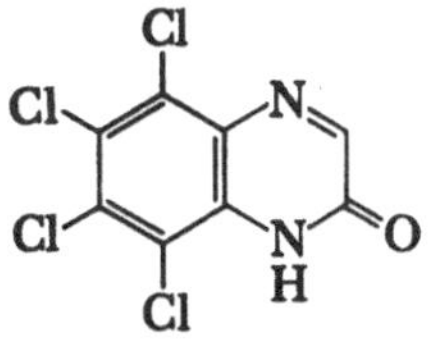

(299)

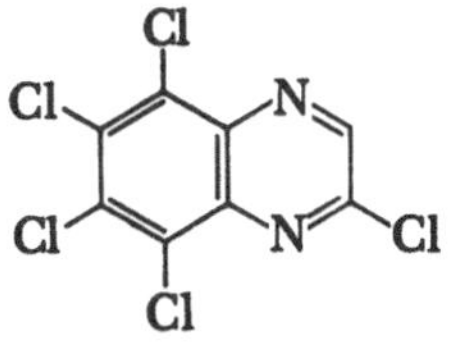

(300)

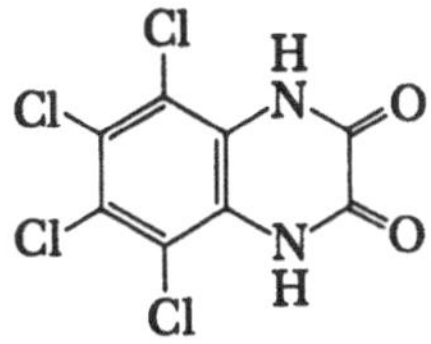

(301)

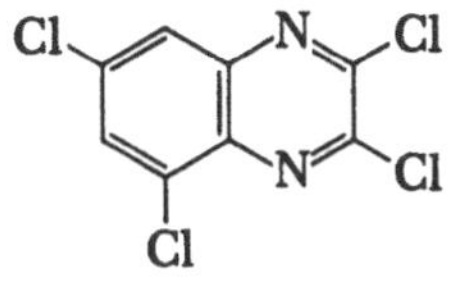

(302)

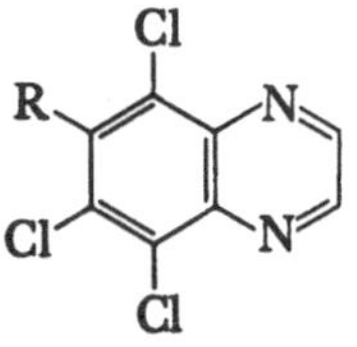

(303)

the corresponding polychlorobenzotriazole (Section 2.18.1) with zinc and acid [643].

Oxidation of 5,6,7,8-tetrachloroquinoxaline with either nitric acid or hydrogen peroxide yields the dione (301) [643], whilst treatment of hexachloroquinoxaline with anhydrous potassium fluoride in an autoclave at high temperatures gives hexafluoroquinoxaline [639-641]. The 2-chlorine atom in the 2,5,6,7,8-pentachloro-compound (300) can be replaced by a methoxyl, ethoxyl, or azido-group in the usual way [643, 644]. It is noteworthy that hexafluoroquinoxaline undergoes attack by nucleophiles successively in the 2-, 3-, and 6-positions [639-641].

Chlorination of the 2-methyl and 2,3-dimethyl derivatives of 5,6,7,8-tetrachloroquinoxaline in acetic acid in the presence of sodium acetate gives the 2-trichloromethyl compound in the former case and the 2-dichloromethyl-3-methyl compound in the latter case [643]. On irradiation, the 2,3-bis(dichloromethyl) compound is formed in the latter case.

Treatment of 5,6,8-trichloro-7-methylquinoxaline with alkali in aqueous ethanol gives 5,8-dichloro-6-methylquinoxaline [645]. 5,6,7,8-Tetrachloroquinoxaline similarly, though less satisfactorily, yields 5,6,8-trichloro- (major product) and 5,8-dichloro-quinoxaline.

Rather surprisingly, the products of these reactions are not accompanied by products arising from nucleophilic substitution. The reactions also occur in aqueous propan-2-ol but not in aqueous methanol or t-butanol alcohol. If ethanol is used, acetaldehyde is formed, and in the case of propan-2-ol acetone is also formed. Hence, the mechanism for these hydrodehalogenation reactions appears to involve transfer of hydride ion from the alkoxide ion, as shown in Scheme (2.25) [645].

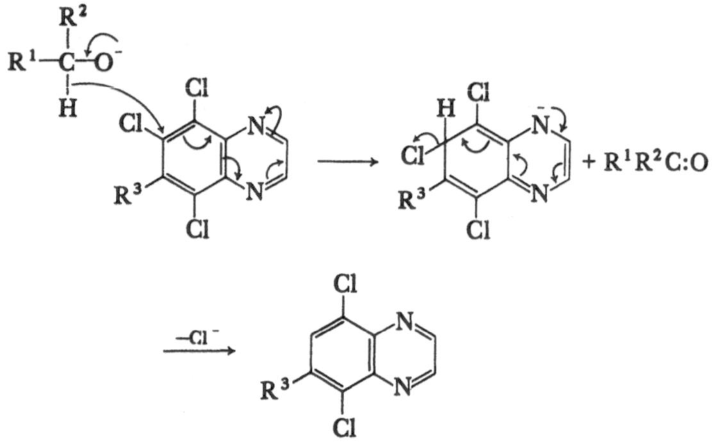

R^1 = Me; R^2 = H or Me; R^3 = Me or Cl

SCHEME 2.25

IR and ^{1}H NMR data and g.l.c. retention ratios have been reported for several polychloroquinoxalines [645].

The adduct (304) formed between tetrachloro-o-benzoquinone and ω-bromostyrene condenses with o-phenylenediamine to give

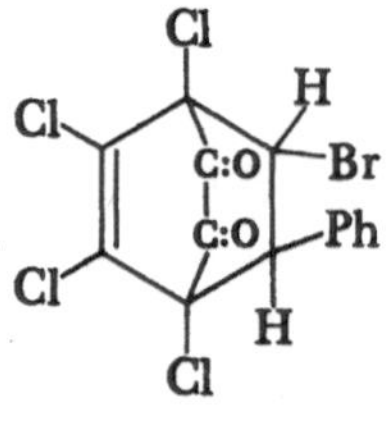

(304) 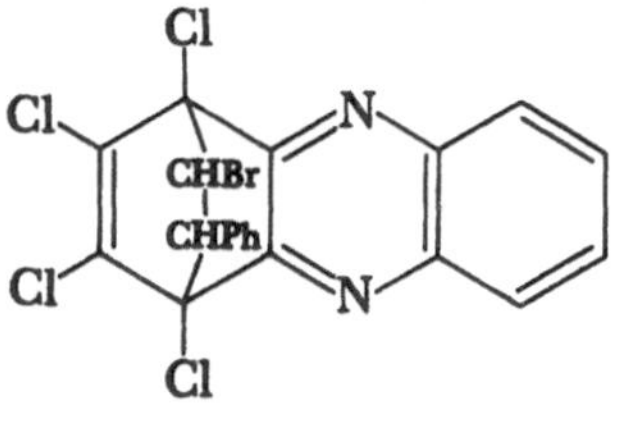   (305)

compound (305) [646]. This loses ω-bromostyrene on being heated to 190° and yields 1,2,3,4-tetrachlorophenazine. This phenazine is obtained also, together with an equivalent amount of 1,2,3,3,4,4,-

hexachloro-3,4-dihydrophenazine, when o-phenylenediamine is condensed with hexachlorocyclohexendione [646].

2.15 OXAZOLES, BENZOXAZOLES, ISOXAZOLES, AND BENZISOXAZOLES

The polychloro-derivatives of this group of compounds are reviewed under the same heading because little is known about them.

Apart from pentachlorobenzoxazole, which may be prepared by exhaustive chlorination of benzoxazole with chlorine at 80-200° in the presence of a halogen carrier such as ferric chloride [647], the other perchloroheterocycles in this group appear to be unknown.

The tetrachlorobenzoxazoles (306: R^1 = Me, Ph, CO_2Et, or α-naphthyl; R^2 = Cl) may be prepared by the action of heat or light on a mixture of tetrachloro-o-benzoquinone diazide and the appropriate nitrile (see also Section 2.3) [37, 648]. These reactions proceed via the 1,3-dipolar species (19) which, when formed at 130° in

(306)   (307)

the presence of phenyl isocyanate, gives tetrachloro-3-phenyl-2-benzoxazolinone (307; R = Ph) [37, 648]. Tetrachloro-2-phenyl-benzoxazole (306: R^1 = Ph; R^2 = Cl) may be prepared also by a cycloaddition reaction between the dipolar species (19) and benzal-methylamine, PhCH:NMe [37, 648]. In this case, the intermediate saturated adduct undergoes loss of methane.

When a mixture of 2-amino-3,4,6-trichlorophenol, nicotinic acid, and polyphosphoric acid is heated, it yields compound (306: R^1 = 3-pyridyl; R^2 = H) [649]. Tetrachloro-2-phenylbenzoxazole (306: R^1 = Ph; R^2 = Cl) can be prepared similarly by heating a mixture of 2-aminotetrachlorophenol and benzoic acid to 200° [37, 648].

Treatment of an o-aminopolychlorophenol with methyl trichloro-acetimidate, $CCl_3C(:NH)OMe$, gives the corresponding 2-trichloro-methylbenzoxazole [650-652]; for example, 2-amino-3,4,6-trichloro-phenol yields compound (306: R^1 = CCl_3; R^2 = H) [650]. With cyanogen bromide in methanol, the same o-aminophenol gives 2-amino-4,5,7-trichlorobenzoxazole (306: R^1 = NH_2; R^2 = H) [653]. With potassium methyl xanthate it yields 4,5,7-trichloro-2-mercapto-benzoxazole (306: R^1 = SH; R^2 = H) [653a].

2,3,5-Trichloro-p-benzoquinonedipivalimide   adds   hydrogen
chloride to give a mixture of tetrachloro-p-phenylenedipivalamide and
2-t-butyltrichloro-6-pivalamidobenzoxazole (306: R^1 = CMe$_3$; R^2 =
NHCOCMe$_3$) [654]. The benzoxazole is formed via protonation of

(308)

the quinonedi-imide and cyclization of the resulting intermediate
(308).

5,7-Dichloro-2-mercaptobenzoxazole   yields   2,5,7-trichloro-
benzoxazole with phosphorus pentachloride in benzene at room
temperature [655], whilst 3,5,6-trichloro-2-benzoxazolinone may be
prepared by the action of chlorine on the sodium salt of 5-chloro-2-
benzoxazolinone-6-sulphonic acid [656, 657].

A large number of tri- and tetra-chlorobenzoxazoles have been
synthesized by the methods already discussed and examined for
biological activity [649-665]. For example, the tetrachlorobenzox-
azole [306: R^1 = CH$_2$SP(:O)(OEt)$_2$; R^2 = Cl], obtained by treating
tetrachloro-2-chloromethylbenzoxazole (306: R^1 = CH$_2$Cl; R^2 = Cl)
with a salt of OO-diethylmonothiophosphoric acid, exhibits pesticidal activity [658], whilst the tetrachloro-2-benzoxazolinones (307;
R = CONHAr), produced from the tetrachloro-2-benzoxazolinone
and an aryl isocyanate (e.g., 3,4-dichlorophenyl isocyanate), exhibit
microbiocidal activity (cf. ref. 659) [660].

The unusual perchloro-oxazolobenzoxazole (309) is obtained
when 2,5-dichloro-3,6-bis(dimethylamino)-p-benzoquinone is chlorinated [70, 71, 253].

It is noteworthy that 5,6,7-trichloro-2,1-benzisoxazole (5,6,7-

(309)   (310)   (311)

trichloroanthranil) (310) may be prepared by reduction of 3,4,5-tri-chloro-2-nitrobenzaldehyde with tin and acid [666], and that 3,5,7-trichloro-1,2-benzisoxazole (311) is obtained when 5,7-dichloro-3-hydroxy-1,2-benzisoxazole is treated with phosphorus oxychloride at 140-150° in the presence of triethylamine [667].

2.16 THIAZOLES, BENZOTHIAZOLES, ISOTHIAZOLES, and BENZOISOTHIAZOLES

Tribromo- [668] and tri-iodo-thiazole [669], and each of the dichlorothiazoles [670-673] are known, but trichlorothiazole appears not to have been reported. Its synthesis should present no problems. Again, we review the polychloro-derivatives of a group of compounds under the same heading because little is known about them.

High yields (70-100%) of pentachlorobenzothiazole may be obtained by exhaustive chlorination of the parent heterocycle, its 2-sulphonic acid, 2-bromo-, 2-chloro-, or 2-mercapto-derivative with chlorine in the presence of antimony trichloride [674], or by the action of heat on pentachlorophenyl isothiocyanate [675]. Other polychlorobenzothiazoles (Table 2.7) may be prepared by the action

TABLE 2.7 Polychlorobenzothiazoles Derived from Phenyl Isothiocyanates [675]

R^1	R^2	R^3	R^4	Yield (%)
Cl	Cl	Cl	Cl	90
Cl	H	Cl	H	50
Cl	Cl	H	Cl	50
Cl	H	Cl	Cl	50
Cl	Cl	CN	Cl	50

of heat on the corresponding polychlorophenyl isothiocyanate [675].

Thiocyanation of 3,4,5-trichloroaniline gives the phenyl thio-cyanate (312) which cyclizes to 2-amino-5,6,7-trichlorobenzo-thiazole (313) [676].

6-Hydroxy-2-phenylbenzothiazole and its 7-amino-derivative undergo photochlorination in acetic acid to give compounds (314) and (315) on work-up [677]. These are reduced by zinc in acetic acid to trichloro-6-hydroxy-2-phenyl- and 4,5-dichloro-6,7-dihydroxy-2-phenyl-benzothiazole, respectively. Presumably, either compound could be converted into tetrachloro-2-phenylbenzothiazole on treatment with a suitable halogenating agent.

(312)   (313)   (314)

(315)   (316)

4,6,7-Trichloro-2-methylbenzothiazole may be prepared from 4,6- or 4,7-dichloro-2-methylbenzothiazole through nitration, reduction, diazotization, and replacement of the diazonium group by chlorine [678].

Polychloro-2-mercaptobenzothiazoles may be prepared from a polychloroaniline through reaction of the amine with sulphur monochloride, followed by alkali (the Herz reaction), and cyclization of the resulting o-aminothiophenol with carbon disulphide [679, 680]; for example, 2,3-dichloroaniline gives a poor yield of 4,5,6-trichloro-2-mercaptobenzothiazole (316; R = SH) (Herz reactions are accompanied by ring chlorination para to the amino-group) [680].

It is noteworthy that 4,6,7-trichloro-2-mercaptobenzothiazolyl-methylchloroarsine and bis(tetrachloro-2-mercaptobenzothiazolyl)-methylarsine are claimed [681] to possess useful insecticidal properties and that compound (316; R = SCH_2SCN), which may be prepared from 4,5,6-trichloro-2-mercaptobenzothiazole by condensation with chloromethyl thiocyanate, exhibits fungicidal activity, as do its side-chain sulphone and sulphoxide [680].

Acrylonitrile yields trichloroisothiazole on being heated with sulphur chlorides or thionyl chloride in pyridine; malononitrile similarly gives dichloro-5-methylisothiazole [682]. The nitriles (317: R^1 = Cl; R^2 = CN) and (317: R^1 = CN; R^2 = Cl) and derivatives of them have been of some industrial interest [683–690]. When ethanolic malononitrile is treated with carbon disulphide in the presence of sodium hydroxide, it yields di(sodiomercapto)methylenemalono-

(317)   (318)

nitrile, $(NC)_2C:C(SNa)_2$, which undergoes cyclization on treatment with chlorine in carbon tetrachloride to give a mixture of the nitrile (317: $R^1 = Cl; R^2 = CN$) (major product) and 3,4-dichloroisothiazolo-[4,5-d]isothiazole (318) [683-686]. The isomeric nitrile (317: $R^1 = CN; R^2 = Cl$) may be prepared by treatment of sodium cyanide with carbon disulphide in dimethylformamide followed by treatment of the resulting sodium cyanodithioformate with chlorine (see refs. 690 and 691 regarding the mechanism of this reaction) [687, 688].

Dichloroisothiazole-4-carbonitrile (317: $R^1 = Cl$; $R^2 = CN$) is substituted by nucleophiles (ammonia, hydrazine and hydrazine derivatives, primary or secondary aliphatic and aromatic amines, alkoxide, and alkylthio-ions) in the 5-position [685, 686, 690]. The structure of the 5-anilino-derivative (317: $R^1 = NHPh; R^2 = CN$) was proved by Raney nickel desulphurization, which gave anilinomethyl-enemalononitrile [686]. Replacement of the second chlorine atom requires more drastic conditions and is accompanied by ring-opened products. For example, treatment of the nitrile (317: $R^1 = Cl; R^2 = CN$) with two equivalents of sodium sulphide followed by alkylation of the product with methyl iodide yields the 3,5-di(methylthio)-derivative together with 1,1-di(methylthio)-2,2-dicyanoethylene [686]. The latter product is believed to arise by attack by sulphide ion on the ring sulphur atom of the nitrile followed by ejection of chloride ion and collapse to the ring-opened product. The nitrile (317: $R^1 = Cl; R^2 = CN$) gives the sulphenyl chloride (317: $R^1 = SCl; R^2 = CN$) through treatment with one equivalent of sodium sulphide, oxidation with iodine, and cleavage of the resulting disulphide with chlorine [683]. The sulphenyl chloride (317: $R^1 = SCl; R^2 = CN$) gives sulphenamides with amines and yields the isothiazoloisothiazole (318) on being heated in benzene [683]. Hydrolysis of the nitrile (317: $R^1 = Cl; R^2 = CN$) yields the corresponding acid (see also ref. 686) [690]. This gives an amide which undergoes a Hofmann degradation to give 4-aminodichloro-isothiazole [690]. This amine also undergoes nucleophilic substitution (e.g., with ammonia) in the 5-position [690].

The nitrile (317: $R^1 = CN; R^2 = Cl$) may be converted into the corresponding acid and various amides by conventional procedures [687, 688].

The isothiazoloisothiazole (318) undergoes ring-cleavage on treatment with an amine to give the corresponding sulphenamide (317: $R^1 = SNR_2; R^2 = CN$); for example, with morpholine, it gives (317: $R^1 = SC_4H_8NO; R^2 = CN$) [683, 684].

Treatment of 2,3,6-trichlorobenzylidene chloride with a mixture of ammonia and sulphur at 150° in benzene gives a 1:1 mixture (72.5%) of 4,5- and 4,7-dichloro-1,2-benzisothiazole [692, 693]. Chlorination of the 4,5-dichloro-isomer gives 3,4,5-trichloro-1,2-benzisothiazole. 3,5,7-Trichloro-1,2-benzisothiazole (319; R = H) may be prepared by the action of phosphorus oxychloride on 5,7-dichloro-3-hydroxy-1,2-benzisothiazole in the presence of triethylamine [694]. The dimethyl compound (319; R = Me) may be prepared similarly. The 5,6-dichloro-derivative of saccharin is converted into 3,5,6-trichloro-1,2-benzisothiazole on treatment with a mixture of phosphorus oxychloride and phosphorus pentachloride [695].

(319)   (320)

Chlorination of 5-amino-1,2-benzisothiazole hydrochloride at room temperature in acetic acid yields the addition compound (320), which is reduced by stannous chloride in acetic acid to 4,6,7-trichloro-5-hydroxy-1,2-benzisothiazole [696].

We are not aware that pentachloro-1,2-benzisothiazole or any tetrachloro-1,2-benzisothiazoles have been prepared or that any nucleophilic substitution reactions have been reported.

It is noteworthy that 3-chloro-1,2-benzisothiazoles may be ring-opened by nucleophiles [697].

2.17   IMIDAZOLES, BENZIMIDAZOLES, PYRAZOLES, AND BENZOPYRAZOLES (INDAZOLES)

2.17.1   Imidazoles

Imidazoles are readily halogenated without catalysts. Aqueous sodium hypochlorite yields 4,5-dichloroimidazole with small amounts of 2,4,5-trichloroimidazole (321; R = H) from imidazole and converts 2-alkylimidazoles into their 4,5-dichloro-derivatives [698, 699]. Chlorination of imidazole and C-alkylimidazoles with N-chlorophthalimide has been reported [700].

(321)

Imidazole [701] and 2,4,5-tribromo- [698, 702, 703] and 2-bromo-4,5-dichloro-imidazole [698] are converted into 2,4,5-trichloro-imidazole (321; R = H) with hydrogen chloride under various conditions; lithium chloride also converts the 2,4,5-tribromo-compound into the 2,4,5-trichloro-compound [704].

Exhaustive chlorination of dimethylaminoacetonitrile (Me_2NCH_2CN) with chlorine at relatively low temperatures yields 2,4,5-trichloro-1-methylimidazole (321; R = Me) [70]. On further chlorination at 180-200°, this compound undergoes ring-cleavage to give tetrachloroethylene-1,2-di(isocyanide dichloride) $(Cl_2C:NCCl_2CCl_2N:CCl_2)$.

At temperatures not exceeding 100°, NN'-dimethyloxamide reacts with phosphorus pentachloride to give a mixture of 5-chloro- and 4,5-dichloro-1-methylimidazole (see Scheme 2.26) [705]. A mixture

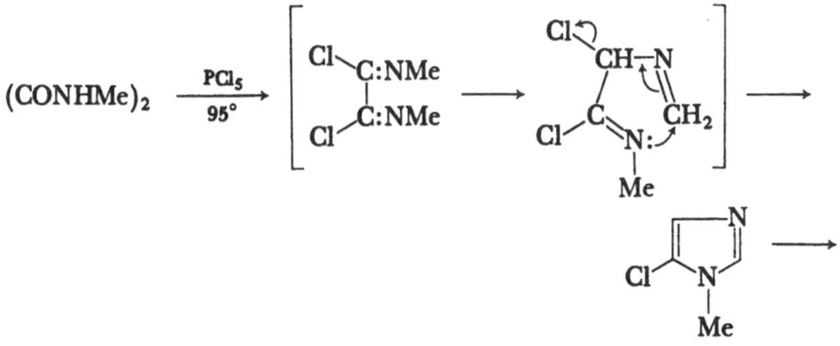

$(CONHMe)_2 \xrightarrow[95°]{PCl_5}$

SCHEME 2.26

of these two compounds and the trichloro-derivative (321; R = Me) is obtained when a larger quantity of phosphorus pentachloride is used in phosphorus oxychloride [705]. This Wallach-type synthesis of imidazoles has been extended to the preparation of other chloro-imidazoles [706, 707]. A mixture of 4,5-dichloro- and tri-chloro-1-methylimidazole (321; R = Me) is obtained on further chlorination of 2-chloro-1-methylimidazole with chlorine in phos-phorus oxychloride [708] or with a mixture of phosphorus pentachloride and phosphorus oxychloride [709]. 1,2-Dialkyl-5-chloroimidazoles similarly give their 4,5-dichloro-derivatives [709]. Chlorination of 4,5-dichloro-1-methylimidazole with chlorine in acetic acid yields N-methylparabanic acid (322) [708].

Only one nucleophilic displacement reaction of a polychloroimid-azole has been reported. This is the reaction of tetrachloroimidazole (321; R = Cl) with cuprous cyanide to give the 1-cyano-derivative, which is claimed [701] to possess herbicidal activity.

Bromination of 4,5-dichloroimidazole yields the 2-bromo-deriva-tive [698, 699]. The trichloro-compound (321; R = H) gives a sodium salt which is alkylated by compounds with the general formula R^1XCH_2Cl (R^1 = alkyl; X = O, S, SO, SO_2). The products (321; R = CH_2XR^1) are claimed [702, 703, 710] to exhibit herbicidal activity. 2,4,5-Trichloroimidazole (321; R = H) exhibits insecticidal and herbicidal activity [107, 704].

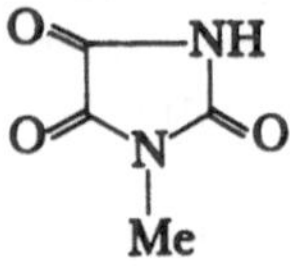

(322)

Nesmeyanov et al. [711] have used ^{35}Cl NQR spectroscopy to study the structures of trichloro-1-methyl- (321; R = Me) and 4,5-dichloro-1(and 2)-methylimidazole.

More interest has been shown in polybromo- and polyiodo-imidazoles than in the corresponding polychloro-compounds. This is also true of pyrazoles (Section 2.17.3).

2.17.2   Benzimidazoles

Several features of benzimidazole chemistry have stimulated a study of the polychloro-derivatives. The first was the discovery that certain N-glycosides of benzimidazole, particularly 5,6-dichloro- (DRB) and 4,5,6-trichloro-1-β-D-ribofuranosylbenzimidazole (323; R = H)

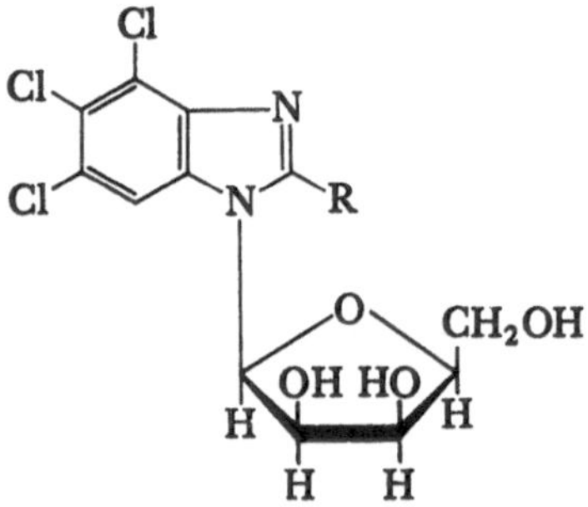

(323)

(TRB), are potent inhibitors of virus multiplication [712-721]. These compounds have been shown also to inhibit the growth of vitamin B_{12}-dependent micro-organisms, such as E. coli [722, 723]. Related to this work are the claims [724-726] that compounds such as 4,5,6-trichloro-2-(4-thiazolyl)-1-β-D-ribofuranosyl-benzimidazole (323; R = 4-thiazolyl) exhibit biocidal activity against a variety of organisms.

Interest in polychlorobenzimidazoles was further stimulated by the discovery that polychloro-derivatives of 2-trifluoromethyl-benzimidazole (TFB) inhibit photosynthesis and are consequently useful as herbicides. Further studies have shown that such compounds uncouple oxidative phosphorylation in mitochondria [107, 727-752]. Tetrachloro-2-trifluoromethylbenzimidazole (324) (TTFB) is the strongest uncoupler in this series of compounds, although the trichloro-derivative (325) is also of considerable interest in this respect. The other trichloro-derivatives and the dichloro-derivatives of TFB are also active uncouplers. The activity of these trifluoromethyl compounds as uncouplers parallels their degree of NH-acidity and is associated with the anion arising by loss of the proton from the ring nitrogen atom [107, 727, 732]. Alkylation at this site results in a loss of activity. Uncouplers of oxidative phosphorylation are generally associated with a broad spectrum of

(324) $R^1 = CF_3$, $R^2 = Cl$

(325) $R^1 = CF_3$, $R^2 = H$

(326) $R^1 = R^2 = H$

(327) $R^1 = H$, $R^2 = Cl$

(328) $R^1 = Me$, $R^2 = Cl$

(329) $R^1 = Et$, $R^2 = Cl$

(330) $R^1 = CCl_3$, $R^2 = Cl$

(331) $R^1 = CO_2 Et$, $R^2 = Cl$

(332) $R^1 = R^2 = Cl$

(333) $R^1 = NHPh$, $R^2 = Cl$

(334) $R^1 = CF_3$, $R^2 = NO_2$

biocidal activity. This is certainly true of the polychloro-2-trifluoromethylbenzimidazoles [107, 727, 729, 734, 737, 738, 753-758]. Derivatives of these compounds with substituents (alkyl [753, 759, 760], $CO_2 R$ [733, 753, 761-765], $SO_2 R$ [766, 767], CONHR [768], $CSNR_2$ [769], OR [769a]) on the ring nitrogen atom or with substituents (alkyl [770], aryl [771], $C_2 F_5$ [753, 754, 756, 769a], $CF_2 Cl$ [772], CCl_3 [770, 773], SH [770], SMe [770], $SO_2 Me$ [770], $CO_2 Et$ [770]) other than a trifluoromethyl group in the 2-position are also of interest in this respect. TTFB (324) is reported to inhibit tumour growth [774].

4(7),5,6-Trichloro- (326) and 4,5,6,7-tetrachloro-benzimidazole (327) are also of interest as biocides [712, 714, 775-777] as are compounds with the general formula (335: n = 1-3; R = OAr, SAr, SOAr, $SO_2 Ar$, $NHSO_2 Ar$, NHCOAr, CONHAr) [778].

Polychlorobenzimidazoles may be prepared either by chlorination of a benzimidazole or from a polychloro-o-phenylenediamine by its treatment with a carboxylic acid or the corresponding anhydride.

$$Cl_n$$

(335)

Chlorination of 2(3H)-benzimidazolone with chlorine in water gives the tetrachloro-derivative (336; R = H); in acetic acid the tautomer (337) is obtained [779]. When a mixture of 3(2H)-benzimidazolone, sodium acetate, dichloroethane, and water is treated with chlorine, the NN-dichloro-compound (338; R = H) is formed, which is converted in hot acetic acid into the tetrachloro-3(2H)-benzimidazolone (336; R = H) [780]. Further chlorination of the last compound in the presence of sodium acetate yields 1,3,4,5,6,7-hexachloro-3(2H)-benzimidazolone (338; R = Cl) [780].

(336)   (337)   (338)

Compound (336; R = H) is converted into its tautomer (337) on treatment with acetic anhydride [779]. With a phosphorus penta-chloride-phosphororus oxychloride mixture at 170°, compounds (336; R = H) and (337) yield pentachlorobenzimidazole (332) [779]. Treatment of 4,5(6),7-trichloro-2(3H)-benzimidazolone with phosphorus oxychloride yields 2,4,5(6),7-tetrachlorobenzimidazole; the 2,4(7),5,6-tetrachloro-compound may be prepared similarly [781]. Hexachlorobenzimidazole may be prepared from the pentachloro-derivative (332) by treatment of an aqueous solution of its sodium salt with chlorine [779]. There is a report [782] that chlorination of 2-mercaptobenzimidazole in acetic acid gives a mixture of tri- and tetra-chlorobenzimidazoles, but these were not identified.

When 2,5(6)-dimethylbenzimidazole is treated with a saturated solution of bleaching powder in aqueous acetic acid, it yields its N-chloro-derivative, which rearranges on being heated in dry benzene to give a product chlorinated in the benzenoid ring [783]. Repetition of this process yields the tetrachloro-derivative (339).

Chlorination of 2-methyl- or 2-ethyl-benzimidazole with chlorine in boiling water gives the corresponding tetrachloro-derivative (328) or (329) respectively [770]. 2-Dichloromethylbenzimidazole [770] or 2-trichloromethylbenzimidazole [781] may be converted similarly into tetrachloro-2-trichloromethylbenzimidazole (330), whilst ethyl benzimidazole-2-carboxylate or its 5(6)-chloro-derivative similarly yields ethyl tetrachlorobenzimidazole-2-carboxylate (331) [770]. Hydrolysis of the ester (331) in hot alkali gives 4,5,6,7-tetrachlorobenzimidazole (327) [770]. Chlorination of 2-trifluoromethylbenzimidazole with chlorine in water or *m*-dichlorobenzene at 120° can be made to give either the trichloro-derivative (325) or the tetrachloro-derivative (324) (TTFB), depending on the reaction

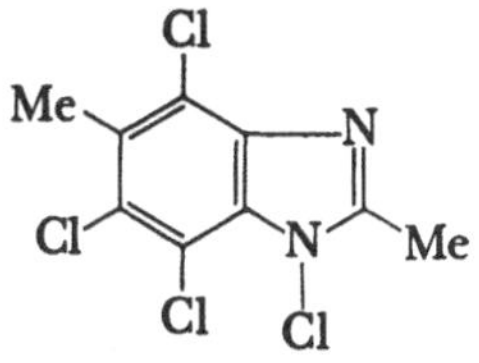

(339)

conditions [727, 728, 731, 757]. Likewise, 5(6)-methyl-2-trifluoromethylbenzimidazole gives its 4,5(6),7-trichloro-derivative [731]. Photochlorination of 2-trifluoromethylbenzimidazole in carbon tetrachloride gives the 1,4,5,6-tetrachloro-derivative whilst similar treatment of TTFB (324) also gives an *N*-chloro-derivative [731].

Chlorination of 5,6-diamino-2-methylbenzimidazole is reported by Fries *et al.* [784] to give the adduct (340; R = H), which is converted

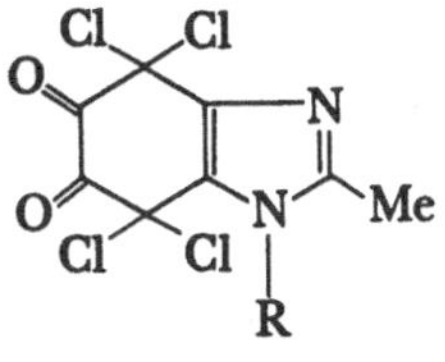

(340)

by zinc and acetic acid into 4,7-dichloro-5,6-dihydroxy-2-methylbenzimidazole. 5,6-Diamino-1,2-dimethylbenzimidazole is reported to give a similar adduct (340; R = Me), which yields 4,7-dichloro-5,6-dihydroxy-1,2-dimethylbenzimidazole on reduction.

Condensation of a polychloro-*o*-phenylenediamine with a carboxylic acid or the corresponding anhydride is probably the preferred route to most polychlorobenzimidazoles. Formic acid yields a 2-unsubstituted compound, whilst acetic acid and its higher homologues give 2-alkyl compounds. Condensation of tetrachloro-*o*-phenylenediamine with trifluoroacetic acid yields 4,5,6,7-tetrachloro-2-trifluoromethylbenzimidazole (TTFB) (324). Trichloroacetic acid similarly gives 2-trichloromethylbenzimidazoles. If an N-mono-substituted-o-phenylenediamine is chosen as the starting material, then the product is a 1- (with formic acid) or 1,2-dialkyl-derivative. Table 2.8 lists some products which have been prepared by these procedures.

Treatment of tetrachloro-o-phenylenediamine with urea gives tetrachloro-2($3H$)-benzimidazolone (336; R = H) and with thio-phosgene or, better, carbon disulphide in dimethylformamide it gives

TABLE 2.8   Polychlorobenzimidazoles Derived from
Polychloro-o-Phenylenediamines

Substituents	Yield^a (%)	Ref.
4(7),5,6-Cl$_3$	—	716, 717, 718, 775
4(7),5,7(4)-Cl$_3$-2-Me	—	785, 781
4,5,6,7-Cl$_4$	95	643, 781
4,5,6,7-Cl$_4$-2-Me	86	643
4,5,6,7-Cl$_4$-2-n-Pr	66	643
4,5,6,7-Cl$_4$-2-n-C$_5$H$_{11}$	75	643
4,5,6,7-Cl$_4$-2-CH$_2$CH$_2$CO$_2$H	54	643
4(7),5(6)-Cl$_2$-2-CCl$_3$	44	755
4(7),5,6-Cl$_3$-2-CCl$_3$	—	773
4(7),5,7(4)-Cl$_3$-2-CCl$_3$	—	773
4(7),5,6-Cl$_3$-2-CClF$_2$	62	772
4(7),5,7(4)-Cl$_3$-2-CClF$_2$	—	772
4,5,6,7-Cl$_4$-2-CClF$_2$	—	772
4(7),5,7(4)-Cl$_3$-2-CF$_3$	70	731
	66	753, 756
4(7),5,6-Cl$_3$-2-CF$_3$	70–90	731, 753, 756
4(7)-Br-5,6,7(4)-Cl$_3$-2-CF$_3$	25	753, 756
4,5,6,7-Cl$_4$-2-CF$_3$	50	753, 756, 643
	60–90	731
	>95	727
4(7),6,7(4)-Cl$_3$-5-F-2-CF$_3$	85	643
4(7),6,7(4)-Cl$_3$-5-Me-2-CF$_3$	72, 62	731, 643
4(7),6,7(4)-Cl$_3$-5-Me-2-C$_2$F$_5$	—	753, 756
4(7),5,6-Cl$_3$-7(4)-NO$_2$-2-CF$_3$	—	753
4(7),5,7(4)-Cl$_3$-6-morpholino-2-CF$_3$	—	753
4,5,6,7-Cl$_4$-1-Et-2-Me	—	786
4,5,6,7-Cl$_4$-1-Ph-2-Me	—	786
4,5,6,7-Cl$_4$-1-Et-2-CF$_3$	59	753, 759
4,5,6,7-Cl$_4$-1-n-Bu-2-CF$_3$	—	753

^{a} Dashes in the Table mean that % yields are not given.

the corresponding thione (cf. ref. 770) [643]. 4,5(6),7-trichloro-2(3H)-benzimidazolone and the corresponding thione may be prepared similarly [781]. Condensation of the diamine with p-nitrobenzaldehyde gives a Schiff's base which yields tetrachloro-2-(p-nitrophenyl)benzimidazole on oxidation with lead tetra-acetate [643].

Reduction of a polychloro-2-nitroacetanilide with sodium dithionite in aqueous sodium hydroxide yields the corresponding polychloro-1-hydroxy-2-methylbenzimidazole (341; R = H or Cl)

(341)

[781]. Polychloro-1-hydroxy-2-trifluoromethyl (and pentafluoroethylbenzimidazoles) may be prepared similarly [769a]. The sodium salts of these 1-hydroxy compounds may be used to introduce various groups in the 1-position (e.g., OMe, OCH_2CO_2H, OCH_2CO_2Et, OCH_2CH_2OH, $OCH_2CH_2NMe_2$, or $OCH_2CH_2SO_3H$) [769a, 781].

There have been few reports of nucleophilic substitution reactions of polychlorobenzimidazoles. The 2-chlorine atom of 2,4,5,6,7-pentachlorobenzimidazole (332) is readily displaced by nucleophiles, for example with aniline, to give the 2-anilino-derivative (333) [779]. With pyridine, a betaine (342) is formed. Treatment of an

(342)   (343)

aqueous solution of the sodium salt of the pentachloro-compound with hydrogen sulphide at 100° under pressure yields tetrachloro-2-mercaptobenzimidazole [770]. This reaction presumably proceeds via the zwitterion (343). With various nucleophiles hexachlorobenzimidazole gives only the pentachloro-compound (332) [779].

Nitration of 4(7),5,6-trichloro-2-trifluoromethylbenzimidazole with a mixture of concentrated nitric and sulphuric acids yields the 7-nitro-derivative (334) [756].

A polychloro-2-methylbenzimidazole condenses with benzaldehyde to give a styryl derivative, which yields the corresponding 2-carbaldehyde on oxidation with osmium tetroxide in the presence of sodium metaperiodate [781]. Direct oxidation of the 2-methyl compounds with selenium dioxide also yields the corresponding aldehyde.

With thionyl chloride, the oxime of a polychlorobenzimidazole-2-carbaldehyde gives the corresponding 2-carbonitrile-derivative [781]. These may be prepared also through treatment of a polychlorobenzimidazole substituted by chlorine in the 2-position with chloromethyl methyl ether in the presence of sodium methoxide, treatment of the N-methoxymethyl compound with sodium cyanide in dimethylformamide, which replaces the 2-chlorine atom by a nitrile group, and removal of the N-methoxymethyl group of the product with pyridine hydrochloride (Scheme 2.27) [781].

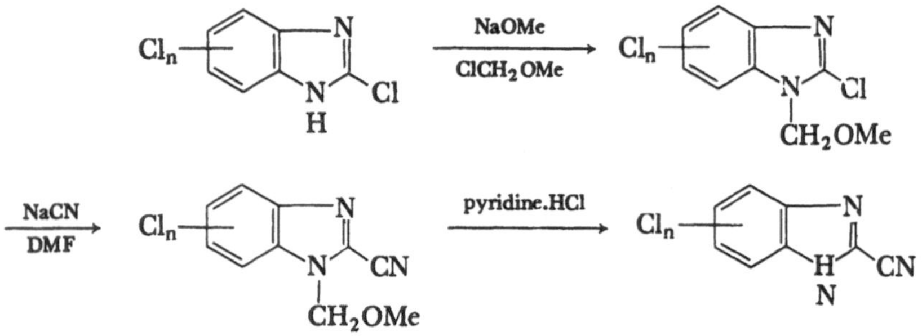

SCHEME 2.27

Tetrachlorobenzimidazole-2-carbonitrile may be prepared also by treatment of tetrachloro-2-trichloromethylbenzimidazole with concentrated aqueous ammonia [781].

Various reactions of polychlorobenzimidazole-2-carbonitriles have been described in a patent [781]. Thus, 2-thiocarbamoyl derivatives may be prepared by treatment of these nitriles with hydrogen sulphide in pyridine in the presence of triethylamine.

It is noteworthy that a polychloro-2-trichloromethylbenzimidazole may be converted into the corresponding 2-trifluoromethyl derivative by treatment with antimony trifluoride in nitrobenzene [755].

Tetrachloro-2-trichloromethylbenzimidazole reacts with 2-mercaptoethylamine to give tetrachloro-2-(2'-thiazolin-2-yl)benzimidazole (344) [781].

Various substituents may be attached in the 1-position of polychlorobenzimidazoles through formation of a sodium salt with a sodium alkoxide or by direct reaction with the reagent (e.g., $ClCO_2R$ or $ArSO_2Cl$) in the presence of triethylamine; the following types of

Cl

Cl

Cl

Cl

(344)

substituents have been introduced by these procedures: various alkyl groups [643, 728, 731, 753, 759, 760, 781, 785], CH_2OMe [781], CH_2CO_2R [781], $CH_2CH_2NR_2$ [781], CH_2CH_2OH [781], $CH_2C_6H_4NO_2\text{-}p$ [731], CO_2R (R = alkyl, aryl, or aralkyl) [733, 753, 761-765, 781], COAr [731], SO_2R (R = alkyl or aryl) [766, 767], and $R_2NC:S$ (NR_2 = dimethylamino, morpholino, etc) [769]. Alkylation of 4(7),5,6-trichloro-2-trifluoromethylbenzimidazole occurs to give exclusively a 1-substituted 4,5,6-trichloro-2-trifluoromethylbenzimidazole [733].

4,5,6,7-Tetrachloro-2-mercaptobenzimidazole may be alkylated with dimethyl sulphate in the presence of base to give either the 2-methylthio-derivative [770] or the 1-methyl-2-methylthio-derivative [779], depending on the amount of reagent used. Similarly, with n-butyl iodide or prop-2-ynl bromide in the presence of anhydrous potassium carbonate, it yields the corresponding 1-alkyl-2-alkylthio-derivative [643]. Treatment of 4,5(6),7-trichloro-2-mercaptobenzimidazole with methyl iodide in the presence of base is reported [781] to give the 2-methylmercapto-compound, which is oxidized to the corresponding sulphone by m-chloroperoxybenzoic acid in methylene chloride.

4,5,6-Trichloro-1-β-D-ribofuranosylbenzimidazole (323; R = H) (TRB) may be prepared from 4(7),5,6-trichlorobenzimidazole (326) by treatment of its 1-chloromercuri-derivative with 1-chloro-2,3,5-tri-O-acetyl-β-D-ribofuranose or tetra-O-acetyl-β-D-ribofuranose, followed by deacetylation of the product with alcoholic ammonia [716-718].

4,5,7-Trichloro-1,2-dimethylbenzimidazole is quaternized on treatment with methyl iodide to give the quaternary iodide (345) which is a precursor of cyanine dyestuffs [785]. Various other cyanine dyes have been prepared similarly [786].

Cl

Cl

Cl   Me

Cl

$\overset{+}{N}Me$ $\bar{I}$

Me

(345)

2.17.3 Pyrazoles

Tetrachloro- and the trichloro-pyrazoles have not been reported [787, 788]. Considerably more interest has been shown in poly-bromo- and polyiodo-pyrazoles.

Direct chlorination of pyrazole itself is complicated [787, 788]. With chlorine in carbon tetrachloride at 0°, or with aqueous sodium hypochlorite at room temperature, it yields 4-chloropyrazole whilst, with the former reagent system at 60°, it gives the 4-chloro-derivative and a trimeric compound to which structure (346) has been assigned [787-790]. An aqueous solution of pyrazole gives compound (347)

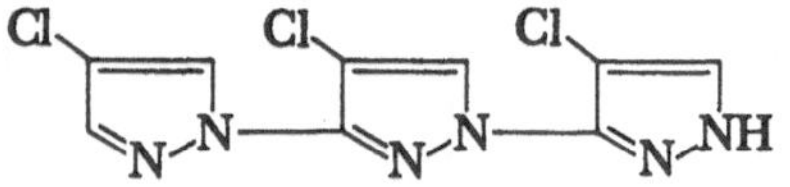

(346)

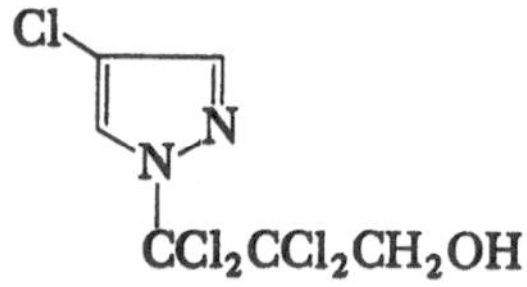

$CCl_2CCl_2CH_2OH$

(347)

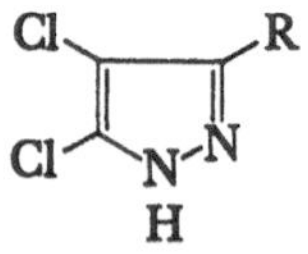

(348)

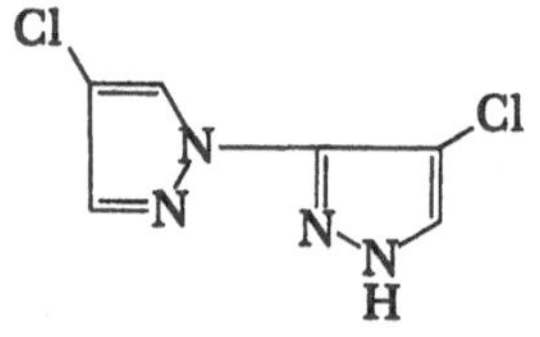

(349)

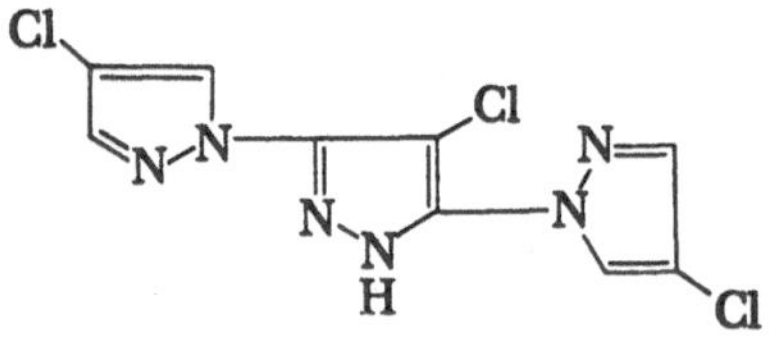

(350)

on treatment with chlorine [789]. The same authors [789] have established that chlorination of 3(5)-methylpyrazole in acetic acid yields the dichloro-derivative (348; R = Me) or 4,5(3)-dichloro-3(5)-trichloromethylpyrazole (348; R = CCl₃), depending on the amount of chlorine used. In carbon tetrachloride, 3(5)-methylpyrazole reacts with chlorine to give its 4-chloro-derivative.

A stepwise chlorination of pyrazoles and 3- or 4-halopyrazoles can be carried out via their silver salts [790]. Thus, chlorination of the silver salt of pyrazole gives the 4-chloro-derivative together with the dimeric compound (349). Similarly, the silver salts of 3(5)- or

4-chloropyrazole yield 3(5),4-dichloropyrazole, the dimer (349), and the trimer (350). Bromination of the silver salt of 3(5),4-dichloropyrazole gives 3(5)-bromo-4,5(3)-dichloropyrazole (348; R = Br). Various chloropyrazoles, which in turn may be prepared by treatment of a pyrazolone with phosphorus oxychloride [791-793], yield polychloropyrazoles when reacted with phosphorus pentachloride. The following conversions are illustrative: 5(3)-Cl-3(5)-Ph to 4,5(3)-Cl_2-3(5)-Ph [791]; 5-Cl-1-Me-3-Ph to 4,5-Cl_2-1-Me-3-Ph [791]; 5(3)-Cl-3(5)-Me to 4,5(3)-Cl_2-3(5)-Me [794, 795]; 3,5-Cl_2-1-Ph to 3,4,5-Cl_3-1-Ph (351) [793].

TABLE 2.9   Polychloro-1-phenylpyrazoles

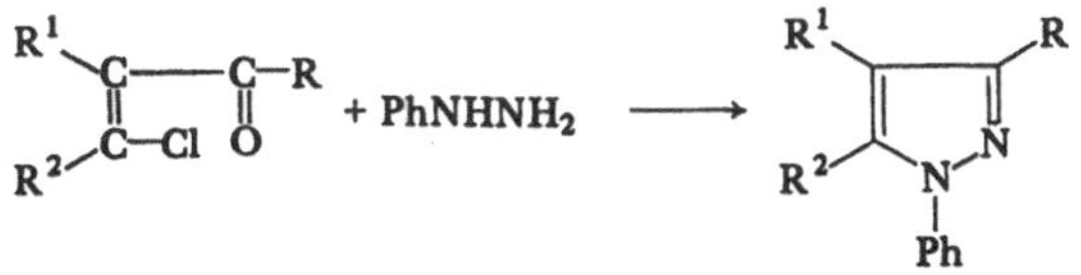

R	R^1	R^2	Yield (%)	Ref.
$CHCl_2$	Cl	Cl	52, 42	798
$CCl:CCl_2$	Cl	Cl	51	796
H	Cl	Cl	4	796
Ph	Cl	Cl	15	796
Me	H	Cl	72	796
$CHCl_2$	Cl	H	38	796
CCl_3	Cl	Cl	38	797
CCl_3	H	Cl	51	797
CCl_3	H	H	61	796

Polychloro-1-phenylpyrazoles may be prepared also by condensation of a conjugated polychlorocarbonyl compound with phenylhydrazine in boiling acetic acid or ether (see Table 2.9 for details) [796-798]. These reactions proceed via cyclization of the phenylhydrazones (352). Perchloropent-1-en-3-one reacts with phenylhydrazine to give Cl_2C:CCl.CO.NHNHPh, which cyclizes to give 4,5-dichloro-1-phenylpyrazol-3-one [796].

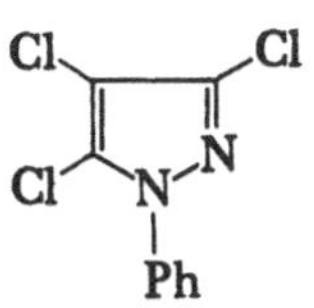

(351)

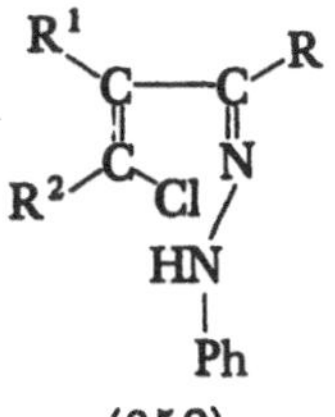

(352)

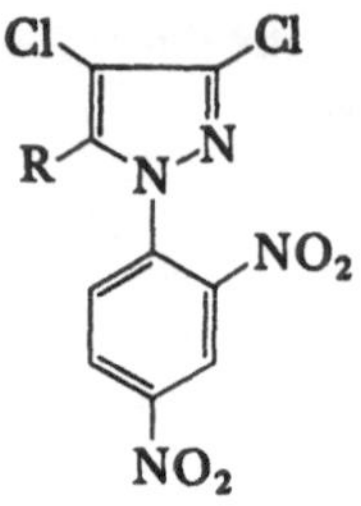

Cl, Cl
R-N-N
NO$_2$
NO$_2$

(353)

Cl, R^2
R^3-N-N
CH$_2$SP(:S)CH$_2$(OR^1)$_2$

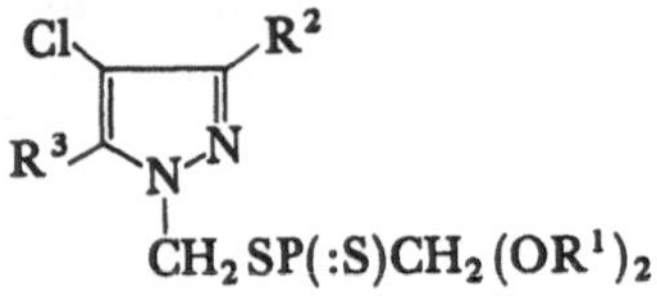

(354)

The *N*-arylpyrazoles (353; R = Me or Ph) may be prepared by treating a refluxing solution of the corresponding Δ^2-pyrazoline with chlorine in chloroform [799].

Compounds (354: R^1 = Me or Et; R^2 and R^3 = Me or Cl) exhibit insecticidal and acaricidal activity [800].

The NMR [801] and UV [802] spectra of a few dichloropyrazoles have been studied.

2.17.4   Benzopyrazoles (Indazoles)

Pentachloroindazoles and hexachloroindazole appear to be unknown.

Chlorination of indazole with chlorine in an acidic medium gives successively 3-chloro-, 3,5-dichloro-, and 3,5,7-trichloro-indazole (355) [803, 804]. 2-Phenylindazole similarly yields its 3,5,7-trichloro-derivative (356) [805]. The structure of this compound was

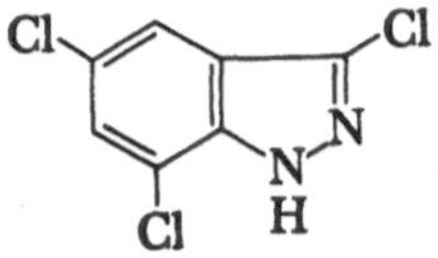

Cl, Cl
N-N
Cl H

(355)

Cl, Cl
N—Ph
N
Cl

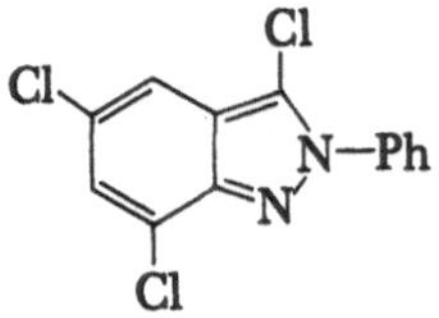

(356)

proved by unambiguous synthesis (see below) as well as by degradation with chromic acid to 3,5-dichloro-2-phenylazobenzoic acid followed by reduction of this compound to 2-amino-3,5-dichlorobenzoic acid. Treatment of 6-aminoindazole with chlorine in the presence of hydrochloric acid gives an adduct which is reduced by zinc and acid to 4,5,7-trichloro-6-hydroxyindazole [806].

Pentachlorobenzonitrile (357; R = Cl) reacts with four equivalents of hydrazine in dimethylformamide at 140° to yield 3-amino-4,5,6,7-tetrachloroindazole (359; R = Cl) via the corresponding intermediate (358; R = Cl); with less hydrazine at room temperature *p*-substitution occurs predominantly [807]. 2,3,5,6-Tetrachlorobenzonitrile

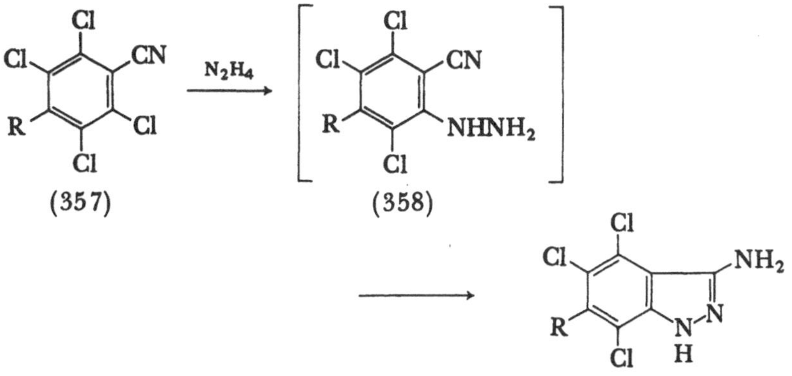

(357)   (358)

(359)

(357; R = H) similarly gives the indazole (359; R = H). The dinitrile (357; R = CN) yields the indazole (359; R = CN) at 80° but substitution by a dimethylamino-group (derived from the solvent) in the benzenoid ring of the indazole occurs at 150° [807].

Cyclization of 5-chloro-2-phenylazobenzoic acid with phosphorus pentachloride, phosphorus oxychloride, or thionyl chloride gives 5,7-dichloro-3-hydroxy-2-phenylindazole, which is converted into 3,5,7-trichloro-2-phenylindazole (356) on further treatment with phosphorus oxychloride [805].

Treatment of the dinitroindazolone (360; R = H) with phosphorus oxychloride at 140° in a sealed tube yields the 3-chlorodinitro-compound whilst, at 180°, both nitro-groups are replaced and the product is 3,5,7-trichloroindazole (355) [808]. The dinitro-2-phenyl-indazolone (360; R = Ph) undergoes analogous reactions [808].

3,4,5,6-Tetrachloroindazole (361; R = H) is reported [107] to possess herbicidal activity, although it is considerably less active than

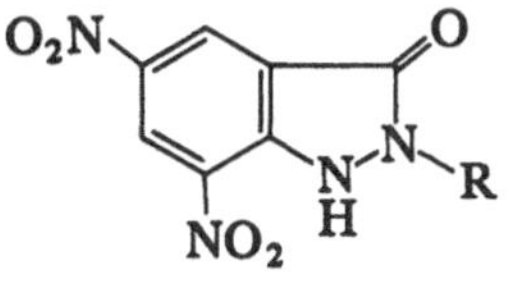

(360)

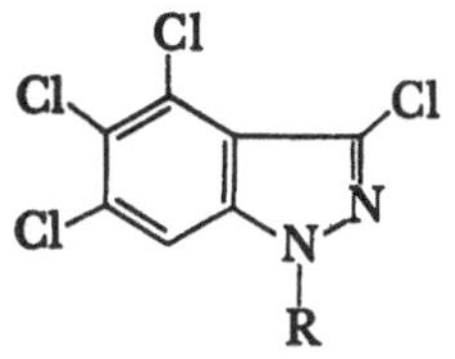

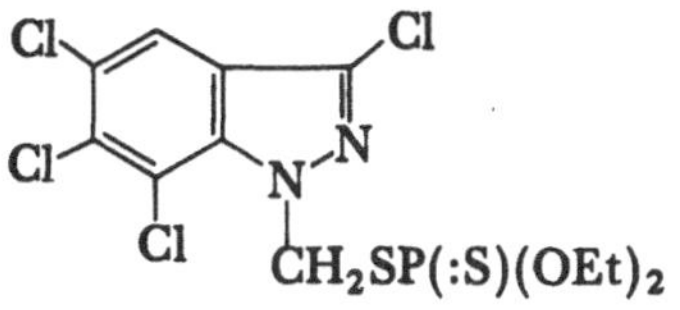

(361)   (362)

polychloro-imidazoles or -benzimidazoles, and compound (362) is claimed [809] to exhibit insecticidal and acaricidal activity. The latter may be prepared by 1-hydroxymethylation of 3,5,6,7-tetra-chloroindazole with formaldehyde followed by conversion of the 1-hydroxymethyl derivative into the 1-chloromethyl compound with thionyl chloride and treatment of this with the ammonium salt of OO-diethyldithiophosphoric acid [$NH_4 SP(:S)(OEt)_2$].

2.18   OTHER POLYCHLOROHETEROAROMATIC COMPOUNDS

In this Section we review a selection of polychloroheteroaromatic compounds which have been studied only briefly; many of them were reported during the preparation of the manuscript.

2.18.1   Nitrogen Heterocycles

We have not reviewed the chemistry of trichloro-1,3,5-triazine (cyanuric chloride) for reasons already given (Section 2.1). Various synthetic uses for cyanuric chloride have been reported recently, however, which are noteworthy. Thus, it can be used to effect the conversion of alcohols into alkyl chlorides [810] or, in the presence of sodium iodide, into the corresponding alkyl iodides [811]. With cyanuric chloride in pyridine, aldoximes are dehydrated to give nitriles at room temperature [812], while benzamide is converted similarly into benzonitrile [813].

The isomeric system, trichloro-1,2,4-triazine (363) (Scheme 2.28), may be prepared by treatment of 5-bromo-6-azauracil (364; R = Br) with phosphorus oxychloride [814-816] or, preferably, a mixture of phosphorus oxychloride and phosphorus pentachloride in the presence of NN-diethylaniline [817]. It is readily hydrolyzed to give 5-chloro-6-azauracil (364; R = Cl) [815] and undergoes nucleophilic

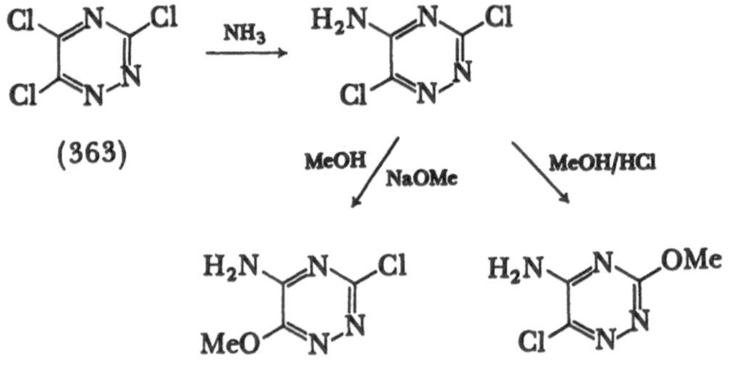

SCHEME 2.28

substitution with an excess of ammonia (Scheme 2.28) or one equivalent of sodium methoxide in the 5-position [816]. Two equivalents of sodium methoxide yield the 5,6-dimethoxy derivative as the major product, together with some of the 3,5-dimethoxy isomer, whilst all three chlorine atoms can be replaced by an excess

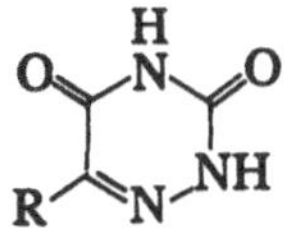

(364)

of this reagent [816]. Dichloro-5-methoxy-1,2,4-triazine gives 5-aminodichloro-1,2,4-triazine on treatment with ammonia [816]. Preferential displacement of the methoxyl group in this case is readily explained by reference to the structures of the transition states for attack in each position (see also Section 2.12) [605].

In acid-catalyzed nucleophilic substitution reactions trichloro-1,2,4-triazine and its derivatives behave differently. For example, with methanolic hydrogen chloride, the perchloro-compound gives the 3,5-dimethoxy derivative (see preceding paragraph) [816]. Further substitution of 5-aminodichloro-1,2,4-triazine can be controlled similarly (Scheme 2.28).

1,2,3,4,5,6-Hexachlorobenzo[α]phenazine (365) may be prepared by condensation of hexachloro-1,2-naphthaquinone with o-phenylenediamine [818].

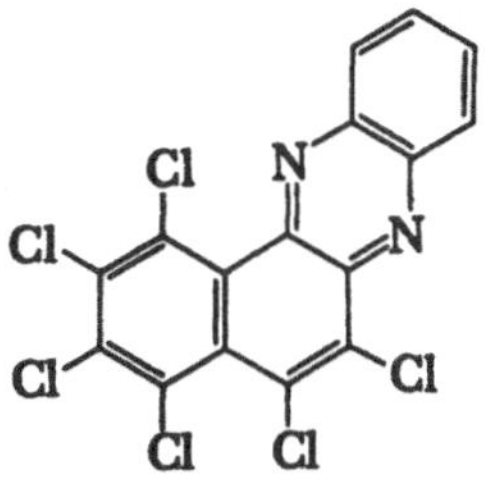

(365)

Exhaustive chlorination of imidazo[1,2-a]pyridine [819], imidazo[1,2-b]pyridazine [820, 821], and s-triazolo[4,3-b]-pyridazine [821] with phosphorus pentachloride in an autoclave at 260-275° gives the corresponding perchloro-compound, (366), (367), and (368), respectively. Chlorination of the s-triazolopyridazine with the same reagent at 280° leads to destruction of the 5-membered ring and gives a mixture of 3,4,6-trichloro- and tetrachloro-pyridazine [821]. 6-Chloro- or 3-bromo-6-chloro-imidazo[1,2-b]pyridazine and 6-chloro- or 3,6-dichloro-s-triazolo[4,3-b]pyridazine also yield the

corresponding perchloro-compound under similar conditions [821].
A study of the chlorination reactions of imidazo[1,2-b]pyridazine
and s-triazolo[4,3-b]pyridazine under milder conditions has led to
the isolation of lower chlorinated products. The order of reactivity
of the ring-positions in the former compound is 3 > 2,7 > 8 > 6
whilst it appears to be 3 > 8 > 7 > 6 in the latter compound [821].
Hexachloroimidazo[1,2-a]pyridine (366) undergoes nucleophilic

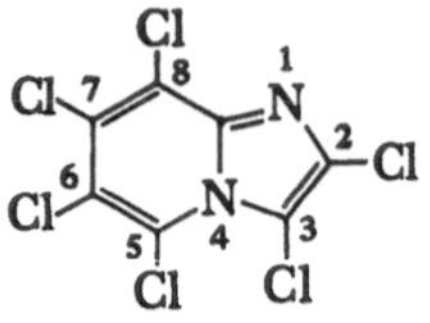

(366)

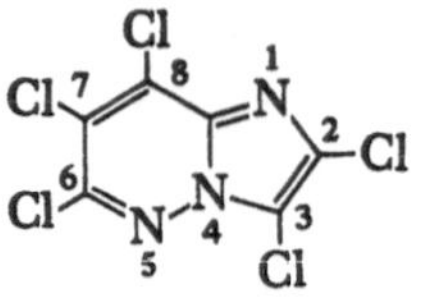

(367)

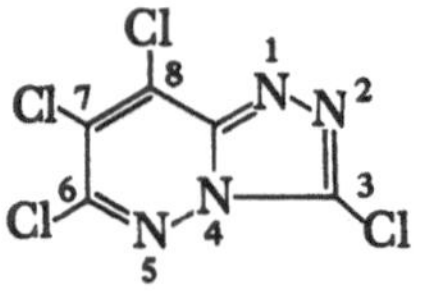

(368)

substitution to give 5-monosubstituted derivatives (e.g., with
methoxide or t-butoxide ions) and 5,7-disubstituted derivatives (e.g.,
with piperidine) [819]. Pentachloroimidazo[1,2-b]pyridazine (367)
[820, 821] and tetrachloro-s-triazolo[4,3-b]pyridazine (368) [821]
each give 8-monosubstituted derivatives (e.g., with dimethylamine,
aniline, hydrazine, and azide or thiophenoxide ions); the former gives
6,8-disubstituted derivatives (e.g., with methoxide or thiophenoxide
ions) [820, 821] and a 6,7,8-trisubstituted derivative with thiophenoxide ion [821]. Catalytic dehydrohalogenation of compounds
(366) [819] and (367) [820] yields the parent heterocycle in each
case, thus excluding skeletal rearrangement during the chlorination
procedures. The structures of the other products were established
similarly.

Chlorination of 3-pyrrolidinopropionitrile or 3-NN-diethylamino-
propionitrile at less than 150° gives hexachloropyrrolo[1,2-a]-
pyrimidine (369); at higher temperatures (150-200°) the former
compound yields polychloropyrimidines (see also Section 2.12)
[250]. Octachloropyrimido[1,2-a]azepine (370) may be prepared
similarly by chlorination of 3-perhydroazepinopropionitrile (94) at
150°; at 250° this nitrile gives tetrachloropyrimidine, pentachloropyridine, and a number of minor chlorinated products (see also
Section 2.12) [250].

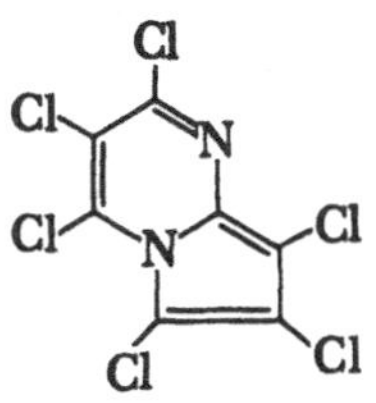

(369)

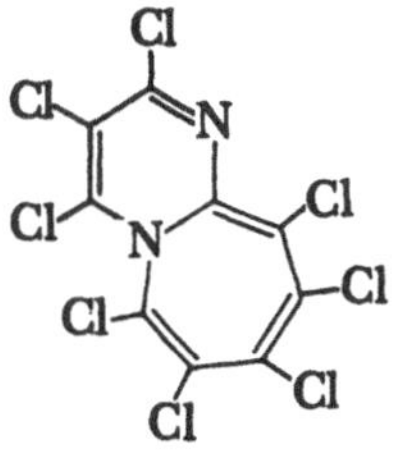

(370)

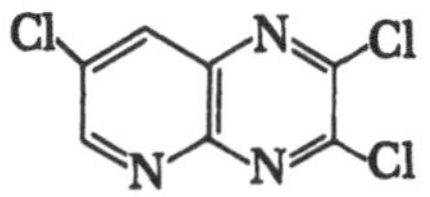

(371)

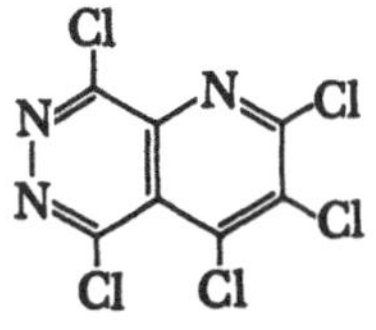

(371a)

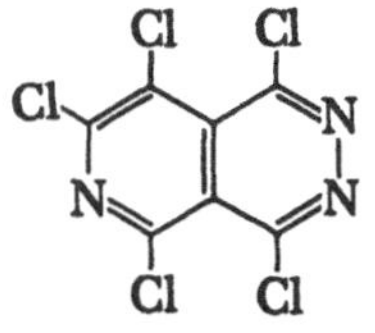

(371b)

2,3,7-Trichloropyrido[2,3-*b*]pyrazine   (2,3,7-trichloro-5-azaquinoxaline) (371) may be prepared by successive condensation of 2,3-diamino-5-chloropyridine with ethyl oxalate and treatment of the 7-chloro-2,3-dihydroxypyrido[2,3-*b*]pyrazine produced with phosphorus oxychloride, or by melting 2,3-dihydroxypyrido[2,3-*b*]-pyrazine with phosphorus pentachloride [822]. It reacts with ethanolic potassium hydroxide to give 7-chloro-2,3-diethoxypyrido-[2,3-*b*]pyrazine. On treatment with an excess of phosphorus penta-chloride in a sealed tube, 5,8-dichloropyrido[2,3-*d*]pyridazine yields its 3,5,8-trichloro-, 2,3,5,8-tetrachloro-, or 2,3,4,5,8-pentachloro-derivative (371a), depending on the reaction temperature and time [822a]. 1,4,8-Trichloro-, 1,4,7,8-tetrachloro-, and 1,4,5,7,8-penta-chloro-pyrido[3,4-*d*]pyridazine (371b) may be prepared similarly [822a]. These compounds are highly reactive towards nucleophiles; in moist air they decompose. With equivalent amounts of other nucleophiles they give complex mixtures and it is only possible to obtain a uniform product if a large excess of the nucleophile is used; for example, compound (371a) reacts with a large excess of sodium methoxide to give 3-chloro-2,4,5,8-tetramethoxypyrido[2,3-*d*]-pyridazine.

   We have referred to the syntheses of 5,6,8-trichloro-2,4-diphenyl-pyrido[3,4-*d*]pyrimidine (135) and 6,7,8-trichloro-2,4-diphenyl-pyrido[3,2-*d*]pyrimidine (138) in Section 2.8.2.

   1,4,5,8-Tetrahydroxypyridazino[4,5-*d*]pyridazine, obtained by treatment of tetraethyl ethylenetetracarboxylate with hydrazine

(372)   (373)

hydrate in refluxing methanol, is converted into the 1,4,5,8-tetra-
chloro-compound (372; R^1 = R^2 = Cl) on being heated with a
mixture of phosphorus pentachloride and phosphorus oxychloride
[823]. With sodium methoxide (in methanol) or sodium thio-
ethoxide, the tetrachloro-compound gives the tetra-substituted
derivatives (372; R^1 = R^2 = OMe or SEt), respectively. With sodium
methoxide in toluene it yields a mixture of two dimethoxy
compounds, (372; R^1 = OMe; R^2 = Cl) and (373; R = OMe), together
with a small amount of the monochlorotrimethoxy derivative. The
structures of these compounds were proved by catalytic dehalogenation. Similarly, hydrolysis with aqueous alkali gives mainly the
dihydroxy compound (373; R = OH) together with a small amount
of its isomer (372: R^1 = OH; R^2 = Cl) [823]. These hydroxy
compounds react with diazomethane to give mixtures of NN- and NO-
dimethyl derivatives.

We have referred to the preparation of trichlorodiaza-indoles (57)
in Section 2.4 [see also Section 2.8.3 for the synthesis of compounds
(191 and 192; X = NH)].

Chlorination of benzotriazole or its 5-chloro-derivative with
boiling *aqua regia* gives 4,5,6,7-tetrachlorobenzotriazole (374; R =
Cl) (90%) [824]. The 5-substituted 4,6,7-trichlorobenzotriazoles

(374)

(374; R = Me [643, 825, 826], Br [825], and F [643]) and
tetrachloro-1(and 2)-methylbenzotriazole [824] may be prepared
similarly. Tetrachlorination of the 1-methyl compound requires more
forcing conditions; 4,5,6-trichloro-1-methylbenzotriazole is readily
isolated under the usual conditions. With *aqua regia*, 4-nitrobenzo-
triazole and its 2-methyl derivative give 4,5,6,7-tetrachlorobenzo-
triazole and its 2-methyl derivative, respectively [826]. Loss of a

nitro-group under such conditions is not uncommon, but more surprising is the loss of a methyl group from 2,5-dimethylbenzotriazole, which also gives tetrachloro-2-methylbenzotriazole [826].

Diazotization of 4,5- or 3,6-dichloro-o-phenylenediamine gives 5,6- or 4,7-dichlorobenzotriazole, respectively [826], whilst treatment of pentachloronitrobenzene with hydrazine gives mainly tetrachloro-1-hydroxybenzotriazole [104, 643].

Polychlorobenzotriazoles are reduced by zinc and acid to polychloro-o-phenylenediamines [643]. Benzotriazole is alkylated and undergoes addition reactions to alkenes to give 1-substituted products but 4,5,6,7-tetrachlorobenzotriazole gives mainly the 2-methyl derivative with dimethyl sulphate, together with a small amount of the 1-methyl isomer, and it undergoes addition reactions with acrylic acid, crotonic acid, acrylonitrile, acrylamide, and benzalacetophenone to give 2-substituted products [824]. With crotonic acid, for example, it gives the acid (375). 4,6,7-Trichloro-5-methylbenzotriazole similarly gives 2-substituted products in alkene addition

(375)

reactions [826]. These differences in behaviour between benzotriazole and its polychloro-derivatives can be attributed to steric hindrance by the 4,7-chlorine atoms. This is supported by the fact that, whereas 4,7-dichlorobenzotriazole gives a 2-substituted product analogous to compound (375) in an addition reaction with crotonic acid, its 5,6-dichloro-isomer gives the isomeric 1-substituted product [826].

The UV spectrum of tetrachloro-1-methylbenzotriazole is analogous to that of benzotriazole, whereas that of the 2-methyl isomer is different [824]. The behaviour of 4,5,6,7-tetrachlorobenzotriazole in its alkylation reactions and the difference between the UV spectra of its 1- and 2-methyl derivatives is attributable to a tautomeric equilibrium, as shown (374; R = Cl). As expected, 4,5,6,7-tetrachlorobenzotriazole is weakly acidic [824].

4,5,6,7-Tetrachlorobenzotriazole (374; R = Cl) and 4,5,7-trichloro-2-methylbenzotriazole exhibit herbicidal properties [107].

Treatment of tetrachloroanthranilic acid with n-butyl nitrite in dioxan is reported [827] to yield a mixture of octachloroacridone (376) (35%) and n-butyl 2,3,4,5-tetrachlorophenyl ether (28%). Similarly, addition of sodium nitrite to a solution of tetrachloro-

(376)

anthranilic acid in acetic acid at 25° gives octachloroacridone (376) (10%) together with 3,4,5,6,2',3',4',5'-octachlorodiphenylamine-2-carboxylic acid (~20%) and 2,3,4,5-tetrachlorophenyl acetate (~40%) [828]. Octachloroacridone (376) arises in these reactions from the reaction of tetrachlorobenzyne with undiazotized tetrachloroanthranilic acid. It can be made also by the action of antimony pentachloride on acridone [829].

2.18.2   Compounds Containing Nitrogen and Sulphur

The synthesis of polychlorobenzothienopyridines (191 and 192; X = S) from 2- or 4-thiophenoxytetrachloropyridines has been described in Section 2.8.3.

Cyclization of the pyridine derivative (377), obtained by condensation of 2,3,6-trichloro-4-mercaptopyridine with chloroacetone, with polyphosphoric acid gives a low yield of 4,6,7-trichloro-3-methylthieno[3,2-c]pyridine (378) [270].

(377)   (378)

There is a report [830] that chlorination of phenothiazine with chlorine in the absence of UV light yields a hexachlorophenothiazine, but an attempt to repeat this work has failed [831]. On UV irradiation, 1,2,3,4,6,7,8,9-octachlorophenothiazine (379) is formed albeit in low yield [830, 831]. The octachloro-compound (379) is most conveniently prepared, however, in 50% yield, when a suspension of a tetrachlorophenothiazine, which is formed on chlorination of phenothiazine with chlorine in nitrobenzene, is further chlorinated [831-833]. Exhaustive chlorination of phenothiazine under these conditions yields an undecachlorodihydrophenothiazine, which may have one of two structures, (380) or (381) [831, 833]. The octachloro-compound (379) may be prepared also by chlorination of N-acetyl- or N-benzoyl-phenothiazine or 2,4,5,7-tetrachlorophenothiazine with sulphuryl chloride [834].

If water is present during the chlorination of phenothiazine with chlorine in nitrobenzene, the product is the sulphoxide (382) [832]. Oxidation of the octachloro-compound (379) with potassium dichromate in boiling acetic acid gives heptachloro-3H-phenothiazin-3-one (383) [832]. The same compound (383) is formed when the undecachlorodihydro-compound, (380) or (381), is treated with acetic anhydride [831, 833].

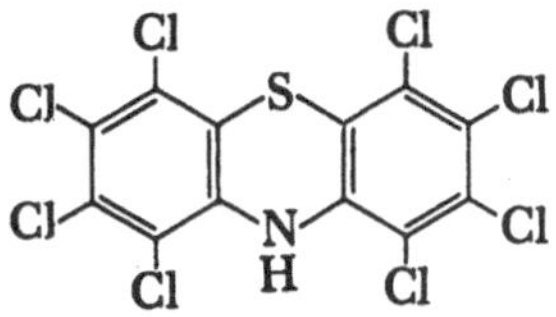

(379) 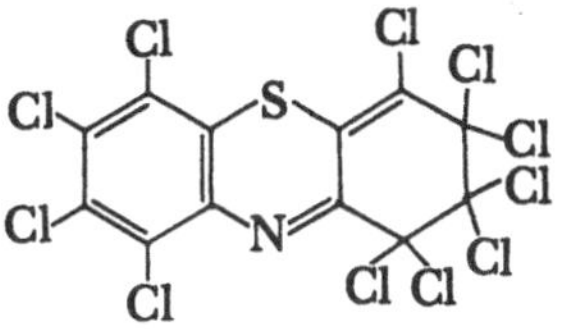   (380)

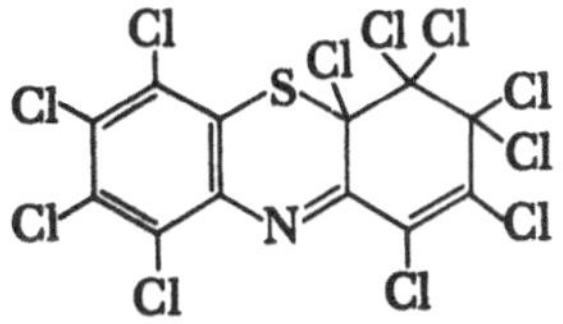

(381) 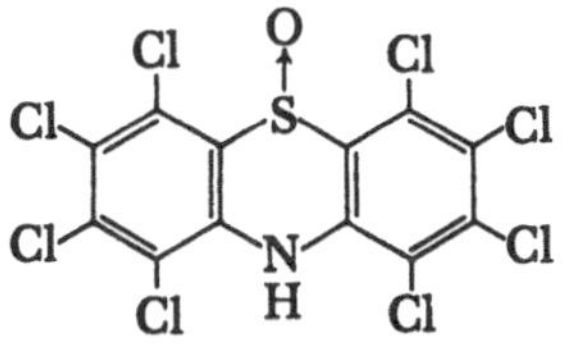   (382)

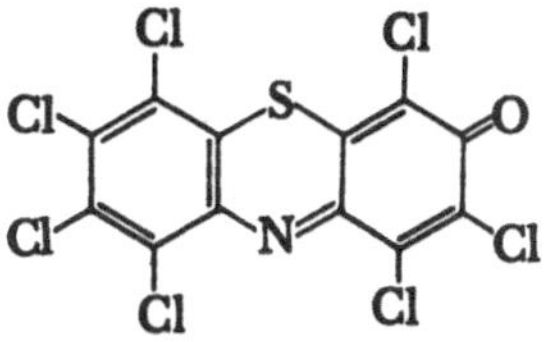

(383) 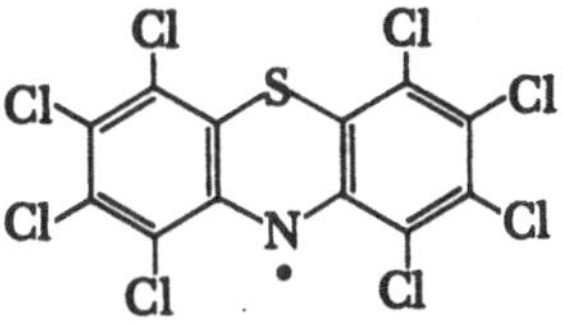   (384)

The undecachlorodihydro-compound undergoes thermal decomposition at 180° with loss of three atoms of chlorine to give the stable octachlorophenothiazinyl radical (384) [831, 835, 836]. A little of the dimer, hexadecachloro-10,10'-biphenothiazinyl, is also formed. The radical (384) can be sublimed at 270° and melts in air at 380° with only slow decomposition. It reacts with aniline by hydrogen abstraction, to give octachlorophenothiazine (379). On being heated in aniline, phenylhydrazine, or a mixture of ethanol and nitrobenzene, the undecachlorodihydro-compound, (380) or (381), similarly gives the octachloro-derivative (379) [831, 833].

Polychlorophenothiazines have been reviewed [831].

Treatment of cyanogen with sulphur dichloride in tetrahydrofuran or dimethylformamide in the presence of tetraethylquaternary ammonium chloride yields 3,4-dichloro-1,2,5-thiadiazole (385) [837]. The same compound is produced when dichloroglyoxime is treated with sulphur dichloride in dimethylformamide [838]. In hydrochloric acid the bisdiazonium salt derived from 2,5-diamino-1,3,4-thiadiazole gives 2,5-dichloro-1,3,4-thiadiazole (386) [839].

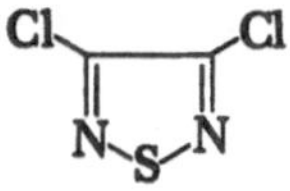

(385) 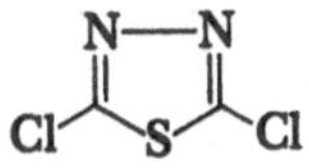   (386)

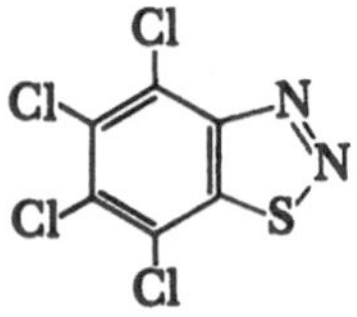

(387)

With sulphur monochloride, 2,3,4,5-tetrachloroaniline is converted into tetrachlorobenzothiazathiolium chloride, which gives tetra-chlorobenzo-1,2,3-thiadiazole (387) on treatment with sodium nitrite in sulphuric acid at 0° [840, 841]. Trichlorobenzo-1,2,3-thiadiazoles and a large number of dichloro-derivatives may be prepared similarly. These compounds are of interest as herbicides, fungicides, and insecticides. A mixture (2:1) of 4,5,7-trichlorobenzo-2,1,3-thiadiazole and 4,5(or 6),7-trichlorobenzo-2,1,3-thiadiazole 1-oxide is obtained when 3,4,6-trichloro-o-benzoquinone dioxime is reacted with sulphur dichloride in benzene at 25° [838].

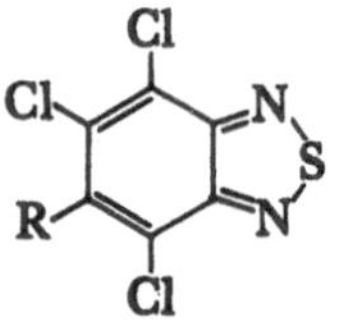

(388)

Benzo-2,1,3-thiadiazole reacts exothermically with chlorine either in the molten state or in solution in various solvents to give an addition compound which, on treatment with alkali, gives the 4,7-dichloro-derivative [842-844]. Chlorination in the molten state at 190-200° in the presence of iron filings gives 4,5,7-trichloro- (388; R = H) or 4,5,6,7-tetrachloro-benzo-2,1,3-thiadiazole (388; R = Cl), depending on the conditions [842-845]. 5-Chlorobenzo-2,1,3-

thiadiazole gives either of these products on chlorination, again depending on the reaction conditions, whilst 4-methylbenzo-2,1,3-thiadiazole gives its 5,6,7-trichloro-derivative. Under similar conditions, the 5-methyl-compound gives a mixture of its 4-chloro- and 4,7-dichloro-derivatives. Other derivatives of benzo-2,1,3-thiadiazole, for example the 5,6-dimethyl compound, behave predictably [844].

A polychlorobenzo-2,1,3-thiadiazole may be prepared also through cyclization of the appropriate polychloro-*o*-phenylenediamine dihydrochloride with thionylaniline or thionyl chloride in benzene in the presence of a base [846-849].

A polychlorobenzo-2,1,3-thiadiazole is reduced by zinc and acid to the corresponding *o*-phenylenediamine [844]. These compounds are readily nitrated [848, 849]; for example, with nitrating mixture at a temperature of 100° or below, 5,6-dichlorobenzo-2,1,3-thiadiazole gives its 4-nitro-derivative whilst the 4,7-dinitro-compound is obtained above 100° [848]. 4,5,7-Trichlorobenzo-2,1,3-thiadiazole similarly gives its 6-nitro-derivative [849]. A number of nucleophilic substitution reactions of mono- and dichlorobenzo-2,1,3-thiadiazoles have been reported [850].

Polychlorobenzo-2,1,3-thiadiazoles show promising herbicidal activity [107, 845-848]. Of particular interest in this respect is the 4,5,7-trichloro-derivative (388; R = H) (Phillips Th-052-H); the 5,7-dichloro-4-methyl derivative also shows strongly herbicidal activity.

2,5,7-Trichlorothiazolo[5,4-*d*]pyrimidine (389) can be prepared from the corresponding tri-hydroxy compound by its treatment with phosphorus oxychloride in the presence of *NN*-dimethylaniline [851]. It reacts with *N*-methylaniline in the 7-position and the product gives the 2-ethylthio-derivative (390) on treatment with ethyl mercaptan in the presence of base.

(389)   (390)

With phosphorus pentasulphide at 180°, perchloronaphthyl-1-isocyanide dichloride (391; R = N:CCl$_2$) yields the corresponding isothiocyanate (391; R = NCS) [852]. With sodium sulphide, however, it gives the naphthothiazole derivatives (392) and (393).

2.18.3 Miscellaneous Compounds

By analogy with the preparation of tetrachlorothiophen from sulphur and perchlorobuta-1,3-diene (Section 2.6.1), treatment of

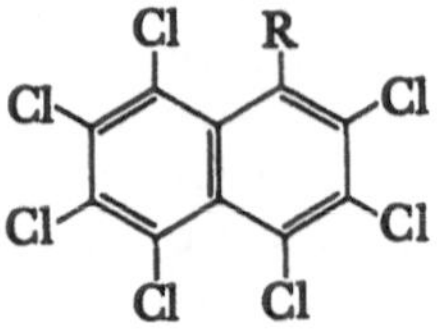

(391)

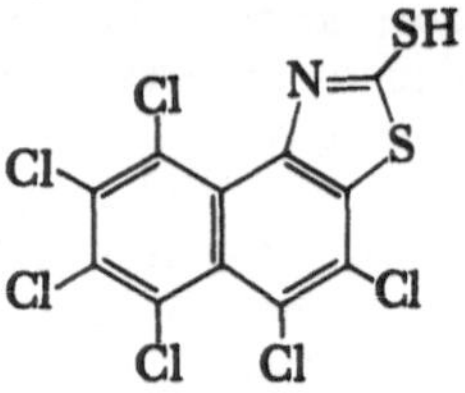

(392)

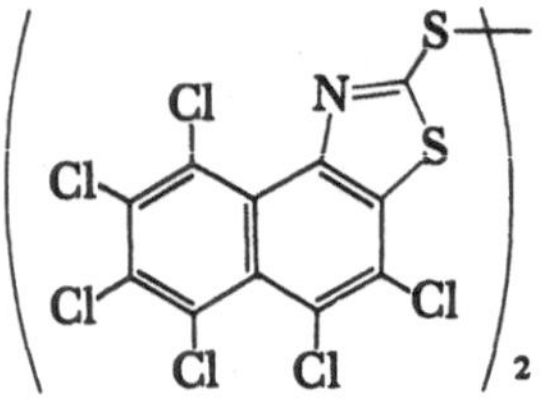

(393)

the butadiene with tellurium or selenium at 250° yields tetrachloro-tellurophen (394; X = Te) and tetrachloroselenophen (394; X = Se),

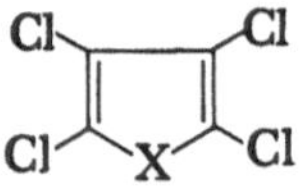

(394)

respectively [853]. We have mentioned the synthesis of a poly-chloroselenophen (77) in Section 2.6.1. Tetrachlorotellurophen was only the second derivative of this ring system to be reported.

Tetrachlorotellurophen (394; X = Te) rapidly absorbs one mole of chlorine to give the hexachloro-derivative (394; X = TeCl$_2$) which is reconverted into the tetrachloro-compound on treatment with sodium hydrogen sulphite [853]. The hexachloro-compound (394; X = TeCl$_2$) is decomposed by aqueous alkali to give 1,2,3,4-tetra-chlorobuta-1,3-diene, probably the *cis-cis*-isomer [853]. With palladium chloride in methanol, tetrachlorotellurophen yields a red-brown crystalline complex, (C$_4$Cl$_4$Te)$_2$PdCl$_2$ [854].

Octachlorodibenzo-1,4-dithiin (395; X = S) and various of its derivatives and the analogous compound (395; X = O) may be prepared by treatment of diphenyl sulphide (or a derivative) or diphenyl ether, respectively, with a mixture of sulphur monochloride and sulphuryl chloride in the presence of aluminium or ferric chloride [167, 855]. Compound (395; X = S) may be prepared also by irradiation of pentachlorobenzenesulphenyl chloride (see also Section 2.7) [166].

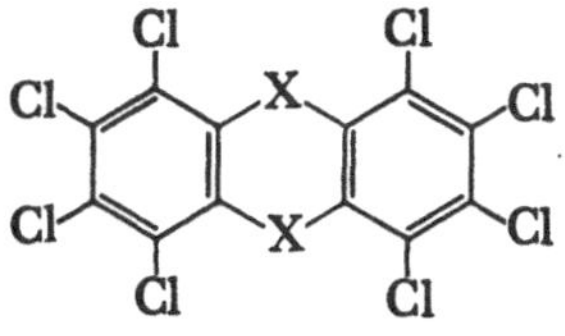

(395)

Despite the fact that octachloronaphthalene is relatively inert to nucleophilic attack, fusion with sulphur or selenium gives 3,4,7,8-tetrachloronaphtho[1,8-cd:4,5-$c'd'$]bis[1,2]dithiole (396) and 3,4,5,6,7,8-hexachloronaphtho[1,8-cd]-1,2-diselenole (397; X = Se), respectively [856]. The dithiole (397; X = S) is obtained when

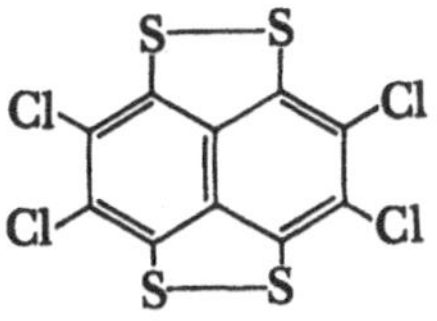

(396)

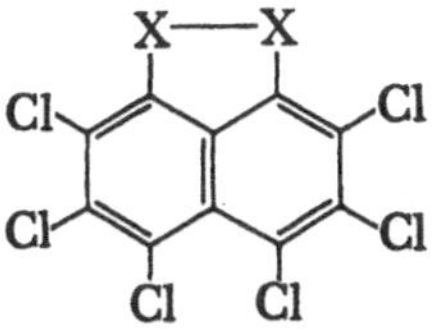

(397)

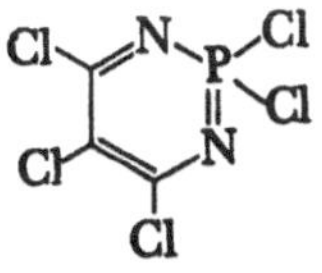

(398)

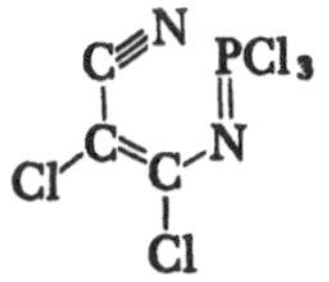

(399)

octachloronaphthalene is treated with sodium disulphide. These interesting reactions are believed to involve bidentate attack, for example with S_2^{2-}, at two peri positions [856].

The interesting heterocycle, 1,1,3,4,5-pentachloro-1,2,6-phosphadiazine (398), may be prepared through reaction of malononitrile with phosphorus pentachloride as well as by treatment of compound (399) with hydrogen chloride [857].

Acknowledgements

The authors would like to thank Drs. M. B. Green and R. D. Bowden, I.C.I. Ltd., Mond Division, Dr. C. D. S. Tomlin, I.C.I. Plant Protection Ltd., and Dr. S. M. Roberts, University of Salford, who read the manuscript and made useful suggestions and contributions, Misses S. M. B. Holt and A. J. Chappell for their patience in the preparation of the manuscript, and Mrs. B. Iddon for help in proof reading.

REFERENCES

1. D. T. CLARK, R. D. CHAMBERS, D. KILCAST, and W. K. R. MUSGRAVE, *J. C. S., Faraday Trans. II*, **68**, 309 (1972).
2. D. B. ADAMS, D. T. CLARK, W. J. FEAST, D. KILCAST, W. K. R. MUSGRAVE, and W. E. PRESTON, *Nature (London)*, *Phys. Sci.*, **239**, 47 (1972).
3. R. C. ELDERFIELD and T. N. DODD, in *Heterocyclic Compounds* (R. C. Elderfield, ed.), Wiley, New York, 1950, Vol. 1, ch. 4, p. 147.
4. O. W. CASS and H. B. COPELIN, U.S. P. 2,430,667 (1947); *Chem. Abs.*, **42**, 2284 (1948).
5. E. I. du Pont de Nemours & Co., Brit. P. 611,851 (1948); *Chem. Abs.*, **43**, 3041 (1949).
6. A. P. DUNLOP and F. N. PETERS, *The Furans*, Reinhold, New York, 1953, ch. 3.
7. J. J. EISCH, *Adv. Heterocyclic Chem.*, **7**, 1 (1966).
8. H. KRZIKALLA and H. LINGE, Ger. P. 932,612 (1955); *Chem. Abs.*, **52**, 17287 (1958).
9. Badische Anilin- & Soda-Fabrik A.-G., Brit. P. 765,281 (1957); *Chem. Abs.*, **51**, 11389 (1957).
10. N. REEVES, W. J. FEAST, and W. K. R. MUSGRAVE, *Chem. Comm.*, 67 (1970).
11. G. MAASS, *Angew. Chem., Internat. Edn.*, **4**, 787 (1965).
12. H. KRZIKALLA, K. MERKEL, and H. LINGE, Ger. P. 921,385 (1954).
13. M. J. MALAGUTI, *Ann. Chim. Phys.* [2], **64**, 275 (1837).
14. M. J. MALAGUTI, *Ann. Chim. Phys.* [2], **70**, 372 (1839) (article commences on p. 337).
15. A. DENARO, *Gazzetta*, **16**, 333 (1886).
16. H. B. HILL and L. L. JACKSON, *Proc. Amer. Acad. Arts Sci.*, **24**, 320 (1889).
17. H. B. HILL and L. L. JACKSON, *Amer. Chem. J.*, **12**, 22 (1890).
18. H. B. HILL and L. L. JACKSON, *Amer. Chem. J.*, **12**, 112 (1890).
19. H. B. HILL and W. S. HENDRIXSON, *Proc. Amer. Acad. Sci.*, **25**, 283 (1889-1890).
20. H. B. HILL and R. W. CORNELISON, *Amer. Chem. J.*, **16**, 188 (1894).
21. R. J. VAN DER WAL, *Iowa State Coll. J. Sci.*, **11**, 128 (1936) [see also, H. GILMAN, R. A. FRANZ, A. P. HEWLETT, and G. F. WRIGHT, *J. Amer. Chem. Soc.*, **72**, 3 (1950)].
22. W. G. LOWE and C. S. HAMILTON, *J. Amer. Chem. Soc.*, **57**, 2314 (1935).
23. H. GILMAN and R. J. VAN DER WAL, *Rec. Trav. Chim.*, **52**, 267 (1933).
24. E. D. WEIL, E. LEON, and J. LINDER, *J. Org. Chem.*, **26**, 5185 (1961).
25. E. LEON, E. D. WEIL, and J. LINDER, U.S. P. 3,158,624 (1964); *Chem. Abs.*, **62**, 9106 (1965): U.S. P. 3,253,982 (1966); *Chem. Abs.*, **65**, 3835 (1966): U.S. P 3,266,883 (1966); *Chem. Abs.*, **65**, 13659 (1966): U.S. P. 3,354,157 (1967); *Chem. Abs.*, **68**, 104960 (1968).
26. A. F. SHEPARD, N. R. WINSLOW, and J. R. JOHNSON, *J. Amer. Chem. Soc.*, **52**, 2083 (1930).
27. H. KRZIKALLA and H. LINGE, *Chem. Ber.*, **96**, 1751 (1963).
28. H. FEICHTINGER and H.-W. LINDEN, *Chem. Ber.*, **102**, 3573 (1969).
29. T. JAWORSKI, *Roczniki Chem.*, **39**, 767 (1965); *Chem. Abs.*, **63**, 8335 (1965).
30. T. JAWORSKI, *Roczniki Chem.*, **43**, 53 (1969); *Chem. Abs.*, **70**, 115033 (1969).

31. S. DORMAL, E. L. DELVAUX, L. E. DILLS, and D. E. H. FREAR, *J. Econ. Entomol.*, 43, 915 (1950).
32. A. N. GRINEV, PAN BON-KHVAR, V. N. FROSIN, and A. P. TERENT'EV, *Zh. Obshch. Khim.*, 26, 561 (1956); *Chem. Abs.*, 50, 13860 (1956).
33. D. BUCKLEY, S. DUNSTAN, and H. B. HENBEST, *J. Chem. Soc.*, 4880 (1957).
34. E. P. FOKIN and E. P. PRUDCHENKO, *J. Org. Chem. U.S.S.R.*, 6, 1258 (1970).
35. E. P. FOKIN and E. P. PRUDCHENKO, *J. Org. Chem. U.S.S.R.*, 6, 90 (1970).
36. E. P. PRUDCHENKO and E. P. FOKIN. *Izv. Sib. Otd. Akad. Nauk. SSSR., Ser. Khim. Nauk.*, 106 (1971); *Chem. Abs.*, 77, 48119 (1972).
37. R. HUISGEN, H. KÖNIG, G. BINSCH, and H. J. STURM, *Angew. Chem.*, 73, 368 (1961).
38. R. HUISGEN, G. BINSCH, and H. KÖNIG, *Chem. Ber.*, 97, 2884 (1964).
39. G. BINSCH, R. HUISGEN, and H. KÖNIG, *Chem. Ber.*, 97, 2893 (1964).
40. H. HEANEY and C. T. McCARTY, *Chem. Comm.*, 123 (1970).
41. P. A. BERKE and W. E. ROSEN, U.S. P. 3,471,537 (1969); *Chem. Abs.*, 72, 12386 (1970).
42. R. RIEMSCHNEIDER and D. LANGE, *Chem. Ber.*, 97, 300 (1964).
43. H. HEANEY, S. V. LEY, A. P. PRICE, and R. P. SHARMA *Tetrahedron Letters*, 3067 (1972).
44. *Europ. Chem. News*, 21, No. 523, 11 (1972).
45. G. L. CIAMICIAN, *Ber.*, 37, 4200 (1904) (relevant section begins on p. 4218).
46. K. SCHOFIELD, *Hetero-aromatic Nitrogen Compounds; Pyrroles and Pyridines*, Butterworths, London, 1967, pp. 77 and 100, and references cited therein.
47. D. G. DURHAM, C. G. HUGHES, and A. H. REES, *Canad. J. Chem.*, 50, 3223 (1972).
48. Kalle and Co., German P. 38,423; *Chem. Zentralblatt*, 423 (1887).
49. Friedländers Fortschritte der TheerFarbenfabrikation, I, 223 (Berlin, 1888).
50. G. L. CIAMICIAN and P. SILBER, *Ber.*, 17, 1743 (1884); 18, 1763 (1885).
51. G. L. CIAMICIAN and P. SILBER, *Gazz. Chim. Ital.*, 14, 356 (1884); 16, 39 (1886).
52. R. ANSCHÜTZ and G. SCHROETER, *Annalen.*, 295, 67 (1897).
53. G. L. CIAMICIAN and P. SILBER, *Ber.*, 16, 2388 (1883); 17, 553 (1884).
54. G. L. CIAMICIAN and P. SILBER, *Gazz. Chim. Ital.*, 13, 403 (1883); 14, 31 (1884).
55. G. MAZZARA, *Gazz. Chim. Ital.*, 32, I, 510 (1902); 32, II, 28 (1902).
56. G. MAZZARA and A. BORGO, *Gazz. Chim. Ital.*, 34, I, 253 (1904); 34, I, 414 (1904).
57. G. MAZZARA and A. BORGO, *Gazz. Chim. Ital.*, 35, I, 477 (1905); 35, II, 19 (1905).
58. A. P. TERENT'EV and L. A. YANOVSKAYA, *J. Gen. Chem. U.S.S.R.*, 21, 307 (1951).
59. U. COLACICCHI, *Atti Accad. Lincei*, 19, II, 645 (1910); *Chem. Abs.*, 5, 1280 (1911).
60. R. A. NICOLAUS and L. MANGONI, *Gazz. Chim. Ital.*, 86, 757 (1956); *Chem. Abs.*, 52, 2833 (1958).
61. R. A. NICOLAUS, *Rass. med. sper.*, 7, Suppl. No. 2 (1960); *Chem. Abs.*, 54, 24644 (1960).

340   B. IDDON AND H. SUSCHITZKY

62. E. BISAGNI, J.-P. MARQUET, and J. ANDRÉ-LOUISFERT, *Bull. Soc. Chim. France*, 637 (1968).
63. J.-P. MARQUET, J. ANDRÉ-LOUISFERT, and E. BISAGNI, *Compt. Rend.*, **265C**, 1271 (1967).
64. H. FISCHER and E. ELHARDT, Z. *Physiol. Chem.*, 257, 61 (1969).
65. D. G. DURHAM and A. H. REES, *Canad. J. Chem.*, 49, 136 (1971).
66. P. HODGE and R. W. RICKARDS, *J. Chem. Soc.*, 495 (1965).
67. G. MAZZARA and A. BORGO, *Gazz. Chim. Ital.*, 35, II, 104 (1905).
67a. H. EL KHADEM, L. A. KEMLER, Z. M. EL-SHAFEI, M. M. A. ABDEL RAHMAN, and S. EL SADANY, *J. Heterocyclic Chem.*, 9, 1413 (1972).
68. P. L. DE BENNEVILLE and H. W. BLESSING, U.S. P. 3,072,673 (1963); *Chem. Abs.*, 58, 12516 (1963).
69. Farbenfabriken Bayer A.-G., Neth. P. Appl. 6,409,122 (1965); *Chem. Abs.*, 63, 8327 (1965).
70. H. HOLTSCHMIDT, E. DEGENER, H.-G. SCHMELZER, H. TARNOW, and W. ZECHER, *Angew. Chem., Internat. Edn.*, 7, 856 (1968).
71. H. HOLTSCHMIDT, *Angew. Chem., Internat. Edn.*, 1, 632 (1962).
72. R. ANSCHÜTZ and C. BEAVIS, *Annalen.*, 295, 29 (1897); 263, 156 (1891).
73. R. ANSCHÜTZ and A. GUENTHER, *Annalen.*, 295, 43 (1897).
74. R. ANSCHÜTZ and J. MEYERFELD, *Annalen.*, 295, 56 (1897).
75. G. B. BONINO, R. MANZONI-ANSIDEI, and P. PRATESI, Z. *Physik. Chem.*, 25B, 348 (1934); *Chem. Abs.*, 28, 5336 (1934).
76. G. L. CIAMICIAN and P. SILBER, *Ber.*, 19, 3027 (1886).
77. G. MAZZARA, *Gazz. Chim. Ital.*, 34, II, 178 (1904).
78. G. MAZZARA, *Gazz. Chim. Ital.*, 34, I, 482 (1904).
79. H. ULRICH, E. KOBER, H. SCHROEDER, R. RÄTZ, and C. GRUNDMANN, *J. Org. Chem.*, 27, 2585 (1962).
80. R. L. HAMILL, H. R. SULLIVAN, and M. GORMAN, *Appl. Microbiol.*, 18, 310 (1969).
81. R. P. ELANDER, J. A. MABE, R. H. HAMILL, and M. GORMAN, *Appl. Microbiol.*, 16, 753 (1968).
82. P. J. MURPHY and T. L. WILLIAMS, *J. Med. Chem.*, 15, 137 (1972).
83. J. GOSTELI, *Helv. Chim. Acta*, 55, 451 (1972).
84. L. L. MARTIN, C.-J. CHANG, H.-G. FLOSS, J. A. MABE, E. W. HAGAMAN, and E. WENKERT, *J. Amer. Chem. Soc.*, 94, 8942 (1972).
85. J. A. ELIX and M. V. SARGENT, *J. Chem. Soc. (C)*, 1718 (1967).
86. D. M. BAILEY and R. E. JOHNSON, *Tetrahedron Letters*, 3555 (1970).
87. K. HATTORI and M. HASHIMOTO, Japan P. 6746 ('67); *Chem. Abs.*, 67, 90667 (1967).
88. R. J. MOTEKAITIS, D. H. HEINERT, and A. E. MARTELL, *J. Org. Chem.*, 35, 2504 (1970).
89. J. C. POWERS, *J. Org. Chem.*, 31, 2627 (1966), and references cited therein.
90. R. V. HEINZELMAN and J. SZMUSKOVICZ, *Prog. Drug Res.*, 6, 75 (1963).
91. J. SZMUSKOVICZ, *J. Org. Chem.*, 29, 178 (1964).
92. E. E. VAN TAMELEN, J. P. YARDLEY, M. MIYANO, and W. B. HINSHAW, *J. Amer. Chem. Soc.*, 91, 7333 (1969).
93. R. J. SUNDBERG, *The Chemistry of Indoles*, Academic Press, New York, 1970.
94. J. M. MUCHOWSKI, *Canad. J. Chem.*, 48, 422 (1970), and references cited therein.
95. J. A. JANKE, Ph.D. Thesis, University of Minnesota (1970); *Diss. Abs.*, 32B, 168 (1971).

96. R. J. BASS, *Tetrahedron*, 27, 3263 (1971).
97. J. BERGMAN, *Acta Chem. Scand.*, 25, 2865 (1971).
98. J. C. POWERS, in *The Chemistry of Heterocyclic Compounds (Vol. 25): Indoles; Part Two* (W. J. Houlihan, A. Weissberger, and E. C. Taylor, eds), Interscience, New York, 1972, ch. V, p. 127.
99. P. G. GASSMAN, G. A. CAMPBELL, and G. MEHTA, *Tetrahedron*, 28, 2749 (1972).
100. R. D. CHAMBERS, R. A. STOREY, and W. K. R. MUSGRAVE, Brit. P. 1,177,628 (1970); *Chem. Abs.*, 72, 100501 (1970).
101. R. D. CHAMBERS, B. IDDON, W. K. R. MUSGRAVE, and R. A. STOREY, unpublished results.
102. Farbenfabriken Bayer A.-G., Belg. P. 673,779 (1964).
103. H. HOLTSCHMIDT and H. TARNOW, Brit. P. 1,135,216 (1968).
104. I. COLLINS, S. M. ROBERTS, and H. SUSCHITZKY, *J. Chem. Soc. (C)*, 167 (1971).
105. W. RIED and P. WIEDEMANN, *Chem. Ber.*, 102, 2684 (1969).
106. S. J. HOLT and P. W. SADLER, *Proc. Royal Soc.*, 148B, 481 (1958).
107. K. H. BÜCHEL and W. DRABER, *Prog. Photosynthesis Res.*, 3, 1777 (1969).
108. W. E. ROSEN, V. P. TOOHEY, and A. C. SHABICA, *J. Amer. Chem. Soc.*, 79, 3167 (1957).
109. W. E. ROSEN, V. P. TOOHEY, and A. C. SHABICA, *J. Amer. Chem. Soc.*, 80, 935 (1958).
110. L. WEITZ, *Ber.*, 17, 792 (1884).
111. H. D. HARTOUGH, in *Thiophene and Its Derivatives* (A. Weissberger, ed.), Interscience, New York, 1952, ch. 7, p. 185.
112. H. ULRICH, E. KOBER, R. RÄTZ, H. SCHROEDER, and C. GRUNDMANN, *J. Org. Chem.*, 27, 2593 (1962).
113. W. STEINKOPF and W. KOHLER, *Annalen.*, 532, 250 (1937).
114. H. L. COONRADT and H. D. HARTOUGH, *J. Amer. Chem. Soc.*, 70, 1158 (1948).
115. H. L. COONRADT, H. D. HARTOUGH, and G. C. JOHNSON, *J. Amer. Chem. Soc.*, 70, 2564 (1948).
116. H. L. COONRADT and H. D. HARTOUGH, U.S. P. 2,492,623 (1949); *Chem. Abs.*, 44, 2567 (1950).
117. H. L. COONRADT and H. D. HARTOUGH, U.S. P. 2,492,624 (1949); *Chem. Abs.*, 44, 2567 (1950).
118. H. L. COONRADT, H. D. HARTOUGH, and H. D. NORRIS, *J. Amer. Chem. Soc.*, 74, 163 (1952).
119. H. D. NORRIS and J. H. McCRACKEN, U.S. P. 2,504,084 (1950); *Chem. Abs.*, 44, 7884 (1950).
120. A. N. AKOPYAN, A. M. SAAKYAN, and Z. A. DZHAUARI, *Arm. Khim. Zh.*, 22, 889 (1969); *Chem. Abs.*, 72, 31527 (1970).
121. H. D. NORRIS and J. H. McCRACKEN, U.S. P. 2,504,085 (1950); *Chem. Abs.*, 44, 7884 (1950).
122. M. HAUPTSCHEIN and V. MARK, U.S. P. 3,364,233 (1968); *Chem. Abs.*, 69, 10350 (1968).
123. H. L. COONRADT and H. D. HARTOUGH, U.S. P. 2,504,068 (1950); *Chem. Abs.*, 44, 7884 (1950).
124. E. R. OSGOOD, L. E. LIMPEL, R. L. ANNIS, and N. J. TURNER, U.S. P. 3,354,179 (1967); *Chem. Abs.*, 68, 104965 (1968).
125. G. McCOY, C. E. INMAN, and G. D. KYKER, U.S. P. 2,914,573 (1959); *Chem. Abs.*, 54, 5430 (1960).
126. G. McCOY, C. E. INMAN, and G. D. KYKER, U.S. P. 2,955,142 (1960); *Chem. Abs.*, 55, 2567 (1961).

127. G. McCOY, C. E. INMAN, and G. D. KYKER, U.S. P. 2,851,464 (1958); *Chem. Abs.*, 53, 3242 (1959).
128. E. J. GEERING, *J. Org. Chem.*, 24, 1128 (1959).
129. E. J. GEERING, U.S. P. 2,900,394 (1959); *Chem. Abs.*, 54, 572 (1960).
130. R. N. HASZELDINE, R. E. BANKS, and J. M. BIRCHALL, Brit. P. 1,069,943 (1967); *Chem. Abs.*, 67, 73518 (1967).
131. R. H. GOSHORN and T. E. DEGER, U.S. P. 3,350,410 (1967); *Chem. Abs.*, 68, 95668 (1968).
132. A. N. AKOPYAN, A. M. SAAKYAN, and Z. A. DZHAUARI, *Arm. Khim. Zh.*, 21, 414 (1968); *Chem. Abs.*, 70, 47195 (1969).
133. YU. A. OL'DEKOP and R. V. KABERDIN, *Vesti Akad. Navuk Belarus. S.S.R., Ser. Khim. Navuk*, 131 (1972); *Chem. Abs.*, 77, 164352 (1972).
134. S. GRONOWITZ and B. MALTESSON, *Acta Chem. Scand.*, 26, 2982 (1972).
135. S. NAKAGAWA, J. OKUMURA, F. SAKAI, H. HOSHI, and T. NAITO, *Tetrahedron Letters*, 3719 (1970).
136. H. BLUESTONE, U.S. P. 3,073,691 (1963); *Chem. Abs.*, 59, 576 (1963).
137. W. MACK, *Angew. Chem., Internat. Edn.*, 6, 1083 (1967).
138. Consortium für Elektrochemische Industrie G.m.b.H., Brit. P. 1,191,088 (1970); *Chem. Abs.*, 73, 77038 (1970).
139. O. SCHERER and F. KLUGE, *Chem. Ber.*, 99, 1973 (1966).
140. W. STEINKOPF, H. JACOB, and H. PENZ, *Annalen.*, 512, 136 (1934).
141. G. B. BACHMAN and L. V. HEISEY, *J. Amer. Chem. Soc.*, 70, 2378 (1948).
142. M. T. RAHMAN, M. R. SMITH, A. F. WEBB, and H. GILMAN, *Organometal. Chem. Syn.*, 1, 105 (1970/71).
143. M. D. RAUSCH, T. R. CRISWELL, and A. K. IGNATOWICZ, *J. Organometal. Chem.*, 13, 419 (1968).
144. M. NILSSON and C. ULLENIUS, *Acta Chem. Scand.*, 25, 2428 (1971).
145. M. R. SMITH and H. GILMAN, *Organometal. Chem. Syn.*, 1, 265 (1970/71).
146. I. HAIDUC and H. GILMAN, *Rev. Roumaine Chim.*, 16, 305 (1971).
147. M. R. SMITH and H. GILMAN, *J. Organometal. Chem.*, 42, 1 (1972).
148. D. S. SETHI, M. R. SMITH, and H. GILMAN, *J. Organometal. Chem.*, 24, C41 (1970).
149. M. R. SMITH and H. GILMAN, *J. Organometal. Chem.*, 37, 35 (1972).
150. M. R. SMITH, M. T. RAHMAN, and H. GILMAN, *Organometal. Chem. Syn.*, 1, 295 (1970/71).
151. J. BURDON, I. W. PARSONS, and J. C. TATLOW, *J. Chem. Soc. (C)*, 346 (1971).
152. D. PILLON and TRINH VAN QUY, Fr. P. 1,563,736 (1969); *Chem. Abs.*, 72, 111289 (1970).
153. TRINH VAN QUY and D. PILLON, Ger. P. Offen., 1,813,194 (1969); *Chem. Abs.*, 72, 21605 (1970).
154. D. PILLON, S. TRINH, and R. CAVIER, *Chim. Therap.*, 5, 32 (1970).
155. D. PILLON and TRINH VAN QUY, Fr. P. 1,563,735 (1969); *Chem. Abs.*, 72, 66797 (1970).
156. TRINH VAN QUY and D. PILLON, Ger. P. Offen. 1,813,195 (1969); *Chem. Abs.*, 72, 12555 (1970).
157. J. F. FREEMAN, J. R. DOVE, and E. D. AMSTUTZ, *J. Amer. Chem. Soc.*, 70, 3136 (1948).
158. O. EBERHARD, *Ber.*, 28, 2385 (1895).
159. W. STEINKOPF, R. LEITSMANN, A. H. MÜLLER, and H. WILHELM, *Annalen.*, 541, 260 (1939).

160. W. STEINKOPF and (in part) H.-J. VON PETERSDORFF, *Annalen.*, 543, 128 (1940).
161. W. STEINKOPF, R. LEITSMANN, and K. H. HOFMANN, *Annalen.*, 546, 180 (1941).
162. G. BARGER and A. J. EWINS, *J. Chem. Soc.*, 93, 2086 (1908).
163. D. HUDSON and B. IDDON, University of Salford, unpublished results.
164. A. H. SCHLESINGER and D. T. MOWRY, *J. Amer. Chem. Soc.*, 73, 2614 (1951).
165. E. ROOS and K. WAGNER, Ger. P. Offen. 1,902,050 (1970); *Chem. Abs.*, 73, 87909 (1970).
166. N. KHARASCH and Z. S. ARIYAN, *Chem. Ind. (London)*, 302 (1965).
167. H. KLUG, Ger. P. 1,222,508 (1966); *Chem. Abs.*, 65, 13727 (1966).
168. W. J. SELL and F. W. DOOTSON, *J. Chem. Soc.*, 71, 1068 (1897).
169. W. J. SELL and F. W. DOOTSON, *J. Chem. Soc.*, 73, 432 (1898).
170. E. AGER and H. SUSCHITZKY, *J. Fluorine Chem.*, 3, 230 (1973/74); *J. C. S., Perkin I*, 2839 (1973).
171. A. J. COPSON, H. HEANEY, A. A. LOGUN, and R. P. SHARMA, *Chem. Comm.*, 315 (1972).
172. G. E. CHIVERS and H. SUSCHITZKY, *J. Chem. Soc. (C)*, 2867 (1971).
173. S. L. BELL, R. D. CHAMBERS, W. K. R. MUSGRAVE, and J. G. THORPE, *J. Fluorine Chem.*, 1, 51 (1971).
174. H. T. HARRISON, U.S. P. 3,288,796 (1966); *Chem. Abs.* 66, 55392 (1967).
175. D. KYRIACOU, U.S. P. 3,668,209 (1972); *Chem. Abs.*, 77, 61827 (1972).
176. J. A. LADD and V. I. P. JONES, *Spectrochim. Acta*, 23A 2791 (1967).
177. J. D. COOK and B. J. WAKEFIELD, *J. Organometal. Chem.*, 13, 15 (1968).
178. J. D. COOK and B. J. WAKEFIELD, *J. Chem. Soc. (C)*, 1973 (1969).
179. J. D. COOK and B. J. WAKEFIELD, *J. Chem. Soc. (C)*, 2376 (1969).
180. S. M. ROBERTS and H. SUSCHITZKY, *Chem. Comm.*, 893 (1967).
181. E. A. C. LUCKEN and C. MAZELINE, in *Nuclear Magnetic Resonance and Relaxation in Solids; Proceedings of the XIIIth Colloque Ampère, University of Leuven, September, 1964* (L. van Goerven, ed.), North Holland Pub. Co., Amsterdam, 1965, p. 235.
182. A. R. KATRITZKY, J. D. ROWE, and S. K. ROY, *J. Chem. Soc. (B)*, 758 (1967).
183. C. D. S. TOMLIN, I.C.I. Plant Protection Ltd., Jealott's Hill, personal communication.
184. R. D. CHAMBERS, J. HUTCHINSON, and W. K. R. MUSGRAVE, *J. Chem. Soc.*, 5634 (1964).
185. F. P. BOER, J. W. TURLEY, and F. P. VAN REMOORTERE, *Chem. Comm.*, 573 (1972).
185a. S. S. T. KING, W. L. DILLING, and N. B. TEFERTILLER, *Tetrahedron*, 28, 5859 (1972).
186. R. T. BAILEY and G. P. STRACHAN, *Spectrochim. Acta*, 26A, 1129 (1970).
187. H. E. MERTEL, in *Pyridine and Its Derivatives* (E. Klingsberg, ed.), Interscience, New York, 1961, Part 2, ch. VI, p. 299.
188. H. S. MOSHER, in *Heterocyclic Compounds* (R. C. Elderfield, ed.), Wiley, New York, 1950, Vol. I, p. 504.
189. YA. N. IVASHCHENKO and S. D. MOSHCHITSKII, *Khim. Prom. Ukr.*, 56 (1967); *Chem. Abs.*, 67, 21785 (1967).
190. R. D. BOWDEN, I.C.I. Mond Division, Runcorn, has computed a list of these. We thank him for allowing us to see this information.

344   B. IDDON AND H. SUSCHITZKY

191. H. J. DEN HERTOG, J. C. M. SCHOGT, J. DE BRUYN, and A. DE KLERK, *Rec. Trav. Chim.*, 69, 673 (1950).
192. H. J. DEN HERTOG, J. P. WIBAUT, and P. H. VAN DER LEY, *Rec. Trav. Chim.*, 51, 387 (1932).
193. J. P. WIBAUT and J. R. NICOLAI, *Rec. Trav. Chim.*, 58, 709 (1939).
194. J. P. WIBAUT and H. J. DEN HERTOG, U.S. P. 1,977,662 (1934); *Chem. Abs.*, 29, 178 (1935).
195. I.C.I. Ltd., Belg. P. 659,475 (1965); *Chem. Abs.*, 64, 2064 (1966): J. A. CORRAN, Brit. P. 1,041,906 (1966): J. A. CORRAN, Ger. P. 1,545,984 (1970); *Chem. Abs.* 73, 45361 (1970).
196. W. H. TAPLIN, U.S. P. 3,420,833 (1969); *Chem. Abs.*, 71, 3279 (1969).
197. R. M. BIMBER, U.S. P. 3,325,503 (1967); *Chem. Abs.*, 68, 68896 (1968).
198. R. D. BOWDEN, Brit. P. 1,276,253 (1972); *Chem. Abs.*, 77, 61830 (1972).
199. R. D. BOWDEN, M. B. GREEN, and T. SEATON, *The Kinetics and Mechanism of the Gas-Phase Chlorination of Pyridine*, paper presented at the International Symposium on Organic Polyhalogen Compounds, University of Salford, April, 1971.
200. W. J. SELL and F. W. DOOTSON, *J. Chem. Soc.*, 73, 442 (1898).
201. W. J. SELL, *J. Chem. Soc.*, 99, 1679 (1911).
202. W. J. SELL, *J. Chem. Soc.*, 87, 799 (1905).
203. W. J. SELL and F. W. DOOTSON, *J. Chem. Soc.*, 75, 979 (1899).
204. W. J. SELL, *J. Chem. Soc.*, 93, 1997 (1908).
205. W. J. SELL, *J. Chem. Soc.*, 93, 2001 (1908).
206. W. J. SELL and F. W. DOOTSON, *J. Chem. Soc.*, 77, 771 (1900).
207. W. J. SELL and F. W. DOOTSON, *J. Chem. Soc.*, 79, 899 (1901).
208. H. JOHNSTON, U.S. P. 3,555,032 (1971); *Chem. Abs.*, 75, 5727 (1971).
209. E. SMITH, U.S. P. 3,538,100 (1970); *Chem. Abs.*, 74, 53542 (1971).
210. W. J. SELL and F. W. DOOTSON, *J. Chem. Soc.*, 77, 233 (1900).
211. W. W. CROUCH and H. L. LOCHTE, *J. Amer. Chem. Soc.*, 65, 270 (1943).
212 W. J. SELL and F. W. DOOTSON, *J. Chem. Soc.*, 77, 1 (1900).
213. R. D. CHAMBERS, J. HUTCHINSON, and W. K. R. MUSGRAVE, *Proc. Chem. Soc.*, 83 (1964); *J. Chem. Soc.*, 3573 (1964).
214. Imperial Smelting Corp. (N.S.C.) Ltd., Belg. P. 660,907 (1965); *Chem. Abs.*, 64, 3568 (1966).
215. R. E. BANKS, R. N. HASZELDINE, J. V. LATHAM, and I. M. YOUNG, *J. Chem. Soc.*, 594 (1965).
216. R. N. HASZELDINE and R. E. BANKS, Brit. P. 1,039,987 (1966); *Chem. Abs.*, 65, 15341 (1966).
217. D. M. W. VAN DEN HAM, *Chem. Ind. (London)*, 730 (1972).
218. R. E. BANKS, R. N. HASZELDINE, and I. M. YOUNG, *J. Chem. Soc. (C)*, 2089 (1967).
219. R. D. CHAMBERS, D. LOMAS, and W. K. R. MUSGRAVE, Brit. P. 1,163,472 (1969); *Chem. Abs.*, 71, 124269 (1969).
220. R. D. CHAMBERS, D. LOMAS, and W. K. R. MUSGRAVE, *Tetrahedron*, 24, 5633 (1968).
221. T. BATKOWSKI, D. TOMASIK, and P. TOMASIK, *Roczniki Chem.*, 41, 2101 (1967); *Chem. Abs.*, 69, 18984 (1968).
222. W. J. SELL, *J. Chem. Soc.*, 93, 1993 (1908).
223. E. T. McBEE, H. B. HASS, and E. M. HODNETT, *Ind. Eng. Chem.*, 39, 389 (1947).
224. C. H. BRETT and E. M. HODNETT, U.S. P. 2,679,453 (1954); *Chem. Abs.*, 48, 9011 (1954).

225. H. JOHNSTON, M. S. TOMITA, F. H. NORTON, and W. H. TAPLIN, Belg. P. 624,800 (1963); *Chem. Abs.*, 61, 1841 (1964).
226. C. T. REDEMANN, Belg. P. 628,486 (1963); *Chem. Abs.*, 60, 15840 (1964).
227. Dow Chem. Co., Brit. P. 957,276 (1964).
228. H. JOHNSTON, F. H. NORTON, and M. S. TOMITA, U.S. P. 3,173,919 (1965); *Chem. Abs.*, 62, 14638 (1965).
229. F. H. NORTON and W. H. TAPLIN, U.S. P. 3,256,167 (1966); *Chem. Abs.*, 65, 8882 (1966).
230. H. JOHNSTON and M. S. TOMITA, U.S. P. 3,224,950 (1965); *Chem. Abs.*, 64, 6623 (1966).
231. Dow Chem. Co., Brit. P. 991,526 (1965); *Chem. Abs.*, 63, 9921 (1965): H. JOHNSTON and M. S. TOMITA, U.S. P. 3,186,994 (1965).
232. P. B. DOMENICO, U.S. P. 3,629,281 (1971); *Chem. Abs.*, 76, 72409 (1972): U.S. P. 3,699,108 (1972); *Chem. Abs.*, 78, 16043 (1973): U.S. P. 3,706,751 (1972); *Chem. Abs.*, 78, 97504 (1973).
232a. F. E. TORBA, U.S. P. 3,705,170 (1972); *Chem. Abs.*, 78, 58253 (1973): U.S. P. 3,711,486 (1973); *Chem. Abs.*, 78, 84271 (1973).
233. F. DOERING and G. BECK, Ger. P. Offen. 2,036,174 (1972); *Chem. Abs.*, 76, 140542 (1972).
234. H. JOHNSTON, U.S. P. 3,686,193 (1972); *Chem. Abs.*, 77, 139827 (1972).
235. B. BOBRAŃSKI, L. KOCHAŃSKA, and A. KOWALEWSKA, *Ber.*, 71, 2385 (1938).
236. C. R. KOLDER and H. J. DEN HERTOG, *Rec. Trav. Chim.*, 72, 285 (1953).
237. H. J. DEN HERTOG, J. MAAS, C. R. KOLDER, and W. P. COMBÉ, *Rec. Trav. Chim.*, 74, 59 (1955).
238. H. N. STOKES and H. VON PECHMAN, *Amer. Chem. J.*, 8, 377 (1886).
239. W. J. SELL and F. W. DOOTSON, *J. Chem. Soc.*, 73, 777 (1898).
240. E. KNUESLI and H. GYSIN, Swiss P. 384,929 (1965); *Chem. Abs.*, 62, 16203 (1965).
241. H. J. DEN HERTOG and J. DE BRUYN, *Rec. Trav. Chim.*, 70, 182 (1951).
242. B. H. CHASE and J. WALKER, *J. Chem. Soc.*, 3548 (1953).
243. S. D. MOSHCHITSKII, G. A. ZALESSKII, A. PAVLENKO, and YA. N. IVASHCHENKO, *Khim. Geterotsikl. Soedin.*, 791 (1970); *Chem. Abs.*, 73, 120466 (1970).
244. C. R. KOLDER and H. J. DEN HERTOG, *Rec. Trav. Chim.*, 72, 853 (1953).
244a. J. A. ELVIDGE and N. A. ZAIDI, *J. Chem. Soc.* (*C*), 2188 (1968).
245. H. J. den HERTOG and J. C. M. SCHOGT, *Rec. Trav. Chim.*, 70, 353 (1951).
246. S. D. MOSHCHITSKII, L. S. SOLUGUB, and YA. N. IVASHCHENKO, *Khim. Geterotsikl. Soedin.*, 1068 (1968); *Chem. Abs.*, 70, 77738 (1969).
247. W. T. FLOWERS, R. N. HASZELDINE, and S. A. MAJID, *Tetrahedron Letters*, 2503 (1967).
248. A. ROEDIG and K. GROHE, *Chem. Ber.*, 98, 923 (1965).
249. O. VON SCHICKH, A. BINZ, and A. SCHULZ, *Ber.*, 69, 2593 (1936).
250. G. BECK, H. HOLTSCHMIDT, and H. HEITZER, *Annalen.*, 731, 45 (1970): Farbenfabriken Bayer A.-G., Fr. P. Demande 2,015,602 (1970); *Chem. Abs.*, 74, 125449 (1971).
251. Farbenfabriken Bayer A.-G., Neth. P. Appl. 6,516,622 (1966); *Chem. Abs.*, 65, 20152 (1966): W. ZECHER, H. TARNOW, and H. HOLT-SCHMIDT, Ger. P., 1,222,918 (1966).

252. H. JOHNSTON and S. H. RUETMAN, Ger. P. Offen, 1,911,023 (1970); *Chem. Abs.*, 74, 31692 (1971).
253. H. HOLTSCHMIDT and W. ZECHER, Belg. P., 622,382 (1962); *Chem. Abs.*, 59, 11534 (1963).
254. T. A. MAGEE, Ger. P. Offen. 2,116,548 (1971); *Chem. Abs.*, 76, 25109 (1972).
255. M. J. MARINAK, U.S. P. 3,532,701 (1970); *Chem. Abs.*, 74, 53558 (1971).
256. H. JOHNSTON, M. J. MARINAK, and S. H. RUETMAN, U.S. P. 3,592,817 (1971); *Chem. Abs.*, 75, 88492 (1971).
257. R. N. HASZELDINE, R. E. BANKS, and J. M. BIRCHALL, U.S. P. 3,359,267 (1967); *Chem. Abs.*, 68, 105015 (1968).
258. R. D. BOWDEN and T. SEATON, Ger. P. Offen. 2,141,632 (1972); *Chem. Abs.*, 77, 5361 (1972).
259. T. JAWORSKI and B. KORYBUT-DASZKIEWICZ, *Roczniki Chem.*, 41, 1521 (1967); *Chem. Abs.*, 69, 2822 (1968).
260. T. JAWORSKI and W. POLACZKOWA, *Roczniki Chem.*, 31, 1337 (1957); *Chem. Abs.*, 52, 11037 (1958).
261. R. G. PEWS, E. B. NYQUIST, and F. P. CORSON, *J. Org. Chem.*, 35, 4096 (1970).
262. A. ROEDIG and G. MÄRKL, *Annalen.*, 659, 1(1962).
263. A. ROEDIG, R. KOHLHAUPT, and G. MÄRKL, *Chem. Ber.*, 99, 698 (1966).
264. A. ROEDIG and G. MÄRKL, *Annalen.*, 636, 1 (1960).
265. A. ROEDIG, G. MÄRKL, W. RUCH, H.-G. KLEPPE, R. KOHLHAUPT, and H. SCHALLER, *Annalen.*, 692, 83 (1966).
266. A. ROEDIG, G. MÄRKL, and H. SCHALLER, *Chem. Ber.*, 103, 1022 (1970).
267. A. ROEDIG, K. GROHE, D. KLATT, and H.-G. KLEPPE, *Chem. Ber.*, 99, 2813 (1966).
268. A. ROEDIG, K. GROHE, and D. KLATT, *Chem. Ber.*, 99, 2818 (1966).
269. E. AGER, B. IDDON, and H. SUSCHITZKY, *Tetrahedron Letters*, 1507 (1969).
270. E. AGER, B. IDDON, and H. SUSCHITZKY, *J. Chem. Soc.* (C), 193 (1970).
271. YA. N. IVASHCHENKO, S. D. MOSHCHITSKII, L. S. SOLOGUB, and G. A. ZALESSKII, *Khim. Geterotsikl. Soedin.*, 963 (1970); *Chem. Abs.*, 75, 48847 (1971): YA. N. IVASHCHENKO, A. V. KIRSANOV, and S. D. MOSHCHITSKII, U.S.S.R. P. 287,942 (1970); *Chem. Abs.*, 74, 125464 (1971).
272. R. ROBERTS, Ger. P. Offen. 1,949,424 (1970); *Chem. Abs.*, 72, 132542 (1970).
273. I.C.I. Ltd., Neth. P. Appl. 6,611,766 (1967); *Chem. Abs.*, 68, 59438 (1968).
274. C. D. S. TOMLIN, J. W. SLATER, and D. HARTLEY, Brit. P. 1,161,492 (1969); *Chem. Abs.*, 71, 91313 (1969).
275. A. NICOLSON, Ger. P. Offen. 2,128,540 (1971); *Chem. Abs.*, 76, 59469 (1972).
276. H. C. FIELDING, Brit. P. 1,198,476 (1970); *Chem. Abs.*, 73, 98807 (1970).
277. R. D. BOWDEN, I.C.I. Mond Division, personal communication.
278. I.C.I. Ltd., Neth. P. Appl. 6,517,158 (1966); *Chem. Abs.*, 65, 18565 (1966).
279. H. C. FIELDING, L. P. GALLIMORE, H. L. ROBERTS, and B. TITTLE, *J. Chem. Soc.* (C), 2142 (1966).

280. G. DIPROSE and R. D. HOWARD, Ger. P. Offen. 2,006,607 (1970); *Chem. Abs.*, 73, 98802 (1970).
281. G. DIPROSE and R. D. HOWARD, Brit. P. 1,234,543 (1971).
282. S. C. CARSON and R. D. HOWARD, Brit. P. 1,272,475 (1972); *Chem. Abs.*, 77, 126439 (1972).
283. F. E. TORBA, Ger. P. Offen. 1,816,685 (1969); *Chem. Abs.*, 72, 12595 (1970).
284. YA. N. IVASHCHENKO, L. S. SOLOGUB, S. D. MOSHCHITSKII, and A. V. KIRSANOV, *J. Gen. Chem. U.S.S.R.*, 39, 1662 (1969).
284a. G. N. SHIBANOV and Y. YA. KAKLYUGIN, U.S.S.R. P. 348,560 (1972); *Chem. Abs.*, 78, 4132 (1973).
285. S. M. ROBERTS and H. SUSCHITZKY, *J. Chem. Soc.* (C), 2844 (1968).
286. S. M. ROBERTS and H. SUSCHITZKY, *J. Chem. Soc.* (C), 1537 (1968).
287. YA. N. IVASHCHENKO, S. D. MOSHCHITSKII, and G. A. ZALESSKII, *Khim. Geterotsikl. Soedin.*, 959 (1970); *Chem. Abs.*, 74, 12962 (1971).
288. H. JOHNSTON and M. S. TOMITA, U.S. P. 3,291,804 (1966); *Chem. Abs.*, 66, 115612 (1967).
289. H. JOHNSTON, U.S. P. 3,364,223 (1968); *Chem. Abs.*, 69, 27254 (1968).
290. I.C.I. Ltd., Neth. P. Appl., 6,516,409 (1966); *Chem. Abs.*, 65, 18564 (1966).
291. C. D. S. TOMLIN, J. W. SLATER, D. HARTLEY, and C. J. CLAYTON, Brit. P. 1,059,990 (1967).
292. L. LEVINE, U.S. P. 3,475,441 (1969); *Chem. Abs.*, 72, 31628 (1970).
293. H. JOHNSTON, U.S. P. 3,549,647 (1970); *Chem. Abs.*, 75, 5715 (1971).
294. L. LEVINE, U.S. P. 3,641,033 (1972); *Chem. Abs.*, 76, 140888 (1972).
295. J. E. DUNBAR and J. W. ZEMBA, U.S. P. 3,674,795 (1972); *Chem. Abs.*, 77, 88316 (1972).
296. E. AGER, B. IDDON, and H. SUSCHITZKY, *J.C.S., Perkin I*, 133 (1972).
297. W. J. SELL and F. W. DOOTSON, *J. Chem. Soc.*, 83, 396 (1903).
298. W. J. SELL, *J. Chem. Soc.*, 101, 1193 (1912).
299. W. J. SELL, *J. Chem. Soc.*, 101, 1945 (1912).
300. D. J. BERRY, B. J. WAKEFIELD, and J. D. COOK, *J. Chem. Soc.* (C), 1227 (1971).
301. E. AGER, G. E. CHIVERS, and H. SUSCHITZKY, *Chem. Comm.*, 505 (1972).
302. E. AGER, G. E. CHIVERS, and H. SUSCHITZKY, *J.C.S., Perkin I*, 1125 (1973).
303. G. C. FINGER, L. D. STARR, D. R. DICKERSON, H. S. GUTOWSKY, and J. HAMER, *J. Org. Chem.*, 28, 1666 (1963).
304. Dow Chem. Co., Brit. P. 957,277 (1964).
305. H. JOHNSTON and M. S. TOMITA, U.S. P. 3,244,722 (1966); *Chem. Abs.*, 65, 8884 (1966).
306. H. JOHNSTON and M. S. TOMITA, Belg. P. 628,487 (1963); *Chem. Abs.*, 61, 1838 (1964).
307. Dow Chem. Co., Brit. P. 957,831 (1964).
308. H. JOHNSTON and M. S. TOMITA, U.S. P. 3,285,925 (1966); *Chem. Abs.*, 66, 46338 (1967).
309. R. M. BIMBER and P. H. SCHULDT, U.S. P. 3,637,716 (1972); *Chem. Abs.*, 76, 126800 (1972).
310. C. E. GRANITO, U.S. P. 3,651,070 (1972); *Chem. Abs.*, 77, 34348 (1972).
311. F. BINNS, S. M. ROBERTS, and H. SUSCHITZKY, *Chem. Comm.*, 1211 (1969).
312. F. BINNS, S. M. ROBERTS, and H. SUSCHITZKY, *J. Chem. Soc.* (C), 1375 (1970).

313. F. BINNS and H. SUSCHITZKY, *Chem. Comm.*, 750 (1970).
314. F. BINNS and H. SUSCHITZKY, *J. Chem. Soc.* (C), 1223 (1971).
315. D. E. BUBLITZ, *J. Heterocyclic Chem.*, 9, 471 (1972).
316. D. E. BUBLITZ, personal communication.
317. S. M. ROBERTS and H. SUSCHITZKY, *J. Chem. Soc.* (C), 1485 (1969).
318. R. D. BOWDEN, M. B. GREEN, and G. T. BROWN, Ger. P. Offen, 2,130,409 (1972); *Chem. Abs.*, 76, 140539 (1972).
319. R. D. BOWDEN, M. B. GREEN, and G. T. BROWN, Ger. P. Offen. 2,127,901 (1972); *Chem. Abs.*, 76, 153615 (1972).
320. B. IDDON, H. SUSCHITZKY, and A. THOMPSON (University of Salford), unpublished results.
321. S. S. DUA and H. GILMAN, *J. Organometal. Chem.*, 12, 234 (1968).
322. S. S. DUA, A. E. JUKES, and H. GILMAN, *Org. Prepns. Proc.*, 1, 187 (1969).
323. I. F. MIKHAILOVA and V. A. BARKHASH, *J. Gen. Chem. U.S.S.R.*, 37, 2662 (1967).
324. R. C. EDMONDSON, A. E. JUKES, and H. GILMAN, *J. Organometal. Chem.*, 25, 273 (1970).
325. S. S. DUA and H. GILMAN, *J. Organometal. Chem.*, 12, 299 (1968).
326. S. S. DUA, R. C. EDMONDSON, and H. GILMAN, *J. Organometal. Chem.*, 27, 33 (1971).
327. S. S. DUA, A. E. JUKES, and H. GILMAN, *J. Organometal. Chem.*, 12, P24 (1968).
328. A. E. JUKES, S. S. DUA, and H. GILMAN, *J. Organometal. Chem.*, 21, 241 (1970).
329. A. E. JUKES, S. S. DUA, and H. GILMAN, *J. Organometal. Chem.*, 24, 791 (1970).
330. YA. N. IVASHCHENKO, S. D. MOSHCHITSKII, and A. K. ELISEEVA, *Khim. Geterotsikl. Soedin.*, 58, (1970); *Chem. Abs.*, 72, 100451 (1970).
331. R. A. FERNANDEZ, H. HEANEY, J. M. JABLONSKI, K. G. MASON, and T. J. WARD, *J. Chem. Soc.* (C), 1908 (1969).
332. R. D. CHAMBERS, J. HUTCHINSON, and W. K. R. MUSGRAVE, *J. Chem. Soc.*, 3736 (1964).
333. R. E. BANKS, J. E. BURGESS, W. M. CHENG, and R. N. HASZELDINE, *J. Chem. Soc.*, 575 (1965).
334. R. D. CHAMBERS, B. IDDON, W. K. R. MUSGRAVE, and L. CHADWICK, *Tetrahedron*, 24, 877 (1968).
335. J. D. COOK, B. J. WAKEFIELD, and C. J. CLAYTON, *Chem. Comm.*, 150 (1967).
336. I. HAIDUC and H. GILMAN, *Rev. Roumaine Chim.*, 16, 597 (1971).
337. D. J. BERRY, J. D. COOK, and B. J. WAKEFIELD, *Chem. Comm.*, 1273 (1969).
338. D. J. BERRY, J. D. COOK, and B. J. WAKEFIELD, *J. C. S.*, *Perkin I*, 2190 (1972).
339. J. D. COOK, N. J. FOULGER, and B. J. WAKEFIELD, *J. C. S.*, *Perkin I*, 995 (1972).
340. S. S. DUA, R. C. EDMONDSON, and H. GILMAN, *J. Organometal. Chem.*, 24, 703 (1970).
341. D. J. BERRY and B. J. WAKEFIELD, *J. Organometal. Chem.*, 23, 1 (1970).
342. J. D. COOK and B. J. WAKEFIELD, *Chem. Comm.*, 297 (1968).
343. C. D. S. TOMLIN and A. S. MANGALJI, Brit. P. 1,241,869 (1971); *Chem. Abs.*, 75, 118235 (1971).
344. J. D. COOK and B. J. WAKEFIELD, *Tetrahedron Letters*, 2535 (1967).

345. J. D. COOK, B. J. WAKEFIELD, H. HEANEY, and J. M. JABLONSKI, *J. Chem. Soc. (C)*, 2727 (1968).
346. R. D. CHAMBERS, F. G. DRAKESMITH, J. HUTCHINSON, and W. K. R. MUSGRAVE, *Tetrahedron Letters*, 1705 (1967).
347. A. E. JUKES, S. S. DUA, and H. GILMAN, *J. Organometal. Chem.*, 12, P44 (1968).
348. J. N. SEIBER, *J. Org. Chem.*, 36, 2000 (1971).
348a. V. D. PARKER, U.S. P. 3,694,332 (1972); *Chem. Abs.*, 77, 164492 (1972).
349. H. OST, *J. Prakt. Chem.* [2], 27, 257 (1883).
350. R. D. CHAMBERS, F. G. DRAKESMITH, and W. K. R. MUSGRAVE, *J. Chem. Soc.*, 5045 (1965).
351. K. J. SAUNDERS, Dip. Tech. Dissertation, University of Salford (1967).
352. E. SMITH, U.S. P. 3,357,984 (1967); *Chem. Abs.*, 68, 105008 (1968).
353. G. E. CHIVERS and H. SUSCHITZKY, *Chem. Comm.*, 28 (1971).
354. A. POLLACK, B. STANOVINIK, and M. TIŠLER, *J. Org. Chem.*, 35, 2478 (1970).
355. Dow Chem. Co., Neth. P. Appl. 6,515,950 (1966); *Chem. Abs.*, 65, 15338 (1966).
356. H. JOHNSTON, Brit. P. 1,103,606 (1968).
357. H. JOHNSTON, U.S. P. 3,296,272 (1967); *Chem. Abs.*, 66, 104903 (1967).
358. H. JOHNSTON, U.S. P. 3,371,011 (1968); *Chem. Abs.*, 69, 59109 (1968).
359. L. S. SOLOGUB, S. D. MOSHCHITSKII, L. N. MARKOVSKII, and YA. N. IVASHCHENKO, *Khim. Geterotsikl. Soedin.*, 1232 (1970); *Index. Chem.*, 40, 172581 (1971).
360. S. D. MOSHCHITSKII, G. A. ZALESSKII, YA. N. IVASHCHENKO, and L. M. YAGULPOL'SKII, *Khim. Geterotsikl. Soedin.*, 1094 (1972); *Chem. Abs.*, 77, 139750 (1972).
361. P. B. DOMENICO, U.S. P. 3,634,436 (1972); *Chem. Abs.*, 76, 99526 (1972).
362. C. D. CRAWFORD, U.S. P. 3,415,832 (1968); *Chem. Abs.*, 70, 57654 (1969).
363. J. BRATT and H. SUSCHITZKY, *Chem. Comm.*, 949 (1972).
364. A. F. PAVLENKO, V. P. AKKERMAN, G. A. ZALESSKII, and YA. N. IVASHCHENKO, *J. Gen. Chem. U.S.S.R.*, 39, 1486 (1969).
365. U. VON GIZYCKI, *Angew. Chem., Internat. Edn.*, 10, 402 (1971).
366. C. B. BARLOW, C. D. S. TOMLIN, G. M. FARRELL, P. F. H. FREEMAN, J. W. SLATER, and C. J. CLAYTON, Ger. P. Offen. 2,139,042 (1972); *Chem. Abs.*, 76, 126795 (1972).
367. C. B. BARLOW and C. D. S. TOMLIN, Ger. P. Offen. 2,143,426 (1972); *Chem. Abs.*, 76, 140858 (1972).
368. R. M. BIMBER and P. H. SCHULDT, U.S. P. 3,637,716 (1972); *Chem. Abs.*, 76, 126800 (1972).
369. YA. N. IVASHCHENKO, S. D. MOSHCHITSKII, and V. P. DANIL'CHENKO, *Ukr. Khim. Zh.*, 37, 474 (1971); *Chem. Abs.*, 75, 63559 (1971).
370. E. AGER and B. IDDON, *Chem. Comm.*, 118 (1970).
371. E. AGER, B. IDDON, and H. SUSCHITZKY, *J. Chem. Soc. (C)*, 1530 (1970): C. D. S. TOMLIN, B. IDDON, and E. AGER, Brit. P. 1,293,909 (1972); *Chem. Abs.*, 78, 58255 (1973).
372. P. B. DOMENICO, U.S. P. 3,635,994 (1972); *Chem. Abs.*, 76, 113078 (1972): U.S. P. 3,692,792 (1972); *Chem. Abs.*, 78, 16051 (1973).
373. L. S. SOLOGUB, S. D. MOSHCHITSKII, YA. N. IVASHCHENKO, and Y. N. LEVCHUK, *Khim. Geterotsikl. Soedin.*, 514 (1972).

373a.S. D. MOSHCHITSKII, L. S. SOLOGUB, YA. N. IVASHCHENKO, and L. M. YAGUPOL'SKII, *Khim. Geterotsikl. Soedin.*, 1634 (1972); *Chem. Abs.*, **78**, 71864 (1973).
374. C. FEST, I. HAMMANN, W. STENDEL, and W. FLUCKE, Brit. P. 1,165,293 (1969); *Chem. Abs.*, 72, 31622 (1970).
375. W. SMITH and G. W. DAVIS, *J. Chem. Soc.*, 41, 412 (1882).
376. Y. C. TONG, *J. Heterocyclic Chem.*, 7, 171 (1970).
377. M. GORDON and D. E. PEARSON, *J. Org. Chem.*, 29, 329 (1964).
378. M. GORDON, H. J. HAMILTON, C. ADKINS, J. HAY, and D. E. PEARSON, *J. Heterocyclic Chem.*, 4, 410 (1967).
379. R. D. CHAMBERS, M. HOLE, B. IDDON, W. K. R. MUSGRAVE, and R. A. STOREY, *J. Chem. Soc.* (C), 2328 (1966).
380. H. TARNOW, H. HOLTSCHMIDT, and O. BAYER, Ger. P. 1,186,859 (1965).
381. H. TARNOW, H. HOLTSCHMIDT, and O. BAYER, Belg. P. 638,861 (1964); *Chem. Abs.*, 62, 7736 (1965).
382. B. RIEGEL, G. R. LAPPIN, B. H. ADELSON, R. I. JACKSON, C. J. ALBISETTI, R. M. DODSON, and R. H. BAKER, *J. Amer. Chem. Soc.*, 68, 1264 (1946).
383. R. A. STOREY, R. D. CHAMBERS, W. K. R. MUSGRAVE, and B. IDDON, Brit. P. 1,151,862 (1969).
384. G. PANGON, *Bull. Soc. Chim. France*, 1993 (1970).
385. G. PANGON, *Bull. Soc. Chim. France*, 1997 (1970).
386. R. SCHÖNBECK and E. KLOIMSTEIN, *Monatsh.*, 99, 15 (1968).
387. M. TIŠLER and B. STANOVNIK, *Adv. Heterocyclic Chem.*, 9, 211 (1968).
388. J. DRUEY, *Angew. Chem.*, 70, 5 (1958).
389. R. D. CHAMBERS, J. A. H. MACBRIDE, and W. K. R. MUSGRAVE, *Chem. Ind. (London)*, 904 (1966).
390. R. D. CHAMBERS, J. A. H. MACBRIDE, and W. K. R. MUSGRAVE, *J. Chem. Soc.* (C), 2116 (1968).
391. R. D. CHAMBERS, J. A. H. MACBRIDE, and W. K. R. MUSGRAVE, Brit. P. 1,163,582 (1969); *Chem. Abs.*, 71, 124495 (1969).
392. D. E. BUBLITZ, U.S. P. 3,466,283 (1969); *Chem. Abs.*, 72, 12756 (1970).
393. D. E. BUBLITZ, U.S. P. 3,637,691 (1972); *Chem. Abs.*, 76, 127006 (1972).
394. R. H. MIZZONI and P. E. SPOERRI, *J. Amer. Chem. Soc.*, 73, 1873 (1951).
395. M. M. ROGERS and J. P. ENGLISH, U.S. P. 2,671,086 (1954); *Chem. Abs.*, 49, 1824 (1955).
396. R. H. MIZZONI and P. E. SPOERRI, *J. Amer. Chem. Soc.*, 76, 2201 (1954).
397. C. J. PENNINO, U.S. P. 2,846,433 (1958); *Chem. Abs.*, 53, 3252 (1959).
398. T. KURAISHI, *Chem. Pharm. Bull. Japan*, 4, 137 (1956).
399. H. FEUER and H. RUBINSTEIN, *J. Org. Chem.*, 24, 811 (1959).
400. P. COAD, R. A. COAD, S. CLOUGH, J. HYEPOCK, R. SALISBURY, and C. WILKINS, *J. Org. Chem.*, 28, 218 (1963): P. COAD and R. A. COAD, *ibid.*, 28, 1919 (1963).
401. J. DRUEY, K. MEIER, and K. EICHENBERGER, *Helv. Chim. Acta*, 37, 121 (1954).
402. T. MAKI and M. TAKAYA, *Yuki Gosei Kagaku Kyokai Shi*, 28, 462 (1970); *Chem. Abs.*, 73, 14789 (1970).
403. K. DURY, *Angew. Chem., Internat. Edn.*, 4, 292 (1965), and references cited therein.
404. C. GRUNDMANN, *Ber.*, 81, 1 (1948).

405. D. T. MOWRY, *J. Amer. Chem. Soc.*, 75, 1909 (1953): U.S. P. 2,628,181 (1953); *Chem. Abs.*, 47, 5065 (1953).
406. T. KURAISHI, *Chem. Pharm. Bull. Japan*, 4, 497 (1956).
407. R. F. HOMER, H. GREGORY, and L. F. WIGGINS, *J. Chem. Soc.*, 2191 (1948).
408. R. N. CASTLE and K. KAJI, *Tetrahedron Letters*, 393 (1962).
409. R. N. CASTLE, K. KAJI, G. A. GERHARDT, W. D. GUITHER, C. WEBER, M. P. MALM, R. R. SHOUP, and W. D. RHOADS, *J. Heterocyclic Chem.*, 3, 79 (1966).
410. Y. MAKI and K. OBATA, *Yakugaku Zasshi*, 83, 819 (1963); *Chem. Abs.*, 60, 1742 (1964).
411. A. ROEDIG and W. WENZEL, *Annalen.*, 728, 1 (1969).
412. H. M. COHEN, *J. Heterocyclic Chem.*, 4, 130 (1967).
413. Ciba Ltd., Brit. P. 962,261 (1964); *Chem. Abs.*, 61, 13460 (1964).
414. H. TARNOW, K. SASSE, and L. EUE, Ger. P. Offen. 1,912,472 (1970); *Chem. Abs.*, 74, 22866 (1971).
415. I. CROSSLAND and H. KOFOD, *Acta Chem. Scand.*, 21, 2131 (1967).
416. R. S. FENTON, J. K. LANDQUIST, and S. E. MEEK, *J. Chem. Soc. (C)*, 1536 (1971).
417. J. K. LANDQUIST, Brit. P. 1,248,094 (1971); *Chem. Abs.*, 75, 151819 (1971).
418. D. W. JOHNSON, V. AUSTEL, R. S. FELD, and D. M. LEVAL, *J. Amer. Chem. Soc.*, 92, 7505 (1970).
419. T. KURAISHI, *Chem. Pharm. Bull. Japan*, 6, 641 (1958).
420. T. KURAISHI and R. N. CASTLE, *J. Heterocyclic Chem.*, 1, 42 (1964).
421. S. SAKO, *Chem. Pharm. Bull. Japan*, 14, 303 (1966).
422. F. YONEDA, T. OHTAKA, and Y. NITTA, *Chem. Pharm. Bull. Japan*, 11, 954 (1963).
423. T. ITAI and S. KAMIYA, *Chem. Pharm. Bull. Japan*, 11, 1059 (1963).
424. T. KURAISHI, *Chem. Pharm. Bull. Japan*, 5, 376 (1957).
425. F. YONEDA, T. OHTAKA, and Y. NITTA, *Chem. Pharm. Bull. Japan*, 14, 698 (1966).
426. K. EICHENBERGER, R. ROMETSCH, and J. DRUEY, *Helv. Chim. Acta*, 39, 1755 (1956).
427. V. G. NYRKOVA, T. V. GORTINSKAYA, and M. N. SHCHUKINA, *J. Org. Chem. U.S.S.R.*, 1, 1711 (1965).
428. M. YANAI, T. KURAISHI, and T. KINOSHITA, *Yakugaku Zasshi*, 81, 708 (1961); *Chem. Abs.*, 55, 23553 (1961).
429. T. NAKAGOME, T. HAYAMA, T. KOMATSU, and Y. EDA, *Yakugaku Zasshi*, 82, 1103 (1962); *Chem. Abs.*, 58, 4559 (1963).
430. V. G. NYRKOVA, T. V. GORTINSKAYA, and M. N. SHCHUKINA, *Zh. Obshch. Khim.*, 34, 3132 (1964); *Chem. Abs.*, 62, 2772 (1965).
431. T. JOJIMA and S. TAMURA, *Agr. Biol. Chem. Japan*, 11, 954 (1963).
432. J. LEDERER, *J. Org. Chem.*, 26, 4462 (1961).
433. T. KURAISHI, *Chem. Pharm. Bull. Japan*, 6, 331 (1958).
434. W. J. HOULIHAN and R. E. MANNING, Brit. P. 1,287,118 (1972); *Chem. Abs.*, 77, 152207 (1972).
435. W. J. HOULIHAN and R. E. MANNING, Brit. P. 1,287,119 (1972); *Chem. Abs.*, 77, 152206 (1972).
435a. W. J. HOULIHAN and R. E. MANNING, U.S. P. 3,579,517 (1971); *Chem. Abs.*, 75, 49118 (1971): U.S. P. 3,683,085 (1972); *Chem. Abs.*, 78, 4270 (1973).
436. J. K. LANDQUIST and S. E. MEEK, *J.C.S., Perkin I*, 2735 (1972).
437. I. CROSSLAND, *Acta Chem. Scand.*, 22, 2700 (1968).

438. R. S. FENTON, J. K. LANDQUIST, and S. E. MEEK, *J.C.S., Perkin I*, 2323 (1972).
439. G. ROSSEELS, *Bull. Soc. Chim. Belges.*, 75, 5 (1966).
440. T. TSUCHIYA, H. ARAI, and H. IGETA, *Tetrahedron Letters*, 3839 (1970).
441. H. IGETA, T. TSUCHIYA, C. OKUDA, and H. YOKOGAWA, *Chem. Pharm. Bull. Japan*, 18, 1340 (1970).
442. T. KURAISHI, *Chem. Pharm. Bull. Japan*, 6, 234 (1958).
443. K. EICHENBERGER, R. ROMETSCH, and J. DRUEY, *Helv. Chim. Acta*, 37, 1298 (1954).
444. D. J. BROWN, in *The Chemistry of Heterocyclic Compounds: The Pyrimidines* (A. Weissberger, ed.), Interscience, New York, 1962, ch. VI, p. 162, and references cited therein.
445. D. J. BROWN, in *The Chemistry of Heterocyclic Compounds (Vol. 26): The Pyrimidines; Supplement I* (A. Weissberger, ed.), Interscience, New York, 1970, ch. VI, p. 110, and references cited therein.
446. R. N. HESLOP, N. LEGG, J. F. MAWSON, W. E. STEPHEN, and J. WARDLEWORTH, Brit. P. 822,047 (1959); *Chem. Abs.*, 54, 6145 (1960).
447. R. BAKER, Brit. P. 822,948 (1959); *Chem. Abs.*, 54, 8094 (1960).
448. R. N. HESLOP, N. LEGG, J. F. MAWSON, W. E. STEPHEN, and J. WARDLEWORTH, U.S. P. 2,935,506 (1960); *Chem. Abs.*, 54, 16853 (1960).
449. E. N. SARANTIS, *Amer. Dyestuffs Reptr.*, 49, No. 22, p. 25 (1960); *Chem. Abs.*, 55, 999 (1961).
450. M. SCHUMACHER, *Melliand Textilber.*, 41, 1548 (1960); *Chem. Abs.*, 55, 9883 (1961).
451. P. DUSSY, J. AMMANN, and W. BOSSARD, Ger. P. 1,112,229 (Appl. 1959); *Chem. Abs.*, 56, 2536 (1962).
452. M. CAPPIONI and A. BARTHOLD, *Tetil-Praxis*, 17, 155 and 255 (1962); *Chem. Abs.*, 57, 8757 (1962).
453. J. SRAMEK, *Textil-Praxis*, 17, 390 (1962); *Chem. Abs.*, 58, 8079 (1963).
454. Ciba Ltd., Belg. P. 609,054 (1962); *Chem. Abs.*, 58, 12707 (1963).
455. J. R. Geigy A.-G., Belg. P. 616,411 (1962); *Chem. Abs.*, 59, 6553 (1963).
456. I.C.I. Ltd., Belg. P. 617,164 (1962); *Chem. Abs.*, 59, 5297 (1963).
457. J. R. Geigy A.-G., Belg. P. 621,038 (1963); *Chem. Abs.*, 59, 4097 (1963).
458. J. BENZ, H. BURKHARD, K. KAEGI, and H. VON TOBEL, Swiss P. 364,854 (1962); *Chem. Abs.*, 59, 15418 (1963).
459. Farbwerke Hoechst A.-G., Belg. P. 628,261 (1963); *Chem. Abs.*, 61, 756 (1964).
460. U. MOECK, *Melliand Textilber.*, 45, 655 (1964); *Chem. Abs.*, 61, 7163 (1964).
461. Sandoz Ltd., Brit. P. 938,078 (1963); *Chem. Abs.*, 61, 10808 (1964).
462. V. D. POOLE, Brit. P. 948,969 (1964); *Chem. Abs.*, 61, 8448 (1964).
463. J. BENZ and W. WEHRLI, Brit. P. 952,068 (1964); *Chem. Abs.*, 61, 13463 (1964).
464. G. D. ANDERSON, G. BOOTH, and V. D. POOLE, Brit. P. 952,619 (1964); *Chem. Abs.*, 61, 13455 (1964).
465. W. STEINEMANN, Ger. P. 1,186,963 (1965); *Chem. Abs.*, 62, 16419 (1965).
466. R. J. HINE and J. R. McPHEE, *J. Soc. Dyers Colourists*, 81, 268 (1965).
467. H. ACKERMANN and H. SEILER, Swiss P. 388,495 (1965); *Chem. Abs.*, 63, 13453 (1965).
468. B. N. PARSONS and E. DRONFIELD, U.S. P. 3,220,793 (1965); *Chem. Abs.*, 64, 8382 (1966).

469. J. R. Geigy A.-G., Brit. P. 1,016,247 (1966); *Chem. Abs.*, **64**, 11374 (1966).
470. A. SUSZER, *Israel J. Chem.*, **4**, 123 (1966).
471. J. BENZ and A. SCHWEIZER, Swiss P. 440,509 (1967); *Chem. Abs.*, **68**, 96806 (1968).
472. M. RUSSOCKI and J. MIELICKI, *Ind. Chim. Belge*, **33**, 449 (1968).
473. U. EINSELE, Ger. P. 1,125,875 (1962); *Chem. Abs.*, **57**, 1123 (1962).
474. E. W. PIETRUSZA and R. PINTER, U.S. P. 3,249,590 (1966); *Chem. Abs.*, **65**, 2373 (1966).
475. E. W. DUCK and B. J. RIDGEWELL, Ger. P. Offen. 1,901,900 (1969); *Chem. Abs.*, **71**, 102912 (1969).
476. J. D. GARFORTH, Ger. P. Offen. 1,902,351 (1969); *Chem. Abs.*, **71**, 103078 (1969).
477. D. A. McGINTY and W. G. BYWATER, *J. Pharmacol. Expt. Therap.*, **84**, 342 (1945).
478. J. A. HENDRY, F. L. ROSE, and A. L. WALPOLE, *Brit. J. Pharmacol.*, **6**, 201 (1951).
479. J. A. HENDRY and R. F. HOMER, *J. Chem. Soc.*, 328 (1952).
480. J. A. HENDRY, R. F. HOMER, and F. L. ROSE, Brit. P. 683,414 (1952); *Chem. Abs.*, **48**, 746 (1954).
481. J. A. HENDRY, R. F. HOMER, and F. L. ROSE, U.S. P. 2,675,386 (1954); *Chem. Abs.*, **48**, 7645 (1954).
482. W. LANGENBECK, H. SCHUBERT, and H. GIESEMANN, *Annalen.*, **585**, 68 (1954).
483. G. F. DEEBEL and P. C. HAMM, U.S. P. 2,879,150 (1959); *Chem. Abs.*, **53**, 15107 (1959).
484. H. GYSIN and E. KNÜSLI, Ger. P. 1,035,398 (1958); *Chem. Abs.*, **54**, 25543 (1960).
485. M. ULRYCHOVÁ-ZELINKOVÁ, *Biol. Plant. Acad. Sci. Bohemoslov.*, **2**, 240 (1960); *Chem. Abs.*, **55**, 6603 (1961).
486. J. ŠKODA, A. ČIHÁK, J. GUT, M. PRYSTAŠ, A. PÍSKALA, C. PÁRKÁNYI, and F. ŠORM, *Coll. Czech. Chem. Comm.*, **27**, 1736 (1962).
487. H. K. GOUCK and G. C. La BRECQUE, *U.S. Dept. Agr.*, *ARS, ARS 33-87*, 8 pp. (1963); *Chem. Abs.*, **59**, 13290 (1963).
488. S. MATSUMOTO, O. SHIOYAMA, and K. MURATA, *Noyaku Seisan Gijutsu*, **9**, 17 (1963); *Chem. Abs.*, **61**, 6300 (1964).
489. H. GERSHON, K. DITTMER, and R. BRAUN, *J. Org. Chem.*, **26**, 1874 (1961).
490. H. GERSHON and R. PARMEGIANI, *Trans. N.Y. Acad. Sci.*, **25**, 638 (1963).
491. H. GERSHON and R. PARMEGIANI, *Appl. Microbiol.*, **11**, 78 (1963).
492. H. GERSHON, R. BRAUN, and A. SCALA, *J. Med. Chem.*, **6**, 87 (1963).
493. H. GERSHON, R. BRAUN, A. SCALA, and R. RODIN, *J. Med. Chem.*, **7**, 808 (1964).
494. H. GERSHON, R. PARMEGIANI, and R. D'ASCOLI, *J. Med. Chem.*, **10**, 113 (1967).
495. H. GERSHON, U.S. P. 3,227,612 (1966); *Chem. Abs.*, **64**, 14901 (1966).
496. H. YUKI, F. SANO, S.-I. TAKAMA, and S. SUZUKI, *Chem. Pharm. Bull. Japan*, **14**, 139 (1966).
497. SPOFA United Pharmaceutical Works, Neth. P. Appl. 6,512,184 (1966); *Chem. Abs.*, **65**, 7970 (1966).
498. E. H. KOBER and R. F. W. RAETZ, U.S. P. 3,259,623 (1966); *Chem. Abs.*, **65**, 8930 (1966).
499. R. BRETSCHNEIDER, W. KLÖTZER, and J. SCHANTL, Swiss P. 397,694 (1966); *Chem. Abs.*, **64**, 19633 (1966).

500. O. E. SCHULTZ and P. WARNECKE, *Arzneim.-Forsch.*, 17, 1060 (1967).
501. K. CULIK, M. HEROLD, J. PALKOSKA, M. VONDRACEK, and J. SKODA, Czech. P. 120,935 (1966); *Chem. Abs.*, 67, 107382 (1967).
502. R. FUSCO, G. LOSCO, and N. TROIANI, Ital. P. 662,501 (1964); *Chem. Abs.*, 65, 20147 (1966).
503. M. W. GOLDBLATT, *Brit. J. Ind. Med.*, 2, 183 (1945).
504. W. JUNGSTAND, P. NEULAND, and R. WEISS, *Z. Chem.*, 5, 156 (1965).
505. J. CHESTERFIELD, J. F. W. McOMIE, and E. R. SAYER, *J. Chem. Soc.*, 3478 (1955).
506. H.-E. KLEINE-NATROP, *Acta Allergologica*, 21, 319 (1966).
507. L. N. SHORT and H. W. THOMPSON, *J. Chem. Soc.*, 168 (1952).
508. S. GOYA, T. TAKAHASHI, and T. OKANO, *Yakugaku Zasshi*, 86, 952 (1966); *Chem. Abs.*, 66, 37208 (1967).
509. T. OKANO, A. TAKADATE, and H. MATSUMOTO, *Yakugaku Zasshi*, 88, 439 (1968); *Chem. Abs.*, 69, 63143 (1968).
510. R. T. BAILEY and D. STEELE, *Spectrochim. Acta*, 25A, 219 (1969).
511. F. M. UBER and R. WINTERS, *J. Amer. Chem. Soc.*, 63, 137 (1941).
512. M. P. V. BOARLAND and J. F. W. McOMIE, *J. Chem. Soc.*, 3722 (1952).
513. T. OKANO and A. TAKADATE, *Yakugaku Zasshi*, 89, 302 (1969); *Chem. Abs.*, 71, 26399 (1969).
514. P. J. BRAY, S. MOSKOWITZ, H. O. HOOPER, R. G. BARNES, and S. L. SEGEL, *J. Chem. Phys.*, 28, 99 (1958).
515. M. J. S. DEWAR and E. A. C. LUCKEN, *J. Chem. Soc.*, 2653 (1958).
516. H. O. HOOPER and P. J. BRAY, *J. Chem. Phys.*, 30, 957 (1959).
516a. E. A. C. LUCKEN, *Tetrahedron*, 19, Suppl. 2, 123 (1963).
517. G. K. SEMIN, T. A. BABUSHKINA, V. P. MAMAEV, and V. P. KRIVOPALOV, *Izv. Sibirsk. Otdl. Akad. Nauk. SSR.*, *Ser. Khim.*, No. 2 [1], 82 (1971).
518. H. WEILER-FEILCHENFELD and E. D. BERGMANN, *Israel J. Chem.*, 6, 823 (1968).
519. S. GABRIEL, *Ber.*, 33, 3666 (1900).
520. E. BÜTTNER, *Ber.*, 36, 2227 (1903).
521. S. GABRIEL and J. COLMAN, *Ber.*, 37, 3657 (1904).
522. H. KAST, *Ber.*, 45, 3124 (1912).
523. S. KAWAI and T. MIYOSHI, *Sci. Papers Inst. Phys. Chem. Research (Tokyo)*, 16, Nos. 306-9, 20-3 (1931); *Chem. Abs.*, 25, 5676 (1931).
524. M.-J. LANGERMAN and C. K. BANKS, *J. Amer. Chem. Soc.*, 73, 3011 (1951).
525. T. MASUDA, *Chem. Pharm. Bull. Japan*, 5, 28 (1957).
526. J. BADDILEY and A. TOPHAM, *J. Chem. Soc.*, 678 (1944).
527. F. E. KING, T. J. KING, and P. C. SPENSLEY, *J. Chem. Soc.*, 1247 (1947).
528. H. BREDERECK, A. BRÄUNINGER, D. HAYER, and H. VOLLMANN, *Chem. Ber.*, 92, 2937 (1959).
529. W. PFLEIDERER and G. NÜBEL, *Annalen.*, 631, 168 (1960).
530. V. H. SMITH and B. E. CHRISTENSEN, *J. Org. Chem.*, 20, 829 (1955).
531. I. WEMPEN and J. J. FOX, *J. Med. Chem.*, 6, 688 (1963).
532. B. H. CHASE, J. P. THURSTON, and J. WALKER, *J. Chem. Soc.*, 3439 (1951).
533. S. J. CHILDRESS and R. L. McKEE, *J. Amer. Chem. Soc.*, 72, 4271 (1950).
534. G. STEFFAN, Ger. P. Offen, 1,933,784 (1971); *Chem. Abs.*, 74, 64255 (1971).
535. D. G. COE, H. N. RYDON, and B. L. TONGE, *J. Chem. Soc.*, 323 (1957).
536. M. M. ROBINSON, *J. Amer. Chem. Soc.*, 80, 5481 (1958).

537. O. GERNGROSS, *Ber.*, 38, 3394 (1905).

538. W. P. PFLEIDERER and H. MOSTHAF, *Chem. Ber.*, 90, 728 (1957).

539. A. VON MERKATZ, *Ber.*, 52B, 869 (1919).

540. J. SHAPIRA, *J. Org. Chem.*, 27, 1918 (1962).

541. A. W. DOX, *J. Amer. Chem. Soc.*, 53, 1559 (1931).

542. Société des usines chimiques Rhône-Poulene, Brit. P. 710,070 (1954); *Chem. Abs.*, 49, 15982 (1955).

543. Z. BUDĚŠINSKÝ, V. BYDŽOVSKÝ, J. KOPECKÝ, A. ŠVÁB, and J. VAVŘINA, *Československ. farm.*, 10, 241 (1961); *Chem. Abs.*, 55, 25973 (1961).

544. R. K. ROBINS, K. L. DILLE, and B. E. CHRISTENSEN, *J. Org. Chem.*, 19, 930 (1954).

545. N. WHITTAKER, *J. Chem. Soc.*, 1565 (1951).

546. L. D. PROTSENKO and YU. I. BOGODIST, *Ukr. Khim. Zh.*, 32, 378 (1966); *Chem. Abs.*, 65, 3869 (1966).

547. YU. I. BOGODIST, U.S.S.R. P. 178,383 (1966); *Chem. Abs.*, 65, 2278 (1966).

548. J. BENZ, U.S. P., 3,075,980 (1963); *Chem. Abs.*, 59, 1660 (1963).

549. YU. I. BOGODIST, U.S.S.R. P. 172,812 (1965); *Chem. Abs.*, 64, 740 (1966).

550. W. KLÖTZER and J. SCHANTL, *Monatsh.*, 94, 1190 (1963).

551. W. SCHOENAUER, Swiss P., 372,679 (1963); *Chem. Abs.*, 60, 13257 (1964).

552. R. C. ELDERFIELD and R. N. PRASAD, *J. Org. Chem.* 25, 1583 (1960).

553. H. GERSHON, *J. Org. Chem.*, 27, 3507 (1962).

554. R. BEHRAND, *Annalen.*, 229, 25 (1885) (article begins on p. 1).

555. J. CLARK and W. PENDERGAST, *J. Chem. Soc.* (C), 2780 (1969).

556. C. KAISER and A. BURGER, *J. Org. Chem.*, 24, 113 (1959).

557. A. A. AROYAN, R. G. MELIK-ORGANDZHANYAN, and L. V. KHAZHAKYAN, *Arm. Khim. Zh.*, 22, 245 (1969); *Chem. Abs.*, 71, 81299 (1969).

558. Ciba Ltd., Belg. P. 645,062 (1964); *Chem. Abs.*, 66, 37942 (1967).

559. G. CIAMICIAN and P. MAGNAGHI, *Ber.*, 18, 3444 (1885); *Gazz. Chim. Ital.*, 14, 173 (1884).

560. W. O. EMERY, *Ber.*, 34, 4178 (1901).

561. Farbenfabriken Bayer A.-G., Fr. P. 1,545,313 (1968); *Chem. Abs.*, 72, 3497 (1970).

562. Farbenfabriken Bayer A.-G., Fr. P. Demande 2,004,698 (1969); *Chem. Abs.*, 72, 111505 (1970).

563. G. BECK, Ger. P. Offen. 2,039,491 (1972); *Chem. Abs.*, 76, 140875 (1972).

564. Farbenfabriken Bayer A.-G., Fr. P. 1,600,587 (1970); *Chem. Abs.*, 74, 100079 (1971).

565. Farbenfabriken Bayer A.-G., Fr. P. 1,545,314 (1968); *Chem. Abs.*, 71, 124469 (1969).

566. Farbenfabriken Bayer A.-G., Fr. P. 1,546,395 (1968); *Chem. Abs.*, 72, 43719 (1970).

567. R. BRADEN, K. FINDEISEN, and H. HOLTSCHMIDT, *Angew. Chem., Internat. Ed.*, 9, 65 (1970).

568. Farbenfabriken Bayer A.-G., Neth. P. Appl. 6,516,622 (1966); *Chem. Abs.*, 65, 20152 (1966).

569. R. D. CHAMBERS, J. A. H. MACBRIDE, and W. K. R. MUSGRAVE, *Chem. Ind. (London)*, 1721 (1966).

570. Farbenfabriken Bayer A.-G., Fr. P. Demande 2,000,995 (1969); *Chem. Abs.*, 72, 55487 (1970).

571. H. HOLTSCHMIDT, H. SCHWARZ, and F. DORING, Brit. P. 1,201,228 (1970).
572. G. BECK, Ger. P. Offen, 2,020,297 (1971); *Chem. Abs.*, **76**, 72540 (1972).
573. D. J. BROWN and T.-C. LEE, *Austral. J. Chem.*, **21**, 243 (1968).
574. H. ACKERMANN and P. DUSSY, *Helv. Chim. Acta*, **45**, 1683 (1962).
575. W. WINKLEMAN, *J. Prakt. Chem.*, 115; 292 (1927).
576. Metal & Thermit Corp., Brit. P. 823,276 (1959); *Chem. Abs.*, **54**, 8868 (1960).
577. H. ACKERMANN, *Helv. Chim. Acta*, **49**, 454 (1966).
578. H. SCHROEDER, E. KOBER, H. ULRICH, R. RÄTZ, H. AGAHIGIAN, and C. GRUNDMANN, *J. Org. Chem.*, **27**, 2580 (1962).
579. E. KOBER and R. RÄTZ, *J. Org. Chem.*, **27**, 2509 (1962).
580. L. D. PROTSENKO and YU. I. BOGODIST, *Ukr. Khim. Zh.*, **32**, 867 (1966); *Chem. Abs.*, **66**, 2531 (1967).
581. YU. I. BOGODIST and L. D. PROTSENKO, U.S.S.R. P. 162,149 (1964); *Chem. Abs.*, **61**, 13327 (1964).
582. YU. I. BOGODIST and L. D. PROTSENKO, *Ukr. Khim. Zh.*, **32**, 1094 (1966); *Chem. Abs.*, **66**, 94988 (1967).
583. R. E. BANKS, D. S. FIELD, and R. N. HASZELDINE, *J. Chem. Soc. (C)*, 1822 (1967).
584. Farbenfabriken Bayer A.-G., Fr. P. 1,545,174 (1968); *Chem. Abs.*, **71**, 124481 (1969).
585. B. A. IVIN, V. I. SLESAREV, and E. G. SOCHILIN, *J. Gen. Chem. U.S.S.R.*, **34**, 4183 (1964).
586. V. G. NEMETS, B. A. IVIN, and V. I. SLESAREV, *J. Gen. Chem. U.S.S.R.*, **35**, 1433 (1965).
587. M. M. BOUDAKIAN, E. H. KOBER, and E. R. SHIPKOWSKI, U.S. P. 3,280,124 (1966); *Chem. Abs.*, **66**, 2582 (1967).
588. H. SCHROEDER, *J. Amer. Chem. Soc.*, **82**, 4115 (1960).
589. C. W. TULLOCK, R. A. CARBONI, R. J. HARDER, W. C. SMITH, and D. D. COFFMAN, *J. Amer. Chem. Soc.*, **82**, 5107 (1960).
590. S. GABRIEL, *Ber.*, **34**, 3362 (1901).
591. F. E. KING and T. J. KING, *J. Chem. Soc.*, 726 (1947).
592. W. R. BOON, *J. Chem. Soc.*, 1532 (1952).
593. J. R. Geigy A.-G., Fr. P. 1,413,722 (1965); *Chem. Abs.*, **64**, 5109 (1966).
594. Ciba Ltd., Fr. P. 1,332,539 (1963); *Chem. Abs.*, **60**, 2981 (1964).
595. H. C. KOPPEL, R. H. SPRINGER, and C. C. CHENG, *J. Org. Chem.*, **26**, 1884 (1961).
596. J. WOJCIECHOWSKI, Pol. P. 48,869 (1964); *Chem. Abs.*, **63**, 18114 (1965).
597. Österreichische Stickstoffwerke A.-G., Austrian P. 174,377 (1953); *Chem. Abs.*, **47**, 12422 (1953).
598. S. GABRIEL and J. COLMAN, *Ber.*, **36**, 3379 (1903).
599. H. J. FISCHER and T. B. JOHNSON, *J. Amer. Chem. Soc.*, **54**, 727 (1932).
600. S. B. GREENBAUM and W. L. HOLMES, *J. Amer. Chem. Soc.*, **76**, 2899 (1954).
601. P. NEWMARK and I. GOODMAN, *J. Amer. Chem. Soc.*, **79**, 6446 (1957).
602. B. H. CLAMPITT and A. P. MUELLER, *J. Polymer Sci.*, **62**, 15 (1962).
603. J. DAVOLL and D. D. EVANS, *J. Chem. Soc.*, 5041 (1960).
604. J. P. HOROWITZ and A. J. TOMSON, *J. Org. Chem.*, **26**, 3392 (1961).
605. W. PFLEIDERER, in *Topics in Heterocyclic Chemistry* (R. N. Castle, ed.), Wiley Interscience, New York, 1969, ch. 3, pp. 67-68.

606. YU. I. BOGODIST, *Ukr. Khim. Zh.*, 32, 1091 (1966); *Chem. Abs.*, 66, 37870 (1967).
607. A. W. DOX, *J. Amer. Chem. Soc.*, 53, 2741 (1931).
608. D. E. O'BRIEN, C. C. CHENG, and W. PFLEIDERER, *J. Med. Chem.*, 9, 573 (1966).
609. YU. I. BOGODIST, *Ukr. Khim. Zh.*, 33, 87 (1967); *Chem. Abs.*, 66, 94989 (1967).
610. R. BROSSMER and E. RÖHM, *Annalen.*, 692, 119 (1966).
611. M. HASEGAWA, *Chem. Pharm. Bull. Japan*, 1, 387 (1953).
611a. E. DYER, T. J. NYCZ, and M. B. LONG, *J. Heterocyclic Chem.*, 9, 1267 (1972).
612. G. W. H. CHEESEMAN and E. S. G. WERSTIUK, *Adv. Heterocyclic Chem.*, 14, 153 (1972).
613. Y. T. PRATT, in *Heterocyclic Compounds* (R. C. Elderfield, ed.), Wiley, New York, 1957, Vol. 6, p. 377.
614. J. M. SAYWARD and J. K. DIXON, U.S. P. 2,442,473 (1948); *Chem. Abs.*, 42, 6861 (1948).
615. J. K. DIXON and A. A. MILLER, U.S. P. 2,540,476 (1951); *Chem. Abs.*, 45, 5726 (1951).
616. A. A. MILLER, U.S. P. 2,573,268 (1951); *Chem. Abs.*, 46, 7594 (1952).
617. G. PALAMIDESSI and F. LUINI, *Farmaco, Ed. Sci.*, 21, 811 (1966).
618. C. G. ALLISON, R. D. CHAMBERS, J. A. H. MACBRIDE, and W. K. R. MUSGRAVE, *J. Chem. Soc. (C)*, 1023 (1970).
619. H. HOLTSCHMIDT and W. ZECHER, Ger. P. 1,179,214 (1961).
620. H. HOLTSCHMIDT and W. ZECHER, Belg. P. 622,381 (1962); *Chem. Abs.*, 59, 11526 (1963).
621. R. D. CHAMBERS, W. K. R. MUSGRAVE, and K. C. SRIVASTAVA, *Chem. Comm.*, 264 (1971).
622. D. KYRIACOU, *J. Heterocyclic Chem.*, 8, 697 (1971).
623. H. JOHNSTON, U.S. P. 3,509,144 (1970); *Chem. Abs.*, 73, 25517 (1970).
624. D. H. HORNE, U.S. P. 3,452,016 (1969); *Chem. Abs.*, 71, 81415 (1969).
625. A. H. GULBENK, U.S. P. 3,471,496 (1969); *Chem. Abs.*, 71, 124489 (1969).
626. L. BERNARDI, G. LARINI, and A. LEONE, Ger. P. 1,178,436 (1964); *Chem. Abs.*, 62, 4039 (1965).
627. G. PALAMIDESSI, L. BERNARDI, and A. LEONE, *Farmaco, Ed. Sci.*, 21, 805 (1966).
628. D. R. CARTER and F. P. BOER, *J. Heterocyclic Chem.*, 9, 335 (1972).
629. F. HALVERSON and R. C. HIRT, *J. Chem. Phys.*, 19, 711 (1951).
630. R. CARBO, *An. Quim.*, 64, 147 (1968); *Chem. Abs.*, 69, 14114 (1968).
631. R. D. CHAMBERS, J. A. H. MACBRIDE, and W. K. R. MUSGRAVE, *Chem. Comm.*, 739 (1970).
632. J. A. H. MACBRIDE, *Chem. Comm.*, 1219 (1972).
633. R. D. CHAMBERS, J. A. H. MACBRIDE, W. K. R. MUSGRAVE, and I. S. REILLY, *Tetrahedron Letters*, 57 (1970).
634. C. G. ALLISON, R. D. CHAMBERS, J. A. H. MACBRIDE, and W. K. R. MUSGRAVE, *Tetrahedron Letters*, 1979 (1970).
635. F. EBEL, W. RUPP, and O. TRAUTH, U.S. P. 2,697,097 (1954); *Chem. Abs.*, 49, 4301 (1955).
636. H. ALBERS, R. OSTER, and H. SCHROEDER, Ger. P. 1,178,052 (1964); *Chem. Abs.*, 61, 16080 (1964).
637. H. WEIDINGER and G. WELLENREUTHER, Brit. P. 927,974 (1963); *Chem. Abs.*, 60, 2987 (1964).
638. D. J. BERRY and B. J. WAKEFIELD, *J. Chem. Soc. (C)*, 642 (1971).

639. C. G. ALLISON, R. D. CHAMBERS, J. A. H. MACBRIDE, and W. K. R. MUSGRAVE, *Chem. Ind. (London)*, 1402 (1968).
640. C. G. ALLISON, R. D. CHAMBERS, J. A. H. MACBRIDE, and W. K. R. MUSGRAVE, *J. Fluorine Chem.*, 1, 59 (1971).
641. C. G. ALLISON, R. D. CHAMBERS, J. A. H. MACBRIDE, and W. K. R. MUSGRAVE, Brit. P. 1,274,445 (1972); *Chem. Abs.*, 77, 48514 (1972).
642. K. SASSE, R. WEGLER, H. SCHEINPFLUG, and H. JUNG, Ger. P. 1,194,631 (1965); *Chem. Abs.*, 63, 8381 (1965).
643. D. E. BURTON, A. J. LAMBIE, D. W. J. LANE, G. T. NEWBOLD, and A. PERCIVAL, *J. Chem. Soc. (C)*, 1268 (1968).
644. D. W. J. LANE and G. T. NEWBOLD, Brit. P. 1,041,011 (1966).
645. D. E. BURTON, D. HUGHES, G. T. NEWBOLD, and J. A. ELVIDGE, *J. Chem. Soc. (C)*, 1274 (1968).
646. L. HORNER and H. MERZ, *Annalen.*, 570, 89 (1950).
647. E. DEGENER, G. UNTERSTENHOEFER, I. HAMMANN, and H. HOLT-SCHMIDT, Ger. P. Offen. 2,059,725 (1972); *Chem. Abs.*, 77, 88480 (1972).
648. R. HUISGEN, G. BINSCH, and H. KÖNIG, *Chem. Ber.*, 97, 2868 (1964).
649. Ciba Ltd., Brit. P. 901,648 (1962); *Chem. Abs.*, 58, 2455 (1963).
650. G. HOLAN, E. L. SAMUEL, B. C. ENNIS, and R. W. HINDE, *J. Chem. Soc. (C)*, 20 (1967).
651. Monsanto Chemicals (Australia) Ltd., Brit. P. 1,087,101 (1967); *Chem. Abs.*, 68, 105184 (1968).
652. Monsanto Chemicals (Australia) Ltd., Brit. P. 1,087,779 (1967); *Chem. Abs.*, 68, 114584 (1968).
653. J. SAM and J. N. PLAMPIN, *J. Pharm. Sci.*, 53, 538 (1964).
653a. L. KATZ and M. S. COHEN, *J. Org. Chem.*, 19, 758 (1954).
654. R. ADAMS and J. M. STEWART, *J. Amer. Chem. Soc.*, 74, 3660 (1952).
655. Farbenfabriken Bayer A.-G., Neth. P. Appl. 6,505,511 (1965); *Chem. Abs.*, 64, 12679 (1966).
656. S. TOYOSHIMA and N. MORISHITA, *Yakugaku Zasshi*, 86, 209 (1966); *Chem. Abs.*, 64, 19586 (1966).
657. S. TOYOSHIMA and N. MORISHITA, Japan P. 68 23,624 (1968); *Chem. Abs.*, 70, 68341 (1969).
658. Farbwerke Hoechst A.-G., Neth. P. Appl. 6,607,822 (1966); *Chem. Abs.*, 68, 21922 (1968).
659. Ciba Ltd., Fr. P. 1,469,297 (1967); *Chem. Abs.*, 67, 90793 (1967).
660. J. W. BAKER and R. E. STENSETH, U.S. P. 3,256,293 (1966); *Chem. Abs.*, 65, 10592 (1966).
661. L. KATZ and M. S. COHEN, *J. Org. Chem.*, 19, 767 (1954).
662. L. KATZ and M. S. COHEN, U.S. P. 2,820,042 (1958); *Chem. Abs.*, 52, 10204 (1958).
663. J. BINDLER and E. MODEL, Ger. P. 1,023,627 (1958); *Chem. Abs.*, 54, 14564 (1960).
664. E. MODEL, J. BINDLER, and R. ZINKERNAGEL, Ger. P. 1,106,927 (1961); *Chem. Abs.*, 56, 15878 (1962).
665. K. GÄTZI and P. MÜLLER, U.S. P. 2,724,678 (1955); *Chem. Abs.*, 50, 10788 (1956).
666. R. C. BERTELSON and W. J. BECKER, *J. Heterocyclic Chem.*, 3, 422 (1966).
667. H. BÖSHAGEN, *Chem. Ber.*, 100, 3326 (1967).
668. M. ROBBA and R. C. MOREAU, *Ann. Pharm. Franc.*, 22, 201 (1964); *Chem. Abs.*, 61, 3087 (1964).
669. G. TRAVAGLI, *Gazz. Chim. Ital.*, 85, 926 (1955); *Chem. Abs.*, 51, 13851 (1957).

670. P. REYNAUD, M. ROBBA, and R. C. MOREAU, *Bull. Soc. Chim. France*, 1735 (1962).
671. G. DAVIDOVICS, G. GARIGOU-LAGRANGE, J. CHOUTEAU, and J. METZGER, *Spectrochim. Acta*, 23A, 1477 (1967).
672. E.-J. VINCENT and J. METZGER, *Compt. Rend.*, 261C, 1964 (1965).
673. E.-J. VINCENT, R. PHAN-TAN-LUU, J. METZGER, and J. M. SURZUR, *Bull. Soc. Chim. France*, 3524 (1966).
674. E. ENDERS and E. DEGENER, Ger. P. 1,168,911 (1964); *Chem. Abs.*, 61, 3073 (1964).
675. E. DEGENER, G. BECK, and H. HOLTSCHMIDT, *Angew. Chem.*, *Internat. Edn.*, 9, 65 (1970).
676. K. PAPKE and R. POHLOUDEK-FABINI, *Pharmazie*, 22, 229 (1967); *Chem. Abs.*, 68, 12586 (1968).
677. K. FRIES and W. BUCHLER, *Annalen.*, 454, 223 (1927).
678. Y. MIZUNO and K. ADACHI, *J. Pharm. Soc. Japan*, 72, 743 (1952).
679. H. D. COSSEY, J. JUDD, and F. F. STEPHENS, *J. Chem. Soc.*, 954 (1965).
680. A. G. M. WILLEMS, A. TEMPEL, D. HAMMINGA, and B. STORK, *Rec. Trav. Chim.*, 90, 97 (1971).
681. M. NAGASAWA, Z. AIKI, and T. MAEDA, Japan P. 8147-8 ('59); *Chem. Abs.*, 54, 2652 (1960).
682. S. NAKAGAWA, J. OKUMURA, F. SAKAI, H. HOSHI, and T. NAITO, *Tetrahedron Letters*, 3719 (1970).
683. W. R. HATCHARD, U.S. P. 3,118,901 (1964); *Chem. Abs.*, 60, 15878 (1964).
684. W. R. HATCHARD, U.S. P. 3,149,107 (1964); *Chem. Abs.*, 63, 7016 (1965).
685. W. R. HATCHARD, U.S. P. 3,155,678 (1964); *Chem. Abs.*, 62, 2778 (1965).
686. W. R. HATCHARD, *J. Org. Chem.*, 29, 660 (1964).
687. E. A. MAILEY, U.S. P. 3,341,547 (1967); *Chem. Abs.*, 68, 114596 (1968).
688. E. A. MAILEY, U.S. P. 3,393,992 (1968); *Chem. Abs.*, 69, 96709 (1968).
689. G. P. VOLPP, U.S. P. 3,375,161 (1968); *Chem. Abs.*, 68, 113581 (1968).
690. K. R. H. WOOLDRIDGE, *Adv. Heterocyclic Chem.*, 14, 1 (1972).
691. H. E. SIMMONS, R. D. VEST, D. C. BLOMSTROM, J. R. ROLAND, and T. L. CAIRNS, *J. Amer. Chem. Soc.*, 84, 4746 (1962).
692. F. BECKE and H. HAGEN, *Annalen*, 729, 146 (1969).
693. Badische Anilin- und Soda-Fabrik A.-G., French P. 1,558,071 (1969); *Chem. Abs.*, 72, 43656 (1970).
694. H. BÖSHAGEN, and W. GEIGER, *Chem. Ber.*, 101, 2472 (1968).
695. I. SEKI and T. MATSUNO, Japan P. 70 14,302 (1970); *Chem. Abs.*, 73, 45499 (1970).
696. K. FRIES, K. EISHOLD, and B. VAHLBERG, *Annalen.*, 454, 264 (1927).
697. D. E. L. CARRINGTON, K. CLARKE, and R. M. SCROWSTON, *J. Chem. Soc. (C)*, 3262, 3903 (1971).
698. A. W. LUTZ and S. DE LORENZO, *J. Heterocyclic Chem.*, 4, 399 (1967).
699. A. W. LUTZ and S. A. DE LORENZO, U.S. P. 3,409,606 (1968); *Chem. Abs.*, 70, 37814 (1969).
700. J. L. IMBACH, R. JACQUIER, and A. ROMANE, *J. Heterocyclic Chem.*, 4, 451 (1967).
701. Shell Internationale Research Maatschappij N.V., Neth. P. Appl. 6,701,756 (1967); *Chem. Abs.*, 68, 11897 (1968).
702. H. RUTZ and K. GUBLER, S. Afr. P. 68 02,643 (1968); *Chem. Abs.*, 71, 38964 (1969).

360   B. IDDON AND H. SUSCHITZKY

703. H. RUTZ and K. GUBLER, S. Afr. P. 68 02,644 (1968); *Chem. Abs.*, 71, 70601 (1969).
704. J. L. WASCO, U.S. P. 3,435,050 (1969); *Chem. Abs.*, 70, 115154 (1969).
705. P. M. KOCHERGIN, *J. Gen. Chem. U.S.S.R.*, 34, 3444 (1964).
706. E. F. GODEFROI, C. A. M. VAN DER EYCKEN, and P. A. J. JANSSEN, *J. Org. Chem.*, 32, 1259 (1967).
707. P. M. KOCHERGIN and R. M. PALEI, *J. Gen. Chem. U.S.S.R.*, 38, 1085 (1968).
708. P. M. KOCHERGIN, *Khim. Geterotsikl. Soedin.*, *Akad. Nauk Latv. SSR.*, 402 (1965); *Chem. Abs.*, 63, 18069 (1965).
709. P. M. KOCHERGIN, *Khim. Geterotsikl. Soedin.*, *Akad. Nauk Latv. SSR.*, 398 (1965); *Chem. Abs.*, 63, 14847 (1965).
710. H. RUTZ and K. GUBLER, S. Afr. P. 68 02,645 (1968); *Chem. Abs.*, 72, 30594 (1970).
711. A. N. NESMEYANOV, D. N. KRAVTSOV, A. P. ZHUKOV, P. M. KOCHERGIN, and G. K. SEMIN, *Dokl. Akad. Nauk SSSR*, 179, 102 (1968); *Chem. Abs.*, 69, 82162 (1968).
712. I. TAMM, *Bull. N.Y. Acad. Med.*, 31, 537 (1955).
713. F. L. HORSFALL, *Bull. N.Y. Acad. Med.*, 31, 783 (1955).
714. I. TAMM, K. FOLKERS, and C. H. SHUNK, *J. Bacteriol.*, 72, 54 (1956).
715. I. TAMM and M. M. NEMES, *Virology*, 4, 483 (1957).
716. Merck & Co., Brit. P. 783,306 (1957); *Chem. Abs.*, 52, 9219 (1958).
717. K. A. FOLKERS and C. H. SHUNK, U.S. P. 2,876,230 (1959); *Chem. Abs.*, 53, 18060 (1959).
718. C. H. SHUNK and K. A. FOLKERS, U.S. P. 2,935,508 (1960); *Chem. Abs.*, 54, 21134 (1960).
719. R. A. BUCKNALL, *Infektionskr.*, *Int. Kongr. Infektionskr. Verhandlungsber.*, 4th, 1966 (Pub. 1967), 775.
720. R. A. BUCKNALL, *J. Gen. Virol.*, 1, 89 (1967), and references cited therein.
721. S. WALTERS, D. C. BURKE, and J. J. SKEHEL, *J. Gen. Virol.*, 1, 349 (1967).
722. G. M. TIMMIS and S. S. EPSTEIN, *Nature*, 184, 1383 (1959).
723. D. M. McNAIR SCOTT, M. L. ROGERS, and C. ROSE, *J. Amer. Chem. Soc.*, 80, 2165 (1958).
724. Chimetron S.a.r.l., Fr. P. 1,476,535 (1967); *Chem. Abs.*, 68, 114917 (1968).
725. Chimetron S.a.r.l., Fr. P. 1,476,537 (1967); *Chem. Abs.*, 68, 105500 (1968).
726. Chimetron S.a.r.l., Fr. P. 1,476,557 (1968); *Chem. Abs.*, 68, 105501 (1968).
727. K. H. BÜCHEL, F. KORTE, and R. B. BEECHEY, *Angew. Chem.*, *Internat. Edn.*, 4, 788 (1965).
728. K. H. BÜCHEL, F. KORTE, A. TREBST, and E. PISTORIUS, *Angew. Chem.*, *Internat. Edn.*, 4, 789 (1965).
729. K. H. BÜCHEL, W. DRABER, A. TREBST, and E. PISTORIUS, *Z. Naturforsch. B*, 21, 243 (1966).
730. R. B. BEECHEY, *Biochem. J.*, 98, 284 (1966).
731. K. H. BÜCHEL, *Z. Naturforsch. B.*, 25, 934 (1970).
732. W. C. ATEN and K. H. BÜCHEL, *Z. Naturforsch. B*, 25, 961 (1970).
733. H. RÖCHLING and K. H. BÜCHEL, *Z. Naturforsch. B*, 25, 1103 (1970).
734. D. E. BURTON, A. J. LAMBIE, J. C. L. LUDGATE, G. T. NEWBOLD, A. PERCIVAL, and D. T. SAGGERS, *Nature*, 208, 1166 (1965).
735. O. T. G. JONES and W. A. WATSON, *Nature*, 208, 1169 (1965).
736. O. T. G. JONES and W. A. WATSON, *Biochem. J.*, 102, 564 (1967).

737. C. C. BLACK and L. MYERS, *Weeds*, 14, 331 (1966).
738. R. L. WILLIAMSON and R. L. METCALF, *Science*, 158, 1694 (1967).
739. D. F. WILSON and R. D. MERZ, *Arch. Biochem. Biophys.*, 119, 470 (1967).
740. M. AVRON, *Curr. Top. Bioenerg.*, 2, 1 (1967).
741. E. C. WEINBACH and J. GARBUS, *Biochem. J.*, 106, 711 (1968).
742. E. A. LIBERMAN, E. N. MOKHOVA, V. P. SKULACHEV, and V. P. TOPALY, *Biofizika*, 13, 188 (1968); *Chem. Abs.*, 68, 84409 (1968).
743. V. P. SKULACHEV, A. A. SHARAF, L. S. YAGUJZINSKY, A. A. JASAITIS, E. A. LIBERMAN, and V. P. TOPALI, *Curr. Mod. Biol.*, 2, 98 (1968).
744. E. A. LIBERMAN and V. P. TOPALY, *Biofizika*, 13, 1025 (1968); *Chem. Abs.*, 70, 34267 (1969).
745. A. P. BABAKOV, V. V. DEMIN, S. D. SOKOLOV, and P. S. SOTNIKOV, *Biofizika*, 13, 1122 (1968); *Chem. Abs.*, 70, 34268 (1969).
746. J. ILIVICKY and J. E. CASIDA, *Biochem. Pharmacol.*, 18, 1389 (1969).
747. A. FINKELSTEIN, *Biochem. Biophys. Acta*, 205, 1 (1970).
748. S. PAPA, K. S. CHEAH, H. N. RASMUSSEN, IN-YOUNG LEE, and B. CHANCE, *Eur. J. Biochem.*, 12, 540 (1970).
749. E. A. LIBERMAN, V. P. TOPALY, and L. M. TSOFINA, *Biofizika*, 15, 69 (1970); *Chem. Abs.*, 72, 128753 (1970).
750. I. M. GLAGOLEVA, E. A. LIBERMAN, and Z. M. KHASHAEV, *Biofizika*, 15, 76 (1970); *Chem. Abs.*, 72, 108864 (1970).
751. E. A. LIBERMAN, V. P. TOPALY, and A. Y. SILBERSTEIN, *Biochem. Biophys. Acta*, 196, 221 (1970).
752. A. V. LEBEDEV and L. I. BOGUSLAVSKII, *Dokl. Akad. Nauk SSSR.*, 189, 1122 (1969); *Chem. Abs.*, 72, 128617 (1970).
753. Fisons Pest Control Ltd., Neth. P. Appl. 6,501,323 (1966); *Chem. Abs.*, 66, 28771 (1967).
754. Fisons Pest Control Ltd., Neth. P. Appl. 6,602,316 (1966); *Chem. Abs.*, 66, 27985 (1967).
755. Fisons Pest Control Ltd., Neth. P. Appl. 6,603,719 (1966); *Chem. Abs.*, 66, 85788 (1967).
756. D. E. BURTON, A. J. LAMBIE, and G. T. NEWBOLD, Brit. P. 1,087,561 (1967); *Chem. Abs.*, 68, 95817 (1968).
757. Shell Internationale Research Maatschappij N.V., Neth. P. Appl. 6,410,413 (1965); *Chem. Abs.*, 63, 9954 (1965).
758. G. M. FARA and K. W. COCHRAN, *Boll. Ist. Sieroterap. Milan*, 42, 630 (1963).
759. Fisons Pest Control Ltd., Fr. P. 1,469,504 (1967); *Chem. Abs.*, 69, 59236 (1968).
760. D. T. SAGGERS, Ger. P. Offen. 1,960,159 (1970); *Chem. Abs.*, 73, 97862 (1970).
761. Fisons Pest Control Ltd., Neth. P. Appl. 6,609,819 (1967); *Chem. Abs.*, 67, 73609 (1967).
762. Fisons Pest Control Ltd., Fr. P. 1,459,782 (1966); *Chem. Abs.*, 67, 54129 (1967).
763. Fisons Pest Control Ltd., Fr. P. 1,513,599 (1968); *Chem. Abs.*, 70, 87803 (1969).
764. Fisons Pest Control Ltd., Fr. P. Addn. 93,257 (1969); *Chem. Abs.*, 72, 21691 (1970).
765. Shell Internationale Research Maatschappij N.V., Neth. P. Appl. 6,605,140 (1966); *Chem. Abs.*, 66, 76008 (1967).
766. Fisons Pest Control Ltd., Neth. P. Appl. 6,610,554 (1967); *Chem. Abs.*, 67, 73610 (1967).

767. G. T. NEWBOLD and A. PERCIVAL, U.S. P. 3,430,259 (1969); *Chem. Abs.*, 70, 96797 (1969).
768. Farbwerke Hoechst A.-G., Fr. P. Demande 2,011,237 (1970); *Chem. Abs.*, 73, 130998 (1970).
769. K. H. BÜCHEL, H. F. W. RÖCHLING and U. HASSERODT, Brit. P. 1,113,999 (1968); *Chem. Abs.*, 69, 96724 (1968).
769a. M. H. FISHER, D. R. HOFF, and R. J. BOCHIS, U.S. P. 3,705,174 (1972); *Chem. Abs.*, 78, 58419 (1973).
770. K. H. BÜCHEL, Z. *Naturforsch. B.*, 25, 945 (1970).
771. W. FRICK, H. HAERLE, T. WENGER, and A. WEISS, Swiss P. 456,236 (1968); *Chem. Abs.*, 70, 12621 (1969).
772. J. H. PARSONS, A. PERCIVAL, and G. T. NEWBOLD, S. Afr. P. 67 07,059 (1968); *Chem. Abs.*, 70, 87812 (1969).
773. Fisons Pest Control Ltd., Neth. P. Appl. 6,609,597 (1967); *Chem. Abs.*, 68, 49603 (1968).
774. D. A. KALBHEN and J. LYNEN, *Arzneim.-Forsch.*, 18, 1506 (1968).
775. Fisons Pest Control Ltd., Neth. P. Appl. 300,883 (1965); *Chem. Abs.*, 64, 5697 (1966).
776. D. W. J. LANE and G. T. NEWBOLD, Brit. P. 1,063,472 (1967); *Chem. Abs.*, 67, 10688 (1967).
777. N. D. MIKHNOVS'KA and O. V. STETSENKO, *Mikrobiol. Zh.* (Kiev), 29, 242 (1967); *Chem. Abs.*, 67, 97905 (1967).
778. Agripat S. A., Neth. P. Appl. 6,611,087 (1967); *Chem. Abs.*, 68, 29699 (1968).
779. H. RÖCHLING, E. FRASCA, and K. H. BÜCHEL, Z. *Naturforsch.*, B, 25, 954 (1970).
780. D. F. KUTEPOV and D. N. KOKHLOV, *J. Org. Chem. U.S.S.R.*, 1, 186 (1965).
781. M. H. FISHER, Ger. P. Offen. 2,047,369 (1971); *Chem. Abs.*, 75, 36036 (1971).
782. W. KNOBLOCH and W. RINTELEN, *Arch. Pharm.*, 291, 180 (1958).
783. E. BAMBERGER and J. LORENZEN, *Annalen.*, 273, 289 (1893) (article begins on p. 269).
784. K. FRIES, E. MODROW, B. RAEKE, and K. WEBER, *Annalen.*, 454, 119 (1927) (article begins on p. 121).
785. A. V. STETSENKO and Yu. I. BOGODIST, *Ukr. Khim. Zh.*, 26, 92 (1960); *Chem. Abs.*, 54, 15938 (1960).
786. G. E. FICKEN, D. J. FRY, and K. J. BANNERT, Brit. P. 1,132,528 (1968); *Chem. Abs.*, 70, 38911 (1969).
787. R. FUSCO, in *The Chemistry of Heterocyclic Compounds: Pyrazoles, Pyrazolines, Pyrazolidines, Indazoles, and Condensed Rings* (A. Weissberger and R. H. Wiley, eds.), Interscience, New York, 1967, Part I, p. 3.
788. A. N. KOST and I. I. GRANDBERG, *Adv. Heterocyclic Chem.*, 6, 347 (1966).
789. R. HÜTTEL, O. SCHÄFER, and G. WELZEL, *Annalen.*, 598, 186 (1956).
790. H. REIMLINGER, A. NOELS, J. JADOT, and A. V. OVERSTRAETEN, *Chem. Ber.*, 103, 1942 (1970).
791. A. MICHAELIS, *Annalen.*, 352, 152 (1907).
792. A. MICHAELIS and K. SCHENK, *Ber.*, 41, 3865 (1908).
793. A. MICHAELIS, *Annalen*, 385, 1 (1911).
794. A. MICHAELIS and A. LACHWITZ, *Ber.*, 43, 2106 (1910).
795. C. MUSANTE, *Gazz. Chim. Ital*, 75, 109 (1945).
796. A. ROEDIG and H.-J. BECKER, *Annalen.*, 597, 214 (1955).
797. H. OHSE and K. PILGRAM, *Tetrahedron Letters*, 1949 (1968).
798. K. PILGRAM and H. OHSE, *J. Org. Chem.*, 34, 1592 (1969).

799. J. ELGUERO and R. JACQUIER, *Bull. Soc. Chim.* France, 610 (1966).
800. J. R. Geigy A.-G., Fr. P. 1,331,721 (1963); *Chem. Abs.*, 60, 1762 (1964).
801. J. ELGUERO, R. JACQUIER, and NGUYEN TIEN DUC HONG CUNG, *Bull. Soc. Chim. France*, 3727 (1966).
802. J. ELGUERO, R. JACQUIER, and NGUYEN TIEN DOC HONG CUNG, *Bull. Soc. Chim. France*, 3744 (1966).
803. K. VON AUWERS and H. LANGE, *Ber.*, 55B, 1139 (1922).
804. L. C. BEHR, in *Pyrazoles, Pyrazolines, Pyrazolidines, Indazoles, and Condensed Rings* (A. Weissberger and R. H. Wiley, eds.), Interscience, New York, 1967, ch. 10, p. 324.
805. P. FREUNDLER, *Bull. Soc. Chim. France* [4], 9, 773 (1911); *Chem. Abs.*, 5, 3684 (1911).
806. K. FRIES and R. WELDERT, *Annalen.*, 454, 314 (1927) (article begins on p. 121).
807. G. BECK, E. DEGENER, and H. HEITZER, *Annalen.*, 716, 47 (1968).
808. J. KENNER and R. CURTIS, *J. Chem. Soc.*, 105, 2717 (1914).
809. M. SAULI, Ger. P. Offen. 2,003,561 (1970); *Chem. Abs.*, 73, 66569 (1970).
810. S. R. SANDLER, *J. Org. Chem.*, 35, 3967 (1970).
811. S. R. SANDLER, *Chem. Ind. (London)*, 1416 (1971).
812. J. K. CHAKRABARTI and T. M. HOTTEN, *Chem. Comm.*, 1226 (1972).
813. A. SENIER, *Ber.*, 19, 310 (1886).
814. P. K. CHANG and T. L. V. ULBRICHT, *J. Amer. Chem. Soc.*, 80, 976 (1958).
815. P. K. CHANG, *J. Org. Chem.*, 26, 1118 (1961).
816. A. PÍSKALA, J. GUT, and F. ŠORM, *Chem. Ind. (London)*, 1752 (1964).
817. B. A. LOVING, C. E. SNYDER, G. L. WHITTIER, and K. R. FOUNTAIN, *J. Heterocyclic Chem.*, 8, 1095 (1971).
818. J. G. E. FENYES, *J. Chem. Soc.* (C), 5 (1968).
819. W. W. PAUDLER, D. J. POKORNY, and J. J. GOOD, *J. Heterocyclic Chem.*, 8, 37 (1971).
820. B. STANOVNIK, *Synthesis*, 424 (1971).
821. B. STANOVNIK, M. TIŠLER, and V. ŽIGON, *Monatsh.*, 103, 1624 (1972).
822. K. WINTERFELD and M. WILDERSOHN, *Arch. Pharm.*, 303, 44 (1970).
822a. M. KRAMBERGER, B. STANOVNIK, and M. TIŠLER, *Croat. Chem. Acta*, 44, 419 (1972).
823. G. ADEMBRI, F. DE SIO, R. NESI, and M. SCOTTON, *J.C.S., Perkin I*, 953 (1972).
824. R. H. WILEY, K. H. HUSSUNG, and J. MOFFAT, *J. Amer. Chem. Soc.*, 77, 5105 (1955).
825. TH. ZINCKE and H. ARZBERGER, *Annalen.*, 249, 371 (1888) (article begins on p. 350).
826. R. H. WILEY and K. F. HUSSUNG, *J. Amer. Chem. Soc.*, 79, 4395 (1957).
827. S. HAYASHI and N. ISHIKAWA, *Chem. Lett.*, 99 (1972); *Chem. Abs.*, 76, 113041 (1972).
828. R. HOWE, *J. Chem. Soc.* (C), 478 (1966).
829. A. ECKERT and K. STEINER, *Ber.*, 47, 2628 (1914); *Chem. Abs.*, 9, 86 (1915).
830. E. RUPPRECHT, Ph.D. Dissertation, University of Munich (1955).
831. C. BODEA and I. SILBERG, *Adv. Heterocyclic Chem.*, 9, 321 (1968).
832. C. BODEA, M. TERDIC, and I. SILBERG, *Annalen.*, 673, 113 (1964).
833. C. BODEA and I. SILBERG, *Rev. Roumaine Chim.*, 9, 425 (1964).

834. AL. SPASOV and N. PANOV, *Godishnik Sofiiskiya Univ. Fiz. Mat.*, 54, 233 (1959/60) (Pub. 1961); *Chem. Abs.*, 56, 11581 (1962).
835. C. BODEA and I. SILBERG, *Nature*, 198, 883 (1963).
836. C. BODEA and I. SILBERG, *Rev. Roumaine Chim.*, 9, 505 (1964).
837. R. D. VEST, U.S. P. 3,115,497 (1963); *Chem. Abs.*, 60, 5512 (1964).
838. K. PILGRAM, *J. Org. Chem.*, 35, 1165 (1970).
839. R. STOLLÉ and K. FEHRENBACH, *J. Prakt. Chem.*, 122, 289 (1929).
840. Shell Internationale Research Maatschappij N.V., Neth. P. Appl. 67 16,077 (1969); *Chem. Abs.*, 71, 91485 (1969).
841. Shell Internationale Research Maatschappij N.V., Fr. P. 1,541,415 (1968); *Chem. Abs.*, 72, 55455 (1970).
842. V. G. PESIN, A. M. KHALETSKII, and C. CHZHI-CHZHUN, *J. Gen. Chem. U.S.S.R.*, 27, 1648 (1957).
843. V. G. PESIN, A. M. KHALETSKII, and V. A. SERGEEV, *J. Gen. Chem. U.S.S.R.*, 33, 935 (1963).
844. V. G. PESIN, V. A. SERGEEV, and A. M. KHALETSKII, *J. Gen. Chem. U.S.S.R.*, 34, 3063 (1964), and references cited therein.
845. H. KOOPMAN, J. J. VAN DAALAN, and J. DAAMS, *Weed Res.*, 7, 200 (1967).
846. N. V. Philips' Gloeilampenfabrieken, Belg. P. 619,371 (1962); *Chem. Abs.*, 59, 10091 (1963).
847. J. DAAMS and H. KOOPMAN, U.S. P. 3,279,909 (1966); *Chem. Abs.*, 66, 28774 (1967).
848. J. J. VAN DAALEN, J. DAAMS, H. KOOPMAN, and A. TEMPEL, *Rec. Trav. Chim.*, 86, 1159 (1967).
849. V. G. PESIN and V. A. SERGEEV, *Khim. Geterotsikl. Soedin.*, 839 (1967); *Chem. Abs.*, 68, 105110 (1968).
850. J. H. DAVIES, E. HADDOCK, P. KIRBY, and S. B. WEBB, *J. Chem. Soc.* (C), 2843 (1971).
851. E. SUZUKI, S. SUGIURA, T. NAITO, and S. INOUE, *Chem. Pharm. Bull. Japan*, 16, 750 (1968).
852. E. KÜHLE B. ANDERS, E. KLAUKE, H. TARNOW, and G. ZUMACH, *Angew. Chem., Internat. Edn.*, 8, 20 (1969).
853. W. MACK, *Angew. Chem., Internat. Edn.*, 4, 245 (1965).
854. K. ÖFELE and E. DOTZAUER, *J. Organometal. Chem.*, 42, C87 (1972).
855. H. KLUG, Ger. P. 1,123,663 (1962); *Chem. Abs.*, 57, 8585 (1962).
856. E. KLINGSBERG, *Tetrahedron*, 28, 963 (1972).
857. V. I. SHEVCHENKO, P. P. KORNUTA, and N. D. BODNARCHUK, *Metody Poluch. Khim. Reaktivov Prep.*, 93 (1969); *Chem. Abs.*, 75, 140803 (1971).
858. D. L. ALDOUS and R. N. CASTLE, in *The Chemistry of Heterocyclic Compounds: Pyridazines* (A. Weissberger, E. C. Taylor and R. N. Castle, eds.), Interscience, New York, 1973, Chap. III, p. 219.

CHAPTER 3

Polychloroaryl Derivatives of Metals and Metalloids

T. Chivers

Department of Chemistry, University of Calgary,
Calgary T2N 1N4, Alberta, Canada

and

B. J. Wakefield

The Ramage Laboratories, Department of Chemistry
and Applied Chemistry, University of Salford,
Salford M5 4WT, Lancashire, England.

3.1   INTRODUCTION

Polyfluoroaryl derivatives of metals and metalloids have been extensively studied over a number of years, and have proved to be of
considerable interest [1]. By comparison, polychloroaryl derivatives
have been neglected. The reasons for this comparative neglect are not
at first sight obvious, as the chloro-compounds are of at least as
much theoretical interest, and the starting materials are cheaper and
more readily available. However, the fluoro-compounds are amenable
to study by ^{19}F n.m.r., while comparable tools have not been
available for the chloro-compounds. The development of simple
methods for preparing polychloroaryl-lithium and -magnesium
derivatives has now opened up the way for the synthesis of a wide
range of other compounds, and physical methods such as nuclear
quadrupole resonance spectroscopy and ^{13}C n.m.r. spectroscopy
may ease the task of structure determination.

In this chapter, an account is given of the preparation, properties
and uses of polychloroaromatic derivatives (including heteroaromatic
derivatives) of the metals and metalloids.† Derivatives of boron,
silicon and phosphorus are included, but sulphur compounds are
covered in Chapters 1 and 2. The spectroscopic properties of the
compounds are discussed together in the concluding section 3.8.

3.2   DERIVATIVES OF THE ALKALI METALS

Polychloroaryl derivatives of the alkali metals other than lithium are
unknown.

3.2.1   Preparation of Polychloroaryl-lithium
Derivatives

The lithium derivatives have not been made by the direct reaction of
lithium with polychloroaromatic compounds, but are readily prepared by metal–hydrogen exchange (metallation) and metal–halogen
exchange reactions. Even chlorobenzene undergoes metallation
ortho to the chlorine atom by organolithium compounds, but
rapid elimination of lithium chloride occurs, leading to benzyne [2].
The presence of more halogen substituents confers greater stability
on *o*-chloroaryl-lithium compounds, and also facilitates metallation,
so that polychloroaryl-lithium compounds may be prepared in
solution from polychloroaromatic compounds and alkyl-lithium
compounds. Some examples are shown in Table 3.1.

---

† Perchloroaromatic derivatives of metals and metalloids are included in the
review by T. Chivers, *Organometal. Chem. Rev. A*, 6, 1 (1970).

TABLE 3.1   Metallation of Polychloroaromatic Compounds by Organolithium Compounds

Substrate	Organolithium compound	Solvent	Product	Yield[a] (%)	Ref.
C_6Cl_5H	Bu^nLi	THF	C_6Cl_5Li	91	3, 4
$1,2,4,5\text{-}C_6Cl_4H_2$	Bu^nLi	THF	H–(ring: Cl Cl / Cl Cl)–Li	48	4
			Li–(ring: Cl Cl / Cl Cl)–Li	27	
$1,2,3,4\text{-}C_6Cl_4H_2$	MeLi	THF	Cl–(ring: Cl H[b] / Cl Cl)–Li	83	4, 5
$1,3,5\text{-}C_6Cl_3H_3$	$3Bu^nLi$[c]	THF	Li–(ring: Cl Li / Cl Li)–Cl	38	6
$1,2,3\text{-}C_6Cl_3H_3$	Bu^nLi	THF	Cl–(ring: Cl[b,d] / Cl)–Li	35	4, 7
Cl,Cl,Cl,Cl-pyridine (Cl at 2,3,5,6; N)	Bu^nLi	Et_2O	3,5-Cl, Li at 4, N-ring (Cl Cl / Cl Cl)	Not stated	8
Cl,Cl,Cl,Cl-pyridine	Bu^nLi	Et_2O	Li at 3, Cl ring	33	9
Cl,Cl,Cl-pyridine	Bu^nLi	Et_2O	Li at 3, Cl ring	ca. 40	10

[a] Highest yield of derivative isolated.
[b] Accompanied by products of metal–halogen exchange. See p. 370.
[c] Smaller proportions of n-butyl-lithium give mono- and di-lithio compounds.
[d] This is the only case where metallation at a position other than *ortho* to chlorine has been reported.

Aryl bromides and iodides readily undergo metal–halogen exchange with organolithium compounds [11]. On the other hand, although the preparation of trichloro-2-thienyl-lithium from tetrachlorothiophen and n-butyl-lithium had been reported as long ago as 1948 [12], the potential of this type of reaction for the synthesis of

TABLE 3.2   Metal–halogen Exchange Reactions Between Organolithium Compounds and Polychlorobenzenes and Pentachlorophenyl Derivatives

Substrate	Organolithium compound	Solvent	Position(s) of metal-halogen exchange[a]	Yield (%)	Ref.
C_6Cl_6	Bu^nLi	Et_2O or THF	1	81	14
C_6Cl_5H	Bu^tLi	THF	3[b]	38	4
$1,2,3,4\text{-}C_6Cl_4H_2$	Bu^tLi	THF	3	85	4
$1,2,3\text{-}C_6Cl_3H_3$	Bu^tLi	THF	2	Not stated	4
C_6Cl_5OMe	Bu^nLi	Et_2O	o (13) m (50) p (36)	Almost quantitative	19
$C_6Cl_5NMe_2$	Bu^nLi	Et_2O	m (60) p (40)	Almost quantitative	19
$C_6Cl_5NC_5H_{10}$	Bu^nLi	Et_2O	m (57) p (43)	Almost quantitative	19
$C_6Cl_5C_6Cl_5$	$2Bu^nLi$	Et_2O	$4,4'\text{-di}^c$	70–80	20
$C_6Cl_5SiMe_3$	Bu^nLi	Et_2O	p^d	44	22

[a] Where more than one product is formed, the proportion of each (%) is given in parentheses.
[b] Accompanied by products of metallation. See text.
[c] Other isomers may also be formed [21].
[d] A small amount of another product was probably the m-isomer [22, 23].

polychloroaryl-lithium compounds was not appreciated until 1966.† In that year, Rausch, Tibbets and Gordon reported the preparation of pentachlorophenyl-lithium from hexachlorobenzene and n-butyl-lithium in diethyl ether or THF [14]. Since then, the reaction has proved to be applicable to a wide range of partly and fully

† Perfluoroaromatic compounds undergo alkylation rather than metal–halogen exchange [11]. Alkylation of polychloroaromatic compounds has been observed only in some nitrogen heterocyclic systems (see p. 372) and in an isolated homocyclic example, tetrachloro-1,3-dicyanobenzene [13].

chlorinated aromatic and heteroaromatic systems. It generally proceeds rapidly and in high yields even at low temperature in ethers, but more slowly in hydrocarbon solvents. Some reactions of polychlorobenzenes and pentachlorophenyl derivatives are shown in Table 3.2. An interesting related example is perchloroferrocene, which with n-butyl-lithium gives an almost quantitative yield of the 2,2'-dilithio-derivative [15].

It may be noted here that the reaction of alkyl-lithium compounds with hexachlorocyclopentadiene gives pentachlorocyclopentadienyl-lithium by a metal–halogen exchange reaction [16]. (A similar product is probably also formed by the reaction of hexachlorocyclopentadiene with lithium metal [17] or lithium aluminium hydride [18].) Although the product may be regarded as a polychloro-aromatic derivative of lithium ('lithium pentachlorocyclopenta-dienide'), it decomposes above 0°.

The mechanism of the metal–halogen exchange reaction is a subject of controversy (see ref. 11 for discussion); but whatever the exact details of the course of the reaction, and whatever the exact nature of the transition state, the position at which exchange occurs in the examples shown in Table 3.2 is simply rationalized in terms of attack by the carbanionic portion of the organolithium compound *on chlorine*. Thus, in the polychlorobenzenes, it is always a chlorine flanked by two other chlorine atoms which is replaced, and in the pentachlorophenyl derivatives the extent of attack at the various positions parallels that of nucleophilic substitution by amines and alkoxides (*cf.* Chapter 1). The lack of *ortho* substitution in the pentachlorophenyl amines shows that co-ordination of the reagent with a substituent does not have a significant directing influence, possibly because the Lewis basicity of these amines is so low that they cannot compete with the solvent ether. An attempt to study the reaction in a non-donor solvent was unsuccessful, but in di-n-butyl

TABLE 3.3   Metallation and Metal–Halogen Exchange in the Reaction of Organolithium Compounds with 1,2,3,4-Tetrachlorobenzene[a] [4]

Organolithium compound	Proportion of product indicated (%)[b]	
	2,3,6-trichloro-phenyl-lithium (I)	2,3,4,5-tetrachloro-phenyl-lithium (II)
Bu^nLi	72	28
Bu^tLi	100	–
PhLi	11	89
MeLi	7	93

[a] In diethyl ether at $-70°$.
[b] Determined by analysis of trimethylsilyl derivatives (see Section 3.5.1).

(I)   (II)

ether pentachloroanisole gave a significantly higher proportion of o-substitution [19].

For the partly chlorinated substrates, metal–halogen exchange is in competition with metallation. The proportions of each reaction depend on the organolithium compound used, and the data in Tables 3.1 and 3.2 are for the reagents recorded as favouring metallation and metal–halogen exchange, respectively. The variation in the products of metallation and metal–halogen exchange in the reactions of organolithium compounds with 1,2,3,4-tetrachlorobenzene is shown in Table 3.3.

TABLE 3.4   Metal–halogen Exchange Reactions of n-Butyl-lithium and Polychloroheterocyclic Aromatic Compounds

Substrate	Solvent	Position(s) of metal–halogen exchange[a]	Yield (%)	Ref.
	Et_2O[b]	2	86	25
	Et_2O[c]	2, 2'-di	Up to 96	26
	Et_2O	4(78), 3(22)[d]	70	27
	Methyl-cyclo-hexane	2(68), 3(16), 4(16)	43	27

TABLE 3.4—continued

Substrate	Solvent	Position(s) of metal–halogen exchange[a]	Yield (%)	Ref.
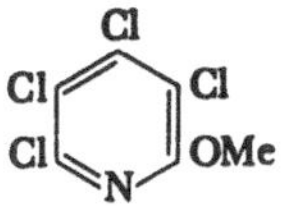 Cl-pyridine with Cl, Cl, Cl, Cl and OMe	Et_2O	4	Up to 84	28
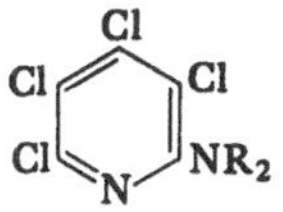 Cl-pyridine with Cl, Cl, Cl, Cl and NR_2	Et_2O	4	88	28
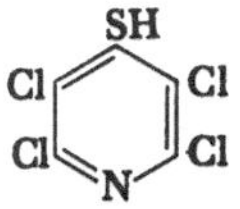 SH-tetrachloropyridine	Et_2O	3(70), 2(30)	75	29
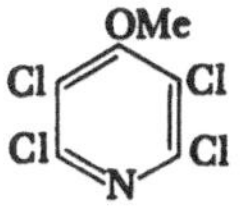 OMe-tetrachloropyridine	Et_2O	3[e]	26	30
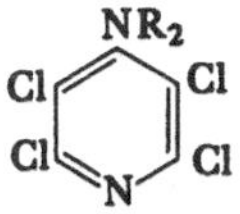 NR_2-tetrachloropyridine	Et_2O	3	Up to 57	31
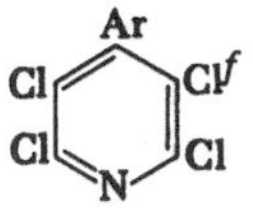 Ar-tetrachloropyridine (Cl[f])	Et_2O	3	36–94	32
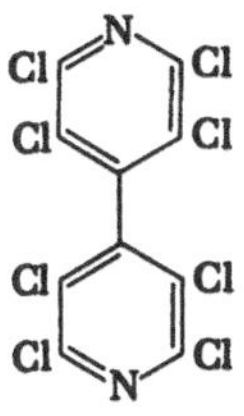 bis(tetrachloropyridyl)	Et_2O	3	70	33

---

[a] Where more than one product is formed, the proportion of each (%) is given in parentheses.

[b] With t-butyl-lithium in THF, a 94% yield is reported[34].

[c] Two molar equivalents of n-butyl-lithium.

[d] Methyl-lithium and phenyl-lithium give almost exclusively the 4-isomer[35,36]. Tetrachloro-4-pyridyl-lithium is also obtained in 76% yield from 4-bromo-tetrachloropyridine and n-butyl-lithium[37,38].

[e] Using t-butyl-lithium. Other alkyl- or aryl-lithium compounds displace the methoxy group, giving 4-alkyltetrachloropyridines[30,32].

[f] Ar = Ph, p-MeC_6H_4, p-$Me_2NC_6H_4$, p-$MeOC_6H_4$, p-$CF_3C_6H_4$.

The usual reaction between perchloroheteroaromatic compounds and organolithium compounds is metal–halogen exchange. One or two examples of alkylation have been reported (for example, with tetrachloropyrazine [24], and as a minor reaction pathway with pentachloropyridine in a hydrocarbon solvent [27]) but in general the perchloroheteroaryl-lithium compounds may be obtained in good to excellent yields.

Some examples of reactions between polychloroheterocyclic aromatic compounds and n-butyl-lithium are recorded in Table 3.4. The position at which exchange occurs in some of the pyridine derivatives is not easy to rationalize. For pentachloropyridine itself, and for the tetrachloro-2-pyridyl derivatives, metal–halogen exchange occurs at the same position(s) as nucleophilic substitution, thus following the pattern established for the pentachlorophenyl derivatives. Two explanations have been offered for the interesting solvent effect with pentachloropyridine [27]. The first explanation is that in ether the organolithium compounds are less associated, and hence less sterically demanding, than in a hydrocarbon solvent [11]. In the latter type of solvent they are thus able to attack the more reactive, but more crowded, 4-position. The second explanation is that although it is so weakly basic, pentachloropyridine is able to co-ordinate with n-butyl-lithium, thus holding the reagent in a position favouring attack at the 2-position, but only in the absence of competition from a basic solvent. The metal–halogen exchange reactions of the tetrachloro-4-pyridyl derivatives are anomalous. Nucleophilic substitution in these compounds occurs at the 2-positions (cf. Chapter 2), whereas metal–halogen exchange occurs at the 3-positions. Indeed, only in the case of the 4-thiolate (and the corresponding oxide [8]) is any exchange at the 2-position observed. It is tempting to explain metal–halogen exchange at the 3-position in terms of co-ordination by the substituents, but it is difficult to account for the complete contrast between, for example, tetrachloro-4-dimethylaminopyridine (III) which undergoes exchange exclusively at the 3-position, and pentachlorodimethylaminobenzene (IV),

NMe$_2$ .... NMe$_2$

Cl — Cl .... Cl — Cl

Cl — Cl .... Cl — Cl

N Cl

(III) .... (IV)

where no exchange *ortho* to the dimethylamino group is observed [19]. Moreover, it does not seem likely that an aryl group could compete with the ether solvent in co-ordination with the butyl-lithium. However, the metal–halogen exchange reaction is known to be reversible [11], and it is conceivable that the products from the tetrachloro-4-pyridyl derivatives result from thermodynamic, rather than kinetic, control. The 3-lithio derivatives could be stabilized by electron-donation by the substituents, which are probably orthogonal to the plane of the pyridine ring.

The synthesis of polychloroaryl-lithium compounds by metallation and metal–halogen exchange is experimentally convenient and utilizes accessible starting materials, but suffers from a tendency to give mixtures of products. An alternative synthesis involves the cleavage of polychloroaryldimethylsilanes by organolithium compounds. This reaction gives clear, colourless solutions of isomer-free products in excellent yields [39].

$$Ar_{Cl}SiMe_2H \xrightarrow[\text{(Ar}_{Cl}\text{ = polychloroaryl)}]{RLi} Ar_{Cl}Li + RSiMe_2H$$

3.2.2   Polychloroaryl-lithium Compounds in Organic Synthesis

Polychloroaryl-lithium compounds undergo most of the 'conventional' reactions of organolithium compounds, and are thus extremely useful intermediates in organic synthesis. Many of their reactions are noted in Chapters 1 and 2. Some typical reactions of pentachlorophenyl-lithium and tetrachloro-4-pyridyl-lithium are summarized in Schemes 3.1 and 3.2. The unusual properties of the

SCHEME 3.1   Some Reactions of Pentachlorophenyl-lithium

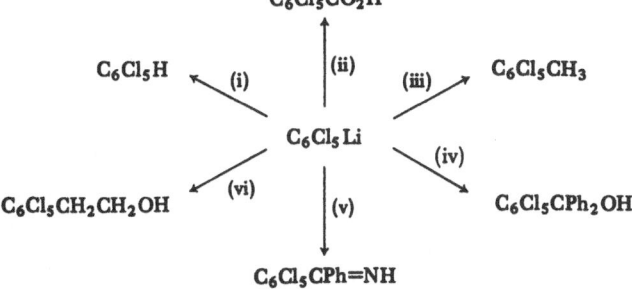

Reagents and references: (i) H_2O [14]; (ii) CO_2, then H_3O^+ [14, 40]; (iii) Me_2SO_4 [41]; (iv) Ph_2CO, then H_3O^+ [14]; (v) PhCN, then H_3O^+ [42]; (vi) $CH_2{-}CH_2$, then H_3O^+ [35].

pentachlorophenyl group (viz. its electronegativity and its large steric demand) mean that modifications of the usual experimental conditions may be required. For example, in converting organolithium compounds into carboxylic acids it is usually necessary to pour a solution of the organolithium compound on to solid carbon dioxide to minimize the formation of ketone [11]. On the other

SCHEME 3.2 Some Reactions of Tetrachloro-4-pyridyl-lithium[a]

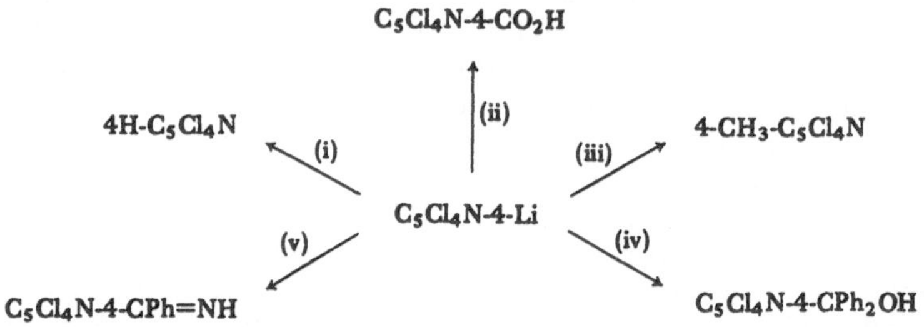
C_5Cl_4N-4-CO_2H
4H-C_5Cl_4N
(ii)
(i)
(iii)
4-CH_3-C_5Cl_4N
C_5Cl_4N-4-Li
(v)
(iv)
C_5Cl_4N-4-CPh=NH
C_5Cl_4N-4-CPh_2OH

Reagents and references: (i) H_2O [27]; (ii) CO_2, then H_3O^+ [27, 40]; (iii) Me_2SO_4 [31]; (iv) Ph_2CO, then H_3O^+ [8]; (v) PhCN, then H_2O [24, 33].

[a] Metal–halogen exchange between pentachloropyridine and *n*-butyl-lithium in diethyl ether gives the 4- and 3-lithio compounds in a ratio of *ca.* 4:1; the products indicated could be contaminated by the 3-isomer in some cases.

hand, pentachlorophenyl-lithium fails to yield pentachlorobenzoic acid under these conditions, but gives an excellent yield when carbon dioxide is passed into a solution of pentachlorophenyl-lithium [40]. In some cases, attempts at reactions may be completely unsuccessful. For example, the reaction of pentachlorophenyl-lithium with dimethylformamide failed to yield any pentachlorobenzaldehyde [44]. The reaction of pentachlorophenyl-lithium with aromatic nitriles is reversible [42]; when an aryl-lithium compound is allowed to react with pentachlorobenzonitrile, pentachlorophenyl-lithium is formed [45]:

$$ArLi + C_6Cl_5CN \rightleftharpoons (Ar)(C_6Cl_5)C{=}NLi \rightleftharpoons ArCN + C_6Cl_5Li$$

The overall reaction is thus formally a metal–cyanide exchange.

When pentachlorophenyl-lithium or tetrachloro-2- or -4-pyridyl-lithium are treated with an excess of an aromatic nitrile, products are obtained derived from reaction with two molecules of the nitrile. With pentachlorophenyl-lithium, addition of the second molecule of nitrile is followed by rapid cyclization, giving a tetrachloroquinazoline [42, 43, 46].† With tetrachloro-4-pyridyl-lithium and

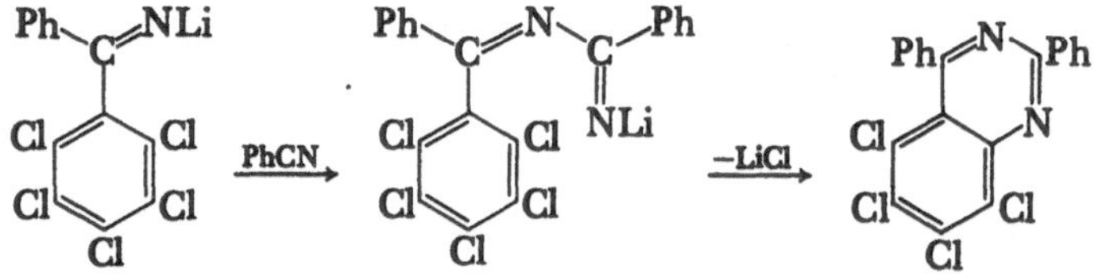
Ph
NLi
Cl
Cl
Cl
Cl
Cl
PhCN
Ph
N
Ph
NLi
Cl
Cl
Cl
Cl
Cl
–LiCl
Ph
N
Ph
N
Cl
Cl
Cl
Cl

† Successive additions of different nitriles lead to mixtures of products, owing to the reversibility of the addition reactions [42].

benzonitrile, cyclization is slower, and besides the 1,3,7-triazanaphthalene (V), the benzoyl-derivative (VI) produced by hydrolysis of the intermediate (VII) is obtained [24, 43].

$$
\text{(V)} \quad \underset{-\text{LiCl}}{\longleftarrow} \quad \text{(VII)} \quad \underset{\text{H}_3\text{O}^+}{\longrightarrow} \quad \text{(VI)}
$$

A classical route to arynes involves the elimination of metal halide from *ortho*-halogenoaryl derivatives of metals [2]. Polychloroaryl-lithium compounds are thus potential precursors for polychloro-arynes. Heaney and his co-workers demonstrated that tetrachloro-benzyne could in fact be conveniently generated by the thermolysis

$$
\longrightarrow \quad \longrightarrow \quad \text{products}
$$

of pentachlorophenyl-lithium [47]; like tetrafluorobenzyne, tetra-chlorobenzyne was an extremely reactive dienophile, and could be trapped by aromatic hydrocarbons in the form of the etheno-naphthalene derivatives such as compound (VIII) (with mesitylene).

$$
\text{C}_6\text{Cl}_5\text{Li} \longrightarrow \quad \longrightarrow \quad \text{(VIII)}
$$

Tetrachlorobenzyne, generated from pentachlorophenyl-lithium, has also been trapped by furan [48], although earlier attempts were unsuccessful [47]. The tetrachloromethoxy- and tetrachloro-dialkylaminophenyl-lithium compounds (see Table 3.2) also furnish arynes; in these cases, however, the presence of electron donating groups renders them almost unreactive towards aromatic hydro-carbons, but they can be trapped with furan [19].

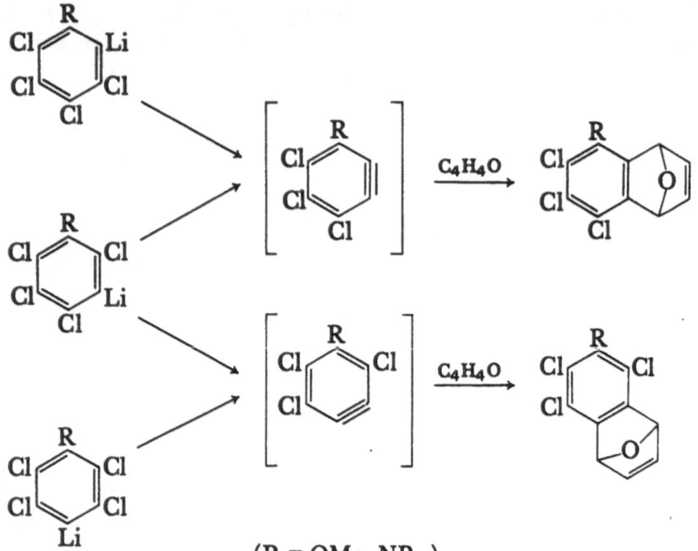

(R = OMe, NR$_2$)

The application of reactions of this type to polychloropyridyl-lithium derivatives has furnished some novel pyridynes, including examples of the elusive 2-pyridynes. Tetrachloro-4-pyridyl-lithium leads to trichloro-3-pyridyne, which may be trapped with aromatic hydrocarbons [49], and 2-substituted trichloro-4-pyridyl-lithium

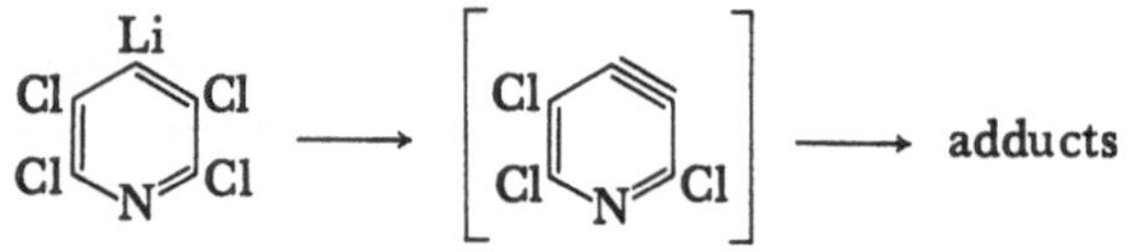

compounds lead to 2-substituted dichloro-3-pyridynes, which may be trapped with furan [28]. Attempts to generate trichloro-2-pyridyne

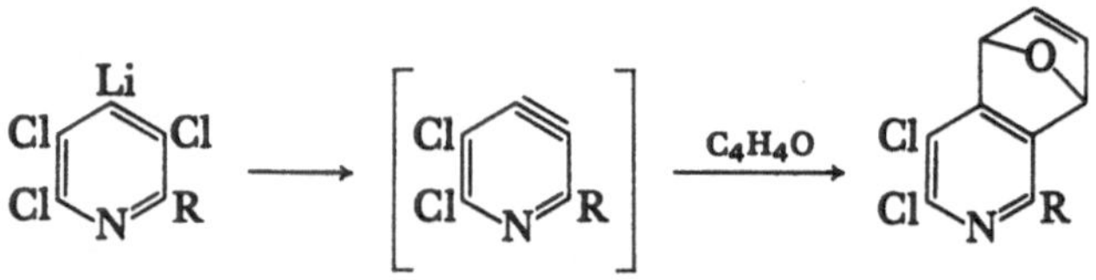

(R = OMe, NR$_2$)

from tetrachloro-2-pyridyl-lithium were unsuccessful [49], and trichloro-2-thienyl-lithium also failed to provide derivatives of dichloro-2-thiophyne [25]. On the other hand, the 4-substituted dichloro-2-pyridynes, from 4-substituted trichloro-3-pyridyl-lithium compounds, were successfully trapped with furan [31]. Pyrolysis of heptachloro-3-lithio-4,4'-bipyridyl in p-di-isopropylbenzene led to the only hydrocarbon adduct of a 2-pyridyne so far isolated [33].

3.2.3 Polychloroaryl-lithium Compounds in Organometallic Synthesis

One of the most general methods for preparing organometallic compounds (particularly σ-bonded compounds) is the reaction of an organic derivative of an electropositive metal with a derivative (usually a halide) of a less electropositive metal. Polychloroaryl-lithium compounds are no exception to this generalization, and are highly versatile intermediates in the synthesis of polychloroaryl derivatives of other metals and metalloids. Examples will be found throughout this Chapter, so that any discussion here would be superfluous.

The preparation of organometallic compounds by the reaction of polychloroaryl-lithium compounds with ligands has been little used. The addition of pentachlorophenyl-lithium to a carbonyl ligand of chromium hexacarbonyl [50] provides a pointer to the types of syntheses which can be envisaged.

$$Cr(CO)_6 \xrightarrow[\text{2. } R_3O^+BF_4^-]{\text{1. } C_6Cl_5Li} C_6Cl_5C{\overset{OR}{\underset{Cr(CO)_5}{\diagdown}}}$$

3.3 DERIVATIVES OF THE GROUP II METALS

Only derivatives of magnesium, cadmium and mercury are known. The Group IIB metals are included here, because in this field, as in so many others, they present little or no evidence of properties associated with the transition metals (see Section 3.7).

3.3.1 Polychloroarylmagnesium Halides

Grignard reagents are much less amenable than organo-lithium compounds to metallation and metal–halogen exchange reactions, so that polychloroarylmagnesium compounds are not normally accessible by these routes. For example, hexachlorobenzene is almost inert to Grignard reagents, and pentachloropyridine [51, 52] and pentachloropyridine N-oxide [53] generally undergo alkylation. However, Grignard reagents can be prepared by the direct reaction between

TABLE 3.5   Preparation of Grignard Reagents from Perchloroaromatic Compounds

Perchloro-aromatic compound[a]	Solvent	Entrainers	Temp.	Position of magnesium in product	Yield (%)	Ref.
C_6Cl_6	Et_2O	$BrCH_2CH_2Br$	reflux	1	65	54
C_6Cl_6	THF	–	reflux	1[b]	High	55 (cf. 56)
C_5Cl_5N	Et_2O	$BrCH_2CH_2Br$	reflux	4	47	27
C_5Cl_5N	Et_2O	–	reflux	4	88	57
C_5Cl_5N	THF	–	$-10°$	4	80	58
C_4Cl_4S	Et_2O	MeBr	reflux	2	17	12 (but see 25, 59)
C_4Cl_4S	THF	–	reflux	2	60	60
C_4Cl_4S	THF	$BrCH_2CH_2Br$	reflux	2	97	61

[a] C_5Cl_5N = pentachloropyridine; C_4Cl_4S = tetrachlorothiophen.
[b] Accompanied by a little 1,4-di-Grignard reagent.

perchloroaromatic compounds and magnesium. The reaction conditions for these reactions have to be carefully controlled, as the conversion of aryl chlorides to Grignard reagents is difficult, and the polychloroarylmagnesium halides, once formed, tend to decompose by the elimination of magnesium halide (see below). Usually, the use of solvents such as tetrahydrofuran and/or 'entrainers' such as 1,2-dibromoethane enable reasonable yields of the reagents to be prepared. Some examples are given in Table 3.5. Polychloroarylmagnesium halides undergo most of the usual reactions of Grignard reagents (see, e.g., Section 2.6.2 and Table 2.3), although they have been less widely employed in organic synthesis than the lithium derivatives. They also resemble the lithium derivatives in furnishing polychloroarynes, in these cases by the elimination of magnesium halide. The Grignard reagents are more stable than the lithium compounds, so that higher temperatures are required for the elimination. For example, pentachlorophenyl-lithium generates tetrachloro-benzyne (trapped as its adduct with the solvent) when heated under reflux with benzene for 2 h, whereas pentachlorophenylmagnesium chloride is stable under these conditions, although it gives the appropriate adducts when heated under reflux in p-xylene or mesitylene [47].

Polychloroarylmagnesium halides again resemble their lithium analogues in their reactions with halides of less electropositive metals to give polychloroaryl derivatives of the metals. Examples are given in the appropriate sections below.

3.3.2 Derivatives of the Group IIB Metals

(a) Synthesis

It is surprising that no polychloroaryl derivatives of zinc have been reported, as bis(pentachlorophenyl)zinc should be easily prepared by the reaction of pentachlorophenyl-lithium or pentachlorophenyl-magnesium halide with zinc chloride; bis(pentafluorophenyl)zinc has been synthesized by this route [62]. A possible alternative route, the thermal decarboxylation of zinc pentachlorobenzoate, was largely unsuccessful, because the temperature required was so high that the desired product decomposed, although under carefully controlled conditions a product apparently consisting largely of pentachloro-phenylzinc chloride was obtained [63]. Bis(pentachlorophenyl)-cadmium has not been prepared by the lithium or Grignard route, but in this case the alternative method was successful, and on heating at 340°, cadmium pentachlorobenzoate gave bis(pentachlorophenyl)-cadmium in 65% yield [63]. The observation that the reaction of trichloro-2-thienylmagnesium halide with cadmium chloride followed by acetyl chloride gave a 10% yield of 2-acetyltrichlorothiophen suggests that a trichloro-2-thienyl cadmium intermediate was formed [61].

Several routes are available for the synthesis of polychloroaryl derivatives of mercury. Bis(pentachlorophenyl)mercury was first

TABLE 3.6   Synthesis of Polychloroarylmercury Derivatives from Polychloroaryl-lithium or -magnesium Compounds

Starting material	Proportion of mercuric chloride[a]	Product	Yield (%)	Ref.
C_6Cl_5MgCl	0.5	$(C_6Cl_5)_2Hg$	50	64
C_6Cl_5Li	0.5	$(C_6Cl_5)_2Hg$	84	14
$4\text{-}LiC_5Cl_4N$	0.5	$(4\text{-}C_5Cl_5N)_2Hg$	12	27
$4\text{-}LiC_5Cl_4N$	1	$(4\text{-}C_5Cl_5N)HgCl^b$	–	27
$2\text{-}LiC_4Cl_3S$	0.5	$(2\text{-}C_4Cl_3S)_2Hg$	40	25
$2\text{-}ClMgC_4ClgS$	1	$2\text{-}ClHgC_4Cl_3S$	88	61

[a] Moles per mole of starting material.
[b] Not obtained pure.

prepared from pentachlorophenylmagnesium halide and mercuric chloride [64], and several analogous syntheses have since been reported, as shown in Table 3.6.

Pentachlorophenylmercury derivatives are also prepared by the pyrolysis of mercury salts of pentachlorobenzoic acid [65]. (Extrusion of sulphur trioxide from mercuric pentachlorophenyl-sulphonate has also been reported [66].) The simplest case is that of mercuric pentachlorobenzoate itself, which on heating in pyridine gives bis(pentachlorophenyl)mercury in 62% yield. This route is particularly useful in the preparation of mixed arylpentachloro-phenylmercury compounds [65].

$$C_6Cl_5CO_2HgPh \rightarrow C_6Cl_5HgPh + CO_2$$

An experimentally convenient adaptation of this method involves heating an alkylmercuric halide with thallous pentachlorobenzoate in pyridine [65].

$$RHgCl + C_6Cl_5CO_2Tl \rightarrow RHgC_6Cl_5 + TlCl + CO_2$$

Finally, pentachlorophenylmercury derivatives may be prepared by direct mercuration of polychlorobenzenes [67, 68]. Vigorous conditions are required for such unreactive systems, but pentachloro-phenylmercuric trifluoroacetate is obtained in high yield when pentachlorobenzene and mercuric trifluoroacetate are heated at 150° during 24 h [68]. Further heating at 240° converts the trifluoro-acetate into bis(pentachlorophenyl)mercury [67, 68]. The more reactive trichlorothiophens are mercurated by mercuric acetate [69].

(b)   Reactions (including Co-ordination Chemistry)

Polychloroaryl derivatives of mercury are characterized by their high melting points, thermal stability and low solubility. For example, bis(pentachlorophenyl)mercury melts at 383° [64], and is only sparingly soluble in organic solvents (it can be re-crystallized satisfactorily only from hot nitrobenzene). In contrast to polyfluoro-aryl derivatives of mercury (see ref. 70), polychloroaryl derivatives are reluctant to form complexes with electron donors. Pentachloro-phenylmercuric chloride [71, 72] and tetrachloro-4-pyridyl-mercuric chloride [27] give isolable complexes with bidentate ligands such as 2,2'-bipyridyl and 1,10-phenanthroline, but no complexes of bis(perchloroaryl)mercury compounds have been obtained (with the possible exception of a bis(pentachlorophenyl)-mercury–3,4,7,8-tetramethyl-1,10-phenanthroline complex [73]). The contrast between the fluorinated and chlorinated compounds is probably not due to differences in electronegativity, since penta-chlorophenyl and pentafluorophenyl groups exert a similar inductive electron-withdrawing effect [74]. It could be due to complex mesomeric effects, to steric hindrance by the o-chlorine atoms, or simply to the low solubility of the uncomplexed compounds.

Only a few comparisons have been made between the reactivity of pentachlorophenylmercury derivatives and that of other organo-mercury compounds. Experiments on the mixed compounds, C_6Cl_5HgR, indicate that the ease of cleavage of organic groups by hydrogen chloride is Ph > C_6Cl_5 > Me [64] and C_6F_5 > C_6Cl_5 [75].

$$C_6Cl_5HgPh \xrightarrow{\text{HCl}} C_6Cl_5HgCl + C_6H_6$$

$$C_6Cl_5HgMe \xrightarrow{\text{HCl}} MeHgCl + C_6Cl_5H$$

$$C_6Cl_5HgC_6F_5 \xrightarrow{\text{HCl}} C_6Cl_5HgCl + C_6F_5H$$

In contrast to its pentafluorophenyl analogue, bis(pentachloro-phenyl)mercury is not rapidly cleaved by iodide in ethanol [76]. This reagent, and other halides, cause pentachlorophenylmercuric chloride to disproportionate to bis(pentachlorophenyl)mercury [77].

$$2C_6Cl_5HgCl + 4X^- \longrightarrow (C_6Cl_5)_2Hg + HgX_4^{2-} + 2Cl^-$$

$$(X = Cl, Br, I)$$

On the other hand, pentachlorophenylmercuric bromide is cleaved to pentachloroiodobenzene by iodine in dimethylformamide or by tri-iodide ion in various solvents [78]. The reaction with iodine follows first-order kinetics, and that with tri-iodide follows second-order kinetics. The suggested mechanisms involve nucleophilic catalysis by the solvent and by tri-iodide, respectively. The catalytic effect of halide on redistribution reactions of polyhalogenoaryl-mercury derivatives is highlighted by the interesting compound, pentachlorophenylpentafluorophenylmercury, $C_6Cl_5HgC_6F_5$. This compound is stable when pure, but rapidly disproportionates in the presence of lithium halide, so that it cannot be prepared by the routes indicated [75].

$$
\begin{array}{ccc}
 & C_6Cl_5Li + C_6F_5HgBr & \\
 & \nearrow \quad \searrow & \\
C_6Cl_5HgC_6F_5 & & (C_6F_5)_2Hg + (C_6Cl_5)_2Hg \\
 & \searrow \quad \nearrow & \\
 & C_6Cl_5HgCl + C_6F_5Li &
\end{array}
$$

Uncatalyzed redistribution reactions of organomercury compounds have been the subject of considerable attention [79]. In general, fairly high temperatures are required for dialkylmercury compounds, but less vigorous conditions for alkylmercuric halides. As a consequence of their high thermal stability, redistribution reactions of pentachlorophenylmercury compounds are readily accomplished, and are useful for preparing unsymmetrical derivatives. Some examples are shown below [64, 75].

$$(C_6Cl_5)_2Hg + R_2Hg \rightarrow RHgC_6Cl_5$$

$$(R = Me, Ph)$$

$$C_6Cl_5HgCl + MeHgC_6F_5 \rightarrow C_6Cl_5HgC_6F_5$$

$$(C_6Cl_5)_2Hg + HgCl_2 \rightarrow 2C_6Cl_5HgCl$$

Although they are so stable, polychloroaromatic derivatives of mercury may possibly function as precursors for arynes. However, it seems very likely that the 'adducts' obtainable may be formed by addition-elimination sequences [2]. As a consequence, the formation of compound (IX) from bis(trichloro-2-thienyl)mercury and tetraphenylcyclopentadienone is not necessarily evidence for the elusive 'thiophyne' [25].

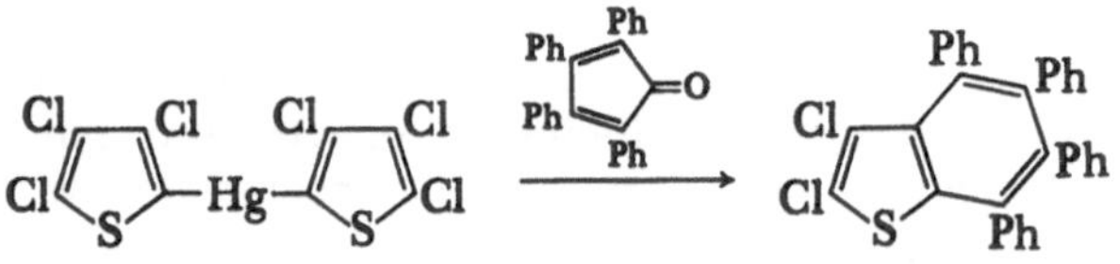

(IX)

3.4   DERIVATIVES OF GROUP IIIA ELEMENTS

Little is known about polychloroaryl compounds of these elements apart from the preparation of pentachlorophenylboron dichloride by the organotin route [80].

$$Me_3SnC_6Cl_5 + 2BCl_3 \rightarrow C_6Cl_5BCl_2 + MeBCl_2 + Me_2SnCl_2$$

Unlike its pentafluorophenyl analogue [81], pentachlorophenylboron dichloride is stable towards disproportionation at temperatures up to 260° [107], and hydrolysis or dimethylamination can be achieved without significant cleavage of the B—C_6Cl_5 bond.

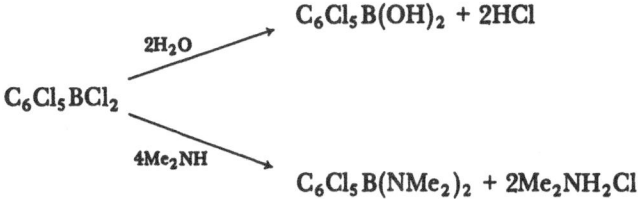

$$C_6Cl_5BCl_2 \xrightarrow{2H_2O} C_6Cl_5B(OH)_2 + 2HCl$$

$$C_6Cl_5BCl_2 \xrightarrow{4Me_2NH} C_6Cl_5B(NMe_2)_2 + 2Me_2NH_2Cl$$

The thermally unstable thallium salt, $C_5Cl_5{}^-Tl^+$, has been prepared by addition of thallium(I) ethoxide to pentachlorocyclopentadiene in pentane at $-78°$ [82]. The IR spectrum shows that the expected D_{5h} symmetry of the $C_5Cl_5{}^-$ ion in this salt is lowered to C_{5v} by interaction with the cation.

3.5 DERIVATIVES OF GROUP IVA ELEMENTS

3.5.1 Silicon

There has been much interest in silicone polymers containing chlorinated phenyl groups because of their improved lubricating properties, greater resistance to acid cleavage and higher thermal stability compared to unsubstituted phenylpolysiloxanes [83]. Certain polychloroarylsilicon chlorides may be prepared by direct chlorination, but by far the most versatile route to polychloroarylsilicon compounds involves the reaction of an organochlorosilane with the appropriate polychloroaryl-lithium or -magnesium reagent.

(a) *Preparation by Direct Chlorination*
    The benzene nucleus bound to silicon may be chlorinated either in the presence of a catalyst, e.g. aluminium chloride, iodine, antimony trichloride, ferric chloride or phosphorus pentachloride at 70-100° or non-catalytically at higher temperatures (350-400°) [84, 85]. Cleavage of the C_6H_5-Si bond accompanies chlorination and is favoured by increasing substitution of organic groups at silicon. With antimony trichloride as catalyst, all of the hydrogen atoms in phenyltrichlorosilane can be substituted with chlorine [86].

$$C_6H_5SiCl_3 \xrightarrow[SbCl_3]{Cl_2} C_6Cl_5SiCl_3$$

p-Tolylsilicon trichloride can be chlorinated selectively in the aromatic ring in the presence of antimony trichloride to give di- and tri-chloro derivatives [87], but only the alkyl group is chlorinated in the presence of free radical initiators.

(b) *Preparation from Polychloroaryl-lithium and -magnesium Reagents*

The use of the recently discovered polychloroaryl-lithium and -magnesium reagents has extended considerably the range of known polychloroarylsilicon compounds. Indeed, many of the polychloroaryl-lithium reagents have been characterized by the preparation of a trimethylsilyl derivative.

The low yields of tetrakis(pentachlorophenyl)silane obtained from the reaction of pentachlorophenylmagnesium chloride and silicon tetrachloride are indicative of the steric requirements of the penta-chlorophenyl group [55]. In fact, dichlorobis(pentachlorophenyl)-silane may be prepared by slow addition of the Grignard reagent to silicon tetrachloride [88], although in contrast, attempts to prepare dichlorobis(pentafluorophenyl)silane by the Grignard route gave only tetrakis(pentafluorophenyl)silane and polymeric residues [89]. Low yields of fluorotris(pentachlorophenyl)silane result from slow passage of silicon tetrafluoride into a solution of the Grignard reagent [88].

The series of compounds with the general formula $C_6Cl_5(SiMe_2)_n$ C_6Cl_5 (n = 1-6) has been prepared from pentachlorophenyl-magnesium chloride and the appropriate chloropoly(dimethyl)silane [90]. Partial replacement of chlorine by pentachlorophenyl in arylchlorosilanes has been achieved using pentachlorophenyl-lithium [91].

$$(n\text{-}1)\ C_6Cl_5Li + Ph_{4-n}SiCl_n \rightarrow (C_6Cl_5)_{n-1}Ph_{4-n}SiCl + (n\text{-}1)LiCl$$

$$(n = 2, 3)$$

Acid hydrolysis of $(C_6Cl_5)_{n-1}Ph_{4-n}SiCl$ gives the corresponding silanol, and $(C_6Cl_5)Ph_2SiCl$ can be reduced to the silane, apparently without significant cleavage of the Si$-C_6Cl_5$ bond.

$$(C_6Cl_5)Ph_2SiCl \underset{Cl_2/CCl_4}{\overset{LiAlH_4}{\rightleftharpoons}} (C_6Cl_5)Ph_2SiH$$

With a liberal excess of dichlorodimethylsilane, pentachlorophenyl-lithium produces (pentachlorophenyl)dimethylsilicon chloride which undergoes reactions typical of a chlorosilane [92].

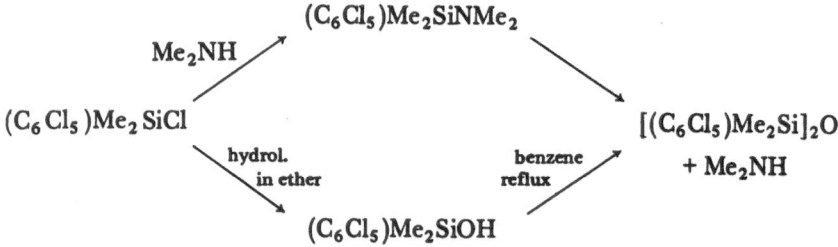

1,3-Bis(pentachlorophenyl)tetramethyldisiloxane has also been prepared from pentachlorophenyl-lithium and 1,3-dichlorotetramethylsiloxane [93].

The bridged siloxane shown below has been obtained by two routes as indicated [93, 94].

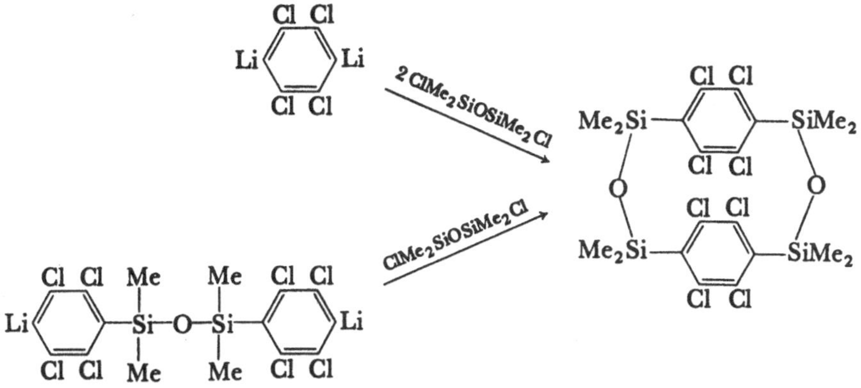

Cl Cl

Li⟨⟩Li

2 ClMe₂SiOSiMe₂Cl →

ClMe₂SiOSiMe₂Cl →

Cl Cl Me   Me Cl Cl

Li⟨⟩–Si–O–Si–⟨⟩Li

Cl Cl Me   Me Cl Cl

Cl Cl

Me₂Si—⟨⟩—SiMe₂

Cl Cl

O   O

Cl Cl

Me₂Si—⟨⟩—SiMe₂

Cl Cl

An interesting preparation of pentachlorophenyl-silicon compounds involves nucleophilic attack on hexachlorobenzene by tris(trimethylsilyl)silyl-lithium or -sodium [95]. In this way both $C_6Cl_5Si(SiMe_3)_3$ and p-$(Me_3Si)_3SiC_6Cl_4Si(SiMe_3)_3$ have been prepared. Similarly, tetrachloro-4-triphenylsilylpyridine can be isolated in low yield from the reaction of triphenylsilyl-lithium with pentachloropyridine, but the major product is hexaphenyldisilane [51]. A number of tetrachloro-4-pyridylsilanes have been prepared by the conventional Grignard procedure [58].

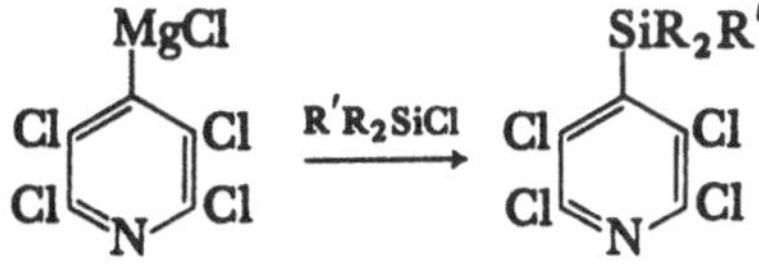

MgCl   SiR₂R'

Cl⟨⟩Cl   R'R₂SiCl →   Cl⟨⟩Cl

Cl⟨N⟩Cl   Cl⟨N⟩Cl

(R = R' = Me; R = Me, R' = H)

(R = Ph, R' = H)

Some 2-silylthiophens have been synthesized from trichloro-2-thienyl-lithium [96] or trichloro-2-thienylmagnesium halide [61].

(c) *Reactions of Polychloroarylsilicon Compounds*

The most characteristic reaction of the silicon-polychloroaryl bond is alkaline cleavage to give the corresponding polychlorobenzene which is easily identified. In this way the relative positions of substituents in polychloroarylsilicon derivatives can be determined. For example, the di-substituted product obtained from the reaction of hexachlorobenzene with two moles of tris(trimethylsilyl)-silyl-lithium yields 1,2,4,5-tetrachlorobenzene on alkaline hydrolysis, clearly indicating p-substitution [95].

Some pentachlorophenylsilicon compounds undergo metal-halogen exchange in preference to $Si-C_6Cl_5$ cleavage with n-butyl-lithium at low temperatures. With $Me_3SiC_6Cl_5$, for example, exchange takes place mainly in the p-position although some m-substitution also occurs as determined by alkaline hydrolysis [22].

The following sequence further illustrates the preparative use of the lithium reagents derived from metal-halogen exchange reactions [94].

In contrast, the lability of the C_6Cl_5-Si bond is demonstrated by the observation that cleavage of the silicon-carbon bond occurs in preference to metal-halogen exchange or $Si-H$ metallation in the reaction of organolithium reagents with pentachlorophenylsilanes of the type $(C_6Cl_5)R_2SiH$ (R = Me, Ph) [22, 91] and related compounds [39]; (see also Section 3.2.1).

$$(C_6Cl_5)R_2SiH \xrightarrow[\text{(2) Me}_3\text{SiCl}]{\text{(1) R'Li}} C_6Cl_5SiMe_3 + R_2R'SiH$$

$$(R' = Me, Ph)$$

Addition of $Me_3SiC_6Cl_5$ to the carbonyl group of benzaldehyde occurs in a fashion reminiscent of the reaction of Grignard or lithium reagents with carbonyl compounds [97].

$$C_6Cl_5SiMe_3 + PhCHO \longrightarrow \begin{array}{c} C_6Cl_5 \\ \diagdown \\ Ph \diagup \end{array} CHOSiMe_3$$

Hydrolysis of the trimethylsiloxy derivative gives the corresponding carbinol. The addition reaction has also been observed for trimethyl-(pentafluorophenyl)silane, but not for trimethylphenylsilane.

The most remarkable reaction of polychloroarylsilicon compounds involves the formation of tetrakis(trimethylsilyl)allene, $(Me_3Si)_2$-$C=C=C(SiMe_3)_2$, on treatment with lithium in tetrahydrofuran in the presence of an excess of chlorotrimethylsilane [98]. This reaction was originally observed for hexachlorobenzene, but higher yields of the allene are formed from either trimethyl(pentachlorophenyl)silane (37%) or tetrachloro-1,4-di(trimethylsilyl)benzene (52%) [99]. It is noteworthy that the allene is also formed from a wide variety of chlorocarbons (including aliphatic) on treatment with lithium and chlorotrimethylsilane in tetrahydrofuran [100, 101]. Recently it has been shown that the tetraene,

$$(Me_3Si)_2C=C=C \begin{array}{c} \diagup SiMe_3 \\ \diagdown \\ Me_3Si \diagup \end{array} C=C=C(SiMe_3)_2$$

is produced in low yield from tetrachloro-1,4-di(trimethylsilyl)-benzene and other aromatic and aliphatic halocarbons [100, 102]. Since the tetraene is easily converted to the allene by lithium and chlorotrimethylsilane

$$(Me_3Si)_2C=C=C(SiMe_3)-C(SiMe_3)=C=C(SiMe_3)_2 \xrightarrow{\text{Li/Me}_3\text{SiCl}}$$

$$2(Me_3Si)_2C=C=C(SiMe_3)_2$$

it may well be an important intermediate in the formation of the allene from halocarbons.

3.5.2   Germanium, Tin and Lead

(a)   *Preparation*

There appear to be no reports in the literature of aryl-germanium or -lead compounds containing more than one chlorine substituent in the aryl group, so this section is devoted entirely to tin derivatives.

Polychloroaryltin compounds are readily prepared from the appropriate organotin halide by the organolithium or Grignard route [22, 55, 80, 103].

$$R_{4-n}SnCl_n + nC_6Cl_5X \rightarrow R_{4-n}Sn(C_6Cl_5)_n + nXCl$$

(R = Me; n = 1) (X = Li, MgCl)

(R = Bu; n = 2)

(R = Ph; n = 1,2,3,4)

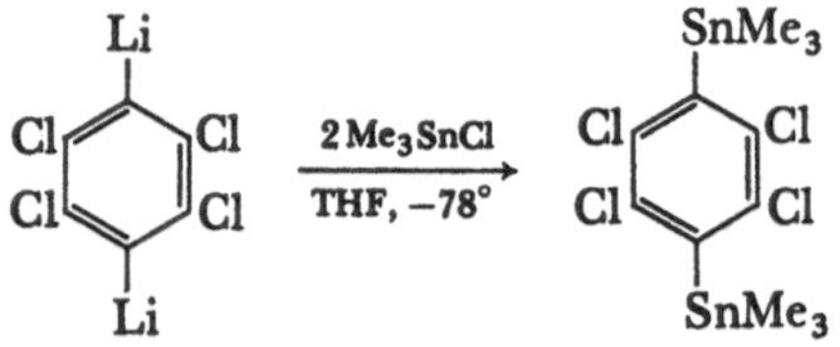

The Diels–Alder reaction between hexachlorocyclopentadiene and trimethyltinacetylenes in refluxing xylene gives rise to polychloro-norbornadienyl derivatives of tin [104]. When 5,5-dimethoxytetra-chlorocyclopentadiene was used for this reaction, however, none of the Diels–Alder adduct was isolated. Instead, aromatization of the adduct occurred to give 1,2-bis(trimethyltin)tetrachlorobenzene.

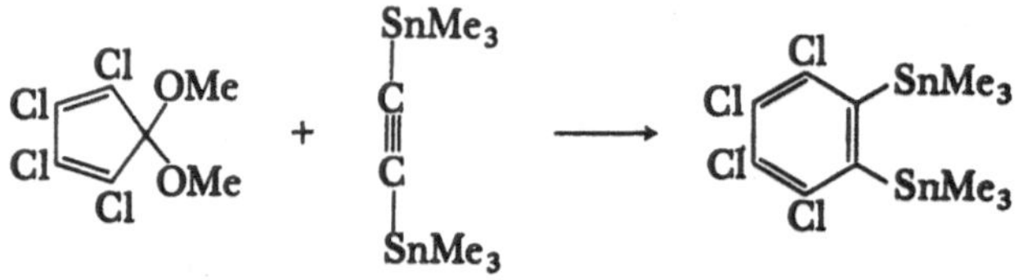

Another product from this reaction, 1-(trimethyltin)-2,3,4,5-tetra-chlorobenzene, probably arises as a hydrolysis product of 1,2-bis(tri-methyltin)tetrachlorobenzene.

(b)   *Reactions*

As mentioned in Section 3.4, the Sn–C_6Cl_5 bond in trimethyl-(pentachlorophenyl)tin is cleaved by the electrophilic reagent boron trichloride to give pentachlorophenylboron dichloride [80]. Nucleophilic cleavage of the $Sn-C_6Cl_5$ bond (e.g. by alkali) occurs very readily (cf. $Si-C_6Cl_5$) to yield pentachlorobenzene. Indeed, trimethyl(pentachlorophenyl)tin can be recrystallized from aqueous ethanol, but in the presence of a trace of fluoride ion it is converted almost quantitatively into pentachlorobenzene [80].

$$Me_3SnC_6Cl_5 \xrightarrow[\text{reflux}]{\text{F}^-/\text{EtOH/H}_2\text{O}} C_6Cl_5H + Me_3SnOH$$

This type of nucleophilic-assisted solvolysis has also been observed for pentafluorophenyltin compounds [105].

The decomposition of trimethyl(pentachlorophenyl)tin at 300° gave a high yield of trimethyltin chloride [80]. Whether the decomposition is unimolecular or bimolecular is uncertain. In contrast, at 230° (2,3,4,5-tetrachlorophenyl)trimethyltin undergoes redistribution to give tetramethyltin and dimethylbis(2,3,4,5-tetrachlorophenyl)tin [104]. Both elimination of trimethyltin halide and disproportionation can occur in the thermal decomposition of (haloaryl)trimethyltin compounds [106], and the relative importance of these two decomposition pathways is temperature-dependent in the case of (2-bromo-3,4,5,6-tetrafluorophenyl)trimethyltin [107].

(2,3,4,5-Tetrachlorophenyl)trimethyltin has been used to prepare a polychlorobiphenyl in a coupling reaction with p-iodotoluene [104]

[119]Sn Mössbauer data for polychloroaryltin compounds are discussed in Section 3.8.4.

3.6   DERIVATIVES OF GROUP VA AND VIA ELEMENTS

Little is known of the polychloroaryl derivatives of these elements, with the exception of phosphorus. Pentachlorophenyldiphenylphosphine and bis(pentachlorophenyl)phenylphosphine have been prepared from pentachlorophenyl-lithium and the appropriate phenylchlorophosphine [14, 108]. Tris(pentachlorophenyl)phosphine was obtained from phosphorus tribromide by the Grignard route [109] or, in low yield, from phosphorus trichloride and pentachlorophenyl-lithium [108].

The   tetrachloro-4-pyridylphosphines, $4\text{-}Ph_2PC_5Cl_4N$, $4\text{-}PhP(C_5Cl_4N)_2$ and $4\text{-}P(C_5Cl_4N)_3$, have been prepared in THF by the Grignard procedure or from the lithium reagent at low temperature [108]. 2-Lithiotrichlorothiophen has been converted to a diphenylphosphino derivative [14, 95].

Metal–halogen exchange occurs in the reaction of n-butyl-lithium with bis(pentachlorophenyl)phenylphosphine, and the dilithium reagent so formed reacts with chlorotrimethylsilane [108].

Indicative of the chemical stability of the C_6Cl_5-P bond is the reported oxidation of pentachlorophenyl- and tetrachloropyridyl-phosphines to the corresponding phosphine oxides by a mixture of sodium dichromate, concentrated sulphuric acid and glacial acetic acid [108].

The ultraviolet spectra of the phosphines and corresponding oxides are discussed in Section 3.8.2.

3.7   TRANSITION METAL DERIVATIVES

3.7.1   Preparation and Properties

The first pentachlorophenyl derivatives of transition metals to be prepared were derivatives of iron, cobalt and nickel, e.g. $(PhEt_2P)_2$-$Co(C_6Cl_5)_2$, made by the Grignard procedure [110, 111].

$$(PhEt_2P)_2CoCl_2 + 2C_6Cl_5MgCl \rightarrow (PhEt_2P)_2Co(C_6Cl_5)_2 + 2MgCl_2$$

Compared to phenylmetal compounds, the pentachlorophenyl derivatives exhibit high thermal stability which has been attributed to the steric and electronic properties of the pentachlorophenyl group (see Section 3.7.2).

The chemistry of pentachlorophenyl transition metal derivatives has recently been extended to include some nickel complexes of the type $(MePh_2P)_2Ni(C_6Cl_5)X$ (X = Cl, Me, C_6F_5) [112] and $(Me_2PhP)_2Ni(C_6Cl_5)X$ (X = Cl, Br, I) [113]. A *trans* structure is assigned to these complexes since a triplet is observed for P–*Me* in the ^1H NMR spectra due to virtual coupling. Green, air-stable complexes of the type $(\pi$-$C_5H_5)(R_2R'P)NiC_6Cl_5$ have been prepared and in solution they show greater aerobic and hydrolytic stability than their phenyl analogues [114]. The lack of success of an attempt

(R = R' = Ph; R = Ph, R' = Me)

to prepare $bipyNi(C_6Cl_5)_2$ is most likely due to steric factors imposed by two pentachlorophenyl groups, since the corresponding palladium compound was obtained from $bipyPdCl_2$ and pentachlorophenyl-lithium [115]. The covalent radius of palladium (1.28 Å) is significantly greater than that of nickel (1.15 Å). Even for the palladium derivative molecular models indicate that the two pentachlorophenyl groups must be oriented perpendicular to the xy plane of the molecule owing to steric constrictions. The high m.p. (335-337°) of $bipyPd(C_6Cl_5)_2$ is indicative of its considerable thermal stability. The palladium derivative *trans*-$(MePh_2P)_2$-$Pd(C_6Cl_5)Cl$ prepared from pentachlorophenylmagnesium chloride was, surprisingly, isolated as a dihydrate [115].

Trichloro-2-lithiothiophen has been used to prepare σ-bonded trichloro-2-thienyl derivatives of nickel, iron and manganese [25].

The trichloro-2-thienyl metal complexes, like their pentachlorophenyl analogues, exhibit enhanced thermal and oxidative stabilities when compared to other σ-bonded transition metal compounds. For example, 2-thienylmanganese pentacarbonyl decomposes in a few minutes when heated to 110° under nitrogen, whereas trichloro-2-thienylmanganese pentacarbonyl was virtually unchanged after 24 hours under the same conditions.

Little use has been made of the reaction between metal carbonyl anions and polychlorinated aromatics or hetero-aromatics for the preparation of polychloroarylmetal compounds. An exception is the reported preparation of disubstituted derivatives from cyanuric chloride and either $Na[Fe(CO)_2(\pi\text{-}C_5H_5)]$ [116] or $NaMn(CO)_5$ [117]. In this connection it is noteworthy that in the reaction of ethylenebis(triphenylphosphine)nickel with 1,2,4-trichlorobenzene, substitution by nickel takes place preferentially at the 2-position [118]. Interestingly, this is the same position at which nucleophilic

substitution by methoxide ion in 1,2,4-trichlorobenzene occurs [119].

There has been considerable interest recently in the preparation of organocopper compounds and in the application of these reagents in organic syntheses. Gilman and coworkers have used three methods to prepare pentachlorophenylcopper and 2,3,5,6-tetrachloropyridyl-copper [120-124].

1. Copper(I) halide (preferably iodide) and tetrachloropyridyl- or pentachlorophenyl-lithium.

2. Copper(I) halide and tetrachloropyridyl- or pentachlorophenyl-magnesium chloride.

3. Lithium dimethylcopper and hexachlorobenzene or penta-chlorobenzene in a THF/ether mixture.

Copper derivatives have also been prepared from trichloro-2-thienyl-lithium [125], dichloro-2,5-dilithiothiophen [26] and the mono- and di-lithio-derivatives of decachlorobiphenyl [21] in re-actions with copper(I) halides, preferably in THF or 1,2-dimethoxy-ethane. When diethyl ether was used as the solvent, the reaction

TABLE 3.7   Some Representative Reactions of Perchloroaryl- and Perchloroheteroaryl-copper Compounds

Copper Compound	Substrate	Product[a]	Ref.
C_6Cl_5Cu	CH_3COBr	$CH_3COC_6Cl_5$ (65)	124
$4\text{-}C_5Cl_4NCu$	C_6H_5COCl	$4\text{-}C_6H_5COC_5Cl_4N$ (66)	124
$C_6Cl_5 \cdot C_6Cl_4Cu$	$n\text{-}C_3H_7COCl$	$n\text{-}C_3H_7COC_6Cl_4 \cdot C_6Cl_5$ (51)	21
C_6Cl_5Cu	$(CH_2)_4(COCl)_2$	$C_6Cl_5CO(CH_2)_4COC_6Cl_5$ (78)	120, 121
$2\text{-}C_4Cl_3SCu$	$(CF_2)_3(COCl)_2$	$2\text{-}C_4Cl_3SCO(CF_2)_3COSC_4Cl_3\text{-}2$ (22)	125
$4\text{-}C_5Cl_4NCu$	C_6H_5I	$4\text{-}C_6H_5C_5Cl_4N$ (55)	123
$2\text{-}C_4Cl_3SCu$	$CH_2{=}CHCH_2Br$	$2\text{-}CH_2{=}CHCH_2C_4Cl_3S$ (62)	125
$2\text{-}C_4Cl_3SCu$	$CHBr{=}CBr_2$	$2\text{-}C_4Cl_3SC{\equiv}CC_4Cl_3S\text{-}2$ (36)	125
$2,5\text{-}C_4Cl_2SCu_2$	I_2	$2,5\text{-}I_2C_5Cl_2S$ (52)	26

[a] Yields are dependent on method of preparation of copper reagent. Optimum yields (%) in parentheses.

between trichloro-2-thienyl-lithium and copper(I) chloride gave hexachloro-2,2'-bithienyl as the principal product.

Like other perchloroarylmetal compounds, pentachlorophenyl-copper and 2,3,5,6-tetrachloropyridylcopper are more thermally stable than their non-halogenated analogues. The copper compounds are less reactive than the corresponding lithium or Grignard reagents towards carbonyl groups, but they arylate many types of organic halide. They have been used in the synthesis of a number of functionally substituted perchlorarenes, including bis(perchloro-aryl)alkynes, perchloroaryl iodides, unsymmetrical biaryls, and per-chloroaryl ketones and diketones, as exemplified in Table 3.7 (see also Scheme 2.12, p. 253).

3.7.2   Structure-stability Relationships

In connection with the observed enhanced thermal and oxidative stabilities of pentachlorophenylmetal compounds, Chatt and Shaw suggested that the electronegative chlorine atoms make an important contribution to stability by withdrawing electrons from the metal atom into the aromatic system, so increasing π-bonding [110]. At the same time the energy gap between the highest occupied bonding orbital and the lowest unoccupied orbital of the metal—carbon bond order will increase with the result that the metal—carbon bond will have greater stability with respect to homolytic dissociation. It is noteworthy that a single crystal X-ray diffraction study of trans-$(PPh_2Me)_2Ni(C_6F_5)(C_6Cl_5)$ shows the $Ni-C_6Cl_5$ distance of 1.905 $\pm$ 0.010 Å to be significantly shorter than the $Ni-C_6F_5$ distance of 1.978 $\pm$ 0.009 Å [126]. Earlier it had been found that the $Ni-C_{aryl}$ bond lengths in $(\pi$-$C_5H_5)Ph_3PNiAr$ [1.919 $\pm$ 0.013 Å (Ar = C_6H_5); 1.914 $\pm$ 0.014 Å (Ar = C_6F_5)] were indistinguishable within experimental error, indicating that although the nickel—carbon bond is greater than unity, π-bonding is not more important for the pentafluorophenyl derivative [127-129]. It is unlikely that the differences in $Ni-C_6Cl_5$ and $Ni-C_6F_5$ bond lengths are steric in origin, but possible reasons for the shorter and therefore stronger $Ni-C_6Cl_5$ bond include:

(a) the trans-effect of a C_6Cl_5 ligand being greater than that of a C_6F_5 ligand.

(b) The $Ni-C_6Cl_5$ σ-bond being stronger than the $Ni-C_6F_5$ σ-bond,

(c) increased $d\pi$-$p\pi$ metal $\rightarrow$ ligand back-donation for the $Ni-C_6Cl_5$ bond relative to the $Ni-C_6F_5$ bond.

The chlorine atoms of the C_6Cl_5 ligand are not coplanar, the deviation being greatest for the m-Cl atoms, presumably as a result of repulsions between adjacent chlorine atoms.

3.8   SPECTROSCOPIC STUDIES

3.8.1   Infrared Spectra

Infrared absorptions characteristic of the pentachlorophenyl group are observed at 670-698, 826-865, 1088-1125 cm^{-1} for a wide variety of C_6Cl_5-metal compounds [14, 112, 115]. For pentachlorophenylmercury derivatives bands at 1516 ± 2, 1328 ± 2, 1297 ± 2 were attributed to $\nu(C-C)$. The 'X-sensitive' mode of the C_6Cl_5X group, attributed to C_6Cl_5-X stretch, was assigned as the band at 845 cm^{-1} and $\nu(C-Cl)$ was found at 677 ± 1 cm^{-1} for the C_6Cl_5-mercury compounds [65, 130].

3.8.2   Ultraviolet Spectra

The ultraviolet spectra of a number of polyhaloaryl-silanes and -phosphines have been reported [108, 131, 132]. Noteworthy features of these spectra are the abnormally high molar absorptivities in comparison with hydrocarbon analogues, and the fact that the benzenoid fine structure of the lower intensity B band (280-310 nm) is still present in pentachlorophenyl derivatives but absent in pentafluorophenyl compounds (260-270 nm).

In the series $C_6Cl_5(SiMe_2)_nC_6Cl_5$ (n = 1-6) the wavelength of the high intensity band at ~215 nm does not vary with increasing chain length, and since a strong absorption is observed for the n = 1 compound this absorption does not originate from an Si—Si bond [131]. The replacement of chlorine by a silyl substituent in hexachlorobenzene causes a bathochromic shift in the B band which is greatest for the compound $(p-C_6Cl_4SiMe_2OSiMe_2)_2$ [132] (see p. 385 for structure). Either deformation of the aromatic nuclei and/or some transannular interaction between the two aromatic rings, as has been found for paracyclophanes [133], may account for this shift.

The molar absorptivities of the band at ~310 nm observed for polychloroarylphosphines of the type $(R_{Cl})_nPPh_{3-n}$ ($R_{Cl} = C_6Cl_5$, C_5Cl_4N; n = 1,2,3) increase in a fairly additive manner along the series, e.g. $R_{Cl} = C_6Cl_5$, n = 1, ϵ = 4510; n = 2, ϵ = 9500; n = 3, ϵ = 14,050, indicating that there is no interaction between the pentachlorophenyl groups [108]. Electron delocalization can occur in these phosphines by $\pi(p{\rightarrow}p)$ donation of the phosphorus lone pair to the ring and by $\pi(p{\rightarrow}d)$ withdrawal of electrons from the ring by the phosphorus d orbitals. In the corresponding phosphine oxides delocalization can only occur by the latter mechanism, as indicated by the lower molar absorptivities observed for polychloroaryl-phosphine oxides compared to the corresponding phosphines.

3.8.3 Mass Spectra

Mass spectrometry is an exceedingly useful tool for studying polychlorinated compounds since chlorine is composed of two stable isotopes, ^{35}Cl and ^{37}Cl, whose relative abundance is in the approximate ratio 3/1. Thus positively charged ions which contain several chlorine atoms give rise to envelopes of peaks which are characteristic of the number of chlorine atoms present in the fragment ion. A very useful table showing the isotope abundance ratios to be expected when a total of up to eight chlorine atoms are present in the ionic species being studied has been compiled by Beynon [134]. In a recent spectroscopic study of polychloroferrocenes, this table has been extended to ionic species containing up to a total of ten chlorines [135]. The fragmentation patterns of a number of polychlorinated compounds $(C_6Cl_5)_3P$ [109], $Me_3SiC_6Cl_5$ [109], $1\text{-}HC_6Cl_4SnMe_3\text{-}2$ [136], o- and $p\text{-}(Me_3Sn)_2C_6Cl_4$ [136], C_6Cl_5HgCl [137] and $(C_6Cl_5)_2Hg$ [137] have been studied by mass spectroscopy. No metastable ions were observed in the spectrum of $(C_6Cl_5)_3P$ but the presence of peaks corresponding to $(C_6Cl_4)_2P^+$, $(C_6Cl_4)_2^+$ and PCl_2^+ indicate that this compound may eliminate either PCl_2 or Cl_2 followed by P. The removal of successive chlorine atoms from the molecular ion is more prevalent than is the corresponding removal of fluorine or hydrogen from $(C_6F_5)_3P$ or $(C_5H_5)_3P$. When only one pentachlorophenyl group is attached to a metalloid atom, as in $Me_3SiC_6Cl_5$, no ions corresponding to $(C_6Cl_4)_2^+$ are found, although fragments which indicate halogen abstraction by silicon are observed.

For the polychloroaryltin compounds $1\text{-}HC_6Cl_4SnMe_3\text{-}2$, 1,2- and $1,4\text{-}(Me_3Sn)_2C_6Cl_4$, ions which result from the migration of chlorine to tin, e.g. $C_2H_6SnCl^+$, $CH_3SnCl_2^+$ and $SnCl^+$, are more abundant than the corresponding fluorine-containing ions in the mass spectra of polyfluoroaryltin compounds, indicative of the weaker C–Cl bonds relative to C–F bonds. Interesting differences in ion abundances are observed for the isomers 1,2- and $1,4\text{-}(Me_3Sn)_2C_6Cl_4$. The loss of CH_3 from the parent ion occurs more readily in the 1,4 isomer, and the ions so formed lose a CH_4 as well as C_2H_6 in the case of the 1,2-isomer, whereas the 1,4-isomer loses CH_3 radical and C_2H_6. The closeness of the tin atoms in the 1,2-isomer may account for this behaviour. The $(CH_3)_3 Sn^+$ ion is more than twice as abundant in the 1,2 isomer. Again the steric effects caused by the proximity of the large tin-methyl groups may encourage loss of $(CH_3)_3 Sn^+$ from $(CH_3)_5 Sn_2 C_6 Cl_4^+$.

The mass spectrum of pentachlorophenylmercuric chloride yields $C_6 Cl_4^+$ as the most intense ion, formed by preferential elimination of $HgCl_2$ $(C_6 Cl_5 HgCl^+ \xrightarrow{-HgCl_2} C_6 Cl_4^+)$ rather than Hg–C bond cleavage. Fragmentation by elimination of $HgCl_2$ is less important for bis(pentachlorophenyl)mercury [137].

The molecular ion is the most intense peak in the mass spectra of polychloroferrocenes, and a major fragmentation pathway is elimination of $FeCl_2$, probably to afford a chlorinated fulvalene [135].

3.8.4 ^{119}Sn Mössbauer Spectra

A number of polychlorophenyltin compounds have been investigated by Mössbauer spectroscopy and the spectral parameters are shown in Table 3.8 [103, 138, 139]. The closeness of the values for the isomer shift, δ, for these compounds indicates that the s-electron density at

TABLE 3.8   Mössbauer Data for Polychloroaryltin Compounds

Tin compound	δ^a	$\Delta^{b,c}$	Ref.
$C_6 Cl_5 SnMe_3$	1.32	1.09 (1.31)	138
$(C_6 Cl_5)_2 SnBu_2$	1.33	1.23	144
$C_6 Cl_5 SnPh_3$	1.27	0.84 $(0.99)^d$	138
$(C_6 Cl_5)_2 SnPh_2$	1.50	1.05 (1.11)	144, 141
$(C_6 Cl_5)_3 SnPh$	1.11	0.80 (0.92)	144, 141
$(C_6 Cl_5)_4 Sn$	1.22	0 (0)	144, 141
$1,4-C_6 Cl_4 (SnMe_3)_2$	1.26	1.10 (1.20)	139
$1,2-C_6 Cl_4 (SnMe_3)_2$	1.25	0.78 (0.85)	139
$1-HC_6 Cl_4 SnMe_3-2$	1.24	0.83	139

^{a} mm sec^{-1} relative to SnO_2; 2.10 mm sec^{-1} has been added to literature values quoted relative to grey tin.
^{b} mm sec^{-1} $(\pm 0.05)^{138,144}$ $(\pm 0.03)^{139}$.
^{c} Δ values in parentheses are for the corresponding fluorocarbon compounds.
^{d} Average of 3 values [138, 140, 141].

the tin nucleus varies little within the series. The following order of decreasing electronegativity of organic groups can be deduced on the basis of isomer shifts for $(C_6X_5)_4Sn$ compounds (X=H, Cl, F):C_6F_5 > C_6Cl_5 > C_6H_5. From the estimated electronegativity values of C_6F_5 (2.68 ± 0.05) and C_6H_5 (2.56 ± 0.05) in these compounds, a value of 2.61 ± 0.05 for the electronegativity of C_6Cl_5 can be interpolated [103, 140]. The resonances observed for Ar_4Sn compounds are single lines, but the signal for $(C_6Cl_5)_4Sn$ is very broad suggesting that there is some deviation from a regular tetrahedral environment due to steric interaction between pentachlorophenyl ligands [103].

In support of the suggestion that the dominant factor in determining the size of the quadrupole splitting, Δ, in tetrahedral organotin compounds is an imbalance in the polarity of the σ-bonds [138], it is found in every case that the value of Δ for a polychloroaryltin compound is less than that of the fluorocarbon analogue. The corresponding aryltin compounds show no resolvable quadrupole splitting. The value of Δ in polychloroaryltin compounds is sensitive to the nature of o-substituents [139]. For example, replacement of an o-chlorine in $Me_3SnC_6Cl_5$ by hydrogen lowers the quadrupole splitting from 1.09 to 0.83. Similarly, 1,4-$(Me_3Sn)_2$-C_6Cl_4 has a value of $\Delta = 1.10$ which falls to $\Delta = 0.78$ in 1,2-$(Me_3Sn)_2C_6Cl_4$, indicating that there is little steric interaction between the $SnMe_3$ groups in the 1,2-isomer, since significant steric interaction between Me_3Sn groups would lead to the introduction of an asymmetry parameter and hence an increased quadrupole splitting compared to 1-$HC_6Cl_4SnMe_3$-2. Also from the Mössbauer data there is no evidence for intramolecular tin-$ortho$-chlorine coordination in pentachlorophenyltin compounds, since 5-coordinate tin would give rise to much larger values of Δ than those observed.

3.8.5   Nuclear Quadrupole Resonance (NQR) Spectra

Since both stable isotopes of chlorine, ^{35}Cl and ^{37}Cl, have a quadrupole moment there is considerable scope for NQR studies of polychloroaryl-metal and -metalloid derivatives. The suggestion that metal-$ortho$-chlorine interactions may be significant in pentachlorophenylmercury compounds has recently been tested by measurements of NQR parameters for C_6Cl_5HgCl, $(C_6Cl_5)_2Hg$, C_6Cl_5HgPh and C_6Cl_5HgMe [142]. The lower frequencies observed for the o-chlorine atoms for three of the compounds can be interpreted in terms of donation of the 3p lone-pair of electrons on chlorine to the empty 6p orbital on mercury. Assuming normal bond angles and distances the Hg—o-Cl distance will be 3.28 Å in pentachlorophenylmercurials, which is only slightly less than the sum of the Van der Waals radii (3.30 Å) [143]. It is interesting to note that abnormal

398 T. CHIVERS AND B. WAKEFIELD

shifts to *higher* frequencies have been observed for the α-chlorine atoms in trichloromethyl- and pentachlorocyclopentadienyl-mercurials, possibly due to 'hyperconjugation' of the empty 6p orbitals with the (fairly polarizable) C–Cl bond [144].

The NQR spectrum of the adduct, $Tl^+C_5Cl_5^-$. 2PhMe, shows a very complicated pattern ($\geqslant 17$ lines), owing to the presence of at least four crystallographically non-equivalent $C_5Cl_5^-$ ions (each with non-equivalent chlorine atoms) per unit cell [82]. The spectrum of decachlororuthenocene shows the expected three lines of intensity 2:2:1. The frequencies are closer to those of C_6Cl_6 than to those of $C_5Cl_5^-$, a fact which is explained by repulsion of the p_z lone pairs of chlorine by filled girdle d orbitals around the metal atom [82]. An X-ray crystal-structure determination of decachlororuthenocene reveals that the chlorine atoms are displaced outward from the nearly planar rings of carbon atoms [145], as expected for this kind of repulsion.

REFERENCES

1. S. C. COHEN and A. G. MASSEY, *Adv. Fluorine Chem.*, 6, 185 (1970).
2. R. W. HOFFMANN, *Dehydrobenzene and Cycloalkynes*, Academic Press, New York, 1967.
3. C. TAMBORSKI, E. J. SOLOSKI, and C. E. DILLS, *Chem. and Ind.*, 2067 (1965).
4. I. HAIDUC and H. GILMAN, *Chem. and Ind.*, 1278 (1968); *idem., Rev. Roum. Chem.*, 16, 907 (1971).
5. I. HAIDUC and H. GILMAN, personal communication.
6. I. HAIDUC and H. GILMAN, *J. Organometal. Chem.*, 12, 394 (1968).
7. I. HAIDUC and H. GILMAN, *J. Organometal. Chem.*, 13, P4 (1968).
8. J. D. COOK, Ph.D. Thesis, Salford (1969).
9. E. AGER, G. E. CHIVERS, and H. SUSCHITZKY, *J. Chem. Soc. Chem. Comm.*, 505 (1972).
10. N. J. FOULGER and B. J. WAKEFIELD, unpublished work.
11. B. J. WAKEFIELD, *The Chemistry of Organolithium Compounds*, Pergamon, Oxford, 1974.
12. G. B. BACHMANN and L. V. HEISEY, *J. Amer. Chem. Soc.*, 70, 2378 (1948).
13. F. H. PINKERTON and S. F. THAMES, *J. Heterocyclic Chem.*, 9, 725 (1972).
14. M. D. RAUSCH, F. E. TIBBETTS, and H. B. GORDON, *J. Organometal. Chem.*, 5, 493 (1966).
15. F. L. HEDBERG and H. ROSENBERG, *J. Amer. Chem. Soc.*, 92, 3239 (1970).
16. R. WEST, *Accts. Chem. Res.*, 3, 130 (1970).
17. E. T. McBEE and D. K. SMITH, *J. Amer. Chem. Soc.*, 77, 389 (1955).
18. R. A. HALLING, Ph.D. Thesis, Purdue (1965); *Diss. Abs.*, 27B, 409 (1966).
19. D. J. BERRY, I. COLLINS, S. M. ROBERTS, H. SUSCHITZKY and B. J. WAKEFIELD, *J. Chem. Soc. (C)*, 1285 (1969).
20. F. BINNS and H. SUSCHITZKY, *J. Chem. Soc. (C)*, 1913 (1971).
21. P. CAUBÈRE and B. GORNY, *J. Organometal. Chem.*, 37, 401 (1972).

22. K. SHIINA, T. BRENNAN, and H. GILMAN, *J. Organometal. Chem.*, 11, 471 (1968).
23. H. GILMAN, personal communication.
24. D. J. BERRY, J. D. COOK, and B. J. WAKEFIELD, *J. Chem. Soc. Perkin I*, 2190 (1972).
25. M. D. RAUSCH, T. R. CRISWELL, and A. K. IGNATOWICZ, *J. Organometal. Chem.*, 13, 419 (1968).
26. M. R. SMITH, Jr. and H. GILMAN, *J. Organometal. Chem.*, 42, 1 (1972).
27. J. D. COOK and B. J. WAKEFIELD, *J. Organometal. Chem.*, 13, 15 (1968).
28. D. J. BERRY, B. J. WAKEFIELD and J. D. COOK, *J. Chem. Soc. (C)*, 1227 (1971).
29. E. AGER, B. IDDON, and H. SUSCHITZKY, *J. Chem. Soc. (C)*, 193 (1970).
30. R. A. FERNANDEZ, H. HEANEY, J. M. JABLONSKI, K. G. MASON, and T. J. WARD, *J. Chem. Soc. (C)*, 1908 (1969).
31. J. D. COOK and B. J. WAKEFIELD, *J. Chem. Soc. (C)*, 1973 (1969).
32. J. D. COOK and B. J. WAKEFIELD, *J. Chem. Soc. (C)*, 2376 (1969).
33. J. D. COOK, N. J. FOULGER, and B. J. WAKEFIELD, *J. Chem. Soc. Perkin I*, 995 (1972).
34. I. HAIDUC and H. GILMAN, *Rev. Roum. Chem.*, 16, 305 (1971).
35. D. J. BERRY and B. J. WAKEFIELD, unpublished observations.
36. I. HAIDUC and H. GILMAN, *Rev. Roum. Chem.*, 16, 597 (1971).
37. S. M. ROBERTS, Ph.D. Thesis, Salford (1969).
38. I. COLLINS, S. M. ROBERTS and H. SUSCHITZKY, *J. Chem. Soc. (C)*, 167 (1971).
39. M. R. SMITH, Jr. and H. GILMAN, *J. Organometal. Chem.*, 37, 35 (1972).
40. D. S. SETHI, M. R. SMITH, Jr., and H. GILMAN, *J. Organometal. Chem.*, 24, C41 (1970).
41. B. J. WAKEFIELD and D. J. WRIGHT, *J. Chem. Soc. (C)*, 1165 (1970).
42. D. J. BERRY and B. J. WAKEFIELD, *J. Organometal. Chem.*, 23, 1 (1970).
43. D. J. BERRY, J. D. COOK, and B. J. WAKEFIELD, *Chem. Comm.*, 1273 (1969).
44. B. J. WAKEFIELD, unpublished work.
45. N. J. FOULGER and B. J. WAKEFIELD, *Tetrahedron Letters*, 4169 (1972).
46. D. J. BERRY and B. J. WAKEFIELD, *J. Chem. Soc. (C)*, 642 (1971).
47. H. HEANEY and J. M. JABLONSKI, *J. Chem. Soc. (C)*, 1895 (1968).
48. G. A. MOSER, F. E. TIBBETS, and M. D. RAUSCH, *Organometal. Chem. Syn.*, 1, 99 (1971).
49. J. D. COOK, B. J. WAKEFIELD, H. HEANEY, and J. M. JABLONSKI, *J. Chem. Soc. (C)*, 2727 (1968).
50. G. A. MOSER, E. O. FISCHER, and M. D. RAUSCH, *J. Organometal. Chem.*, 27, 379 (1971).
51. S. S. DUA and H. GILMAN, *J. Organometal. Chem.*, 12, 299 (1968).
52. YA. N. IVASHCHENKO, S. D. MOSHCHITSKII, and A. K. YELISEEVA, *Khim. Geterotsikl. Soedinenii*, 58 (1970).
53. F. BINNS and H. SUSCHITZKY, *J. Chem. Soc. (C)*, 1223 (1971).
54. D. E. PEARSON and D. COWAN, *Org. Synth.*, 44, 78 (1964).
55. H. GILMAN and S. Y. SIM, *J. Organometal. Chem.*, 7, 249 (1967).
56. S. D. ROSENBERG, J. J. WALBURN, and H. E. RAMSDEN, *J. Org. Chem.*, 22, 1606 (1957).
57. I. F. MIKHAILOVA and V. A. BARKHASH, *Zhur. Obshchei Khim.*, 37, 2792 (1967).
58. S. S. DUA and H. GILMAN, *J. Organometal. Chem.*, 12, 234 (1968).

59. W. STEINKOPF, H. JACOB, and H. PENZ, *Annalen*, 512, 136 (1934).
60. D. PILLON, S. TRINH, and R. CAVIER, *Bull. Chim. Therap.*, 32 (1970).
61. M. T. RAHMAN, M. R. SMITH, Jr., A. F. WEBB, and H. GILMAN, *Organometal. Chem. Synth.*, 1, 105 (1970-71).
62. J. G. NOLTES and J. W. G. VAN DEN HURK, *J. Organometal. Chem.*, 1, 377 (1964).
63. M. WEIDENBRUCH and S. BÖKE, *Chem. Ber.*, 103, 510 (1970).
64. F. E. PAULIK, S. I. E. GREEN, and R. E. DESSY, *J. Organometal. Chem.*, 3, 220 (1965).
65. G. B. DEACON and P. W. FELDER, *J. Chem. Soc.* (*C*), 2313 (1967).
66. P. G. COOKSON and G. B. DEACON, *J. Organometal. Chem.*, 27, C9 (1971).
67. G. B. DEACON and F. B. TAYLOR, *Inorg. Nuclear Chem. Letters*, 5, 477 (1969).
68. R. J. BERTINO, G. B. DEACON, and F. B. TAYLOR, *Austral. J. Chem.*, 25, 1645 (1972).
69. W. STEINKOPF and W. KOHLER, *Annalen*, 532, 250 (1937).
70. A. J. CANTY and G. B. DEACON, *Austral. J. Chem.*, 24, 489 (1971).
71. A. J. CANTY, G. B. DEACON, and P. W. FELDER, *Inorg. Nuclear Chem. Letters*, 3, 263 (1967).
72. A. J. CANTY and P. W. FELDER, *Austral. J. Chem.*, 21, 1757 (1968).
73. G. B. DEACON and P. W. FELDER, *Austral. J. Chem.*, 19, 2381 (1966).
74. W. A. SHEPPARD, *J. Amer. Chem. Soc.*, 92, 5419 (1970).
75. R. D. CHAMBERS and D. J. SPRING, *J. Organometal. Chem.*, 31, C13 (1971).
76. G. B. DEACON, *J. Organometal. Chem.*, 9, P1 (1967).
77. G. B. DEACON, *J. Organometal. Chem.*, 12, 389 (1968).
78. I. P. BELETSKAYA, L. V. SAVINYKH, V. N. GULYACHKINA, and O. A. REUTOV, *J. Organometal. Chem.*, 26, 23 (1971).
79. K. MOEDRITZER, *Advan. Organometal. Chem.*, 6, 171 (1968).
80. T. CHIVERS and B. DAVID, *J. Organometal. Chem.*, 13, 177 (1968); T. CHIVERS and B. DAVID, *J. Organometal. Chem.*, 10, P35 (1967).
81. R. D. CHAMBERS and T. CHIVERS, *J. Chem. Soc.*, 3933 (1965).
82. G. WULFSBERG and R. WEST, *J. Amer. Chem. Soc.*, 94, 6069 (1972).
83. V. BAZANT, V. CHVALOVSKY, and J. RATHOUSKY, *Organosilicon Compounds*, Vol. 1, p. 275, Publishing House of the Czechoslovak Academy of Sciences, Prague, 1965.
84. V. A. PONOMARENKO, A. D. SNEGOVA, M. R. PITINA, and A. D. PETROV, *Doklady Akad. Nauk S.S.S.R.*, 135, 339 (1960); *Chem. Abstr.*, 55, 12328 (1961).
85. B. J. AYLETT and I. A. ELLIS, *J. Chem. Soc.*, 3415 (1960).
86. A. YA. YAKUBOVICH and G. V. MOTSAREV, *Doklady Akad. Nauk S.S.S.R.*, 99, 1015 (1954); *Chem. Abstr.*, 50, 217 (1956).
87. G. V. MOTSAREV and A. YA. YAKUBOVICH, *Zhur. Obshchei. Khim.*, 27, 1318 (1957).
88. F. W. G. FEARON and H. GILMAN, *J. Organometal. Chem.*, 6, 577 (1966).
89. L. A. WALL, R. E. DONADIO, and W. J. PUMMER, *J. Amer. Chem. Soc.*, 82, 4846 (1960).
90. P. J. MORRIS and H. GILMAN, *J. Organometal. Chem.*, 11, 463 (1968).
91. P. J. MORRIS, F. W. G. FEARON, and H. GILMAN, *J. Organometal. Chem.*, 9, 427 (1967).
92. I. HAIDUC and H. GILMAN, *J. Organometal. Chem.*, 13, 257 (1968).
93. I. HAIDUC and H. GILMAN, *J. Organometal. Chem.*, 14, 73 (1968).
94. I. HAIDUC and H. GILMAN, *J. Organometal. Chem.*, 14, 79 (1968).

95. H. GILMAN and K. SHIINA, *J. Organometal. Chem.*, 8, 369 (1967).
96. M. R. SMITH, Jr. and H. GILMAN, *Organometal. Chem. Synth.*, 1, 265 (1971).
97. A. F. WEBB, D. S. SETHI, and H. GILMAN, *J. Organometal. Chem.*, 21, P61 (1970).
98. D. BALLARD, T. BRENNAN, F. W. G. FEARON, K. SHIINA, I. HAIDUC, and H. GILMAN, *Pure and Applied Chem.*, 19, 449 (1969).
99. K. SHIINA and H. GILMAN, *J. Amer. Chem. Soc.*, 88, 5367 (1966).
100. T. BRENNAN and H. GILMAN, *J. Organometal. Chem.*, 11, 625 (1968).
101. D. H. BALLARD and H. GILMAN, *J. Organometal. Chem.*, 12, 237 (1968).
102. D. H. BALLARD, T. BRENNAN, and H. GILMAN, *J. Organometal. Chem.*, 22, 583 (1970).
103. M. CORDEY-HAYES, R. D. W. KEMMITT, R. D. PEACOCK and G. D. RIMMER, *J. Inorg. Nucl. Chem.*, 31, 1515 (1969).
104. D. SEYFERTH and A. B. EVNIN, *J. Amer. Chem. Soc.*, 89, 1468 (1967).
105. R. D. CHAMBERS and T. CHIVERS, *J. Chem. Soc.*, 4782 (1964); R. D. CHAMBERS and T. CHIVERS, *Proc. Chem. Soc.*, 108 (1963).
106. D. SEYFERTH and A. B. EVNIN, *J. Amer. Chem. Soc.*, 89, 952 (1967).
107. T. CHIVERS, unpublished observations.
108. S. S. DUA, R. C. EDMONDSON, and H. GILMAN, *J. Organometal. Chem.*, 24, 703 (1970).
109. J. M. MILLER, *J. Chem. Soc.* (*A*), 828 (1967).
110. J. CHATT and B. L. SHAW, *J. Chem. Soc.*, 1718 (1960).
111. J. CHATT and B. L. SHAW, *J. Chem. Soc.*, 285 (1961).
112. M. D. RAUSCH and F. E. TIBBETTS, *Inorg. Chem.*, 9, 512 (1970).
113. J. R. MOSS and B. L. SHAW, *J. Chem. Soc.* (*A*), 1793 (1966).
114. M. D. RAUSCH, Y. F. CHANG, and H. B. GORDON, *Inorg. Chem.*, 8, 1355 (1969).
115. M. D. RAUSCH and F. E. TIBBETTS, *J. Organometal. Chem.*, 21, 487 (1970).
116. R. B. KING, *Inorg. Chem.*, 2, 531 (1963).
117. R. B. KING, *J. Amer. Chem. Soc.*, 85, 1922 (1963).
118. D. R. FAHEY, *J. Amer. Chem. Soc.*, 92, 402 (1970).
119. G. M. KRAAY, *Rec. Trav. Chim.*, 49, 1082 (1930).
120. S. S. DUA, A. E. JUKES, and H. GILMAN, *J. Organometal. Chem.*, 12, P24 (1968).
121. S. S. DUA, A. E. JUKES, and H. GILMAN, *Organometal. Chem. Synth.*, 1, 87 (1970-71).
122. A. E. JUKES, S. S. DUA, and H. GILMAN, *J. Organometal. Chem.*, 12, P44 (1968).
123. S. S. DUA, A. E. JUKES, and H. GILMAN, *Org. Prep. and Procedures*, 1(3), 187 (1969).
124. A. E. JUKES, S. S. DUA, and H. GILMAN, *J. Organometal. Chem.*, 21, 241 (1970).
125. M. R. SMITH, Jr., M. T. RAHMAN, and H. GILMAN, *Organometal. Chem. Synth.*, 1, 295 (1971).
126. M. R. CHURCHILL and M. V. VEIDIS, *Chem. Comm.*, 1099 (1970).
127. M. R. CHURCHILL, T. A. O'BRIEN, M. D. RAUSCH, and Y. F. CHANG, *Chem. Comm.*, 992 (1967).
128. M. R. CHURCHILL and T. A. O'BRIEN, *J. Chem. Soc.* (*A*), 2970 (1968).
129. M. R. CHURCHILL and T. A. O'BRIEN, *J. Chem. Soc.* (*A*), 266 (1969).
130. G. B. DEACON and P. W. FELDER, *Austral. J. Chem.*, 20, 1587 (1967).
131. H. GILMAN and P. J. MORRIS, *J. Organometal. Chem.*, 6, 102 (1966).

402   T. CHIVERS AND B. WAKEFIELD

132. I. HAIDUC, I. HAIDUC, and H. GILMAN, *J. Organometal. Chem.*, 11, 459 (1968).
133. D. J. CRAM, R. C. HELGESON, D. LOCK, and L. A. SINGER, *J. Amer. Chem. Soc.*, 88, 1325 (1966).
134. J. H. BEYNON, *Mass Spectrometry and its Applications to Organic Chemistry*, p. 298, Elsevier Publishing Co., Amsterdam, 1960.
135. L. SMITHSON, A. K. BHATTACHARYA, and F. L. HEDBERG, *Org. Mass Spectrometry*, 4, 383 (1970).
136. T. CHIVERS, G. F. LANTHIER, and J. M. MILLER, *J. Chem. Soc. (A)*, 2556 (1971).
137. S. C. COHEN, *J. Chem. Soc. (A)*, 632 (1971).
138. R. V. PARISH and R. H. PLATT, *J. Chem. Soc. (A)*, 2145 (1969); R. V. PARISH and R. H. PLATT, *Chem. Comm.*, 1118 (1968).
139. T. CHIVERS and J. R. SAMS, *J. Chem. Soc. (A)*, 928 (1970); T. CHIVERS and J. R. SAMS, *Chem. Comm.*, 249 (1969).
140. M. CORDEY-HAYES, *J. Inorg. Nucl. Chem.*, 26, 2306 (1964).
141. H. A. STÖCKLER and H. SANO, *Trans. Faraday Soc.*, 64, 577 (1968).
142. V. I. BREGADZE, T. A. BABUSHKINA, O. YU. OKHLOBYSTIN, and G. K. SEMIN, *Teor. Eksp. Khim.*, 3, 547 (1967).
143. R. WEST and G. WULFSBERG, private communication.
144. M. CORDEY-HAYES, R. D. W. KEMMITT, R. D. PEACOCK, and G. D. RIMMER, *J. Inorg. Nucl. Chem.*, 31, 1515 (1969).
145. H. A. STOCKLER and H. SANO, *Trans. Faraday Soc.*, 64, 577 (1968).
145. G. M. BROWN, F. L. HEDBERG, and H. ROSENBERG, *J. Chem. Soc. Chem. Comm.*, 5 (1972).

CHAPTER 4

Polychloroaromatics and Heteroaromatics of Industrial Importance

M. B. Green

Imperial Chemical Industries Ltd., Mond Division, Runcorn, Cheshire

403

4.1   POLYCHLOROBENZENES—INTRODUCTION

Chlorination of benzene was one of the first manufacturing processes
for an organic chemical to be carried out industrially. Production was
started in England in 1909 by the United Alkali Company (now part
of Imperial Chemical Industries Limited) [1] and in the United
States in 1915 by the Dow Chemical Company [2]. Up to that time
the 'heavy' chemical industry had been mainly concerned with
production of inorganic chemicals, particularly acids and alkalies.
Widespread introduction of manufacture of sodium hydroxide by
electrolysis of brine led to production of large amounts of chlorine
for which, at that time, there were no commercial outlets,
particularly as manufacture of bleaching powder was becoming
obsolescent. Petrochemicals were still a thing of the future and the
only organic raw materials which were available in large amounts
were those derived from coal-tar, such as benzene. It was thus logical
to look to chlorination of these products as a possible profitable use
of the excess of chlorine to which industry was irrevocably tied as a
result of the natural association of one atom of sodium with one
atom of chlorine.

Chlorination of benzene is, therefore, of historical interest as representing the start of the heavy organic chemicals industry. It was first studied by Jungfleisch in 1868 [3] but it was not until around 1910 that the need to find an outlet for unwanted chlorine provided the economic stimulus to its speculative introduction as an industrial process. The venture was successful because a major use was discovered for monochlorobenzene, namely, hydrolysis to give phenol for manufacture of picric acid to meet the demand for explosives in World War I. Annual world production of dichloro-benzenes is now about 200,000 tonnes. It is this interaction of chemistry, economics and history which provides the basis for

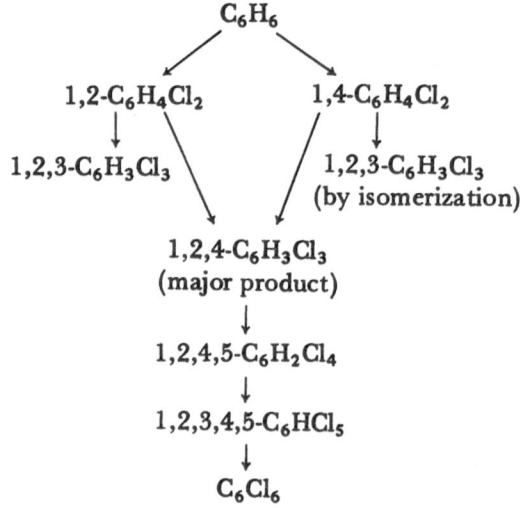

C_6H_6

$1,2\text{-}C_6H_4Cl_2$ $1,4\text{-}C_6H_4Cl_2$

$1,2,3\text{-}C_6H_3Cl_3$ $1,2,3\text{-}C_6H_3Cl_3$
(by isomerization)

$1,2,4\text{-}C_6H_3Cl_3$
(major product)

$1,2,4,5\text{-}C_6H_2Cl_4$

$1,2,3,4,5\text{-}C_6HCl_5$

C_6Cl_6

Fig. 4.1   Industrial catalytic chlorination of benzene

successful technology, which can be summarized as 'the right product at the right time'. There is an interesting parallel in chlorinated heterocyclic compounds (Section 4.18) where specula-tive industrial production of cyanuric chloride, about a century after the basic chemical discovery, resulted in two major groups of commercial products—the 'Procion' dyes and the triazine herbicides. The very recent introduction of economical manufacturing process for pentachloropyridine—a compound which was first discovered in 1897—is currently stimulating a similar search for profitable products based on this intermediate.

The process of chlorination of benzene is simple, since all hydrogen atoms can be replaced progressively, and its mechanism and kinetics have been studied by a number of workers [4-6]. Benzene and chlorine are passed over a heated catalyst and, since the rate of chlorination falls drastically as successive atoms of chlorine are introduced, it is possible to carry out the process stepwise (Fig. 4.1). Various isomers are produced in differing proportions, thus, the main dichlorobenzenes formed are the 1,2- and 1,4- isomers, while the main trichlorobenzene is the 1,2,4- isomer. The usual catalysts are iron, aluminium, tin or antimony chlorides or metallic iron or aluminium [7, 8] and the reaction is, at least in the first stages, highly exothermic. Industrially, chlorination of benzene is carried out mainly to obtain monochlorobenzene and 1,2- and 1,4-dichlorobenzenes. The usual catalyst is ferric chloride and the process is operated on a batch or continuous basis [9, 10]. If dichlorobenzenes are specifically required, reaction temperature is raised to 150-190°. The two dichlorobenzenes are separated by fractional distillation or, less usually, by crystallizing out the solid 1,4- isomer, filtering it off and washing it with methanol. If only the 1,4- isomer is required, either a specific orientating additive such as benzene sulphonic acid is added to the ferric chloride catalyst [11], or the crude mixture of dichlorobenzenes is chlorinated further when the 1,2- isomer chlorinates much more rapidly than the 1,4- isomer, giving rise to an easily separable mixture of 1,4-dichlorobenzene and 1,2,4-trichlorobenzene [12]. If 1,3-dichlorobenzene is required, this is obtained by isomerizing the 1,2- or 1,4-dichlorobenzenes by heating at 120° under pressure with aluminium chloride [13].

Further chlorination of dichlorobenzenes gives successively tri-, tetra-, penta-, and hexachlorobenzenes as reaction temperature is progressively raised. The main trichlorobenzene produced is the 1,2,4- isomer together with some 1,2,3- and 1,3,5- isomers. For industrial purposes the mixture is generally not separated. 1,2,4-trichlorobenzene is also manufactured by dehydrochlorination of the unwanted α and β isomers of hexachlorocyclohexane formed in production of the γ isomer which is the widely-used insecticide lindane. Tetrachlorobenzene is manufactured mainly for hydrolysis to 2,4,5-trichlorophenol for production of the herbicide 2,4,5-T. Hexachlorobenzene is the final product of chlorination, and continuous processes for its manufacture have been described [14] and it is used for manufacture of pentachlorophenol. It is also produced industrially by reaction of hexachlorocyclohexane with chlorine under pressure at about 200° [15, 16] or by oxidation of hexachlorocyclohexane with air at 450-550° over a copper chloride catalyst [17], and it has recently become available as a by-product from manufacture of carbon tetrachloride.

4.2   USES OF 1,2-DICHLOROBENZENE

4.2.1   Solvent

1,2-Dichlorobenzene is used as a solvent in a number of chemical processes, for instance, in production of toluene di-isocyanate from which urethanes are obtained for expanded polyurethanes, bonding agents, etc. [18].

4.2.2   Engine Cleaning Formulations

1,2-Dichlorobenzene is used to make cleaning formulations for aircraft and transport engines generally in conjunction with the mixture of cresols known as 'cresylic acid' [19]. These formulations are very powerful removers of the carbon and gummy deposits from lubricants, which are not easily removed by other means [20, 21]. In addition, they strip off paints and stoved enamels, including those not affected by alkalies, and facilitate decarbonizing by removing the lacquers from the layers of carbonized oil so that residual porous deposits can be readily removed by light brushing.

4.2.3   Heat Transfer Medium

1,2-Dichlorobenzene has been used in the range 150-250° as a heat transfer medium as it requires pressures much lower than those needed for steam at the same temperature. It is also used as a coolant for magnetic coils [22].

4.2.4   Paints and Paint Stripping

1,2-Dichlorobenzene has been used as a constituent of metal polishing compounds, paints, paint strippers and surface coatings [18].

4.2.5   Fuel Combustion Additive

1,2-Dichlorobenzene has been used as a source of chlorine in high temperature combustion zones [18].

4.2.6   Sewer Cleaning

1,2-Dichlorobenzene has been used in sewers and septic tanks where it is effective in eliminating sulphides and preventing accumulation of grease and other odiferous material [18].

4.2.7   Insecticide Formulations

1,2-Dichlorobenzene is used as a component in eradicant formulations against wood-beetle infestations in buildings and furniture. It is claimed to have some insecticidal action in its own right but is non-persistent so is generally used with a long-lasting insecticide such as benzene hexachloride [18].

4.2.8   Water Weed Control

In ponds and lakes which are not used for drinking, washing or bathing and in which there are no fish, 1,2-dichlorobenzene can be used for water-weed control by spraying through an underwater nozzle [18].

4.2.9   Chemical Intermediate

1,2-Dichlorobenzene is used in the synthesis of many dyestuffs, pharmaceuticals and pesticides, via processes such as nitration, sulphonation, etc. Hydrolysis gives successively 2-chlorophenol, a cheap industrial biocide for fungus control, and pyrocatechol, an antioxidant. Reaction with ammonia gives 2-chloroaniline (Fast Yellow G Base) which is used for manufacture of azo dyestuffs [18]. A substantial proportion of the total production of 1,2-dichlorobenzene is used to make 3,4-dichloroaniline for pesticides and dyestuffs.

4.2.10   Industrial Hazards of 1,2-Dichlorobenzene

Although the flashpoint of 1,2-dichlorobenzene is 67.8° it has still to be treated as an inflammable substance and the usual precautions against fire observed in its handling. It is not corrosive to common metals, such as aluminium, magnesium, zinc, copper, iron or their alloys. Mild steel or cast iron vessels are suitable for storage. It has been suggested that repeated breathing of air containing more than 50 ppm v/v may affect the liver and kidneys. At ordinary temperature the vapour pressure is so low (1 mm Hg at 20°) that dangerous concentrations of vapour are unlikely to occur in well-ventilated conditions, but special precautions are needed if the solvent is heated, and extraction equipment must be provided. A vapour concentration above 25 ppm is undesirably high and the limit suggested by the American Conference of Government Industrial Hygienists in 1968 to avoid lachrymation was 50 ppm.

4.3   USES OF 1,4-DICHLOROBENZENE

4.3.1   Deodorant

1,4-Dichlorobenzene is widely used in block form as a deodorant in lavatories, and other confined places [23-25]. It can be cast molten by heating in an enamelled or aluminium pan and running into suitably designed moulds. More usually, the freshly-ground dichlorobenzene is compressed into blocks, and perfumes such as 0.5-1.0% w/w pine oil and 0.05% w/w spike lavender are often added. The blocks can be coloured by incorporating either wax soluble dyes of the 'Waxoline' range or pigment type dyes of the 'Monastral' range. For use in lavatories 2.5-5.0% of a bleaching agent such as 1,3-dichloro-5,5-dimethylhydantoin may be added [26].

4.3.2   Clothes Moth Control

1,4-Dichlorobenzene is widely used as a fumigant for control of the larvae of the clothes moth and is recommended by the U.S. Department of Agriculture and the British Wool Industries Research Association [27]. It has no residual activity so does not 'moth-proof', but it is lethal to the clothes moth in its larval and pupal stages. Egg laying and subsequent hatching may not be prevented but feeding by the larvae is almost immediately inhibited. At 15° 1000 litres of air contain about 5 g 1,4-dichlorobenzene vapour, and this concentration will arrest immediately the activities of fully grown and ravenous larvae and subsequently kill them. To maintain an effective fumigant concentration it is obviously essential to use 1,4-dichlorobenzene in a confined space such as a closed cupboard or to contain the garments in polythene bags. The compound has no harmful effects on fabrics or furs and is not toxic to man or animals in the concentrations needed for moth control. The vapour can, however, taint foodstuffs.

4.3.3   Other Uses

About 70% of the 1,4-dichlorobenzene produced is used for deodorant blocks and clothes moth control. It has very minor application as an extreme pressure lubricant [28] and in production of grinding wheels, and is used as an intermediate for manufacture of dyestuffs, pharmaceuticals and pesticides [29].

4.3.4   Industrial Hazards of 1,4-Dichlorobenzene

1,4-Dichlorobenzene is practically non-inflammable and is non-corrosive. Exposure to very high concentrations of the vapour can cause anaesthesia but irritation of the eyes and nose becomes intolerable long before a dangerous concentration is established. A maximum concentration of 75 ppm has been recommended for continuous working. If 1,4-dichlorobenzene is held in contact with the skin for long periods it produces a burning sensation but no serious dermatitic effects.

4.4   USES OF HIGHER CHLORINATED BENZENES

The commercial liquid mixture of trichlorobenzenes is used as a solvent for oil-soluble dyes, as a degreasing agent, as a dielectric fluid, as a lubricating oil-additive and as a termite exterminant. Pure 1,2,4-trichlorobenzene finds use as a coolant in electrical installations [30] and in glass tempering [31].

Tetrachlorobenzenes have few industrial uses. They have been suggested as components of fire-resisting compositions and as setting-point depressants in transformer oils [32]. The main use of 1,2,4,5-tetrachlorobenzene is for hydrolysis to 2,4,5-trichlorophenol.

Hexachlorobenzene has some use as a wood preservative and in seed-dressing mixtures [33] but its main application is in manufacture of pentachlorophenol, a wood preservative and industrial biocide.

4.5   POLYCHLORONAPHTHALENES–INTRODUCTION

Naphthalene can be chlorinated under the same conditions as benzene, namely, with a catalyst such as ferric chloride [34]. As with benzene, all the hydrogen atoms can, with increasing difficulty, be replaced by chlorine. The number of possible isomers is large, e.g. two monochloro-, ten dichloro-, fourteen trichloro-, etc. Many of these have not yet been isolated in a pure state. The products used industrially are merely chlorinated naphthalenes with a specified chlorine content.

4.6   USES OF POLYCHLORONAPHTHALENES

Chlorinated naphthalenes are another example of the time-lag between chemical discovery and industrial exploitation since they were first made by Laurent in 1833 [35], re-examined by Fischer in 1878 [36] but were not manufactured until Aylsworth [37-39] observed in 1909 that they were useful protective materials for impregnating wood, paper and fabrics. They have high specific inductance capacity, low power loss factor and high electrical resistance and are consequently very suitable for imparting electrical insulating properties, e.g. in covering fabrics for cables. As with chlorobenzene, the success of this speculative industrial venture was a result of military demands in World War I.

The chlorinated naphthalenes have a pronounced dermatitic action and their vapours are toxic to the liver [40, 41]. They have, therefore, been largely replaced by more acceptable polymeric materials and their production has been largely discontinued. As they have no other important commercial applications they are now obsolescent in an industrial sense. There is a small outlet as high-boiling solvents, plasticisers, insecticides and as components in lubricating and cutting oils, protective coatings and underwater paints [42-44], but even these uses are dying out as more suitable and safer alternatives become available.

4.7 POLYCHLOROBIPHENYLS—INTRODUCTION

Biphenyls, terphenyls and higher polyphenyls can be chlorinated in a similar way to benzene using a catalyst such as ferric chloride, and batch and continuous processes have been described [45-49]. No attempt is made industrially to isolate any specific isomers; the chlorination is normally carried on to a specified chlorine content and the mixed product purified by distillation under reduced pressure and sometimes also by slurrying in the molten state with Fullers earth and filtering [50].

The industrial value of the chlorinated biphenyls resulted from their great thermal stability, chemical inertness and non-flammability coupled with excellent electrical properties. Commercial manufacture was started in 1929 and they have been made throughout the world and very widely applied. However, now it has been realized that these compounds present a much greater environmental hazard than DDT and the polychlorinated insecticides because they are used on a larger scale and are much more widely distributed. Therefore, they have either been banned by law or their manufacture discontinued voluntarily. Thus, this class of compound has lost much of its industrial importance as a result of increased public concern about environmental pollution, although world production in 1971 still amounted to 50,000 tonnes.

4.8 USES OF POLYCHLOROBIPHENYLS

4.8.1 Electrical Uses

Commercial chlorinated biphenyl mixtures have a dielectric constant of 6.0 at 1000 cycles/25° and this can be increased to 7.0 by isomerization techniques [51-55]. This property, together with their high resistivity, high dielectric strength, very low power loss factor, non-flammability and chemical inertness enabled electrical equipment such as transformers to be greatly reduced in size for the same capacity and voltage. They have, therefore, been widely used as transformer fluids, as sealants for electrical equipment, as impregnants for condensers and for insulating materials, and as plasticizers for cable-covering materials such as neoprene, rubber and polyvinyl-chloride [56-60].

4.8.2   Plastics

The chlorinated biphenyls have been extensively used in combination with asphalt, chlorinated rubber, styrene–butadiene co-polymers and other plastics as protective coatings for wood, metal and concrete [61-63]. In combination with dioctyl phthalate they have been one of the most ubiquitous plasticizers for polyvinylchloride articles of all types in industrial and domestic use [61, 64].

4.8.3   Paints

Lower chlorinated biphenyls have been used to give flexibility to paint films, and higher chlorinated materials to give hardness. In nitrocellulose lacquers they are used to increase weather resistance, adhesion and lustre. They are compatible with most normal paint constituents [65-68].

4.8.4   Lubricants

The resinous lower chlorinated biphenyls are used as lubricants under conditions of high pressure or temperature or in underwater applications.

4.8.5   Heat Transfer Media

The chlorinated biphenyls have been used as inert and non-flammable heat transfer media in applications where uniform temperature is required [69].

4.8.6   Other Uses

Chlorinated biphenyls have been used as constituents in wax polishes [70] and in combination with substances such as antimony oxide or barium sulphate to prepare fire-resisting paints or to manufacture fireproof fibreboard by impregnation [71].

They have also been incorporated into sealing compounds for wood, canvas, etc. to prevent biodeterioration and in encapsulation processes for carbonless copying papers [72].

4.8.7   Industrial Hazards of Chlorinated Biphenyls

The vapours of chlorinated biphenyls are toxic and the maximum safe concentration is 0.5-1.0 mg per litre, and these are the maximum allowable concentrations for exposure during an eight hour working day. They are not pronouncedly skin irritant although repeated or prolonged contact with skin can produce dermatitic effects.

4.8.8 Environmental Hazards of Chlorinated Biphenyls

In recent years it has become apparent that polychlorinated benzenes, naphthalenes, biphenyls etc and their polychlorinated cycloalkyl analogues are very persistent in the environment because of their inertness to chemical attack, micro-organisms and natural processes of decay. It has also become apparent that, because of their strong lipophilic character, they tend to accumulate in the fatty tissues of living organisms, some of which can specifically extract and concentrate them. Although there is no evidence that this has had any deleterious or toxic effects on man, there is increasing evidence that it can be harmful to birds and to marine life. Mounting public concern with all forms of environmental pollution has resulted in the discontinuance of use of chlorinated biphenyls in situations where it could produce environmental contamination, e.g. as a plasticizer or in paints, polishes, etc., on the principle of 'better safe than sorry'.

An international agreement in 1973 by the 23 member countries of the Organization for Economic Cooperation and Development controls production of polychlorinated biphenyls and limits their use to sealed systems such as transformers.

It has now been shown that the toxic effects of polychlorinated biphenyls are mainly due to the presence of small amounts of chlorinated dibenzofuran derivatives [73]. These by-products are present in differing amounts according to the manufacture (Fig. 4.2).

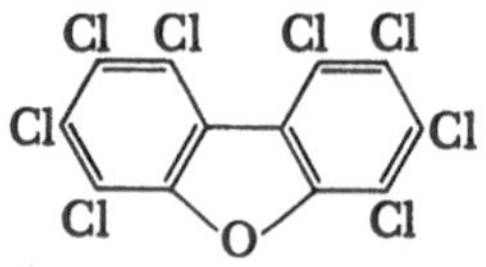

Fig. 4.2 Polychlorinated dibenzofuran. A toxic impurity in commercial polychlorinated biphenyl

It is noteworthy that there is evidence that the polychlorinated heteroaromatics do not suffer from the same disadvantages because they are, in fact, attacked by micro-organisms. Moreover, most of them undergo photochemical decomposition and their different lipophilic/hydrophilic balance does not result in accumulation in fatty tissues. Also, there is no evidence that any toxic by-products analogous to the dibenzofurans are formed during their production.

4.9 POLYCHLOROPHENOLS—INTRODUCTION

Apart from the polychlorinated hydrocarbons the only poly-chlorinated phenyl compounds which are produced on a significant scale are the polychlorophenols [74]. Two other compounds should

also be mentioned, namely, 3,4-dichloroaniline which is manufactured in the U.S.A. on a scale of several thousand tons per year,
by nitration of 1,2-dichlorobenzene followed by separation and
reduction of the appropriate isomer for conversion into the
herbicides diuron, linuron, propanil and related compounds; 2,5-
dichloroaniline which is an intermediate for the dyestuffs Pigment
Red 2 and Direct Green 12 and is made in the U.S.A. on a scale of
about 400 tons/year (Fig. 4.3).

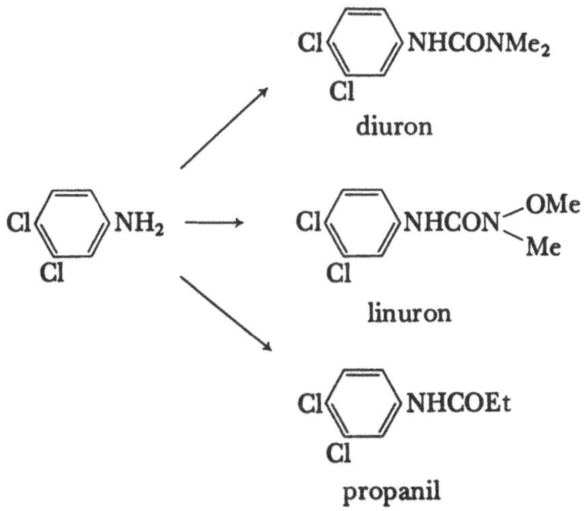

Fig. 4.3   Uses of 3,4-dichloroaniline

The utility of the polychlorophenols resides almost entirely in
their biocidal properties which make them useful as comparatively
cheap antiseptics, disinfectants, repellants, fungicides, wood preservatives, mould inhibitors, etc. Although their activities are, in
general, lower than those of many of the more complex modern
pharmaceutical and industrial biocides, they can be used in situations
where cost is a critical factor (Table 4.1).

A very large industrial use of 2,4-dichlorophenol and 2,4,5-
trichlorophenol, is to manufacture the corresponding phenoxyacetic
acids, 2,4-D and 2,4,5-T which are used on a very large scale all over

TABLE 4.1   Bacteristatic Activity of Chlorophenols Against *E. Typhosa* in
Absence of Organic Matter

2,4-Dichlorophenol	13.3 (phenol = 1)
2,5-Dichlorophenol	10.2
2,4,5-Trichlorophenol	18.1
2,4,6-Trichlorophenol	22.6

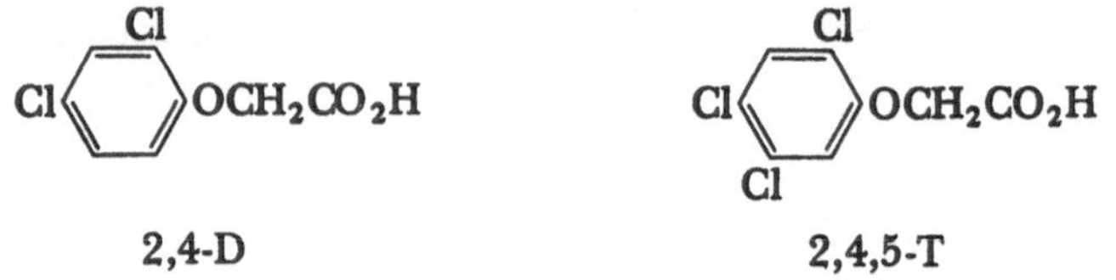
Cl
Cl
OCH₂CO₂H
2,4-D
Cl
Cl
OCH₂CO₂H
Cl
2,4,5-T

Fig. 4.4 Phenoxyacetic acid herbicides

the world as herbicides [75-77] (Fig. 4.4). These compounds will be discussed later.

The most usual method of manufacture of polychlorinated phenols is by direct chlorination of a phenol in the presence of a catalyst such as ferric chloride or aluminium chloride. Successive replacement of the hydrogen atoms in phenol itself gives mainly 2- and 4-chlorophenols, 2,4- and 2,6-dichlorophenols, 2,4,6-trichlorophenol, 2,3,4,6-tetrachlorophenol and pentachlorophenol (Fig. 4.5).

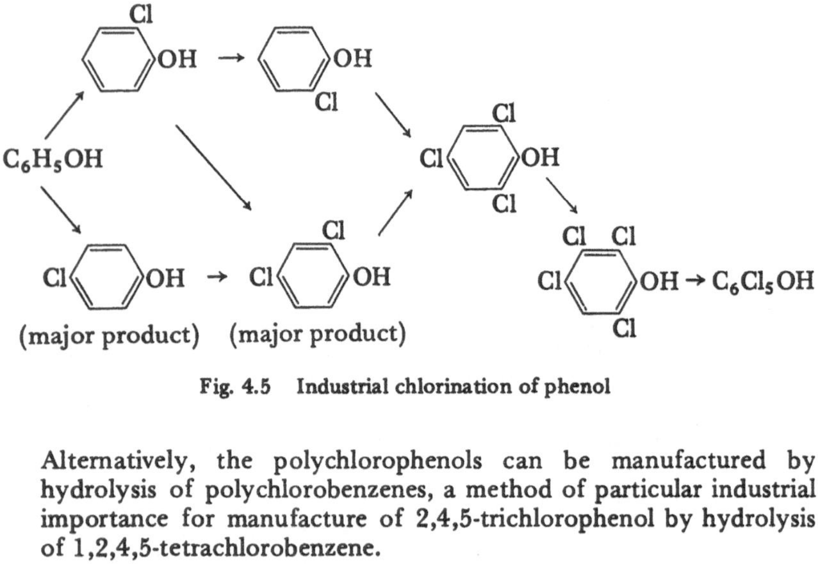
C₆H₅OH
(major product)
(major product)
C₆Cl₅OH

Fig. 4.5 Industrial chlorination of phenol

Alternatively, the polychlorophenols can be manufactured by hydrolysis of polychlorobenzenes, a method of particular industrial importance for manufacture of 2,4,5-trichlorophenol by hydrolysis of 1,2,4,5-tetrachlorobenzene.

4.10 USES OF POLYCHLOROPHENOLS

TABLE 4.2 Estimated Production of Chlorophenols in the U.S.A. in 1970

2,4-Dichlorophenol	17,000 tonnes
2,4,5-Trichlorophenol	9000 tonnes
Tetrachlorophenol	6000 tonnes
Pentachlorophenol	28,000 tonnes

4.10.1   2,4-Dichlorophenol

2,4-Dichlorophenol is mainly manufactured by chlorinating molten phenol at 80-100° until a product with melting point 34-36° is obtained. Under controlled conditions, very little 2,6-dichlorophenol or trichlorophenols are produced [74].

Two derivatives of 2,4-dichlorophenol which have some commercial value are 2,4-dichlorophenyl benzene sulphonate, which is used as an acaricide for fruit trees, and 2,2-dihydroxy-3,5,3',5'-tetrachlorodiphenylmethane which is used as a moth-proofing agent and seed disinfectant.

By far the major industrial use of 2,4-dichlorophenol is manufacture of the herbicide 2,4-dichlorophenoxyacetic acid (2,4-D) by condensation of 2,4-dichlorophenol with chloroacetic acid in sodium hydroxide solution. 2,4-Dichlorophenol has also found some use as an antiseptic [78, 79].

4.10.2   2,4,5-Trichlorophenol

2,4,5-Trichlorophenol is mainly manufactured by continuous hydrolysis of 1,2,4,5-tetrachlorobenzene in a coil reactor at 160° with methanolic sodium hydroxide. Its major use is to manufacture the herbicide 2,4,5-trichlorophenoxyacetic acid (2,4,5-T) by condensation with chloroacetic acid and the closely related herbicide 2,4,5-trichlorophenoxypropionic acid.

A minor use of the alkali metal salts of 2,4,5-trichlorophenol is as fungicides and wood preservatives [80, 81].

4.10.3   2,4,6-Trichlorophenol

2,4,6-Trichlorophenol is manufactured by direct chlorination of phenol. It is used as a cheap general antiseptic [82], a wood preservative [80], a glue preservative [83], and as an antifungal agent for textile treatment [84].

4.10.4   2,3,4,6-Tetrachlorophenol

2,3,4,6-Tetrachlorophenol is manufactured by further chlorination of 2,4,6-trichlorophenol. It is used as a preservative for latex [85], wood [86] and leather [87].

4.10.5   Pentachlorophenol

Pentachlorophenol is manufactured by hydrolysis of hexachlorobenzene [88, 89] or by further chlorination of mixtures of chlorinated phenols [90, 91]. In the latter process chlorine is passed into phenol at 105° until the melting point of the product is 95° and 3-4 atoms of chlorine have been introduced into the molecule. The temperature is then raised gradually, keeping it 10° above the melting point of the reaction mass until five atoms of chlorine have been introduced. In the chlorination process, catalyst concentration is very critical and is generally 0.0075 mole anhydrous aluminium chloride per mole of phenol.

Hydrolysis of hexachlorobenzene is generally carried out under pressure at 125-170° in methanolic sodium hydroxide for about two hours. The methanol is partially distilled off, water is added, the solution filtered to remove insoluble material then acidified to give pentachlorophenol.

Pentachlorophenol is widely used as an industrial biocide, especially for wood preservation [92], and it is particularly useful in this connection as a good fungicide which is also effective against termites, which are a main cause of wood deterioration in the sub-tropics. Its low solubility and high persistence give protection for many years. It has come under criticism from an environmental view-point because of this great persistence, and its use may therefore decrease. In general it is too toxic for agriculture and horticulture although it has found a limited use for weed control in cotton and for the 'stale seed-bed' technique in sugar beet [93].

If pentachlorophenol is introduced into water as a concentrated solution of the sodium salt it is effective as a molluscicide at a few parts per million. It is strongly absorbed by, and is very toxic to, snail eggs but it is also toxic to fish and to some vegetation, and the concentrate presents great handling hazards for the operators. Aquatic snails are adapted to make good molecular contact with a great volume of water in order to obtain sufficient oxygen and they are therefore very susceptible to highly lipophilic substances, and this is why a polychlorinated compound is particularly effective, since the chlorine atoms increase lipoid solubility.

4.11 OTHER POLYCHLOROAROMATICS—INTRODUCTION

Apart from the various uses already described for the poly-chlorinated benzenes and a limited use of polychlorinated phenols and polychlorinated anilines as intermediates for production of dyestuffs, most of the industrial uses claimed in the patent literature for polychlorinated phenyl compounds are as intermediates for the manufacture of pharmaceuticals and pesticides. This particular claim is generally made for any new organic compound and little purpose would be served in citing all such references since many would be misleading as implying real commercial utility where, in fact, there is none. The only compounds which will be referred to are those which have actually been put on to the market and have achieved some degree, albeit a small one, of commercial success.

There appear to be no pharmaceuticals based on polychlorinated phenyl compounds which have made any notable commercial impact. On the other hand, there are many pesticides of this type, including a number which are extremely important in present day horticulture and agriculture and which are manufactured commercially on a very large scale. The accompanying tables (4.3-4.6) list the chemical structures, common names (those marked * are accepted common names of the British Standards Institution), the trade name, the originating company and the patent reference of all polychlorinated phenyl pesticides which have been put on to the market. In the following paragraphs comments will be made on those which are of outstanding importance.

It is interesting to consider why polychlorinated compounds have found extensive application as pesticides but little as pharmaceuticals. To be effective as pesticides, compounds generally need a high lipoid solubility or, more precisely, a particular lipoid/water partition, which is generally much higher than that desirable in pharmaceuticals. Introduction of chlorine atoms into a molecule increases lipoid/water partition but it also increases toxicity, which is a desirable effect for a pesticide, the purpose of which is to kill, but a very undesirable effect for a pharmaceutical which is generally aimed at modifying some biochemical or physiological process with minimum damage to the patient. The one exception is, perhaps, externally applied bactericides and it is significant that it is in this area that polychlorinated compounds, particularly polychlorinated phenols, find their sole pharmaceutical utility, and these will be commented on later.

Many herbicides fall into the general classification of compounds with an acid hydrogen atom and, in general, high activity is associated with a narrow range of Hansch π values and pKa's. Chlorine atoms tend to increase the π value and also to raise pKa by virtue of their electron-withdrawing effect. It is this combination of properties which makes chlorine a particularly effective substituent in herbicides and accounts for the fact that, in general, the derivatives of polychlorinated benzenes have much more usefulness as herbicides than as fungicides, acaricides or insecticides.

4.12 POLYCHLOROPHENOXYACETIC ACID HERBICIDES

The introduction of the phenoxyacetic acid herbicides in the early 1940's was the starting point of the very rapid expansion of chemical weed control during the past thirty years and marks the beginning of the pesticide industry in a fine-chemical sense. Before that time weed-killers had been mainly inorganic compounds, such as sodium

TABLE 4.3   Herbicides

Structure	Common name	Trade name	Originating company	Patent ref.
A. Derivatives of chlorinated phenoxyalkanoic acids				
Cl Cl⟨C$_6$H$_3$⟩OCH$_2$CO$_2$H	2,4-D*	—	Amchem	(139)
Cl Cl⟨C$_6$H$_2$⟩OCH$_2$CO$_2$H Cl	2,4,5-T*	—	Amchem	(140, 141)
Cl Cl⟨C$_6$H$_3$⟩OCHCO$_2$H Me	dichlorprop*	—	Boots	(142)
Cl Cl⟨C$_6$H$_2$⟩OCHCO$_2$H Cl Me	fenoprop*	Silvex	Dow	(143, 144)
Cl Cl⟨C$_6$H$_3$⟩O(CH$_2$)$_3$CO$_2$H	2,4-DB*	—	ICI	(145)
Cl Cl⟨C$_6$H$_2$⟩O(CH$_2$)$_3$CO$_2$H Cl	2,4,5-TB	—	May & Baker	(146)
B. Derivatives of chlorinated phenols				
Cl Cl Cl⟨C$_6$⟩OH Cl Cl	PCP	—	Dow	(147)
Cl Cl⟨C$_6$H$_3$⟩-O-⟨C$_6$H$_4$⟩NO$_2$	nitrofen*	FW 925	Rohm & Haas	(148, 149)
Cl NC⟨C$_6$H$_2$⟩OH Cl	chloroxynil	—	May & Baker	(150)
Cl Cl⟨C$_6$H$_3$⟩OCH$_2$CH$_2$OSO$_3$H	2,4-DES*	—	Henkel & Cie	(151, 152)

TABLE 4.3—*continued*

Structure	Common name	Trade name	Originating company	Patent ref.
Cl, Cl, Cl ring $OCH_2CH_2O_2CCCl_2Me$	erbon	Baron	Dow	(153)
Cl, Cl ring OCH_2CSNH_2	—	Bayer 50870	Bayer	(154)
[Cl, Cl ring OCH_2CH_2O]$_3$P	2,4-DEP	Galone	Uniroyal	(155)
Cl, Cl ring OP $\overset{\parallel}{S}$ $\begin{smallmatrix}OMe\\NHCHMe_2\end{smallmatrix}$	DMPA	Zytron	Dow	(156)

C. Derivatives of chlorinated benzoic acids

Structure	Common name	Trade name	Originating company	Patent ref.
Cl Cl, Cl ring CO_2H	TBA	—	Heyden Newport	(157, 158)
Cl OMe, Cl ring CO_2H	dicamba*	Mediben . Banvel D	Velsicol	(159–161)
Cl OMe, Cl Cl ring CO_2H	tricamba*	Metriben	Velsicol	(162)
H_2N Cl, Cl ring CO_2H	chloramben*	Amiben	Amchem	(163)
O_2N Cl, Cl ring CO_2H	—	Dinoben	Amchem	(164)
MeO_2C Cl Cl, Cl Cl ring CO_2Me	chlorthal methyl*	Dacthal	Diamond Shamrock	(165, 166)

TABLE 4.3—*continued*

Structure	Common name	Trade name	Originating company	Patent ref.
D. Derivatives of chlorinated alinines				
MeO₂C⟨Cl Cl / Cl Cl⟩COSMe	OCS 21944	Glenbar	Velsicol	(167)
Cl⟨Cl⟩NHCOEt	propanil*	Rogue	Rohme & Haas	(168)
Cl⟨Cl⟩NHCOC=CH₂ (Me)	chloranocryl*	Dicryl	FMC	(169)
Cl⟨Cl⟩NHCOCH(CH)₂Me (Me)	—	Karsil	Du Pont	(170)
Cl⟨Cl⟩NHCOCH⟨CH₂/CH₂⟩	cypromid*	Clobber	Spencer	(171)
Cl⟨Cl⟩NHCO₂Me	swep	—	FMC	(172, 173)
Cl⟨Cl⟩NHCONMe₂	diuron*	Karmex	Du Pont	(174–176)
Cl⟨Cl⟩NHCONOMe (Me)	linuron*	Lorox	Du Pont	(177–180)
Cl⟨Cl⟩NHCON(CH₂)₃Me (Me)	neburon*	Kloben	Du Pont	(181)
E. Miscellaneous chlorophenyl derivatives				
⟨Cl Cl / Cl⟩CH₂CO₂H	chlorfenac*	Trifen	Heyden Newport	(182)
⟨Cl Me / Cl⟩CH₂OCOMe	—	ACS 93	Velsicol	(183–186)

TABLE 4.3—*continued*

Structure	Common name	Trade name	Originating company	Patent ref.
Cl — CN (with two Cl)	dichlobenil*	Casoron	Philips	(187)
Cl — $CSNH_2$ (with two Cl)	chlorthiamid*	Prefix	Shell	(188)
Cl — $CSNHCH_2OH$ (with two Cl)	–	TH 073	Philips	(189)
Cl — CNH_2HCl, NH (with two Cl)	–	7585	Shell	(190)
Cl — $CONHCMe_2C{\equiv}CH$ (with two Cl)	pronamide*	Kerb	Rohm & Haas	(191, 192)
Cl — $CH{=}NNHCO$—phenyl (with two Cl)	–	Bayer 58119	Bayer	–
Cl Cl — $CH_2N{=}$cyclohexyl (with three Cl)	–	THC 1626	Tenneco	–
Cl — $CH_2OCONHMe$ (with two Cl)	dichlormate*	Rowmate	Union Carbide	(193)
Cl —N ring NMe, O,O (with two Cl)	–	Tunic	Velsicol	(194)
Cl —N–N ring CMe_3 (with two Cl)	–	Ronstar	Rhodia	(195)

* Indicates accepted common name of the British Standards Institution.

TABLE 4.4   Fungicides

Structure	Common name	Trade name	Originating company	Patent ref.
C_6Cl_6	HCB	–	–	–
C_6Cl_5OH	PCP	–	Dow	(196)
	chloranil	Spergon	Uniroyal	(197)
	tecnazene*	Fusarex	Bayer	–
$C_6Cl_5NO_2$	quintozene	–	I.G.Farb	(198)
	TCNA	–	I.G.Farb	(199)
	dichloran*	Allisan	Boots	(200)
	chloroneb*	Demosan	Du Pont	(201)
	chlorothalonil	Daconil	Diamond Shamrock	(202)
$C_6Cl_5CH:NOH$	–	Minokol	Sumitomo	(203, 204)
	–	PH 50–82	Philips	(205)
	chlorani-formethan	Imugan	Bayer	(206)

TABLE 4.4—*continued*

Structure	Common name	Trade name	Originating company	Patent ref.
[structure: Cl OH HO Cl substituted diphenylmethane with CH₂ bridge, Cl Cl Cl Cl]	hexachloro-phene	—	—	(207)
[structure: Cl Cl benzene OCOCH₂CH₂CH₂N piperidine Me]		Pipron	Eli Lilly	(208)
[structure: Cl Cl Cl benzene OEt OPO phenyl O]		H 0034	Hokka	—
$C_6Cl_5OP[NMe_2]_2$ with O		—	—	Philips	(209)
[structure: naphthoquinone with O, O, Cl Cl]	dichlone*	Phygon	Uniroyal	(210, 211)
[structure: Cl Cl benzene N succinimide O O]		—	Ohric	Sumitomo	(212)
[structure: Cl Cl Cl Cl benzene fused ring O O]		—	Rabcide	Sumitomo	(213)
[structure: Cl Cl benzene N O O Me O Me]	dichlozoline	Sclex	Hokko	(214)

TABLE 4.4—*continued*

Structure	Common name	Trade name	Originating company	Patent ref.
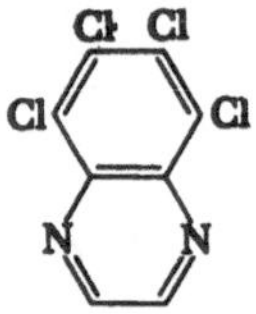	chlorquinox	Lucel	Bayer	(215)
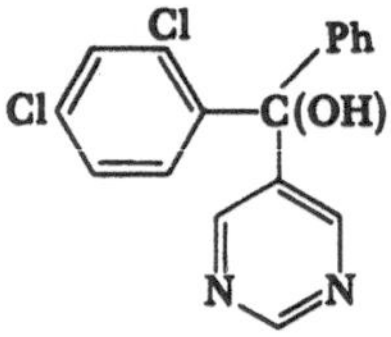	triarimol*	Elanocide	Eli Lilly	(216)

TABLE 4.5 Acaricides

Structure	Common name	Trade name	Originating company	Patent ref.
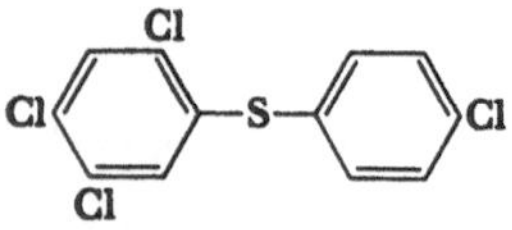	tetrasul*	Animert	Philips	(217)
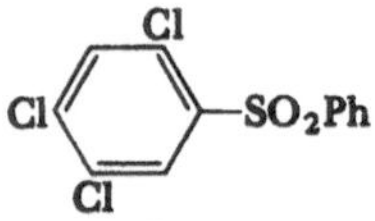	tetradifon*	Tedion	Philips	(218, 219)
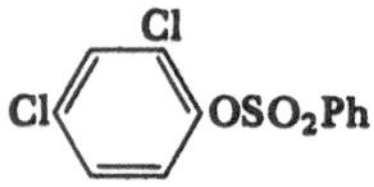	—	Genite	Allied Chem	(220)
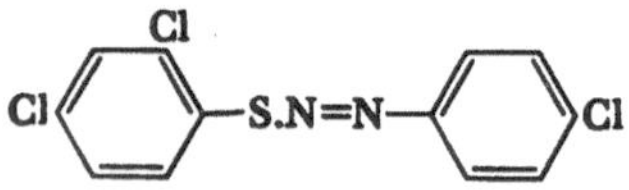	chlorfensulphide	—	Nippon Soda	(223)
$C_6Cl_5N=CClCCl_3$	—	Bayer 58733	Bayer	(221)
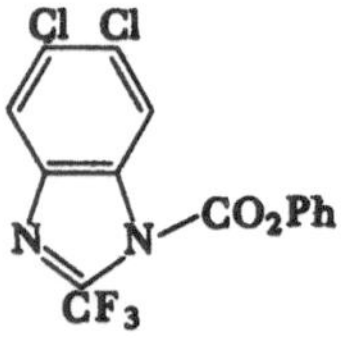	fenazaflor	Lorozal	Fisons	(222)

TABLE 4.6　Insecticides

Structure	Common name	Trade name	Originating company	Patent ref.
Cl—C₆H₃(Cl)—OP(OEt)₂ (=S)	dichlofenthion*	Nemacide	Virginia-Carolina	(224)
Cl—C₆H₂(Cl)(Cl)—OP(OEt)(Et) (=S)	trichloronate*	Agritox	Bayer	(225)
Cl—C₆H₂(Cl)(Cl)—OP(OMe)₂ (=S)	fenchlorphos*	Korlan	Dow	(226–228)
I—C₆H₂(Cl)(Cl)—OP(OMe)₂ (=S)	iodofenphos*	Nuvanol N	Ciba	(229)
Br—C₆H₂(Cl)(Cl)—OP(OEt)₂ (=S)	bromophos-ethyl*	Nexagan	Boehringer	(230, 231)
Br—C₆H₂(Cl)—OP(OCH₃)(Ph) (=S)	leptophos	Phosrel	Velsicol	(232)
Cl—C₆H₃(Cl)—OP(OEt)(Ph) (=S)	—	S-Seven	Takeda	(233)
MeS—C₆H₂(Cl)(Cl)—OP(OEt)₂ (=S)	chlormercaptofen	—	Bayer	(234)
Cl—C₆H₃(Cl)—C(CHCl)=... OP(OEt)₂	chlorfenvinphos*	Birlane	Shell	(235–239)
Cl—C₆H₂(Cl)(Cl)—C(CHCl)=... OP(OMe)₂	tetrachlorvinphos*	Gardona	Shell	(240, 241)
C₆H₃(Cl)(Cl)—SCHSP(OEt)₂ (=S)	phenkapton*	Phentol	Geigy	(242)

(Structures are aromatic ring systems with substituents as shown; S denotes the $P{=}S$ thiophosphoryl group and O the $P{=}O$ phosphoryl group.)

chlorate, and had been unselective in their effects. The phenoxy-acetic acids introduced the concept of 'selectivity', that is, the ability to kill one species of plant but not another. Their discovery arose from a war-time search for compounds which could be used to destroy cereal crops, particularly rice, and thus starve an enemy out of occupied islands, but, by chance, what was in fact discovered were chemicals which would kill other plants but not cereals. The significance of this war-time discovery in the immediate post-war period, when labour was scarce and maximum food production essential, was that these new herbicides were cheap, easy and safe to use. If this had not been so, it is doubtful whether chemical control of weeds would have become adopted so quickly by farmers. Their introduction paved the way for subsequent acceptance by farmers of more expensive herbicides which required more skill in application.

It is important to emphasize the effect of the introduction of these herbicides not only on the subsequent growth of the pesticides industry but also on farming practices. Rotation of crops has been traditional but the widespread introduction of chemical herbicides has led to a rapid tendency towards monoculture, that is, cultivation of the same crop year after year. In the U.K., for instance, barley is now grown for five or more years in succession. In the U.K. and U.S.A. 80% of the total acreage of cereal crops are now sprayed with herbicides. This has changed the character of the countryside by eliminating the fallow years in which fields were grazed, and has not been without its ecological consequences. Also, new weed problems have arisen to replace the older ones; competitive grass weeds such as wild oat, which are not affected by the phenoxyacetic acids, are now a major threat to optimum cereal production, a direct consequence of the abandonment of the grazing and root-crop years.

TABLE 4.7   Estimated Use of Phenoxyacetic Acid Herbicides in the U.S.A.
in 1970

2,4-D	30,000 tonnes
2,4,5-T	5 000 tonnes

(Total usage of all herbicides, 150,000 tonnes)

4.12.1   2,4-D (2,4-Dichlorophenoxyacetic Acid)

2,4-D is manufactured by chlorination of phenol to 2,4-dichloro-phenol which is then reacted with sodium monochloroacetate in water [94, 95]. Its major use is for selective control of broad-leaved weeds in cereals, but it has also been used in a variety of other applications such as weed-control around fruit trees [96] and in lawns and turf [97].

4.12.2   2,4,5-T (2,4,5-Trichlorophenoxyacetic Acid)

2,4,5-T is manufactured by hydrolysis of 1,2,4,5-tetrachlorobenzene to give 2,4,5-trichlorophenol which is then reacted with sodium monochloroacetate, generally in a higher alcoholic solvent such as amyl alcohol [98]. Its particular utility lies in its effective control of woody plants and it has been widely used on grasslands to maintain pasture [99]. Recently it has been shown that some

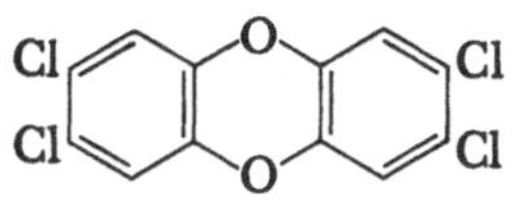

Fig. 4.6 2,3,7,8-Tetrachlorodibenzo-1,4-dioxin. A toxic impurity in commercial 2,4,5-T.

commercial samples contain a by-product 2,3,7,8-tetrachlorodibenzo-1,4-dioxin, which is highly teratogenic (causing malformations in embryos) and its use has been restricted in some countries, but it has also been shown that, by careful control of the production process, this impurity can be completely eliminated [100] (Fig. 4.6).

4.12.3   Dichloroprop (2,4-Dichlorophenoxypropionic Acid)

Dichloroprop is manufactured from 2,4-dichlorophenol and sodium α-chloro-propionate. It has a wider spectrum of activity than 2,4-D and, in particular, will control the two common cereal weeds, chickweed and cleavers, which are resistant to 2,4-D [101].

4.12.4   2,4-DB (2,4-Dichlorophenoxybutyric Acid)

2,4-DB is manufactured by reaction of 2,4-dichlorophenol with butyrolactone. The phenol is first reacted with sodium hydroxide in a mixture of n-butanol and nonane, the water distilled out, butyrolactone added and the mixture heated at 160-165° [102].

2,4-DB is converted in plants by a process of biochemical β-oxidation into 2,4-D, but there are wide differences in the abilities of various species to effect this oxidation. In particular, many leguminous species [103] such as clover and lucerne [104] are not sensitive to 2,4-DB. Thus, although 2,4-DB does not extend the range of weeds which can be controlled, it does extend the range of crops which can be treated, e.g. cereals undersown with clover [105].

4.13   OTHER HERBICIDES BASED ON POLYCHLOROAROMATICS

Two other classes of compounds listed have substantial commercial importance—the chlorinated benzoic acids, especially TBA, dicamba and chloramben, and the chlorinated ureas, especially diuron and linuron. Two more recent discoveries, the closely related compounds dichlobenil and chlorthiamid, are developing quite rapidly.

TABLE 4.8   Estimated Use of Other Herbicides in the U.S.A. in 1970

TBA	2000 tonnes
Dicamba	500 tonnes
Chloramben	1500 tonnes
Diuron	1500 tonnes
Linuron	500 tonnes
Propanil	4000 tonnes

4.13.1   TBA (2,3,6-Trichlorobenzoic Acid)

TBA is manufactured by monochlorination of toluene, separation of the 2-chloro-toluene and further chlorination of this to a mixture of 2,3,6- and 2,4,5-trichlorotoluenes which is oxidized with nitric acid to a mixture of the corresponding benzoic acids which are then separated by selective extraction at controlled pHs (Fig. 4.7). Direct chlorination of benzoyl chloride gives only a very small yield of the required isomer.

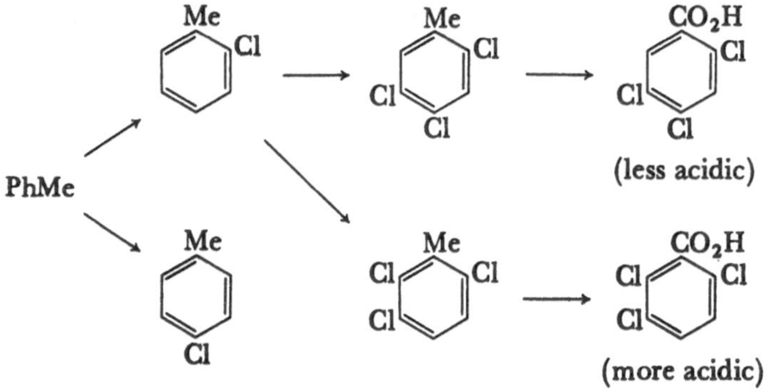

Fig. 4.7   Manufacture of TBA

TBA was first introduced in the U.S.A. in 1954 where it found limited use as a total weed-killer because it is very phytotoxic and non-selective. It can be used for total weed control, especially of deep-rooted perennial weeds, and is particularly effective when applied to foliage in late summer [106]. Its selective use, particularly in admixture with the phenoxyacetic acid herbicides to extend the range of activity of the latter especially to control of cleavers, chickweed and mayweed, was a U.K. discovery [107].

4.13.2   Dicamba (3,6-Dichloro-2-methoxybenzoic Acid)

Dicamba is manufactured by hydrolysis of 1,2,4-trichlorobenzene with aqueous sodium hydroxide under pressure to give 2,5-dichloro-phenol which is then reacted with carbon dioxide to give 3,6-dichloro-2-hydroxybenzoic acid, and this compound is methylated with methanolic hydrogen chloride and the product hydrolyzed to give 3,6-dichloro-2-methoxybenzoic acid (Fig. 4.8).

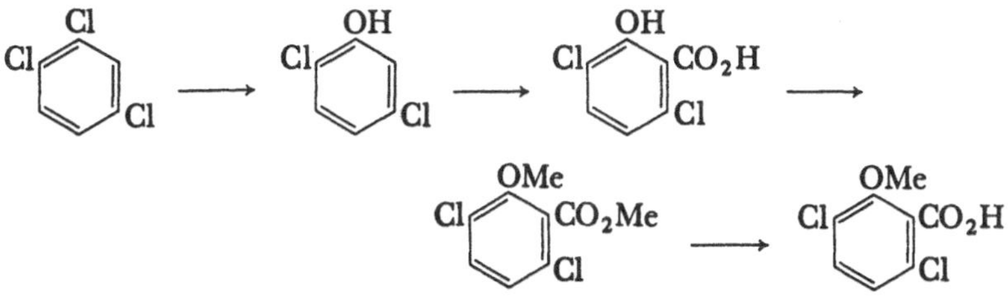

Fig. 4.8   Manufacture of dicamba

Dicamba is somewhat more selective than TBA and it is particularly useful in forestry applications as it controls bracken for several years after one application [108]. Like TBA it can be used to extend the range of usefulness of phenoxyacetic acid and herbicides especially against polygonum and spurrey, and this selective use in cereals was also a U.K. discovery [109].

4.13.3   Chloramben (3-Amino-2,5-dichlorobenzoic Acid)

Chloramben was first introduced in 1961 and has been mainly used in the U.S.A. especially for weed control in soya beans, which are a major crop in that country. It is manufactured by nitration of 2,5-dichlorobenzoic acid, followed by reduction.

4.13.4   Dichlobenil (2,6-Dichlorobenzonitrile)

This compound, which is closely related to the chlorinated benzoic acids, was introduced in 1960. It is manufactured by chlorination of 2,6-dichlorotoluene, hydrolysis to the aldehyde and formation and dehydration of the oxime (Fig. 4.9). It inhibits germination of weed seeds and has been used to control weeds in fruit trees [110] and in dormant bush fruits such as blackcurrants [111], raspberries [112]

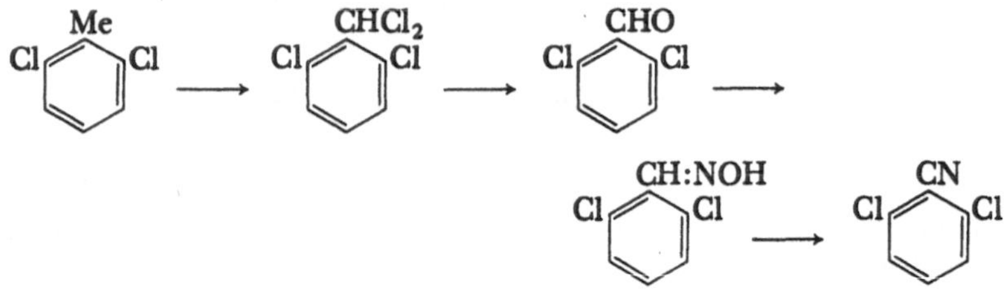

Fig. 4.9   Manufacture of dichlobenil

and gooseberries [113], and also to control water weeds. It can also be used to control most annual and perennial weeds for a whole season by application in early spring [114].

4.13.5   Chlorthiamid (2,6-Dichlorobenzthioamide)

This compound is so rapidly broken down in soil and plants to dichlobenil that it has almost exactly the same range of applications.

4.13.6   Diuron (3-(3,4-Dichlorophenyl)-1,1-dimethylurea)

Diuron was introduced in the late 1940's and is manufactured by reaction of 3,4-dichloroaniline with phosgene and anhydrous hydrogen chloride at 70-75° to give 3,4-dichlorophenyl isocyanate which is then reacted with dimethylamine at 25° to give diuron (Fig. 4.10).

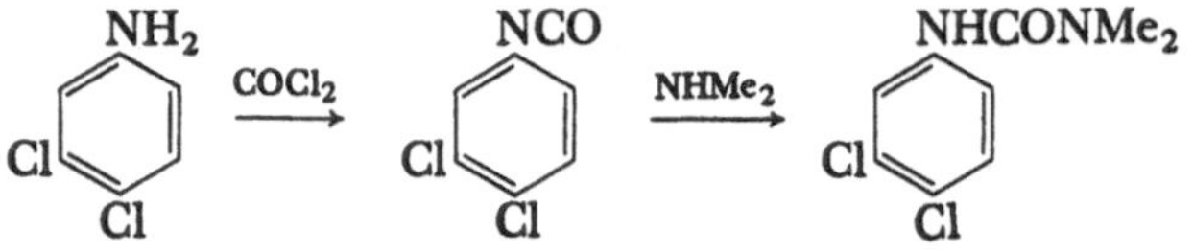

Fig. 4.10   Manufacture of diuron

Diuron is a very persistent non-selective herbicide which has been widely used for general weed control on industrial properties, roads, railways, etc. It can also be used for weed control around fruit trees [115] and as a pre-emergent herbicide in asparagus beds [116]. It is effective against algae but, because of its persistence, it cannot be used in water for domestic purposes or irrigation, and it is toxic to fish [117].

4.13.7   Linuron (3-(3,4-Dichlorophenyl)-1-methoxy-1-methylurea)

Linuron is manufactured in a similar way to diuron but using *N*-methylhydroxylamine instead of dimethylamine in the final stage. It is used as a pre-emergent herbicide to control annual weeds in potatoes [118], carrots [119], parsnips [120], linseed [121], flax [121], parsley [122] and celery [123]. Its use in spring-sown cereals is hazardous since its margin of selectivity for these crops is not great [124].

4.13.8   Propanil (*N*-(3,4-Dichlorophenyl)-propionamide)

Propanil is manufactured by reaction of 3,4-dichloroaniline with propionyl chloride. It is outstandingly effective against barnyard grass which is one of the most troublesome weeds in rice paddies.

4.14   FUNGICIDES BASED ON POLYCHLOROAROMATICS

None of the fungicides derived from polychlorinated benzenes has made a substantial commercial impact in agriculture. Hexachlorobenzene has been used in North America for seed treatment of wheat against bunt, but is not effective against other seed-borne diseases. The sodium salt of pentachlorophenol has been very widely used for timber preservation since 1936 but its high toxicity and skin-irritant and sternutatory properties make it unsuitable for general use as an agricultural fungicide. Technazene has been mainly used for control of the dry rot of potato caused by Fusarium. Quintazene was introduced in the 1930's but has only recently become widely used as a soil fungicide and for control of Rhizoctonia. It has high persistence and low mammalian toxicity. Chloranil was introduced 25 years ago and is still used in some seed-dressing applications. Dichloran was introduced in 1959 and has been successfully used to control Botrytis, especially on lettuce. It also has high persistence and low mammalian toxicity. A number of the other fungicides listed, e.g. chloroneb and dichlone, have found limited commercial uses, particularly in the U.S.A.

4.15   ACARICIDES BASED ON POLYCHLOROAROMATICS

Of the acaricides listed the two of outstanding commercial importance are tetrasul and tetradifon, and, of these, tetradifon is much more widely used (Fig. 4.11). It was introduced in 1954 and is effective against all stages and eggs of phytophagous mites, and it is very persistent. It has a very low toxicity both to mammals and to insects. However, its long reign as the most important acaricide for

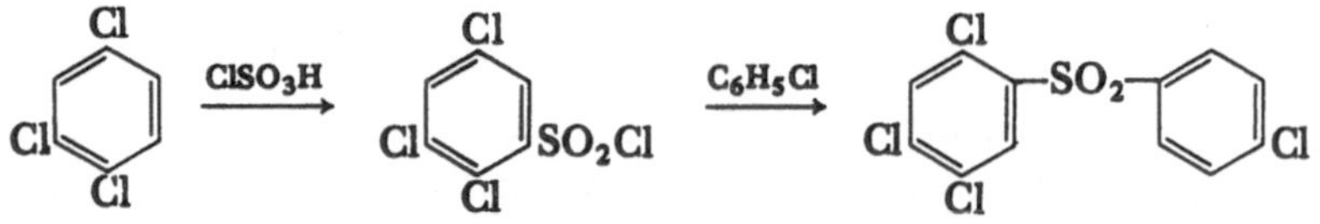

Fig. 4.11   Manufacture of tetradifon

fruit trees is rapidly coming to an end as development of resistance is now widespread, and it is probable that this market will be taken over by one of the newer compounds such as fenazaflor.

4.16   INSECTICIDES BASED ON POLYCHLOROAROMATICS

Of the insecticides listed only three have achieved substantial commercial success. Fenchlorphos, introduced in 1954, is a systemic insecticide but has such a low mammalian toxicity that it has been proposed for treatment of dairy cattle against internal and external pests (Fig. 4.12). As an insecticide it is not very effective against lepidopterous larvae but is useful against flies and sap-feeding insects.

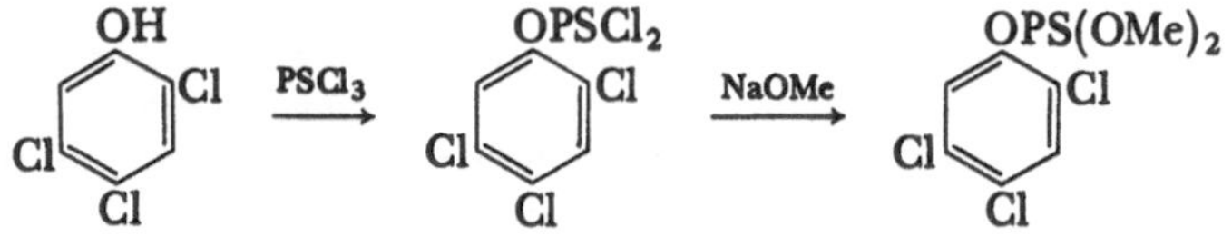

Fig. 4.12   Manufacture of fenchlorphos

Phenkapton has a fairly wide spectrum of activity but has a much higher mammalian toxicity. Chlorfenvinphos also has a fairly high mammalian toxicity but is accepted for use on brassicae, root crops and cereals.

4.17   ANTISEPTICS BASED ON POLYCHLOROAROMATICS

The polychlorinated bis-(hydroxyphenyl)-alkanes have achieved considerable commercial success as externally applied antibacterials. In particular, hexachlorophene [bis-(3,5,6-trichloro-2-hydroxyphenyl)-methane] acts as a bacteristatic and slowly bactericidal agent [125], and its activity is thought to be due to its power to chelate iron in the bacterial cell and thus inhibit its essential enzyme systems [126] (Fig. 4.13). It is retained in trace quantities by the skin and, because of its high activity, it reduces the bacterial flora considerably [127-129]. However, it is far more active against Gram-positive than against Gram-negative organisms [130, 131] and it is inactivated by

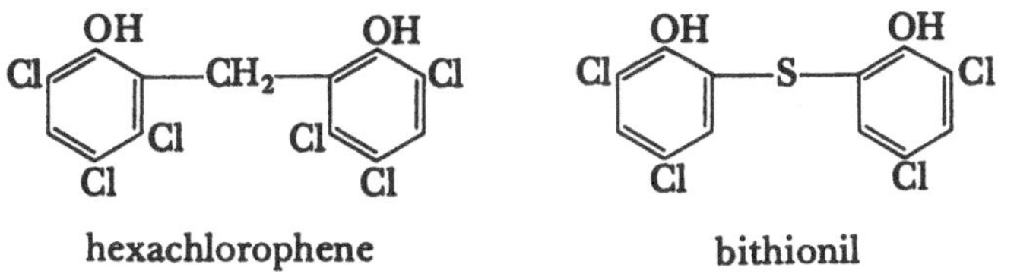

OH   OH   OH   OH

Cl ⬡ —CH₂— ⬡ Cl   Cl ⬡ —S— ⬡ Cl

Cl   Cl   Cl   Cl   Cl   Cl

hexachlorophene   bithionil

Fig. 4.13   Antiseptics

blood so it is not suitable for applications to wounds or for pre-operative sterilization [132].

It has been used extensively as an additive to soaps, cosmetics, etc. and has found a major use in cosmetic deodorants to inhibit the cutaneous bacteria which cause perspiration odour by decomposition of sweat [133, 134]. Recently hexachlorophene has come under criticism because of an alleged possibility that it could be absorbed and cause brain damage. There has been an adverse public reaction because of its widespread use in baby creams, and it has consequently been withdrawn in a number of countries.

A related compound which has very similar antiseptic properties is bithionil (*bis*-(3,5-dichloro-2-hydroxyphenyl)-sulphide) [135, 136] (Fig. 4.13).

TABLE 4.9   Antibacterial Activity of Hexachlorophene and Bithionil

Compound	Organism	Minimum inhibitory concentration
Hexachlorophene	*S. aureus*	$1:10^6$
Hexachlorophene	*E. coli*	$1:10^4$
Bithionil	*S. aureus*	$1:10^6$
Bithionil	*S. typhosa*	$1:10^3$

An examination of over 200 phenyl urea and phenyl thiourea derivatives led to discovery of a number of compounds with high bacteristatic activity [137, 138] (inhibitory concentrations for *S. aureus* as low as $1:10^7$), in particular, 3,3',4-trichlorocarbanilide, 3,4,4'-trichlorocarbanilide, 3,3',4,4'-tetrachlorocarbanilide, 3,3',4,5,5'-pentachlorocarbanilide and 3,4,4'-trichlorothiocarbanilide. Of these, 3,4,4'-trichlorocarbanilide (TCC) has found commercial use as an additive to soap. However, it is not very effective against Gram-negative bacteria or pathogenic fungi.

4.18   POLYCHLOROHETEROAROMATICS–INTRODUCTION

There are, in the patent literature, many claims for various polychlorinated heteroaromatic compounds as intermediates for the production of commercially useful materials. British patent law demands that an invention shall be not only new, but new and useful. However, a very slight amount of utility is sufficient to support a patent. Jessel in Otto vs. Linford [243] said 'As to this question of utility, very little will do'. It is also clear that commercial success is not a requisite. Halsbury in Badische Anilin and Soda Fabrik vs. Levinstein [244] said: 'The element of commercial pecuniary success has, it appears to me, no relation to the question of utility in patent law generally'. The degree of utility which is required to obtain a valid patent is, therefore, of a totally different order from that required to sell a product profitably and successfully. It is common practice, when a new organic compound is discovered, to patent its method of preparation with a vague statement that: 'This compound is useful as an intermediate for the preparation of pharmaceuticals and pesticides', in order to satisfy the utility requirement. Cynics have suggested that an insect could be squashed to death if sufficient of any compound were piled on top of it.

In this chapter, therefore, only those derivatives of polychlorinated heteroaromatic compounds will be mentioned which have, as far as the author can ascertain, actually become the basis for viable and established commercial products. There are, in fact, very few such products. The reason for this is not hard to find—they are too expensive for fields of application where cost considerations are critical. For instance, they could hardly be used as intermediates for polymers unless these had some very exceptional properties, nor for uses such as surface-coating of textiles which are destined for a highly competitive consumer goods market. Biologically active compounds normally command considerably higher prices than those whose commercial applications depend on purely physical effects, so it is not surprising that it is towards pharmaceuticals and pesticides that most thought has been directed in seeking commercial outlets for polychlorinated heteroaromatic compounds. It appears likely that until cheaper processes are evolved for their manufacture, the economically viable future of this class of compounds will lie in these fields.

A major exception is cyanuric chloride which has become the basis for three large industrial applications—the triazine herbicides, the triazine dyes and the optical whitening (fluorescent brightening) agents (Fig. 4.14). The reason why such large and successful businesses have been built upon cyanuric chloride is that it is currently available at a cost which brings its derivatives into the right price for use in textiles, etc. It is readily produced by trimerization of cyanogen chloride by passage in the vapour phase over an activated carbon catalyst, and cyanogen chloride is itself made cheaply either from hydrogen cyanide and chlorine in aqueous media or from

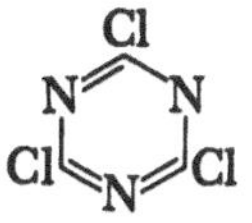

cyanuric chloride

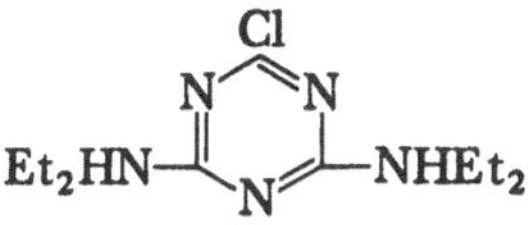

triazine herbicide

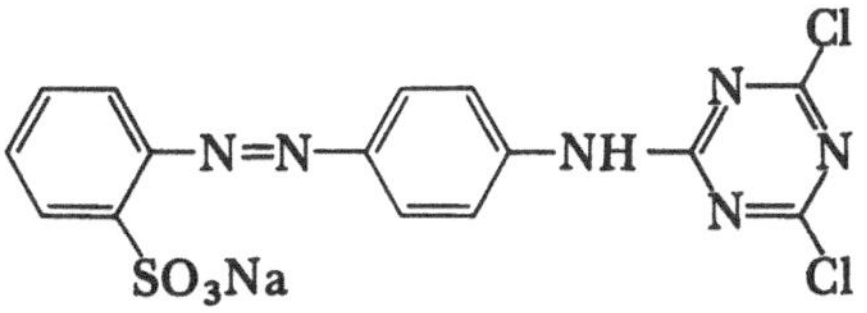

triazine dyestuff

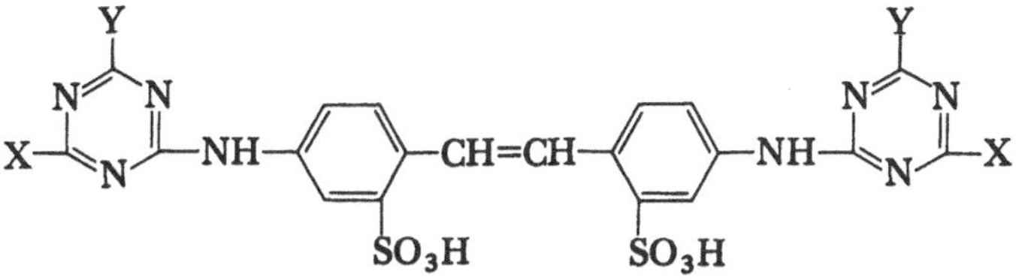

optical whitening agent

Fig. 4.14   Commercial products from cyanuric chloride

sodium cyanide liquor and chlorine in nozzle reactors (Fig. 4.15). World production of cyanuric chloride is currently of the order of 20,000 tonnes/year. None is produced in the U.K., and the main centres of production are the U.S.A. (about 5000 tonnes/year), Europe (about 10,000 tonnes/year) and Japan (about 5000 tonnes/year). World demand has been rising steadily for the past ten years and imports into the U.K. have risen from about 400 tonnes in 1960 to about 1800 tonnes in 1970.

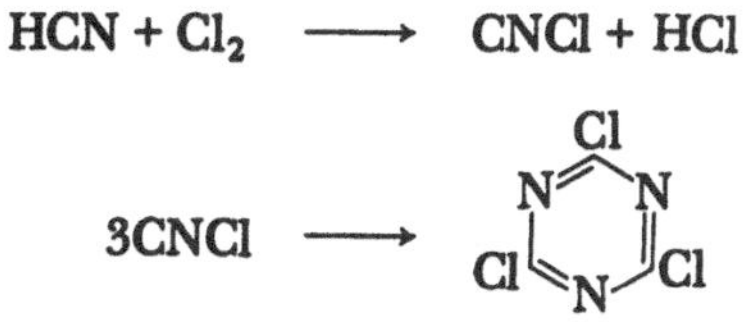

$$HCN + Cl_2 \longrightarrow CNCl + HCl$$

$$3CNCl \longrightarrow$$

Fig. 4.15   Manufacture of cyanuric chloride

Commercial outlets for cyanuric chloride are entirely limited to its use as an intermediate for the production of compounds containing the s-triazine ring structure. However, it does not necessarily provide the only route of commercial importance to s-triazines, nor is it by any means the most economical intermediate. Its value for the formation of triazine ring compounds derives from the ease with which the chlorine atoms can be made to react with substances containing active hydrogen atoms. Condensation products are formed with elimination of one, two or three molecules of hydrogen chloride per mole of cyanuric chloride, and, in many cases, replacement of the chlorine atoms can be effected stepwise by control of the reaction conditions, particularly temperature. For instance, aminolysis of cyanuric chloride leads in turn to 2-amino-4,6-dichloro-*s*-triazine (at 0°), 2,4-diamino-6-chloro-*s*-triazine (at 30-50°) and triamino-*s*-triazine (at 90-100°) (Fig. 4.16).

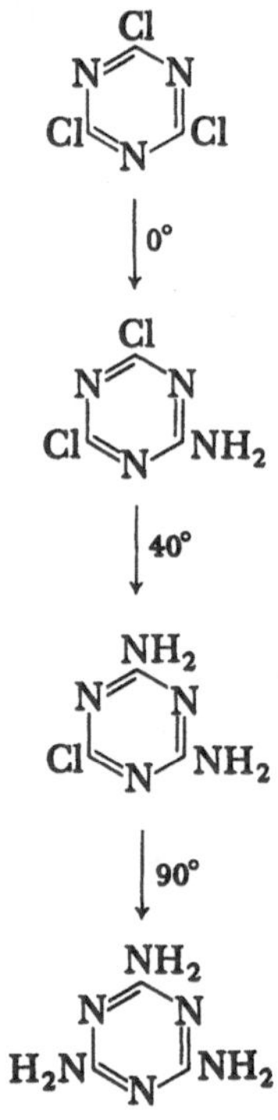

Fig. 4.16   Reaction of cyanuric chloride with ammonia

The history of the development of successful industries based on cyanuric chloride is an example of the way in which basic chemistry must await the 'enabling' technology which will make its commercial application economically viable. Cyanuric chloride was first described in 1830 and, during the next 100 years, its chemistry was fairly thoroughly investigated. But application to the production of commercially useful products had to await the decision of a chemical manufacturer to set up a fairly large-scale plant to manufacture cyanuric chloride at a price which would encourage chemical industry to investigate its potential as an intermediate for profitable products. Chemical industry produces chemicals, but the consumer buys the effects of chemicals, and this is the difficult bridge which has to be crossed by technological research in industry. The commercial availability of cyanuric chloride prompted the investigation of its possibilities as an intermediate for crop protection chemicals which led to the discovery and development of the triazine herbicides.

Pentachloropyridine is currently at the stage where development of economical processes for large-scale manufacture may similarly stimulate wider investigation of its potential as an intermediate for commercial products. The standard laboratory method of preparation of this compound has been by reaction of pyridine with phosphorus pentachloride [245], which is an extremely unpleasant process and one which could hardly be visualized as a method of manufacture. In a series of papers between 1898 and 1903 Sell and Dootson [246] described the attempted reaction of pyridine with chlorine in the vapour phase at fairly high temperatures but obtained complex mixtures of products together with large amounts of tars and charred materials. They did, in fact, isolate and characterize a range of di-, tri-, tetra- and penta-chloropyridines but these remained laboratory curiosities in the absence of economical manufacturing processes.

In the last decade it has been discovered that pyridine can be conveniently chlorinated in the vapour phase at about 500° using either a fluidized bed of an inert substance such as silica [247] or an empty tube under conditions of turbulent flow [248] (Fig. 4.17). Both of these methods are expedients to ensure thorough mixing and to avoid formation of 'hot spots' which occur very readily as the reaction is highly exothermic, and which lead to extensive breakdown of the pyridine ring and formation of large amounts of tarry and charred materials. Decomposition can be further reduced by addition of a diluent such as carbon tetrachloride. With suitable design of equipment and operating procedures reaction is very clean and gives high yields of pure pentachloropyridine. Pentachloropyridine is, therefore, potentially a reasonably priced chemical intermediate. Nevertheless, its present cost of production is still too high for it to be useful for anything other than preparation of biologically active compounds, but this situation may well change when scale of production increases and cost falls, thus providing an incentive for development of other commercial uses, as happened with cyanuric chloride.

As far as the author has been able to ascertain, there are few marketed commercial products based on any polychlorinated hetero-

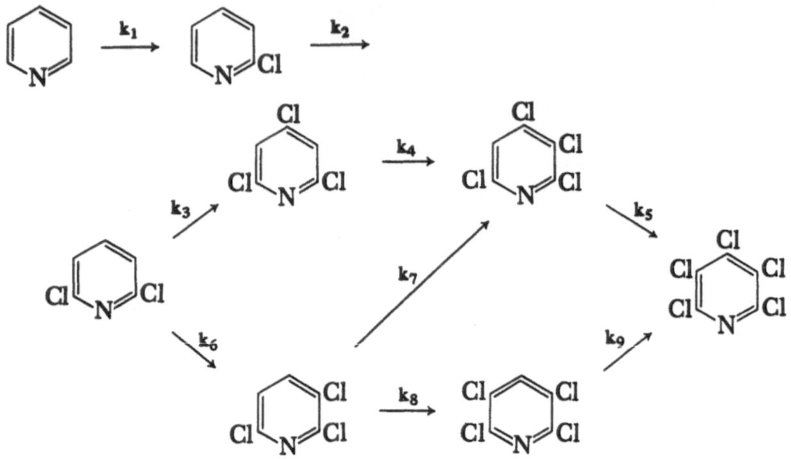

Second order $k_1 = 10^4$ $k_2 = 2 \times 10^3$ $k_3 = 5.1$

Rate constants $k_4 = 0.16$ $k_5 = 0.14$ $k_6 = 1.57$

Litres moles^{-1} sec^{-1} $k_7 = 0.08$ $k_8 = 0.23$ $k_9 = 0.07$

R. D. Bowden, Imperial Chemical Industries Limited, unpublished work.

Fig. 4.17 Chlorination of pyridine at 500°

aromatic compounds other than cyanuric chloride, tetrachloro-pyrimidine and the polychlorinated picolines and pyridine, despite the number of claims in patents relating to such compounds. Their commercial exploitation must await the development of manu-facturing processes based on cheap petrochemical C_1 and C_2 fragments and cheap sources of nitrogen such as ammonia or urea. That such possibilities exist is exemplified by the observation that carbon tetrachloride and ammonia when heated for several days in a sealed tube at 150° give cyanuric chloride and pentachloropyridine [249] (Fig. 4.18).

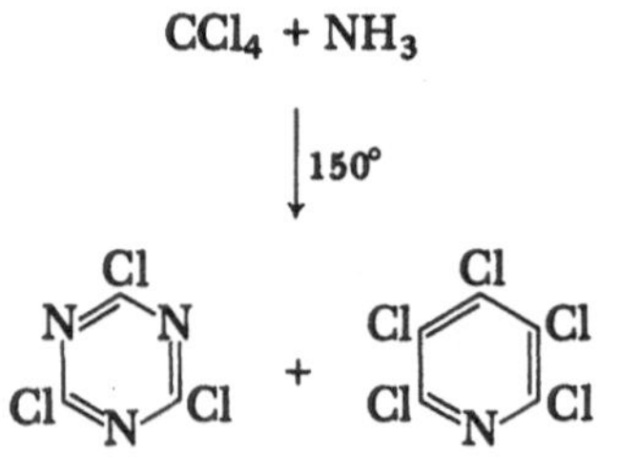

$$CCl_4 + NH_3$$

150°

Fig. 4.18 Reaction of carbon tetrachloride with ammonia

4.19 POLYMERS BASED ON POLYCHLOROHETEROAROMATICS

It has already been pointed out that polychlorinated heteroaromatic compounds are, in general, too expensive to be considered as intermediates for commercial polymers. Even cyanuric chloride is not cheap enough for this application. Melamine (triamino-s-triazine) has considerable commercial importance as a component of melamine resins but, although it can be made from cyanuric chloride and ammonia, it is much more economically manufactured from calcium cyanamide via dicyandiamide.

There have been a number of claims in the patent literature for polymers derived from cyanuric chloride. Acetyl cellulose and cyanuric chloride are said to form a fibre-forming and filament-forming polymer [250]. Polyalkylene polyamines and cyanuric chloride are claimed to give anion-exchange resins [251]. The triallyl ester of cyanuric acid (from cyanuric chloride and allyl alcohol) is reported to polymerize to a hard but brittle cross-linked polymer with good clarity and high heat-distortion temperature, and it is said that the monomer can be used to impregnate glass and then polymerized to give mats with high flexural strength and thermal stability [252]. Similar properties have been claimed for cross-linked polymers from triallyl cyanurate and other vinyl monomers such as styrene and methacrylates [253]. Transesterification of triallyl cyanurate with glycols is said to give a variety of products in the form of gels, rubbery solids or viscous oils [254]. There is, however, no evidence that any of these products have made any substantial commercial impact.

Polychlorinated heteroaromatic compounds are, therefore, unlikely to make any inroad into the polymer market because a wide range of acceptable cheaper alternatives is available.

4.20 DYESTUFFS BASED ON POLYCHLOROHETEROAROMATICS

Colour has always been a major element in social life and world trade in dyestuffs and organic pigments is currently of the order of £300 million per year. The simple traditional dyeing process is to immerse the material in a solution of the dyestuff which is absorbed onto the fibre and held there by ionic bonding to active sites. The problem is that these ionic bonds are relatively easily broken and the dyestuff is consequently removed by repeated washing and laundering. The former method of overcoming this difficulty was to change the form of the dye once it had become attached to the fibre so that it became much less soluble, e.g. by oxidation, by treatment with a heavy metal salt, by reaction with a diazonium salt, etc.

For many years a target of the dyestuffs chemist was to produce a dyestuff which would form a covalent bond with active sites and thus become chemically a part of the fibre itself. In 1895, Cross and Bevan [255] produced coloured esters of cellulose by a process which was quite impracticable industrially. In 1923 patents appeared for dyestuffs derived from cyanuric chloride for fast dyeing of wool [256], and, in 1925, the first dye which was known to react with the fibre was described by Günther [257]. Wool dyes containing a chloroacetylamino group to react with the fibre were introduced in 1929 but were not of much commercial value because of lack of fastness [258]. In 1930 Haller and Hekendorn attempted to devise a practical dyeing process utilizing the reaction between cyanuric chloride and cellulose but failed to develop a water-based process so made no commercial impact [259]. This reaction was later looked at in greater detail by Warren, Reid and Hamalainen in 1952 and their publication revived interest in the subject [260]. The major technological advance was made by Rattee and Stephen in 1954 when they solved the problem of fast dyeing of cellulosic fibres by a simple and practical process [261-263]—a further example of the way in which basic chemistry must await the solution of technological problems before it can be commercially exploited.

The first three 'Procion' fibre-reactive dyestuffs based on cyanuric chloride were introduced in 1956 and this new concept made an immediate impact on dyestuff users throughout the world with the result that dyestuffs of this type are now used very widely. They are still the most versatile and commercially advantageous reactive dyestuffs and, because the reactive group can be attached to many types of dyestuff chromophores, they can be used to produce very bright or very dull shades of almost every colour by conventional methods of dyeing and textile printing. In addition, new highly-productive techniques for their use have been evolved to give the vast modern range of multi-coloured fabrics and novel designs by polychromatic dyeing processes [264, 265]. The number of 'Procion' dyes has grown from 3 in 1956 to 79 in 1970, that is, a new product every two months. Their discovery triggered off intense research activity and similar dyestuffs have been produced by other manufacturers with the result that the total number of products of this type based on cyanuric chloride available commercially now exceeds 200 [266]. Their advent has enabled the fashion trade to do what was hitherto impossible, namely, to reconcile high fastness and brilliance and clarity of shade with versatility and economy in application—all vital factors in the economic viability of the rapidly expanding mass markets for fashion wear.

The earliest 'Procion' dyes† ('Procion M') were dichloro-s-triazinyl

† 'Procion M' and 'Procion H' are registered trade-marks of Imperial Chemical Industries Limited.

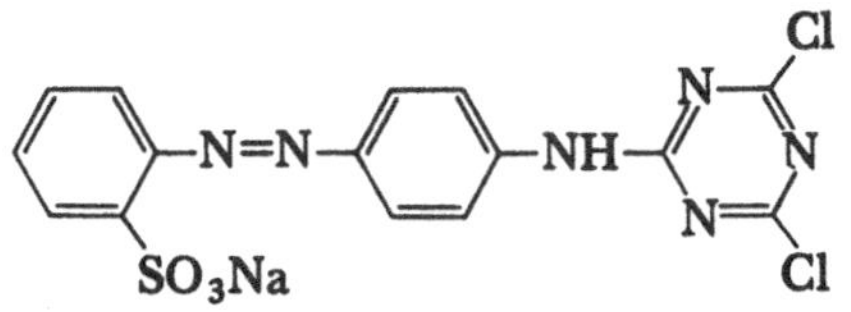

Fig. 4.19   Early Procion dyestuff

derivatives formed by reaction of cyanuric chloride with one molecule of a dyestuff (generally an azo or anthraquinone compound) one of the earliest examples being the compound shown in Fig. 4.19 [267]. They are still the most useful dyestuffs for some dyeing processes but they are somewhat unstable in textile printing pastes and so the rather less reactive 'Procion H' range was introduced [268-271], which made reactive dyestuffs a technically attractive proposition for the textile printer. These are monochloro-s-triazinyl derivatives containing one molecule of the dyestuff (generally an azo, anthraquinone or copper phthalocyanine compound) and one other substituent (Fig. 4.20). Variation of this substituent enables the dyestuffs chemist to control such properties as affinity, solubility and reactivity and, with the experience gained over the past 15 years, it is now possible to 'tailor' all new dyestuffs for their specific end-use so that appropriate 'Procion H' dyes find important uses in textile printing and dyeing applications. Manufacture involves reaction of cyanuric chloride with an amino group

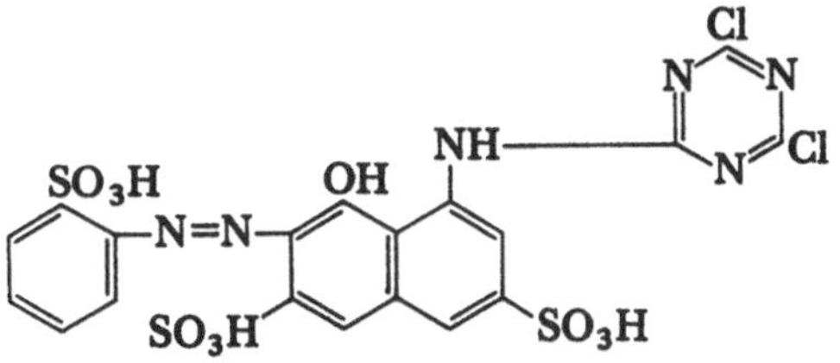

Procion M-type dyestuff

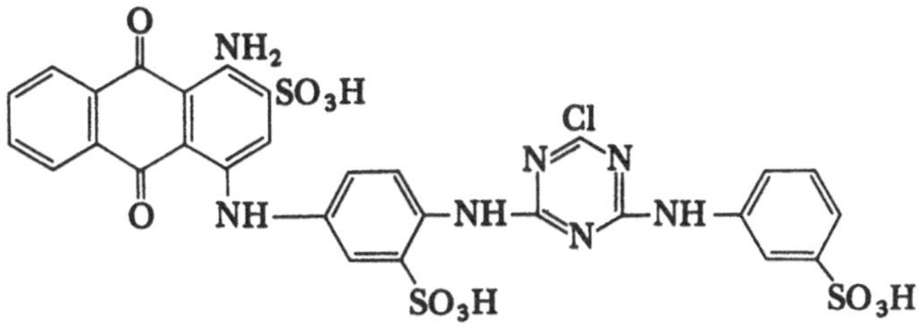

Procion H-type dyestuff

Fig. 4.20   Procion dyestuffs

on a dyestuff or dyestuff intermediate under neutral aqueous conditions at 0-10° [272]. Very extensive studies have been carried out on the precise way in which the triazinyl dyestuff reacts with the cellulose fibre and of the physical chemistry of the reaction [273]. It is not appropriate to review in this book the hundreds of papers on the technology of application of triazinyl dyestuffs. Once the

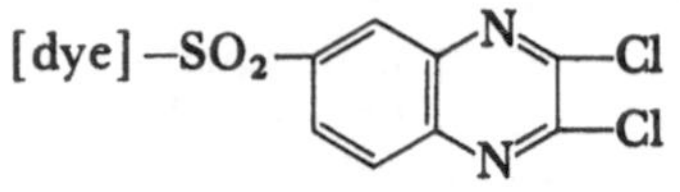

(ref. 274) 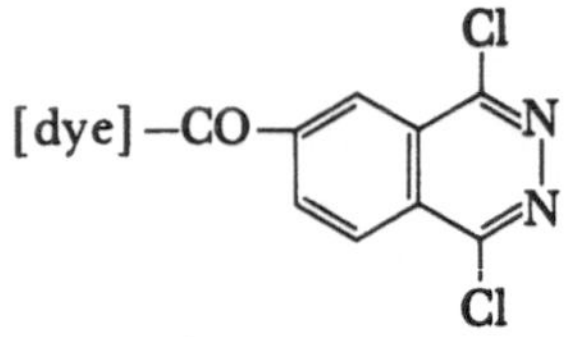   (ref. 274)

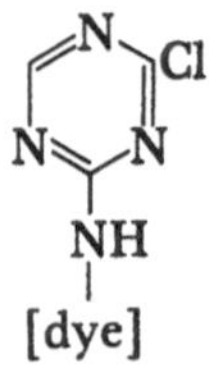

(ref. 275) 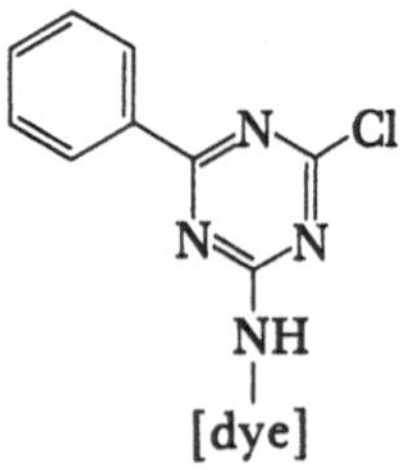   (ref. 276)

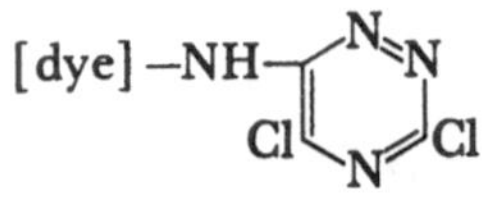

(ref. 277) 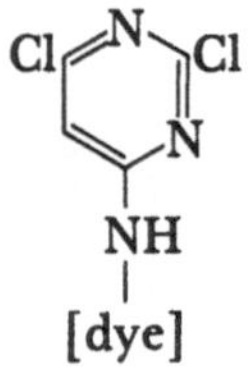   (refs. 278–281) 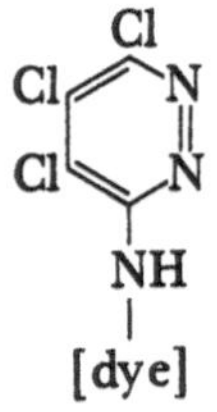   (ref. 282)

Fig. 4.21   Systems proposed for reactive dyes

principle of the dyestuff-fibre reaction under practical application conditions had been established by introduction of the 'Procion' dyes, an intensive search for alternatives to cyanuric chloride was made all over the world and, in particular, practically every conceivable polychlorinated heterocyclic compound was looked at in this context. Some examples are given in Fig. 4.21. The patent literature contains hundreds of claims for utility but, because of economic considerations and lack of availability of intermediates, very few have achieved any commercial significance, and little purpose would be served in citing all the references. The only polychlorinated heterocyclic compounds apart from cyanuric chloride which have given rise to a commercial range of reactive dyestuffs are tetrachloropyrimidine [283] 5-chloro-2,4,6-trifluoro-pyrimidine [284]; 4,5-dichloro-6-methyl-2-methylsulphonyl-pyrimidine [285]; 2,4-dichloro-5-chlorocarbonylpyrimidine [286]; 2,3-dichloro-6-chlorocarbonylquinoxaline [287] and 4,5-dichloro-6-pyridazonyl-1-propionoyl chloride [288] (Fig. 4.22). Triazinyl dye-stuffs are not major users of cyanuric chloride in a tonnage sense—in

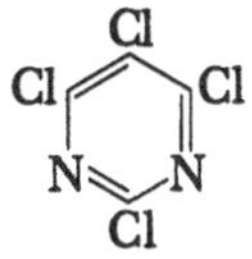

tetrachloropyrimidine

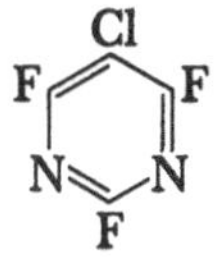

5-chloro-2,4,6-trifluoropyrimidine

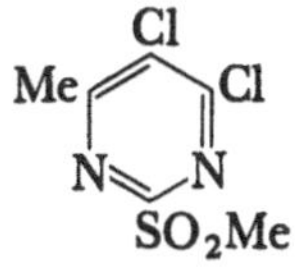

4,5-dichloro-6-methyl-
2-methylsulphonylpyrimidine

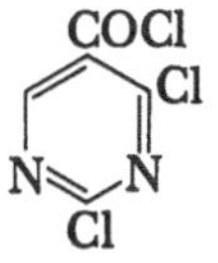

2,4-dichloro-5-
chlorocarbonylpyrimidine

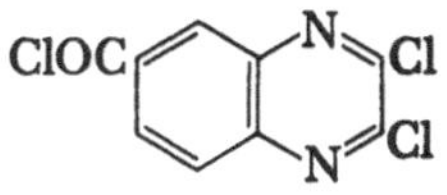

2,3-dichloro-6-
chlorocarbonylquinoxaline

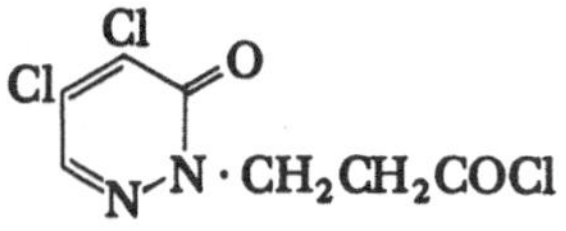

4,5-dichloro-6-pyridazonyl
-1-propionoyl chloride

Fig. 4.22   Intermediates for commercial reactive dyestuffs

1970 the amount of cyanuric chloride used for this purpose was about 500 tonnes in the UK, 300 tonnes in Europe, 150 tonnes in the U.S.A. and 400 tonnes in the rest of the world. They do, however, constitute a very high value market in terms of final products, and one which is steadily increasing.

4.21   OPTICAL WHITENING AGENTS BASED ON POLYCHLOROHETEROAROMATICS

The principle of optical bleaching was first described by Krais [289] in 1928 but industrial use of optical brightening began about 10 years later. Optical whitening agents (fluorescent brighteners) are substances which are added to washing powders and detergents to give an illusion of whiteness and the 'bright clean' look which is much prized by housewives. They are also extensively used to improve the appearance of new paper, leather and textile goods prior to sale [290-293]. They depend for their action on absorption of short wavelength solar radiation which is re-emitted as a bluish fluorescence which counteracts any residual grey look which results from incomplete removal of soiling from the material during the washing process. An essential requirement is that they should be colourless on the fabric, that is, that they should not absorb any light in the visible part of the spectrum.

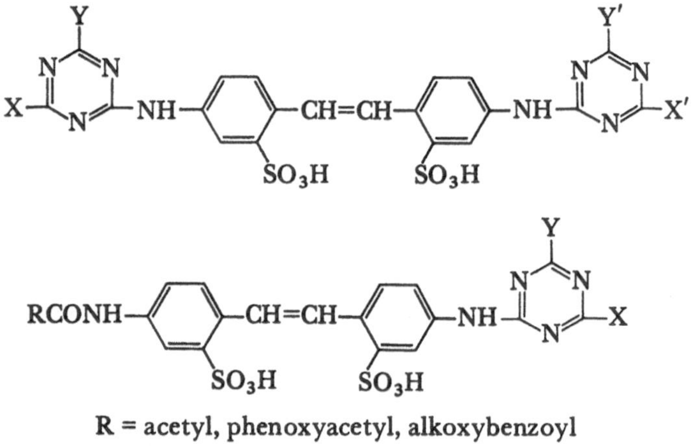

R = acetyl, phenoxyacetyl, alkoxybenzoyl

Fig. 4.23 Typical triazine optical whiteners

The most versatile optical whitening agents so far are produced by reaction of cyanuric chloride with highly unsaturated molecules containing a conjugated benzenoid system. A typical product is that obtained by reaction of 2 molecules of cyanuric chloride with one molecule of 4,4'-diaminostilbene-2,2'-disulphonic acid (Fig. 4.23). There is a range of optical whitening agents of this type, and, in general, the substituents X and X' are the same, as are Y and Y'. Thus X and X' may be anilino groups and Y and Y' hydroxyl groups [294]. More commonly, X, X', Y, and Y' are all substituted amino groups such as those derived from anilines, aminobenzene sulphonic acids, aliphatic or heterocyclic amines, etc. [295-305]. In some cases, X and X' are substituted amino groups and Y and Y' are alkoxy, hydroxyalkoxy, phenoxy or thio groups [306, 307] or chlorine [308]. These *bis*-triazinyl compounds are mainly used on cotton and paper.

A second type of optical whitening agents are the mixed acyl triazinyl derivatives of 4,4'-diaminostilbene-2,2'-disulphonic acid [309-311] and these are claimed to have a better affinity for wool than the *bis*-triazinyl compounds. Another type is that in which the sulphonic acid group in the stilbenedisulphonic acid has been replaced by a group such as alkyl, alkoxy, or sulphonyl [312], and it is claimed that these compounds give more natural white effects.

Although many new types of optical whitening agents have been introduced, those based on cyanuric chloride still command the major share of the market because they are cheap and because they have good solubility and substantivity. For paper treatment they are the only products which are inexpensive enough for practical use. Because of the constant competitive struggle amongst manufacturers to offer even more superlative grades of whiteness and brightness the market for optical whitening agents has been increasing at about 18% per year over the past ten years, and virtually every commercial washing powder and detergent now contains them. This growth is likely to diminish as the detergent market becomes saturated and, although the typical amount of optical whitening agent in a detergent in the U.K. has been increased from 0.1% to 0.3%, and in the U.S.A. to as much as 0.5%, it is unlikely to be increased further. However, the market for non-textile uses, especially for paper, is expected to grow. The usage of cyanuric chloride in 1970 for production of optical whitening agents was about 1500 tonnes in the U.K., 3000 tonnes in Europe and 4500 tonnes in the U.S.A.

4.22   HERBICIDES BASED ON CYANURIC CHLORIDE

The herbicidal effects of derivatives of *s*-triazines were discovered in 1952 [313, 314]. The commercially useful compounds are all derived from cyanuric chloride by replacement of two of the active chlorine atoms by alkylamino groups which can be the same, as in simazine, or different, as in atrazine. A second generation of triazine herbicides has been developed by replacement of the remaining chlorine atom with a methoxy, methylthio or azido group, as exemplified by prometryne [315, 316] (Fig. 4.24). Table 4.10 lists the chemical structures, common names (those marked * are accepted common names of the British Standards Institution), the originating company and the patent reference of all triazine herbicides which have been put onto the market.

Replacement of only one chlorine atom in cyanuric chloride with an alkylamino group gives compounds which are very active as contact herbicides, but these have never been exploited commercially because of their instability [317, 318]. Replacement of either one or

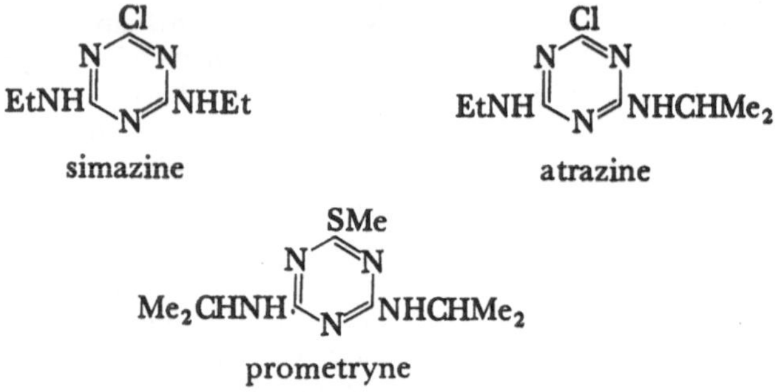

Cl
N⤬N
EtNH⟋N⟍NHEt
simazine

Cl
N⤬N
EtNH⟋N⟍NHCHMe$_2$
atrazine

SMe
N⤬N
Me$_2$CHNH⟋N⟍NHCHMe$_2$
prometryne

Fig. 4.24   Triazine herbicides

two of the chlorine atoms in cyanuric chloride with an arylamino group gives compounds which are inactive as herbicides [315].

The triazine herbicides are readily manufactured by successive replacement of the chlorine atoms in cyanuric chloride and there is no difficulty in introducing differing groups as reaction of the first chlorine atom can take place very readily below room temperature whereas replacement of the second requires moderate heating and

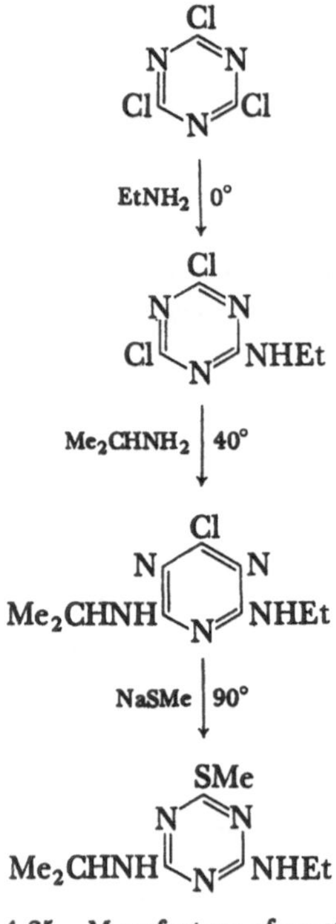

Cl
N⤬N
Cl⟋N⟍Cl

EtNH$_2$ | 0°

Cl
N⤬N
Cl⟋N⟍NHEt

Me$_2$CHNH$_2$ | 40°

Cl
N⤬N
Me$_2$CHNH⟋N⟍NHEt

NaSMe | 90°

SMe
N⤬N
Me$_2$CHNH⟋N⟍NHEt

Fig. 4.25   Manufacture of ametryne

TABLE 4.10   Triazine Herbicides—General Formula

$$\begin{array}{c} R_1 \\ N \diagup \diagdown N \\ R_3 \diagdown_N \diagup R_2 \end{array}$$

Common name	R_1	R_2	R_3	Originating company	Patent reference
Simazine*	Cl	NHEt	NHEt	Ciba-Geigy	(343–346)
Atrazine*	Cl	NHEt$_2$	NHCHMe$_2$	Ciba-Geigy	(343, 347)
Sebuthylazine*	Cl	NHEt	NHCHMeEt	Ciba-Geigy	(343, 348)
Terbuthylazine*	Cl	NHEt	NHCMe$_3$	Ciba-Geigy	(343, 348)
Trietazine*	Cl	NHEt	NEt$_2$	Ciba-Geigy	(343, 348)
Propazine*	Cl	NHCHMe$_2$	NHCHMe$_2$	Ciba-Geigy	(343, 347)
Ipazine*	Cl	NEt$_2$	NHCHMe$_2$	Ciba-Geigy	(343, 347)
Chlorazine*	Cl	NEt$_2$	NEt$_2$	Ciba-Geigy	(343, 347)
Simeton*	OCH$_3$	NHEt	NHEt	Ciba-Geigy	(343)
GS 14254	OCH$_3$	NHEt	NHCH$_2$CHMe$_2$	Ciba-Geigy	(343)
GS 14259	OCH$_3$	NHEt	NHCMe$_3$	Ciba-Geigy	(343)
Atraton*	OCH$_3$	NHEt	NHCHMe$_2$	Ciba-Geigy	(343, 349)
Prometon*	OCH$_3$	NHCHMe$_2$	NHCHMe$_2$	Ciba-Geigy	(343, 349)
Desmetryne*	SCH$_3$	NMEe	NHCHMe$_2$	Ciba-Geigy	(350)
Simetryne*	SCH$_3$	NHEt	NHEt	Ciba-Geigy	(350)
Ametryne*	SCH$_3$	NHEt	NHCHMe$_2$	Ciba-Geigy	(349, 350)
Terbutryne*	SCH$_3$	NHEt	NHCMe$_3$	Ciba-Geigy	(350–352)
Prometryne*	SCH$_3$	NHCHMe$_2$	NHCHMe$_2$	Ciba-Geigy	(349, 350)
Methoprotryne*	SCH$_3$	NHCHMe$_2$	NH(CH$_3$)$_3$OMe	Ciba-Geigy	(353, 354)
Aziprotryne	SCH$_3$	NHCHMe$_2$	N$_3$	Ciba-Geigy	(355)
Cyprazine	Cl	NHCHMe$_2$	NHC$_3$H$_5$	Gulf Oil	(356)
SD 15417	Cl	NHMe	NHCMe(C$_3$H$_7$)CN	Shell	(357)
SD 15418	Cl	NHEt	NHCMe$_2$CN	Shell	(358)
WL 9385	N$_3$	NHEt	NHCMe$_3$	Shell	(359)

* Indicates accepted common name of the British Standards Institution.

replacement of the third needs a fairly high temperature (Fig. 4.25). This reaction has been extensively studied and manufacturing processes have been developed which give very high yields of pure products [319-323].

The triazine herbicides generally have low solubilities (e.g. simazine 5 ppm) and are, on the whole, not effective by application to the leaves of plants, although there are considerable variations within the series and, in particular, the 2-methoxy- and 2-methylthio-compounds have much more contact activity than the 2-chloro-compounds.

The triazines act slowly when applied to the soil and are very persistent, not so much because of low solubility but because of adsorption onto the organic matter in soil. Their original use was for total weed control on industrial and non-agricultural land and they are still widely used for this purpose because of their long-lasting effects. It was later found that maize is very resistant to the triazine herbicides because the plant sap contains an active principle—the glucose derivative of 2,4-dihydroxy-7-methoxy-1,4-benzoxazine-3-one- which can remove the Cl atom very readily by prior conversion of one of the ring nitrogen atoms to the N-oxide, and thus destroy the herbicidal activity [324] (Fig. 4.26). Maize is a major crop in the U.S.A. and in many other parts of the world, and a very high proportion of this crop is now treated with atrazine, which is the favoured compound, although simazine is used preferably in wet districts.

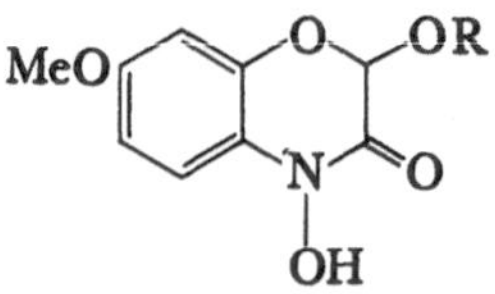

R = glucose
Fig. 4.26   2,4-Dihydroxy-7-methoxy-1,4-benzoxazine-3-one

Some fruit crops are resistant to the triazine herbicides, particularly raspberries and blackcurrants, and compounds such as simazine can also be used for weed control in orchards of large trees, presumably because the roots of these are at a lower level and the herbicide tends to stay in the upper soil layers and not leach downwards to an appreciable extent.

Nearly all s-triazine compounds of the above types show some herbicidal activity, but there are differences in selectivity which can be commercially exploited. The main limitation on commercial development is the high cost of toxicological and ecological testing which is required to obtain Government approval for marketing, which makes it uneconomic to develop commercially all the variants which could have specific advantages for minor crops.

The triazines are the second most important and widely used herbicides in the world after the phenoxyacetic acids with a market value approaching £100 million and the total annual usage of cyanuric chloride to make them is probably of the order of 10,000 tonnes. It has been suggested that use of triazine herbicides may decline because of their persistence and residual toxicity in the soil but there are few signs of this happening, and a continuing growth rate of 10% per year during the next ten years seems likely.

4.22.1   Simazine (2-Chloro-4,6-bis-ethylamino-s-triazine)

Simazine is outstandingly selective towards maize, but, because of its low solubility, is not very effective in dry conditions. Because it is strongly bound to colloidal particles in the soil it is a safe herbicide for deep-rooted crops and is used in late winter or early spring to control germinating weeds around fruit trees [325], although plums and cherries are less tolerant of large doses than are apples and pears [326]. It can also be safely used as pre-emergent herbicide in bush fruits such as raspberries [327], gooseberries [328], and black-currants [329]. Other crops which have been successfully treated with simazine include sugar-cane, pineapples and asparagus [330]. It is very widely used at high application rates for total weed control on industrial land, railways, paths, etc. Its strong adsorption may result in inadequate control of weeds in soils which contain a high proportion of organic matter.

4.22.2   Atrazine (2-Chloro-4-ethylamino-6-isopropylamino-s-triazine)

Atrazine can be used more effectively than simazine when there is a lack of water and it is the pre-emergent herbicide of choice for maize except in very wet districts [331]. The total cultivated area devoted to maize is about 32×10^6 hectares in the U.S.A. and about 8×10^6 hectares in Europe and a high proportion of this crop is treated with atrazine as a pre-emergent herbicide. The total amount of atrazine used as a herbicide in the U.S.A. in 1970 was about 8000 tonnes, and a considerable further amount was used for non-agricultural purposes, such as industrial weed control for which atrazine is preferable to simazine if there is a high proportion of grasses.

4.22.3   Trietazine (2-Chloro-4-ethylamino-6-diethylamino-s-triazine)

Trietazine is mainly useful as a pre-emergent herbicide in peas, potatoes and tobacco.

4.22.4   Propazine (2-Chloro-4,6-*bis*-isopropylamino-s-triazine)

Propazine is a useful post-emergent selective herbicide in carrots and celery, but has now been largely superseded by prometryne.

4.22.5   Ipazine (2-chloro-4-diethylamino-6-isopropylamino-s-triazine)

Ipazine is particularly useful as a post-emergent herbicide in cotton.

4.22.6   Atraton (2-Methoxy-4-ethylamino-6-isopropylamino-s-triazine)

4.22.7   Prometon (2-Methoxy-4,6-bis-isopropylamino-s-triazine)

The main use of these two triazine herbicides is for control of very deep-rooted perennial grasses.

4.22.8   Desmetryne (2-Methylthio-4-methylamino-6-isopropylamino-s-triazine)

Desmetryne is used as a post-emergent herbicide in brussels sprouts [332], cabbage [333], cauliflower [334] and kale [335] but application at precisely the right stage of growth is vital to avoid crop damage.

4.22.9   Ametryne (2-Methylthio-4-ethylamino-6-isopropylamino-s-triazine)

Ametryne is used for pre-emergent control of weeds in potatoes [336].

4.22.10   Terbutryne (2-Methylthio-4-ethylamino-6-tert-butylamino-s-triazine)

Terbutryne is the safest triazine for use as a pre-emergent herbicide in barley, wheat, oats and rye.

4.22.11   Prometryne (2-Methylthio-4,6-bis-isopropylamino-s-triazine)

Prometryne can be used safely for post-emergent weed control in carrots [337], celery [338], leeks [339], and onions [340] and as a pre-emergent herbicide in peas [341] and potatoes [342].

4.23   OTHER CROP PROTECTION PRODUCTS BASED ON POLYCHLOROHETEROAROMATICS

Although there are a large number of claims in the patent literature for pesticidal activity of various derivatives of polychlorinated heterocyclic compounds very few have, as yet, actually been marketed, although there is currently a large amount of research activity in this field which can be expected eventually to yield commercially useful crop protection products. Those which have appeared on the market are listed in Table 4.11 which gives their chemical structures, common names (those marked * are accepted common names of the British Standards Institution), trade names, the type of use, the originating company and the patent reference.

4.23.1   Pyriclor (2,3,5-Trichloro-4-hydroxypyridine)

Pyriclor was introduced as a herbicide in 1965. It controls annual and perennial grasses and most broad-leaved weeds and has a long residual effect [379]. However, its activity against grasses is much higher than its activity against broad-leaved weeds so its main

TABLE 4.11   Polychlorinated Heterocyclic Pesticides

Formula	Use†	Common name	Trade name	Originating company	Patent reference
(structure: pyriclor)	H	pyriclor*	'Daxtron'	Dow	(360)
(structure: picloram)	H	picloram*	'Tordon'	Dow	(361–363)
(structure: pyridinitril)	F	pyridinitril*	—	Merck	(364–367)
(structure: Torelle)	I		'Torelle'	Dow	(368)
(structure: chlorpyriphos)	I	chlorpyriphos	'Dursban'	Dow	(369)
(structure: anilazine)	F	anilazine*	'Dyrene'	Esso	(370)
(structure: chlorquinox)	F	chlorquinox	'Lucel'	Fisons	(371)
(structure: tetrachlorothiophen)	N	tetrachloro-thiophen	—	Pennwalt	(372, 373)

† H = herbicide; F = fungicide; I = insecticide; N = nematicide.
* Indicates accepted common name of the British Standards Institutions.

commercial use has been for control of the former [375]. It has been used for total weed control in industrial situations and alongside roads [376] and, in this context, it is quite effective against a number of tropical weeds [377]. Its high activity against grasses makes possible its use for selective control of this type of weed in a number of crops, particularly sugar-cane and cotton. Its use in sugar-cane does not reduce sugar yield [378], and, when used for cotton defoliation, it is very effective in preventing re-growth [379].

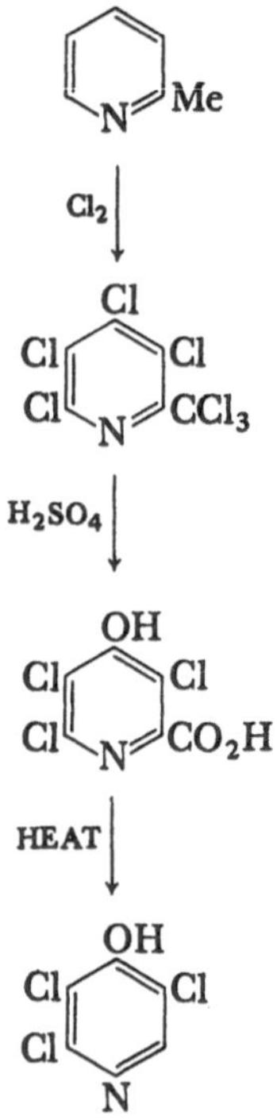

Fig. 4.27 Manufacture of pyriclor

It can be used as a pre-emergent herbicide in peanuts [380], and promising results have also been reported with potatoes [381, 382] and tobacco [383]. It has been used successfully as a pre-ploughing herbicide for control of quack grass (Agropyron repens) in maize [384], but its performance in rice appears to be variable and it sometimes causes crop damage [385, 386]. Its use has also been suggested to control grass weeds in irrigation canals [387].

Its mode of action appears to be inhibition of chloroplast division and growth [388]. Some doubt about its toxic effects on mammals has resulted in a considerable restriction of its commercial use. It is manufactured by liquid phase chlorination of 2-picoline to give tetrachloro-2-trichloromethylpyridine which is then hydrolyzed to 3,5,6-trichloro-4-hydroxypyridine-2-carboxylic   acid,   which decarboxylates quite readily to pyriclor [389, 390] (Fig. 4.27).

4.23.2   Picloram (4-Amino-3,5,6-trichloropyridine-2-carboxylic Acid)

Picloram was introduced in 1963 and it soon became apparent that it is one of the most active and persistent herbicides ever discovered [391]. It kills vegetation of practically every type, and crops such as beans, potatoes, peas, cotton, tobacco, beet and peanuts are highly sensitive to it, although cereals have some resistance [392]. Its use as a selective herbicide has, therefore, been attempted only in maize [393] and wheat [394], but the timing of application is critical to avoid crop damage.

Its main commercial use arises from the fact that it will control deep-rooted perennials [395], woody plants [396], bracken [397-400] and forest trees [401-403], and few herbicides have this type of activity. It has been used quite extensively for forest

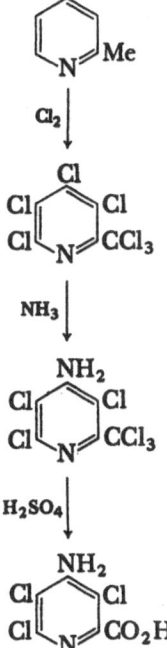

Fig. 4.28   Manufacture of picloram

management [404] and for defoliation of tropical forests by aerial spraying [405, 406] but its use for this purpose in a military context has brought some criticism. There is a very extensive literature on its use for control of weedy plants of various kinds including oak [407], prickly pear [408], nutgrass [409], juniper [410], whitebrush [411], backbrush [412], rangebrush [413], maple [414], cypress [415], elm [416], etc.

It is manufactured by liquid phase chlorination of 2-picoline to give tetrachloro-2-trichloromethylpyridine which is then reacted with ammonia to give 4-amino-3,5,6-trichloromethylpyridine which is hydrolyzed with sulphuric acid to picloram [417, 418] (Fig. 4.28).

4.23.3   Pyridinitril (2,6-Dichloro-3,5-dicyano-4-phenylpyridine)

This compound was discovered in 1964 and has been introduced commercially as an agricultural fungicide. It is manufactured by reaction of cyanacetamide with benzaldehyde according to the method of Day and Thorpe [419] to give 3,5-dicyano-6-hydroxy-4-phenylpyrid-2-one which is then treated with a chlorinating agent such as phosphorus pentachloride or thionyl chloride [420] (Fig. 4.29). Preparation of the same compound as a dyestuff intermediate was reported at about the same date [421].

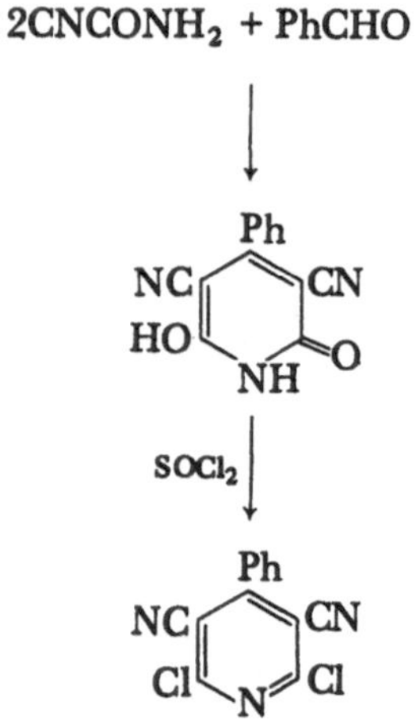

$$2CNCONH_2 + PhCHO$$

Fig. 4.29   Manufacture of pyridinitril

Pyridinitril has a broad spectrum of activity against fungi such as Venturia inequalis, Cladosporium friluum, Botrytis cinerea, Plasmopara viticola, Phytophora species and Alternaria species, but its effectiveness against the mildews is considerably less [422]. It is a good general foliar fungicide which has a relatively long-term effect and is well tolerated by plants [423]. In particular, it is effective against apple scab with little phytotoxicity [424].

It was subsequently shown that pyridinitril has a pronounced synergistic effect with captan (*N*-trichloromethyl tetrahydrophthalimide) which is a well-known and widely used fungicide [425, 426]. It is mainly used commercially, as a mixture with captan, typically as a dusting powder containing 25% captan and 15% pyridinitril.

A large range of analogous 2,6-dichloro-3,5-dicyano-4-alkylthio-pyridines have been reported to have similar fungicidal activity [427].

4.23.4 Chlorpyriphos (3,5,6-Trichloropyridyl-2-*o*,*o*-diethylphosphorothioate)

This compound was introduced in 1965 as a wide-spectrum insecticide with moderate residual activity [428, 429]. Its metabolism has been studied with the ^{14}C-labelled compound [430], and

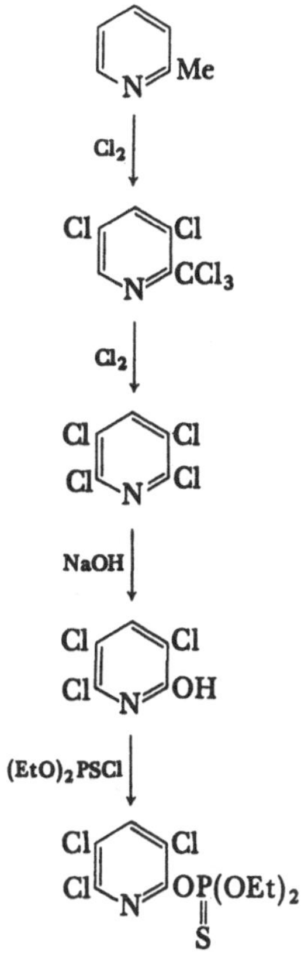

Fig. 4.30   Manufacture of chlorpyriphos

it has been shown that continuous feeding of 100 ppm in the diet to chickens produces no adverse toxic effects [431]. Its major commercial uses have been for industrial and domestic cockroach control [432] and for control of ear tick in cattle. For cockroaches it is very effective, being active against the American, German and brown-banded species [433]. For cattle, dipping in 0.1% emulsion of the compound gives 90% control of the ear tick after one month [434], and a dip vat containing 0.11% of the compound gave protection for 42 weeks [435]. The major strains of cattle tick in all tropical countries are now resistant to most of the organophosphorus insecticides in current use but are still fairly susceptible to chlorpyriphos, and this has created a market for this product.

For mosquito control the compound has been less successful as it appears not to have a sufficiently long-lasting action [436-438]. It has also been used to control clinch-bugs [439], boll-weevils [440] and sugar-cane froghopper [441].

The compound was selected for commercial development from a large number of phosphates and phosphorothioates of hydroxy-halopyridines [442]. It is manufactured by hydrolysis of 2,3,5,6-tetrachloropyridine to 3,5,6-trichloro-2-hydroxypyridine which is then reacted with o,o-diethylthiophosphoryl chloride [443, 444] (Fig. 4.30). The economic viability of the process depends on the technology which has been developed to obtain the required tetrachloropyridine in high yield by manipulation of chlorination conditions and catalysts.

4.23.5   Anilazine (2-(2-Chloroanilino)-4,6-dichloro-s-triazine)

This compound, which is related to the triazine herbicides is, however, a commercial fungicide which, at the application rates required for effective fungal control, does not damage the plants [445]. It is manufactured by direct condensation of cyanuric chloride with 2-chloroaniline [446].

It is interesting to note that the corresponding 4-chloroanilino compound is also a fungicide but has less activity and greater phytotoxicity than anilazine [447].

4.24   VETERINARY PESTICIDES BASED ON POLYCHLOROHETEROAROMATICS

The one commercially important veterinary pesticide derived from polychlorinated heterocyclic compounds is clopidol (3,5-dichloro-2,6-dimethyl-4-hydroxypyridine). This compound, which was discovered in 1966, introduced a new concept of coccidiosis protection in poultry in that it acts at the sporozoite stage of the parasite, which is very early in its life cycle [448]. The mode of action has been studied by a number of workers [449] and it has been shown that the compound is active against the sporozoites only in vivo, not in vitro [450].

Various groups of workers [451-454] have reported excellent results with clopidol, obtaining 98% control of the eight common species of the parasite Eimeria with 0.0125% w/w of the compound in the feed, without any adverse effects on weight gain and feed efficiency [455], and also without reducing the quality or quantity of eggs [456]. It has, however, been shown that continuous medication is necessary to prevent upsurge of infection [457, 458]. There has been some suggestion that strains of Eimeria resistant to the compound may be appearing [459].

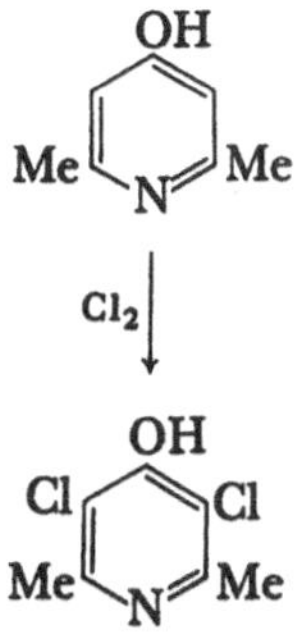

Fig. 4.31. Manufacture of clopidol

It has been reported that up to the end of 1968, over 300 million broiling chickens had been reared in the U.K. with clopidol as a supplement in their diet [460]. Imports into the U.K. in 1970 were about 200 tonnes and total production in 1970 is estimated to have been about 500 tonnes (entirely in the U.S.A.). The compound is manufactured by liquid phase chlorination of 4-hydroxy-2,6-dimethylpyridine [461] (Fig. 4.31).

4.25   INDUSTRIAL BIOCIDES BASED ON POLYCHLOROHETEROAROMATICS

The only derivative of a polychlorinated heterocyclic compound which has significant commercial importance as an industrial biocide is tetrachloro-4-methylsulphonylpyridine which was discovered in 1964. It is manufactured by reaction of pentachloropyridine with methyl mercaptan to give the 4-methylthio compound which can then be oxidized with hydrogen peroxide [462-464 or, more economically, with chlorine [465] to give the 4-methylsulphonyl compound (Fig. 4.32). A very large number of 4-alkylsulphonyl pyridines were originally tested for biological activity but tetra-chloro-4-methylsulphonylpyridine was the outstanding one and was selected for commercial development.

Wolf and Bobalek [466] have described the antimicrobial performance of this compound as an industrial preservative. Their results indicate that it has a wide spectrum of activity against

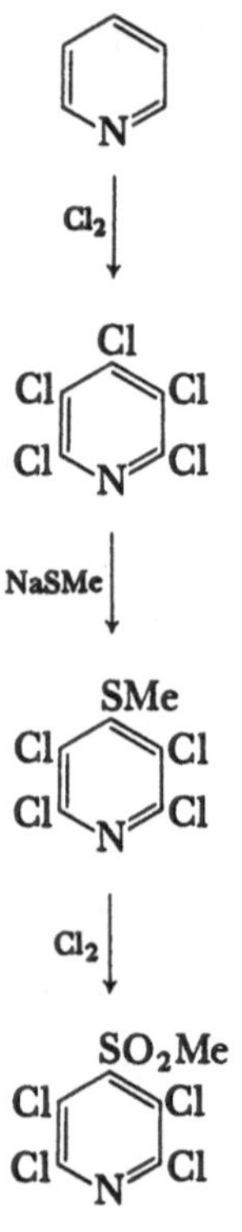

Fig. 4.32   Manufacture of tetrachloro-4-methylsulphonyl pyridine

micro-organisms which cause deterioration in industrial systems. It is particularly active as a fungistat, but, as a bacteristat, it is more active against gram-positive than against gram-negative bacteria. Its fungistatic activity and persistence are said to give it outstanding performance as a mildew-killer in exterior paints, and it also appears to have useful effects as a preservative in asphalt, caulking compositions, joint cements, paper, plastics, cutting fluid emulsions and textiles. Trueblood and Anderson [467] made a special study of its effect in paints and found that, for short-term exposures of 6-10 weeks, it was more effective than organo-tin or organo-mercury fungicides. Incorporation of the compound during the manufacture of polyurethanes gives a product with high resistance to fungal and bacterial attack as evidenced by lack of cracking on prolonged exposure [468]. It has also been reported that addition of the compound to pentachlorophenol has a synergistic effect for wood preservation [469].

Total production of this product, which is manufactured entirely in the U.S.A., was estimated at about 1000 tonnes in 1970.

4.26   RODENTICIDES BASED ON POLYCHLOROHETEROAROMATICS

The only derivative of a chlorinated heteroaromatic compound which has been used commercially as a rodenticide is crimidine, 2-chloro-4-dimethylamino-6-methylpyrimidine, manufactured from 2,4-dichloro-6-methylpyrimidine [470] (Fig. 4.33).

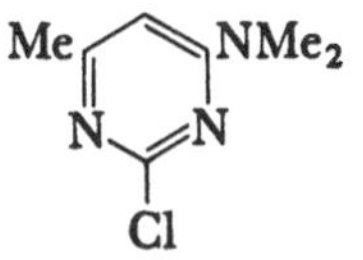

$$Me \underset{N \quad N}{\overset{NMe_2}{\bigcirc}} Cl$$

Fig. 4.33   Crimidine

For many years warfarin has been the most widely used and effective rodenticide, but spread of resistance to this compound has stimulated search for other products. Derivatives of polychlorinated heterocyclic compounds have been looked at fairly widely in this context, but the only compounds for which any effectiveness is claimed are derivatives of tetrachloro-4-hydroxy-pyridine, particularly the carbamate [471].

4.27   PHARMACEUTICALS BASED ON POLYCHLOROHETEROAROMATICS

As far as the author is aware, no derivatives of polychlorinated heterocyclic compounds have found any significant commercial utility as pharmaceuticals despite the large number of claims in the patent literature. The reason, as was indicated for the poly-chlorinated benzenes, is that a large number of chlorine atoms in the molecule is undesirable in a chemotherapeutic substance because they tend to increase toxicity and to produce an undesirably high lipoid/water partition.

Some years ago the aziridino derivatives formed by reaction of chloropyrimidines and chlorotriazines with ethylene imine received some attention as potential tumour-inhibiting agents but the only commercial product which arose from this work was tretamine, the derivative from cyanuric chloride [472-479] (Fig. 4.34). This compound has been used as a neoplastic suppressant in certain restricted types of leukaemia and sarcoma.

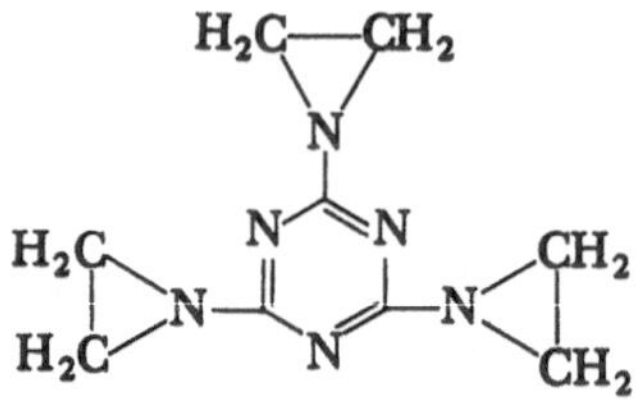

Fig. 4.34 Tretamine

Various polychloropyridazinones have been investigated as anti-tubercular drugs, e.g. 4,5,6-trichloro-2-(4-aminophenyl)-3-pyridazin-one [480, 481] and 6-amino-4,5-dichloro-2-methyl-3-pyridazinone [482, 483]. The 2,6-dichloro-4-pyrimidinylaminophenyl ether of 2-diethylaminoethanol has been claimed as a hypocholesteremic agent [484].

REFERENCES

1. D. W. F. HARDIE, *A History of the Chemical Industry in Widnes*, Imperial Chemical Industries Limited, 155, (1950).
2. M. CAMPBELL and H. HATTON, *Herbert H. Dow: Pioneer in Creative Chemistry*, p. 114, Appleton-Century-Crofts Inc., New York, 1951.
3. E. JUNGFLEISCH, *Ann. Chem. Phys.*, [4], 15, 215 (1868).
4. J. B. COHEN and P. HARTLEY, *J. Chem. Soc.*, 87, 1360 (1905).
5. M. F. BOURION, *Ann. Chim.*, 14, 215 (1920).
6. R. B. MACMULLIN, *Chem. Eng. Progress*, 44, 183 (1948).
7. V. MERZ and W. WEITH, *Ber.* 10, 747 (1877).
8. J. R. MARES, U.S. Pat. No. 2,111,866 (1938).
9. A. J. BRUNJES and M. J. P. BOGART, U.S. Pat. No. 2,395,777 (1942).
10. A. J. BRUNJES and M. J. P. BOGART, U.S. Pat. No. 2,470,336 (1949).
11. Société d'Ugine, French Pat. No. 1,209,222 (1960).
12. N. BENNETT, Brit. Pat. No. 440,205 (1934).
13. B. O. PRAY, U.S. Pat. No. 2,819,321 (1958).
14. B. H. NICOLAISEN, U.S. Pat. No. 2,777,003 (1958).
15. Badische Anilin und Soda-Fabrik AG, Ger. Pat. No. 973,588 (1960).
16. Farbwerke Hoechst, Ger. Pat. No. 1,100,610 (1961).
17. R. T. FOSTER, Canadian Pat. No. 544,274 (1953).
18. *Orthodichlorobenzene*, Imperial Chemical Industries Limited, Mond Division Technical Service Note TS/B/2280/E (1970).
19. *Orthodichlorobenzene for Engine Cleaning and Decarbonising*, Imperial Chemical Industries Limited, Mond Division Technical Service Note TS/B/2181 (1970).

20. M. R. BLAND, U.S. Pat. No. 2,964,429 (1960).
21. *J. Commerce*, New York, 26 (1955).
22. A. G. TENNER and D. DE KLERK, *Rev. Sci. Instruments*, **28**, 206 (1957).
23. L. SCHWAREZ, *Sanitary Products: Their Manufacture, Testing and Use*, p. 189, Maenair-Dorland, New York, 1943.
24. Progil SA, Belgium Pat. No. 601,358 (1961).
25. H. E. PARKER, Brit. Pat. No. 806,726 (1957).
26. *'Maskador'* in *Deodorant Blocks*, Imperial Chemical Industries Limited, Mond Division Technical Service Note TS/B/2276 (1970).
27. *'Maskador'* for *Clothes Moth Control*, Imperial Chemical Industries Limited, Mond Division Technical Service Note TS/B/2277 (1970).
28. E. A. EVANS and J. S. ELLIOTT, Brit. Pat. No. 538,159 (1939).
29. *Paradichlorbenzene*, Imperial Chemical Industries Limited, Mond Division, Technical Service Note TS/B/2275 (1970).
30. J. FORD, U.S. Pat. No. 2,139,945 (1938).
31. C. J. PHILLIPS, U.S. Pat. No. 2,198,738 (1936).
32. F. M. CLARK and W. M. KUTZ, U.S. Pat. No. 2,169,872 (1939).
33. J. S. HOWDEN and E. M. SMYE, Canadian Pat. No. 590,736 (1957).
34. P. B. COURSEY, BIOS Final Report No. 1936 (1947).
35. A. LAURENT, *Ann. Chim. Phys.*, [2], **52**, 275 (1833).
36. E. FISCHER, *Ber.*, **11**, 735 (1878).
37. J. W. AYLSWORTH, U.S. Pat. No. 1,111,289 (1913).
38. J. W. AYLSWORTH, U.S. Pat. No. 914,222 (1909).
39. J. W. AYLSWORTH, U.S. Pat. No. 914,223 (1909).
40. A. T. JONES, *J. Ind. Hyg. Toxicology*, **23**, 290 (1941).
41. M. R. MAYERS and A. R. SMITH, *Ind. Bull. NY State Dept. Labour*, **21**, 30 (1942).
42. R. ENGELHARDT, Ger. Pat. No. 567,261 (1931).
43. E. P. BREAKEY and A. C. MILLER, *J. Econ. Entomology*, **29**, 820 (1936).
44. J. DOFAY, French Pat. No. 1,268,907 (1960).
45. R. L. JENKINS, U.S. Pat. No. 1,892,397 (1932).
46. R. L. JENKINS, U.S. Pat. No. 1,892,398 (1932).
47. R. THERMET and L. PARVI, U.S. Pat. No. 3,029,295 (1962).
48. R. L. JENKINS and J. A. SIKARSKI, U.S. Pat. No. 1,892,400 (1932).
49. J. W. J. FAY and J. H. RICHARDS, *BIOS Final Report*, No. 893.
50. F. M. CLARK, U.S. Pat. No. 1,994,302 (1935).
51. H. I. WEINGARTEN, U.S. Pat. No. 3,038,107 (1962).
52. H. I. WEINGARTEN, U.S. Pat. No. 3,065,297 (1962).
53. H. I. WEINGARTEN, *J. Org. Chem.*, **26**, 4347 (1961).
54. H. I. WEINGARTEN, *J. Org. Chem.*, 27, 2024 (1962).
55. H. I. WEINGARTEN, U.S. Pat. No. 2,977,516 (1961).
56. J. H. YOUNG, U.S. Pat. No. 1,836,147 (1932).
57. E. W. TROELANDER and W. C. WILSON, U.S. Pat. No. 2,219,157 (1938).
58. F. M. CLARK and J. H. KOENIG, U.S. Pat. No. 2,130,264 (1938).
59. W. C. HAYMAN, U.S. Pat. No. 2,141,910 (1938).
60. J. G. FORD and C. F. HILL, U.S. Pat. No. 2,158,281 (1939).
61. *Aroclor Plasticizers*, Monsanto Co. Tech. Bulletin No. PL-306 (1960).
62. *Aroclor Resins*, Monsanto Co. Tech. Bulletin No. PL-311 (1962).
63. *Aroclor Fire-retarding plasticizers*, Monsanto Co. Tech. Bulletin No. CS-14 (1960).
64. *Aroclor Compounds*, Monsanto Co. (1960).
65. H. O. BOWRON, *Paint Technology*, 2, 25 (1937).

66. H. A. GARDNER and G. G. SWORD, *National Paint Association Sci. Sec. Circular* No. 55, 100 (1938).
67. R. L. JENKINS and R. N. FOSTER, *Ind. Eng. Chem.*, 23, 1362 (1931).
68. E. KLEIN, U.S. Pat. No. 2,077,700 (1937).
69. W. J. DAVIS and P. G. BENIGNUS, *Chem. Eng. Progress*, 59, 39 (1963).
70. G. W. WOOD, *Manufacturing Chemist*, 19, 99 (1948).
71. R. G. QUINN, U.S. Pat. No. 2,030,653 (1936).
72. B. K. GREEN and R. W. SANDBERG, U.S. Pat. No. 2,548,366 (1951).
73. *European Chemical News*, 21, No. 523,11 (1972).
74. R. T. FOSTER and N. BENNETT, U.S. Pat. No. 2,440,602 (1948).
75. E. J. KRAUS and J. W. MITCHELL, *Botan. Gaz.*, 108, 301 (1947).
76. A. J. NORMAN, *Botan. Gaz.*, 107, 475 (1946).
77. H. E. THOMPSON, C. P. SWANSON, and A. G. NORMAN, *Botan. Gaz.*, 107, 476 (1946).
78. E. J. ORDAL, *Proc. Soc. Exp. Biol. Med.*, 47, 387 (1924).
79. I. G. Farbenindustrie AG, Brit. Pat. No. 356,192 (1931).
80. F. W. NORTON, U.S. Pat. No. 2,304,013 (1942).
81. J. F. MILLS, U.S. Pat. No. 1,955,080 (1934).
82. I. G. Farbenindustrie AG, French Pat. No. 709,788 (1931).
83. E. L. BRITTON and L. E. MILLS, U.S. Pat. No. 1,946,057 (1934).
84. British Dyestuffs Corporation, Brit. Pat. No. 259,690 (1926).
85. Monsanto Chemical Co., Brit. Pat. No. 530,836 (1940).
86. G. H. ELLIS, U.S. Pat. No. 2,161,654 (1939).
87. J. F. RICHARDSON, *Hide Leather Shoes*, 99, 28 (1940).
88. F. B. SMITH and J. E. LIVAK, U.S. Pat. No. 2,107,650 (1938).
89. I. ROSEN, U.S. Pat. No. 2,812,366 (1957).
90. W. C. STOESSER, U.S. Pat. No. 2,131,259 (1938).
91. F. J. SHELTON, U.S. Pat. No. 2,947,790 (1960).
92. J. SPROULE, *Pest. Technology*, 41, 3 (1960).
93. J. FRYER and R. MAKEPEACE, *Weed Control Handbook Vol. II*, Oxford, 40 (1970).
94. R. T. FOSTER and N. BENNETT, U.S. Pat. No. 2,440,602 (1948).
95. T. SANDERS and J. PRESCOTT, Ind. Eng. Chem., 51, 974 (1959).
96. J. FRYER and R. MAKEPEACE, *Weed Control Handbook Vol. II*, Oxford, 116 (1970).
97. Ibid., 171
98. E. E. GILBERT, U.S. Pat. No. 2,830,083 (1958).
99. J. FRYER and R. MAKEPEACE, *Weed Control Handbook Vol. II*, Oxford, 187 (1970).
100. *Chem. Eng. News*, 48, 60 (1970).
101. J. FRYER and R. MAKEPEACE, *Weed Control Handbook Vol. II*, Oxford, 17 (1970).
102. J. N. HOGSETT, U.S. Pat. No. 3,076,025 (1963).
103. J. FRYER and R. MAKEPEACE, *Weed Control Handbook Vol. II*, Oxford, 19 (1970).
104. Ibid., 146.
105. Ibid., 16.
106. Ibid., 208.
107. Ibid., 20.
108. Ibid., 200.
109. Ibid., 20.
110. Ibid., 116.
111. Ibid., 118.
112. Ibid., 122.
113. Ibid., 120.
114. Ibid., 206.

115. Ibid., 115.
116. Ibid., 111.
117. Ibid., 223.
118. Ibid., 96.
119. Ibid., 61.
120. Ibid., 85.
121. Ibid., 77.
122. Ibid., 83.
123. Ibid., 71.
124. Ibid., 270.
125. W. GUMP, U.S. Pat. No. 2,250,408 (1941).
126. J. B. ADAMS and M. HOBBS, *J. Pharm. & Pharmacol.*, 10, 507 (1958).
127. G. M. COMPEAU, *J. Am. Pharm. Asscn. Sci. Ed.*, 49, 1574 (1960).
128. E. F. TRAUB, C. A. NEWHALL, and J. R. FULLER, *Surg. Gynecol. Obstet.*, 79, 205 (1944).
129. E. F. TRAUB, C. A. NEWHALL, and J. R. FULLER, *Arch. Dermatol. Syphilis*, 52, 385 (1945).
130. P. B. PRICE and A. BENNETT, *Surgery*, 24, 542 (1948).
131. G. V. SEASTONE, *Surg. Gynecol. Obstet.*, 84, 355 (1947).
132. J. M. MILLER and D. A. JACKSON, *Military Med.*, 127, 576 (1962).
133. E. G. KLARMANN, E. S. WRIGHT, and V. A. SHTERNOV, *Am. J. Pharm.* 122, 5 (1950).
134. E. G. KLARMANN, *Acta Dermato-Venereol*, 37, 59 (1957).
135. W. B. SHELLEY, H. J. HURLEY, and A. C. NICHOLS, *Arch. Dermatol. Syphilis*, 68, 430 (1953).
136. R. B. SHUNARD, D. J. BEAVER, and M. C. HUNTER, *Soap, Sanit. Chemicals*, 29, 34 (1953).
137. K. M. WOOD and S. H. HOPPER, *J. Am. Pharm. Assoc. Sci. Ed.*, 47, 317 (1958).
138. D. J. BEAVER, D. P. ROMAN, and P. F. STOFFEL, *J. Am. Chem. Soc.*, 79, 1236 (1957).
139. F. D. JONES, U.S. Pat. No. 2,390,491 (1945).
140. F. D. JONES, Brit. Pat. No. 598,072 (1948).
141. F. D. JONES, Brit. Pat. No. 598,105 (1948).
142. H. A. STEVENSON and R. F. BROOKES, Brit. Pat. No. 822,199 (1959).
143. B. M. WILLIAMS, U.S. Pat. No. 2,749,360 (1956).
144. M. M. WRIGHT, U.S. Pat. No. 2,792,295 (1957).
145. G. W. KITCHINGMAN and A. C. TUCKER, Brit. Pat. No. 804,565 (1958).
146. B. J. HEYWOOD, U.S. Pat. No. 2,866,816 (1958).
147. W. C. STOESSER, U.S. Pat. No. 2,131,259 (1938).
148. H. F. WILSON and D. H. McRAE, Brit. Pat. No. 974,475 (1963).
149. H. F. WILSON and D. H. McRAE, U.S. Pat. No. 3,080,225 (1963).
150. May and Baker Limited, French Pat. No. 1,375,311 (1964).
151. Henkel and Cie. AG, Brit. Pat. No. 798,476 (1958).
152. W. GÜNDEL and H. LINDEN, U.S. Pat. No. 2,852,548 (1958).
153. H. F. BRUST and H. O. SEINKBEIL, U.S. Pat. No. 2,754,324 (1956).
154. W. SCHAEFER, R. WEGLER, L. EUE, and H. HUCK, Belg. Pat. No. 639,645 (1964).
155. W. D. HARRIS and A. W. FELDMAN, U.S. Pat. No. 2,828,198 (1958).
156. J. K. LEASURE, U.S. Pat. No. 3,074,790 (1963).
157. T. A. GIRARD, E. P. DiBELLA, and H. SIDI, U.S. Pat. No. 2,848,470 (1958).
158. N. TISCHLER, U.S. Pat. No. 3,081,162 (1963).
159. Hooker Chemical Corporation, Brit. Pat. No. 900,561 (1962).
160. Velsicol Chemical Corporation, Brit. Pat. No. 901,553 (1962).

161. S. B. RICHTER, U.S. Pat. No. 3,013,054 (1961).
162. S. B. RICHTER, U.S. Pat. No. 3,013,055 (1961).
163. S. R. McLANE, J. R. BISHOP, and H. P. RAMAN, U.S. Pat. No. 3,014,063 (1958).
164. H. P. RAMAN, U.S. Pat. No. 3,174,999 (1965).
165. R. F. LINDEMANN, U.S. Pat. No. 2,923,634 (1960).
166. E. ZINN, C. E. ENTEMANN, J. H. PERKINS, and W. N. WHEELER, U.S. Pat. No. 3,052,712 (1962).
167. Velsicol Chemical Corporation, Brit. Pat. No. 1,005,049 (1965).
168. J. W. NEMEC and R. S. COOK, Belg. Pat. No. 619,660 (1963).
169. J. D. POLLARD and E. F. ORWELL, U.S. Pat. No. 3,200,150 (1965).
170. M. J. FIELDING and D. L. GOODARD, U.S. Pat. No. 3,108,038 (1963).
171. R. P. NEIGHBORS and J. R. HOPKINS, French Pat. No. 1,356,497 (1964).
172. J. R. WILLARD and K. P. DORSCHNER, U.S. Pat. No. 3,382,280 (1961).
173. J. R. WILLARD and K. P. DORSCHNER, Brit. Pat. No. 903,766 (1961).
174. C. W. TODD, Brit. Pat. No. 691,403 (1953).
175. C. W. TODD, Brit. Pat. No. 692,589 (1953).
176. C. W. TODD, U.S. Pat. No. 2,655,444 (1953).
177. Farbwerke Hoechst AG, Brit. Pat. No. 950,254 (1962).
178. R. S. JOHNSON, Brit. Pat. No. 982,344 (1963).
179. R. W. LUCKENBAUGH, U.S. Pat. No. 3,095,299 (1963).
180. R. W. LUCKENBAUGH, U.S. Pat. No. 3,288,580 (1964).
181. C. W. TODD, U.S. Pat. No. 2,655,444 (1953).
182. N. TISCHLER, U.S. Pat. No. 2,977,212 (1961).
183. S. RICHTER, Brit. Pat. No. 1,101,396 (1966).
184. S. RICHTER, U.S. Pat. No. 3,334,125 (1966).
185. S. RICHTER, U.S. Pat. No. 3,356,737 (1966).
186. S. RICHTER, U.S. Pat. No. 3,393,064 (1966).
187. H. KOOPMAN and J. DAAMS, U.S. Pat. No. 3,027,248 (1962).
188. J. YATES, Brit. Pat. No. 987,253 (1965).
189. N. V. PHILIPS, French Pat. No. 1,362,906 (1964).
190. Shell Research Limited, Belg. Pat. No. 626,351 (1963).
191. B. W. HORROM, A. J. CROVETTI, and K. L. VISTE, Brit. Pat. No. 1,209,068 (1969).
192. B. W. HORROM, A. J. CROVETTI, and K. L. VISTE, U.S. Pat. No. 3,535,098 (1969).
193. R. A. HERRETT and R. V. BERTHOLD, U.S. Pat. No. 3,399,048 (1965).
194. Ciba Limited, French Pat. No. 1,469,297 (1967).
195. R. BOESCH and J. METIVIER, French Pat. No. 1,394,774 (1965).
196. W. C. STOESSER, U.S. Pat. No. 2,131,259 (1938).
197. W. P. TER HORST, U.S. Pat. No. 2,349,771 (1944).
198. A. STEINDORFF, K. PFAFF, M. ERLENBACH, and W. STAUDER-MANN, Ger. Pat. No. 682,048 (1939).
199. R. E. THORNSON and L. GOLDMAN, U.S. Pat. No. 3,081,224 (1963).
200. N. G. CLARK, H. A. STEVENSON, R. F. BROOKES, and A. F. HAMS, Brit. Pat. No. 845,916 (1960).
201. E. I. Dupont de Nemours and Co., U.S. Pat. No. 3,265,564 (1964).
202. R. D. BATTERSHELL and N. BLUESTONE, French Pat. No. 1,397,521 (1965).
203. Sumitomo Chemical Co., Jap. Pat. No. 69/10,775 (1969).
204. Sumitomo Chemical Co., Brit. Pat. No. 1,164,556 (1969).
205. N. V. PHILIPS, Dutch Pat. No. 66/11,689 (1968).
206. H. MALTZ, F. GREWE, A. DORKEN, and H. KASPERS, Brit. Pat. No. 1,123,850 (1968).

207. W. GUMP, U.S. Pat. No. 2,250,408 (1941).
208. Eli Lilly and Co., Belg. Pat. No. 647,351 (1964).
209. C. J. SCHOOT, M. J. KOOPMANS, and B. G. VAN DEN BOS, U.S. Pat. No. 3,038,924 (1962).
210. W. P. TER HORST, U.S. Pat. No. 2,302,384 (1943).
211. W. P. TER HORST, U.S. Pat. No. 2,349,771 (1944).
212. Sumitomo Chemical Co., Dutch Pat. No. 68/17,250 (1970).
213. Sumitomo Chemical Co., Jap. Pat. No. 69/32,592 (1971).
214. F. ETAT, Brit. Pat. No. 982,940 (1956).
215. K. SASSE, R. WEGLER, H. SCHEINPFLUG, and H. JUNG, Ger. Pat. No. 1,194,631 (1965).
216. J. D. DAVENPORT, R. E. HACKLER, and H. M. TAYLOR, Brit. Pat. No. 1,218,623 (1969).
217. J. H. UHLENBROEK and J. MELTZER, U.S. Pat. No. 3,054,719 (1962).
218. J. MELTZER and H. O. HUISMAN, Brit. Pat. No. 761,837 (1956).
219. J. MELTZER and H. O. HUISMAN, U.S. Pat. No. 2,812,281 (1957).
220. E. E. GILBERT, U.S. Pat. No. 2,618,583 (1952).
221. H. G. SCHMELZER, E. DEGENER, H. TARNOW, H. HOLTSCHMIDT, and G. UNTERSTENHOFER, Brit. Pat. No. 1,151,627 (1968).
222. Fisons Limited, French Pat. No. 1,459,782 (1966).
223. K. ISHII, R. SAKIMOTO, S. KANO, and T. TANIGUCKI, Jap. Pat. No. 63/156/8 (1963).
224. W. P. BOYER, U.S. Pat. No. 2,761,806 (1956).
225. G. SCHRADER, Ger. Pat. No. 1,099,525 (1959).
226. C. L. MOYLE, Brit. Pat. No. 699,064 (1953).
227. C. L. MOYLE, U.S. Pat. No. 2,599,515 (1952).
228. C. L. MOYLE, U.S. Pat. No. 2,599,516 (1952).
229. Ciba Limited, Brit. Pat. No. 1,057,609 (1966).
230. C. H. BOEHRINGER, Brit. Pat. No. 956,343 (1963).
231. C. H. BOEHRINGER, U.S. Pat. No. 3,275,718 (1963).
232. S. B. RICHTER, U.S. Pat. No. 3,459,836 (1969).
233. C. HAVUKAWA and K. KONISHI, Jap. Pat. No. 65/4,750 (1960).
234. G. SCHEGK and G. SCHRADER, Ger. Pat. No. 1,117,110 (1957).
235. R. WHETSTONE and D. HARMAN, Brit. Pat. No. 775,085 (1960).
236. R. WHETSTONE and D. HARMAN, U.S. Pat. No. 2,888,648 (1960).
237. R. WHETSTONE and D. HARMAN, U.S. Pat. No. 2,956,073 (1960).
238. E. E. GILBERT, J. A. OTTO, and E. J. RUMANOWSKI, U.S. Pat. No. 3,003,916 (1961).
239. R. WHETSTONE and D. HARMAN, U.S. Pat. No. 3,116,201 (1963).
240. D. D. PHILIPS and L. F. WARD, U.S. Pat. No. 3,102,842 (1963).
241. D. D. PHILIPS and L. F. WARD, Brit. Pat. No. 972,006 (1963).
242. K. GAETZ, Ger. Pat. No. 1,137,006 (1962).
243. *Otto v Linford, Law Times Reports (new series)*, London, 46, 35, 41.
244. *Badische Anilin and Soda Fabrik v Levinstein*, Reports of Patent Cases, London, 4, 449, 462.
245. W. J. SELL and F. W. DOOTSON, *J. Chem. Soc.*, 1082 (1897); 432 (1898).
246. W. J. SELL and F. W. DOOTSON, *J. Chem. Soc.*, 777 (1898); 205 (1899); 979 (1899); 233 (1900); 771 (1900); 899 (1900); 396 (1903).
247. Imperial Chemical Industries Limited, Brit. Pat. No. 1,041,906 (1965).
248. W. H. TAPLIN, U.S. Pat. No. 3,251,848 (1966).
249. R. HASZELDINE, R. E. BANKS, and J. M. BIRCHALL, U.S. Pat. No. 3,359,267 (1967).
250. Soc. Chem. Ind. Basle, Brit. Pat. No. 393,914 (1933).
251. J. R. DUDLEY, U.S. Pat. No. 2,467,523 (1949).

468　　　　　　　　　　　M. B. GREEN

252. E. L. KROPA, U.S. Pat. No. 2,510,503 (1950).
253. J. G. ERICKSON and W. M. THOMAS, U.S. Pat. No. 2,580,901 (1952).
254. J. R. DUDLEY, *J. Am. Chem. Soc.*, **73**, 2986 (1951).
255. C. F. CROSS and E. J. BEVAN, *Res. Cellulose*, **34** (1895).
256. Soc. Chem. Ind. Basle, Brit. Pat. No. 209,723 (1923).
257. F. GÜNTHER, U.S. Pat. No. 1,567,731 (1925).
258. I. G. Farbenind AG, Brit. Pat. No. 341,461 (1929).
259. R. HALLER and A. HECKENDORN, U.S. Pat. No. 1,886,480 (1932).
260. J. WARREN, J. D. REID, and E. HAMALAINEN, *Text. Res. J.*, **22**, 584 (1952).
261. F. R. ALSBERG, I. D. RATTEE, W. E. STEPHEN, W. J. MARSHALL, R. W. SPEKE, and C. D. WESTON, Brit. Pat. No. 797,946 (1958).
262. F. R. ALSBERG, R. N. HESLOP, I. D. RATTEE, W. E. STEPHEN, W. J. MARSHALL, R. W. SPEKE, and C. D. WESTON, Brit. Pat. No. 798,121 (1958).
263. I. D. RATTEE, *Research*, **12**, 15 (1959).
264. *Procion Dyestuffs in Textile Dyeing*, Imperial Chemical Industries Limited (1962).
265. *Procion Dyestuffs in Textile Printing*, Imperial Chemical Industries Limited (1961).
266. F. R. ALSBERG, H. G. CONNOR, W. F. LIQUORICE, and S. W. MILNE, *Internat. Dyer & Textile Printer*, **141**, 151 (1969).
267. C. PRESTON and A. S. FERN, *Chimia*, **15**, 177 (1961).
268. R. C. SENN, O. A. STAMM, and H. ZOLLINGER, *Milliand Textilber.*, **44**, 261 (1963).
269. R. C. SENN and H. ZOLLINGER, *Helv. Chim. Acta.*, **44**, 261 (1963).
270. J. WEGMANN, *J. Soc. Dyers Colourists*, **76**, 205 (1960).
271. T. L. DAWSON, *J. Soc. Dyers Colourists*, **80**, 134 (1964).
272. W. E. STEPHEN and C. G. TILLEY, Brit. Pat. Nos. 836,248; 837,124; 838,728 (1960).
273. T. L. DAWSON, A. S. FERN, and C. PRESTON, *J. Soc. Dyers Colourists*, **76**, 210 and references cited therein (1960).
274. K. KLEB, E. SIEGEL, and K. SASSE, *Angew. Chem.*, **3**, 408 (1964).
275. Ciba Limited, Belgian Pat. No. 592,148 (1959).
276. J. EISELE, W. FERKIEL, J. DEHNERT, O. TRAUTH, and H. WEIDINGER, German Pat. No. 1,067,404 (1957).
277. Sandoz Limited, Belgian Pat. No. 607,999 (1960).
278. R. N. HESLOP, N. LEGG, J. F. MAWSON, W. E. STEPHEN, and J. WARDLEWORTH, U.S. Pat. No. 2,935,506 (1960).
279. Badische Anilin und Soda-Fabrik, French Pat. No. 1,194,043 (1957).
280. Sandoz Limited, Belgian Pat. No. 572,944 (1957).
281. Farbenfabriken Bayer AG, Belgian Pat. No. 572,973 (1957).
282. Ciba Limited, Belgian Pat. No. 604,068 (1960).
283. H. ACKERMANN and P. DUSSY, *Milliand Textilber*, **42**, 1167 (1961).
284. M. CAPPONI, E. METZGER, and A. GIAMARA, *Am. Dyestuff Reptr.*, **50**, 505 (1961).
285. M. CAPPONI, *Am. Dyestuff Reptr.*, **53**, 913 (1964).
286. W. BECKMANN and D. HILDEBRAND, *J. Soc. Dyers Colourists*, **81**, 11 (1965).
287. K. KLEB, E. SIEGEL, and K. SASSE, *Angew. Chem.*, **3**, 408 (1964).
288. A. J. HALL, *Textile World*, **114**, 82 (1964).
289. P. KRAIS, *Milliand Textilber.*, **10**, 468 (1928).
290. E. ÜHLEIN, *Optische Aufheller, Garmische-Partenkirchen (1959)*.
291. W. WINCOR, *Textil-Rundschau*, **14**, 316 (1959).
292. R. S. LONG, *Proc. Perkin Centennial*, 411 (1956).

293. E. ALLEN, *Am. Dyestuff Reptr.*, **48**, 27 (1959).
294. J. EGGERT and B. WENDT, U.S. Pat. No. 2,171,427 (1937).
295. E. KELLER and B. ZWEIDLER, U.S. Pat. No. 2,473,475 (1943).
296. H. HAUSERMANN, U.S. Pat. No. 2,762,801 (1953).
297. W. W. WILLIAMS and W. E. WALLACE, U.S. Pat. No. 2,618,636 (1948).
298. W. E. WALLACE, U.S. Pat. No. 2,595,030 (1948).
299. W. E. WALLACE and W. W. WILLIAMS, U.S. Pat. No. 2,658,064 (1950).
300. W. W. WILLIAMS and H. B. FREYERMUTH, U.S. Pat. No. 2,700,665 (1950).
301. H. GOLD and S. PETERSON, U.S. Pat. No. 2,764,582 (1950).
302. F. FLECK, U.S. Pat. No. 2,766,239 (1953).
303. D. W. HEIN, U.S. Pat. No. 2,671,784 (1952).
304. D. A. W. ADAMS and R. H. WILSON, U.S. Pat. No. 2,667,458 (1949).
305. R. H. WILSON, U.S. Pat. No. 2,612,501 (1949).
306. W. W. WILLIAMS and B. H. FREYERMUTH, U.S. Pat. No. 2,713,046 (1951).
307. B. W. ROTTSCHAEFER and A. F. PLUE, U.S. Pat. No. 2,703,801 (1951).
308. M. E. CHIDDIX, S. K. HESSE, and M. R. WILLIAMS, U.S. Pat. No. 2,809,999 (1949).
309. E. KELLER and R. ZWEIDLER, U.S. Pat. No. 2,526,668 (1947).
310. R. ZWEIDLER and H. HAUSERMANN, U.S. Pat. No. 2,539,766 (1948).
311. H. HAUSERMANN, U.S. Pat. No. 2,694,064 (1949).
312. H. HAUSERMANN, U.S. Pat. No. 2,762,802 (1954).
313. A. GAST, E. KNÜSLI, and H. GYSIN, *Experientia*, **11**, 107 (1955).
314. A. GAST, E. KNÜSLI, and H. GYSIN, *Experientia*, **12**, 146 (1956).
315. H. GYSIN and E. KNÜSLI, *Advances in Pest Control Research*, Vol. 3, p. 289, New York, (1960).
316. E. KNÜSLI, *Phytiat–Phytopharm.* **7**, 81 (1958).
317. H. KOOPMAN and J. DAAM, *Rec. Trav. Chim.* **77**, 235 (1958).
318. J. R. Geigy SA, French Pat. No. 1,135,848 (1956).
319. O. DIELS, *Ber.* **32**, 700 (1899).
320. A. W. HOFMANN, *Ber.* **18**, 2766 (1885).
321. W. M. PEARLMANN and C. K. BANKS, *J. Am. Chem. Soc.* **70**, 3726 (1948).
322. J. T. THURSTON, *J. Am. Chem. Soc.*, **73**, 2981 (1951).
323. I. H. WITT and C. S. HAMILTON, *J. Am. Chem. Soc.*, **67**, 1078 (1945).
324. J. L. HILTON, L. L. JANSEN, and H. M. HULL, *Ann. Rev. Plant Physiol.*, **14**, 353 (1963).
325. J. FRYER and R. MAKEPEACE, *Weed Control Handbook*, Vol. 2, Oxford, 114 (1970).
326. Ibid., 121.
327. Ibid., 121.
328. Ibid., 119.
329. Ibid., 117.
330. Ibid., 111.
331. Ibid., 78.
332. Ibid., 53.
333. Ibid., 54.
334. Ibid., 70.
335. Ibid., 73.
336. Ibid., 93.
337. Ibid., 61.
338. Ibid., 71.
339. Ibid., 75.
340. Ibid., 83.

341. Ibid., 89.
342. Ibid., 93.
343. J. R. GEIGY SA, Brit. Pat. No. 814,947 (1959).
344. J. YATES, Brit. Pat. No. 987,253 (1965).
345. M. B. WEED, U.S. Pat. No. 3,022,150 (1960).
346. J. S. NEWCOMER, E. D. WEIL, E. DORFMAN, J. LINDER, and K. J. SMITH, U.S. Pat. No. 3,154,396 (1963).
347. H. GYSIN and E. KNÜSLI, U.S. Pat. No. 2,891,855 (1958).
348. T. GRAUER, U.S. Pat. No. 3,376,302 (1963).
349. H. GYSIN and E. KNÜSLI, U.S. Pat. No. 2,909,420 (1958).
350. J. R. Geigy SA, Brit, Pat. No. 814,948 (1959).
351. J. R. Geigy SA, Brit, Pat. No. 978,249 (1963).
352. H. GYSIN and E. KNÜSLI, U.S. Pat. No. 3,145,208 (1961).
353. H. GYSIN and E. KNÜSLI, Brit. Pat. No. 927,348 (1964).
354. H. GYSIN and E. KNÜSLI, U.S. Pat. No. 3,326,914 (1967).
355. Ciba Limited, Brit. Pat. No. 1,093,376 (1965).
356. R. P. NEIGHBORS and L. S. PHILLIPS, U.S. Pat. No. 3,451,802 (1969).
357. Shell Research Limited, Dutch Pat. No. 6,715,520 (1967).
358. Shell Research Limited, Dutch Pat. No. 6,709,805 (1967).
359. P. A. GABBOTT and G. E. BARNSLEY, *J. Sci. Fd. Agric.*, **19**, 16 (1968).
360. H. A. JOHNSTON and M. S. TOMITA, Brit. Pat. No. 1,031,083 (1964).
361. H. A. JOHNSTON and M. S. TOMITA, Brit. Pat. No. 957,831 (1963).
362. H. A. JOHNSTON and M. S. TOMITA, U.S. Pat. No. 3,285,925 (1966).
363. H. T. HARRISON, U.S. Pat. No. 3,288,796 (1960).
364. G. MOHR, K. NIETHAMMER, S. LUST, and G. SCHNEIDER, Brit. Pat. No. 988,310 (1964).
365. G. MOHR, K. NIETHAMMER, S. LUST, and G. SCHNEIDER, Brit. Pat. No. 1,148,810 (1968).
366. G. MOHR, K. NIETHAMMER, S. LUST, and G. SCHNEIDER, U.S. Pat. No. 3,284,293 (1964).
367. G. MOHR, K. NIETHAMMER, S. LUST, and G. SCHNEIDER, U.S. Pat. No. 3,468,895 (1964).
368. R. H. RIGTERINK, French Pat. No. 1,360,901 (1964).
369. R. H. RIGTERINK, U.S. Pat. No. 3,244,586 (1964).
370. Ethyl Corporation, French Pat. No. 1,061,791 (1953).
371. Fisons Pest Control Limited, Belgian Pat. No. 631,044 (1962).
372. R. E. PLUMP, U.S. Pat. No. 2,651,579 (1953).
373. M. J. JANES, and D. J. CROWLEY, U.S. Pat. No. 2,690,413 (1954).
374. M. J. HURAUX and H. M. LAWSON, *Symp. New Herbicides*, Paris, 261 (1965).
375. E. R. LANING, *Proc. West. Weed Contr. Conf.*, **20**, 46 (1965).
376. A. J. CLAPHAM, S. KENYON, and R. S. BELL, *Proc. North East. Weed Contr. Conf.*, **21**, 395 (1967).
377. C. E. ROMERO, L. S. JEFFERY, and M. REVELS, *Rev. Inst. Colomb. Agropecuen.*, **4**, 99 (1969).
378. W. N. L. DAVIES, *Proc. Brit. West Indies Sugar Technol.*, **1**, 38 (1966).
379. R. E. FRANS and C. W. HAGUE, *Arkansas Agr. Expt. Sta. Bull.*, No. 746 (1969).
380. G. A. BUCHANAN, R. DICKENS, E. R. BURNS, and R. M. McCORMICK, *Proc. South. Weed Sci. Soc.*, **22**, 122 (1969).
381. O. H. FRICKE, *Proc. North East. Weed Contr. Conf.*, **23**, 108 (1969).
382. H. J. MURPHY and M. J. GOVEN, *Proc. North East. Weed Contr. Conf.*, **23**, 122 (1969).
383. A. D. WORSHAM, *Weed Sci.*, **18**, 648 (1968).
384. K. P. BUCHHOLTZ, *Weed Sci.*, **16**, 439 (1970).

385. J. C. MOOMAND and D. S. KIRN, *Weed Res.*, 8, 163 (1968).
386. W. L. CHANG, *Nung Yeh Yeu Chiu*, 17, 15 (1968).
387. J. M. HODGSON, *Weed Sci.*, 16, 465 (1968).
388. S. SAWAMURA and W. T. JACKSON, *Cytologia*, 33, 545 (1968).
389. R. J. MARTIN, U.S. Pat. No. 3,249,419 (1966).
390. Dow Chemical Co., French Pat. No. 1,393,606 (1965).
391. J. W. HAMAKER, H. JOHNSTON, R. J. MARTIN, and C. T. REDEMANN, *Sci.*, 141, 363 (1963).
392. P. M. RITTY, *Proc. Brit. Weed Contr. Conf. Glasgow*, 23 (1964).
393. W. R. ARNOLD and P. N. SANTELMANN, *Proc. South. Weed Sci. Soc.*, 18, 56 (1965).
394. F. VERNIE, J. LHOSTE, and A. CASANOVA, *Weed Res.*, 6, 322 (1966).
395. E. E. HEIKES, *Down Earth*, 20, 9 (1964).
396. E. R. LANING, *Down Earth*, 19, 3 (1963).
397. G. L. HODGSON and C. DONALDSON, *Proc. Brit. Weed Contr. Conf.*, 360 (1966).
398. D. S. C. ERSKINE, *Proc. Brit. Weed Contr. Conf.*, 488 (1966).
399. J. FARNWORTH and G. M. DAVIES, *Proc. Brit. Weed Contr. Conf.*, 493 (1966).
400. B. J. MITCHELL, *Proc. Brit. Weed Contr. Conf.*, 498 (1966).
401. H. A. BRADY, *Proc. South. Weed Sci. Soc.*, 22, 245 (1969).
402. F. E. PEEVY, *Proc. South. Weed Sci. Soc.*, 22, 251 (1969).
403. R. W. BOVEY, *Weed Sci.*, 17, 538 (1965).
404. C. J. LICHY, *Proc. South. Weed Sci. Soc.*, 18, 303 (1965).
405. R. W. BOVEY, C. C. DOWLER, and J. D. DIAZ-COLON, *Weed Sci.*, 17, 285 (1969).
406. C. C. DOWLER, F. H. TSCHIRLEY, R. W. BOVEY, and H. L. MARTIN, *Weed Sci.*, 18, 164 (1970).
407. R. W. BOVEY, H. L. MARTIN, and J. R. BAUR, *Weed Sci.*, 17, 373 (1969).
408. G. A. WICKS, C. R. FENSTER, and O. C. BURNSIDE, *Weed Sci.*, 17, 408 (1969).
409. H. CARDENAS, P. CABRERA, and A. FERNANDEZ, *Agron. Trop. Venezuela*, 19, 41 (1964).
410. C. P. REID and W. HURTT, *Plant Physiol.*, 44, 1393 (1969).
411. R. E. MEYER, T. E. RILEY, H. L. MARTIN, and M. G. MERKLE, *Tex. Agr. Expt. Sta. Misc. Publ.*, No. 930 (1969).
412. E. J. PETERS and J. F. STRITZKE, *U.S. Dept. Agr. Tech. Bull.*, No. 1400 (1969).
413. P. T. TUELLER and R. A. EVANS, *Weed Sci.*, 17, 233 (1969).
414. K. L. CARVELL, *Down Earth*, 24, 17 (1968).
415. M. CORNET, R. VIDAL, and C. R. JOURNEES, *Etud. Herbic. Conf. Columa*, 1, 248 (1967).
416. B. KIRBY, P. STRYKES, and P. SANTELMANN, *J. Range Manage.*, 20, 158 (1967).
417. H. A. JOHNSTON and M. S. TOMITA, Belgian Pat. No. 628,487 (1963).
418. H. A. JOHNSTON and M. S. TOMITA, U.S. Pat. No. 3,285,925 (1966).
419. J. N. E. DAY and J. F. THORPE, *J. Chem. Soc.*, 117, 1465 (1920).
420. G. MOHR, K. NIETHAMMER, S. LUST, and G. SCHNEIDER, German Pat. No. 1,182,896 (1964).
421. C. SOZNOWSKI and J. WALENS, Polish Pat. Nos. 47,071; 47,072; 47,892; 49,292 (1964).
422. G. MOHR, D. ERDMANN, S. LUST, and G. SCHNEIDER, *Meded. Rijksfac. Landbouwwetensch. Gent*, 33, 1293 (1968).
423. S. LUST, *Meded. Rijksfac. Landbouwwetensch. Gent*, 34, 787 (1969).

424. R. W. BYRDE and C. W. HARPER, *Long Ashton Agr. Hort. Res. Sta. Ann. Rep.*, 151 (1966).
425. E. MERCK, AG, French Pat. No. 1,557,693 (1969).
426. G. MOHR, K. NIETHAMMER, S. LUST, and G. SCHNEIDER, South African Pat. No. 6,707,305 (1968).
427. G. MOHR, K. SCHUERER, and S. LUST, South African Pat. No. 6,801,931 (1968).
428. W. K. WHITNEY, J. L. HARDY, and A. E. DOVY, *J. Econ. Entomol.*, **58**, 1043 (1965).
429. R. H. RIGTERINK and E. E. KENAGA, *J. Agr. Food. Chem.*, **14**, 394 (1966).
430. G. N. SMITH, *Down Earth*, **22**, 3 (1966).
431. J. C. SCHLINKE, *J. Econ. Entomol.*, **63**, 319 (1970).
432. J. M. GRAYSON, *Pest Control*, **34**, 12 (1966).
433. K. WHITNEY, R. P. HARRISON, and R. G. HOWE, *Pest Control*, **35**, 25 (1967).
434. R. O. DRUMMOND, J. H. WHETSTONE, and S. E. ERNST, *J. Econ. Entomol.*, **60**, 1021 (1967).
435. R. O. DRUMMOND, S. E. ERNST, J. L. TREVINE, and O. H. GRAHAM, *J. Econ. Entomol.*, **61**, 467 (1968).
436. M. S. MULLER, R. L. METCALF, and A. H. GEIB, *Mosquito News*, **26**, 236 (1966).
437. U. E. BRADY, D. W. MEIFORT, and G. C. LE BRECQUE, *J. Econ. Entomol.*, **59**, 1522 (1966).
438. P. D. LUDWIG and J. C. McNEIL, *Mosquito News*, **26**, 344 (1966).
439. B. L. COLLIER and C. E. DIETER, *Down Earth*, **21**, 3 (1965).
440. R. L. WALKER and E. CANTU, *U.S. Dept. Agr. ARS*, No. 33-122 (1967).
441. D. W. FEWKES and D. A. BUXO, *Proc. Brit. West Indies, Sugar Technol.*, **1**, 172 (1966).
442. R. H. RIGTERINK, French Pat. No. 1,360,901 (1964).
443. C. FEST, I. HAMMANS, W. STENDEL, and W. FLECKE, Brit. Pat. No. 1,165,293 (1969).
444. R. H. RIGTERINK, South African Pat. No. 6,800,893 (1969).
445. C. N. WOLF, *Sci.*, **121**, 61 (1955).
446. Ethyl Corporation, French Pat. No. 1,668,029 (1954).
447. H. GYSIN, *Advances in Pest Control Research, Vol. 3*, p. 309, New York, 1960.
448. O. SIEGMANN, *Acta. Vet. Brno*, **38**, 47 (1969).
449. J. F. RYLEY, *J. Parasitol.*, **53**, 1151 (1967).
450. D. L. LONG and B. J. MILLARD, *Expt. Parasitol.*, **23**, 331 (1968).
451. N. VIAENE, A. DEVOS, L. SPANOGHE, and J. VAN IMPE, *Vlaams Diergeneesk. Tijdschr.*, **36**, 395 (1967).
452. W. M. REID and R. N. BREWER, *Poultry Sci.*, **46**, 638 (1967).
453. L. KEPPENS and N. REYNTENS, *Rev. Agr. Brussels*, **22**, 727 (1969).
454. R. N. BREWER and L. M. KOWALSKI, *Expt. Parasitol.*, **28**, 64 (1970).
455. B. L. STOCK, G. T. STEVENSON, and T. A. HYMAS, *Poultry Sci.*, **46**, 485 (1967).
456. O. C. BUCEK, *Poultry Sci.*, **48**, 2173 (1969).
457. P. L. LONG and B. J. MILLARD, *Vet. Rec.*, **81**, 11 (1967).
458. C. C. NORTON and L. P. JOYNER, *Vet. Rec.*, **82**, 317 (1968).
459. D. K. McLOUGHLIN, *Expt. Parasitol.*, **28**, 129 (1970).
460. F. G. BROWN, *Soc. Chem. Ind. Monograph No. 33*, 120 (1969).
461. Dow Chemical Co., Dutch Pat. No. 6,409,766 (1965).
462. Dow Chemical Co., Dutch Pat. No. 6,515,950 (1966).
463. H. JOHNSTON, U.S. Pat. No. 3,296,272 (1967).

464. H. JOHNSTON, U.S. Pat. No. 3,371,011 (1968).
465. C. D. CRAWFORD, U.S. Pat. No. 3,415,832 (1968).
466. P. A. WOLF and F. J. BOBALEK, *Applied Microbiol.*, 1376 (1967).
467. R. C. TRUEBLOOD and K. ANDERSON, *J. Paint Technol.*, 39, 650 (1967).
468. O. C. ELMER, U.S. Pat. No. 3,531,433 (1970).
469. F. L. BROWN, Brit. Pat. No. 1,144,278 (1969).
470. K. WESTPHAL, U.S. Pat. No. 2,219,858 (1941).
471. K. L. LAPHAM and P. J. SHEA, U.S. Pat. No. 3,544,677 (1970).
472. J. A. HENDRY, F. L. ROX, and A. L. WALPOLE, *Brit. J. Pharmacol.*, 84, 342 (1945).
473. J. A. HENDRY and R. F. HOMER, *J. Chem. Soc.*, 328 (1952).
474. J. A. HENDRY, R. F. HOMER, and F. L. ROSE, Brit. Pat. No. 683,414 (1952).
475. J. A. HENDRY, R. F. HOMER, and F. L. ROSE, U.S. Pat. No. 2,675,386 (1954).
476. J. A. HENDRY and R. F. HOMER, *J. Chem. Soc.*, 328 (1952).
477. J. A. HENDRY, F. L. ROSE, and A. L. WALPOLE, *Nature*, 165, 993 (1950).
478. J. H. BURCHENAL, C. C. STOCK, M. L. CROSSLEY, and C. P. RHOADS, *Cancer Res.*, 10, 207 (1950).
479. J. H. BURCHENAL, C. C. STOCK, M. L. CROSSLEY, and C. P. RHOADS, *Cancer Res.*, 10, 208 (1950).
480. DAI-ICHI SEIYAKU, Japanese Pat. No. 6,903,357 (1969).
481. DAI-ICHI SEIYAKU, Japanese Pat. No. 6,822,309 (1968).
482. HODOGAYA, Japanese Pat. No. 6,905,398 (1969).
483. HODOGAYA, Japanese Pat. No. 6,808,954 (1968).
484. American Cyanamid Co., Brit. Pat. No. 1,034,538 (1965).

Author Index

In this Author Index the first figure cited is the page in which the reference appears. The figure in parentheses indicates the reference number.

King, S.S.T., 224 (185a)
King, T.J., 284 (527), 292
(527, 591), 293, 294
(591)
Kinoshita, T., 279 (428)
Kirby, B., 456 (416)
Kirby, P., 335 (850)
Kirmse, W., 123 (339)
Kirn, D.S., 454 (385)
Kirpal, A., 17 (90)
Kirpal, C., 46 (284)
Kirsanov, A.V., 236 (271),
238 (271, 284), 239
(284), 246 (271, 284),
272 (284)
Kitchingman, G.W., 420
(145)
Klarmann, E.G., 435 (133,
134)
Klatt, D., 236 (267, 268),
238 (268), 254, 260,
270 (267)
Klauke, E., 335 (852)
Kleb, K., 444 (274), 445
(287)
Klee, W.E., 40 (268)
Klein, E., 413 (68)
Kleine-Natrop, H.-E., 283
(506)
Kleppe, H.G., 232-234
(265), 236, 254, 260,
270 (267)
Klerk, D. de, 408 (22)
Klingsberg, E., 337 (856)
Kloimstein, E., 275, 277-
279 (386)
Klötzer, W., 283 (499),
285, 294 (499, 550)
Klug, H., 221, (167), 336
(167, 855)
Kluge, F., 217 (139)
Knobloch, W., 316 (782)
Knusli, E., 228 (240),
283 (484), 447 (313-
316), 449 (347, 349,
352-354)

Kober, E.H., 283 (498),
291 (498, 587), 293
(498)
Kober, E., 211 (79), 214,
216, 217, 220 (112),
291 (578, 579), 293
(579)
Kochanska, L., 227 (235)
Kochergin, P.M., 313
(705, 707-709), 314
(711)
Koenig, J.H., 412 (58)
Kofod, H., 277, 279, 280
(415)
Kohler, W., 214, 216,
217, 220, 221 (113),
380 (69)
Kohlhaupt, R., 66 (305),
67 (305, 306), 69
(305), 232, 233 (263,
265), 234 (263)
Kohlrausch, K.W.F., 21,
180 (150)
Kokhlov, D.N., 316 (780)
Kolder, C.R., 227 (236,
237), 228 (236, 237,
244), 240 (236, 244),
254 (244), 269 (237,
244)
Komatsu, T., 279 (429)
König, H., 203, 204 (37-
39), 307 (37, 648)
Konishi, K., 427 (233)
Koopman, H., 334 (845),
335 (845, 847, 848),
423 (187), 447 (317
Koopmans, M.J., 425
(209)
Kopecký, J., 285, 295
(543)
Kopelman, R., 21 (127,
130)
Koppel, H.C., 292, 294
(595)
Kornuta, P.P., 337 (857)
Korte, F., 34 (256), 315,

317 (727, 728), 318
(727), 321 (728)
Korybut-Daszkiewicz, B.,
231 (259)
Kost, A.N., 322 (788)
Kowalewska, A., 227
(235)
Kowalski, L.M., 459 (454)
Kraay, G.M., 392 (119)
Kraft, J., 11 (61)
Krafft, L.F., 6, 14 (29),
19 (93)
Krais, P., 445 (289)
Kramberger, M., 329
(822a)
Kraus, E.J., 416 (75)
Kravtsov, D.N., 314 (711)
Krishnan, K.S., 21 (160)
Krivopalov, V.P., 283
(517)
Kropa, E.L., 441 (252)
Krukonis, A.P., 43 (271a),
128 (342)
Kruse, C.W., 23 (180)
Krynitsky, J.A., 19 (96,
101)
Krzikalla, H., 200 (8, 12),
201 (27)
Kuchinke, E., 19 (98)
Kuhlberg, A., 4 (16, 17),
14, 42, 57 (16)
Kühle, E., 335 (852)
Kulka, M., 6, 27 (36)
Kunze, H., 17 (90)
Kuraishi, T., 276 (398,
406), 278 (406, 419,
420), 279 (398, 424,
428, 433), 280 (398),
281 (424), 282 (442)
Kurbatow, A., 9 (52)
Kusuda, F., 19 (104), 45
(104, 280), 173 (280)
Kutepov, D.F., 316 (780)
Kutz, W.M., 411 (32)
Kyker, G.D., 215 (125-
127)

Subject Index

Derivatives are listed either under the parent structure or under
"per- chloro-" if fully chlorinated.